REGIONAL LANDSCAPES OF THE UNITED STATES AND CANADA

FOURTH EDITION

Stephen S. Birdsall
John W. Florin
University of North Carolina at Chapel Hill

John Wiley & Sons, Inc.
New York • Chichester • Brisbane • Toronto • Singapore

ACQUISITIONS EDITOR Barry Harmon
PRODUCTION MANAGER Katharine Rubin
DESIGNER Pete Noa
COVER PHOTO Reginald Wickham
PRODUCTION SUPERVISOR Sandra Russell
MANUFACTURING MANAGER Lorraine Fumoso
COPY EDITOR Marjorie Shustak
PHOTO RESEARCHER Jennifer Atkins
PHOTO RESEARCH MANAGER Stella Kupferberg
ILLUSTRATION Edward Starr

Recognizing the importance of preserving what has been written, it is a policy of John Wiley & Sons, Inc. to have books of enduring value published in the United States printed on acid-free paper, and we exert our best efforts to that end.

Library of Congress Cataloging in Publication Data:

Birdsall, Stephen S.

Regional landscapes of the United States and Canada / Stephen S. Birdsall, John W. Florin.—4th ed.

Includes index

ISBN 0-471-61646-X

1. United States—Description and travel—1981- 2. Canada—Description and travel—1981- 3. United States—Economic conditions—1981- 4. Canada—Economic conditions—1945- I. Florin, John William. II. Title.

E169.04.B57 1991

917.3—dc20

91-14787

CIP

Printed in the United States of America

10 9 8 7 6 5 4 3

Printed and bound by R.R. Donnelley & Sons, Inc.

Acknowledgements

We continue to find great pleasure in the willingness of friends, colleagues, and even people we did not know to take time to help us make this a better book. With each edition, the list of those to whom we are in debt grows longer. Even when we did not adopt their suggestions, people who sent us their thoughts about the last edition helped improve this new version by forcing us to see with another's eyes what we had written. We appreciate deeply the encouragement and the assistance we have received over the years.

In addition to the many reviewers who contributed to improvements in earlier editions. we want to thank those who made helpful comments on our efforts for this fourth edition. Giving their time and thoughts were Robert Bacon, Marvin Baker, James Davis, Lee Dexter, John J. Ford, William Imperatore, Daniel Jacobson, J. Harold Leaman, Alan Lew, Darrell Napton, Thomas Orf, Paul Phillips, Leon Pitman, Paul Salley, Thomas Schmidlen, George Schnell, Gail Sechrist, R. Hal Sharp, and Dennis Spetz. Each of these reviewers added their insights and reactions to the many who advised us on previous editions. All offered tangible assistance through their reviews of all or some of the text. We also want to express appreciation to the staff at *American Demographics,* who efficiently provided us with an advance copy of the U.S. metropolitan population data used in the regional chapter tables. Margo L. Price deserves special thanks for taking time from her other writing to add two excellent new case studies to this edition and helping to improve the text in many other ways as well.

We do not always recognize that the advice we are given is good advice. Any weaknesses in this edition are a result of our own shortcomings, not those of the many people who have recommended ways to improve it.

The staff at John Wiley & Sons have been patient and helpful, even when we missed deadlines or sent the wrong material first. We are especially indebted to our editor, Barry Harmon, who took over partway through the revision for this edition and pushed us gently but firmly to completion. We also want to thank the production team at John Wiley & Sons for their efficient and effective assistance. A deeply felt thank you to Sandra Russell, Katy Rubin, Stella Kupferberg, Jennifer Atkins, Edward Starr, Pete Noa, Marjorie Shustak, Cynthia Michelsen, and all the others at Wiley who performed their jobs so well.

S.S.B.
J.W.F.

Preface to the Fourth Edition

As we undertake each new edition of this text, we face the challenge of how to retain what we (and others) think is good and valuable while incorporating changes that are significant and substantial. Some changes are relatively easy: Replace outdated statistics with more recent figures. Other changes depend more on judgments about timing: Have long-term trends in a region progressed sufficientlly to require the total reorganization of a chapter, or does the original organization remain valid? And still other changes yield something entirely new. Will this new material be a contribution to or a distraction from the overall presentation? We hope that in each case we have found a proper balance in this fourth edition.

Following some debate, we decided to delay revision of the third edition until we could incorporate data from the 1990 United States Census of Population. We understand that this decision created difficulties during the last several years for many who used the text, and for that we apologize. As it turned out, technical problems in the census led to delays in publishing even early U.S. population figures, and this forced us to retain some estimates rather than actual counts. However, we incorporated 1990 U.S. population data as late as May 1991 in an attempt to be as current as possible. We did not think we could wait for the 1991 Canadian Census, as well.

A new chapter has been added to an early portion of the book where we address human and physical patterns across the two countries in a systematic, as opposed to a regional, manner. The new chapter offers a look ahead to the end of the century and beyond, suggesting five issues that will have repercussions on the geography of the United States and Canada. As we say in the chapter itself, we are not attempting an exhaustive presentation of the geographic problems our two countries face. Nor do we want to begin an argument that our selection is the "best" according to some criterion. The five issues selected are large, serious, and (we hope) will provide instructors and students with an appreciation of the range of national policy issues with which geographers are concerned.

Based on feedback from instructors who had used our book, we doubled the number of case studies. We believed from the first that students could benefit from a brief demonstration of how the broad themes and concepts discussed in the text were also apparent at a scale closer to their personal experience. Our goal, then, is to provide students a detailed look at a place within a region. Case studies have been added to several chapters in which none existed. Some chapter have two case studies, each touching on different themes in a region.

The "Additional Readings" lists at the end of each chapter were revised. We continue to include in these short lists a mix of readings about the subjects covered in the chapter. We also make it a point to include a few popular selections along with more academic writing.

In our experience, this book is a text frequently used in the first or second geography course taken by students at the college level. We have worked hard to keep the writing clear and interesting. Complicated and important issues do not have to be described in obscure language to make them seem more complicated or important than they are. We continue to try to improve our communication skills and hope all who are familiar with earlier editions will approve of the changes.

We cannot and will not try to make this text appeal to all who teach and study the geography of the United States and Canada. But to those moved to read it through, we extend an open invitation to let us know what you think, regardless of the balance between roses and thorns.

Stephen S. Birdsall
John W. Florin

Preface to Second Edition

It is in the nature of all instructors to want to modify available texts; most instructors have their own teaching orientations and research interests. We believe that student interest and understanding is increased by interweaving factual information with the distinctive geographic approach. This is hardly a new notion to professional geographers. Factual information is important. Our primary emphasis, however, remains on the character of a region: its "flavor" as defined on the basis of regional themes and on the specifically geographic contributions that have helped form the region's distinctiveness.

One of the ways that geographic perspective can be shown in a regional text is through the application of organizational ideas that are basic to the discipline. Several dozen fundamental geographic concepts have been fused into the regional discussions of this book. Each is presented where it will enhance students' understanding of both the region and the concept. Many of the concepts appear in more than one chapter, and most can be applied to other sections of the book as well.

We believe that a regional text should be organized within a regional framework. We have chosen to define the major regions of the United States and Canada thematically as well as locationally. In doing so, we are simply making explicit the difference between a region and a territory. A territory is a portion of the earth's surface, but a region is a portion of the earth's surface that is different from other areas because of the features found there. Those who look in each chapter for a rigid topical sequence applied to a territory—for example, the physical environment, the agriculture, economy, population, urbanization, and so forth—will be disappointed. There are general similarities in the subjects treated in many regions, of course, but this does not depend on an arbitrary structuring of topics. Much of the material in each regional chapter is built around the themes that we have selected as most important in defining the region's character. The identifying themes in most chapters stem from the cultural rather than the physical environment, although the interaction of the two landscapes is usually strong. The major focus, then, is on human patterns and the interaction between these patterns and the physical environment.

Some of our regions are contained within political boundaries; most are not. A few follow the general pattern of physiographic or climatic variations; others are more fully defined by the sprawl and interaction of their human geography. Some portions of the United States and Canada have been identifiable as distinct regions since the earliest stages of European development; others are only recently formed or are still in the process of development and, therefore, remain vague in outline and functional character. None of the regions that we define should be viewed as permanent. A complex pattern such as that in any large region is always in flux with the region's character and boundaries changing. It is possible that several of the regions described in this book may disappear through a merging process, whereas others may be created as subregional characteristics become stronger.

The thematic regions that we use are not spatially exclusive—their territories overlap. The continental manufacturing core, for example, overlaps spatially with the continent's agricultural core region, the Canadian national core, and the continent's major urbanized region, called Megalopolis. Northern Indiana is located in both the manufacturing and agricultural core regions and is discussed in both regional chapters.

This territorial overlap of regions takes two forms. For instance, for northern Indiana, the important characteristics of two different types of regions must be considered. In other instances, a territory may lie in parts of two regions primarily because the boundary between them is indistinct. In the latter example, the thematic elements of both regions blend together; they are not at all separated by a sharp line. This complexity is reflected in Figure 1-5, our basic regional map. Some students may find this map difficult to read because of the illustration of regional overlaps. We have tried to ease this problem by eliminating the peripheral, or "transitional," zones around each region. Students should use the regional maps at the beginning of each chapter for more specific locations. Throughout the book we make every effort to make our discussions of overlapping thematic regions especially clear as well as to emphasize the advantages of this approach.

This is a thematic regional geography of North America. On the basis of political and cultural patterns, it treats only the United States and Canada; Mexico and the Caribbean countries are not included. The term "Anglo-America" has also been discarded because of the strong non-Anglo cultural influences in many sections of both countries. French culture in Canada and the strong Spanish influence in the American Southwest are only two examples of this cultural diversity. The British heritage is dominant in both countries, but non-British cultures have modified and enriched this heritage sufficiently to render the term "Anglo-America" unnecessarily misleading to the unwary.

Stephen S. Birdsall
John W. Florin

Contents

CHAPTER 1

Themes and Regions

This is a book about two of the largest and wealthiest countries in the world. One, the United States, is the world's most powerful nation, both economically and militarily. It has had a dominant influence on the world for decades. That position of strength has diminished somewhat in recent years, but not enough to rob the country of international preeminence. The other, Canada, even larger than the United States in land area, has only a tenth as many people. The sparseness of its population, coupled with a large potential resource availability, has led many to anticipate that Canada's already significant world economic influence will grow in the future.

Yet this is not a book about international power, nor is it really about the wealth of nations. Instead, our goal is to discuss and, to some degree, evaluate the geography of these two countries. Although we will look at their physical geography, our central interest is not landforms, climate, soils, or vegetation but rather the human imprint on the landscape. The people in these technologically advanced countries are the prime creators of their own environment, and it is the cultural environment that interests us most. The specific elements of this environment are diverse, and their number is almost endless, from the pattern of land use in a city, to the distribution of racial or ethnic groups, to the style of fences that farmers use to enclose their land.

This does not mean that the physical environment will, or can, be ignored. In fact, in some instances it will hold a central role in the text. There are several reasons for this. One is that the physical environment often has a significant influence on people's activities in the landscape. One factor in the importance of New York City is certainly its location on one of the world's finest natural harbors. Southern Florida's long growing season and mild winters enable it to be a leader in the United States in the production of such crops as oranges, limes, and sugarcane. You can surely add many other examples.

Still, Florida's mild climate does not automatically mean that it will be a supplier of oranges; and New York City's harbor, taken by itself, is only one of many important reasons for the city's growth. The physical environment is one of the elements that influences human activities, but it does not in itself determine those activities. In general, the more advanced the level of technology, the greater the leeway that a population has in dealing with the land. For Canada and the United States, this means that the possible variations in activity for any particular environment are great. If one simply looked at its climate, the valley around Great Salt Lake in central Utah should be a barren wasteland. The valley is, however, a productive agricultural region and the home of several sizable urban areas, largely because the people of Utah wanted this to be and had the technology to make the necessary changes.

It is obviously impossible to cover all the material that might fit into a geography of the two countries. To do so would require many volumes. We have chosen to limit that unmanageable complexity by subdividing the United States and Canada into a number of areas of distinctive character, or regions. We believe that each of these regions has a special identity that

The landscape of the United States and Canada, rich with diversity, provides the best way to view the regional diversity of the two countries. This Iowa farm scene suggests the abundance of American agriculture, the rectangular land pattern of the mid-section of both countries, and the relative flatness of much of the Deep North (USDA) has developed out of a small number of interacting elements. We have used these elements to form the themes around which each of the regional chapters is organized.

A FEW BASIC THEMES

There are a few general cultural patterns in the two countries that are important because they serve to give the countries an identity distinct from most of the rest of the world. These identifying themes are often so common or universal within the United States and Canada that most of their residents view them as simply a part of the normal nature of things. Their influence cuts across regional and political boundaries, and in many cases ignores major differences in the physical environment. These general themes indicate what appear to be elements that are especially characteristic of the ways that Canadians and Americans have organized their national territories. In this manner, they define the character of the entire continent.

Urbanization

Citizens of both the United States and Canada hold dearly to their rural roots, and there continues to be a widespread attitude that life in a rural setting is somehow good, whereas life in an urban environment is less good, if not downright bad. Millions of Americans, most of them urbanites, prefer to consider their country as a basically rural place. They seem to believe that this rurality sets them apart from Europeans and provides our two North American countries with a basic national vigor that has been lost in Europe. Few politicians seeking elective office in either country would have their chances damaged by a personal claim to a rural background.

There is no longer much justification for this view of rural dominance in either country. Although most of the countryside is rural by definition—a matter that may distort the residential image of the two countries to the casual visitor—most of the people simply do not live in most of the country. By their own definitions, over 70 percent of the residents of Canada and the United States live in urban areas. It has been predicted that by the year 2000, over 75 percent of the U.S. population will live in 30 urban regions, each numbering over 1 million people (Figure 1-1). The U.S. farm population numbered only about 5 million in 1989, and that percentage has declined steadily since the first national census in 1790, when over 90 percent of all Americans were farmers. Over 40 percent of the population of the United States lives in Metropolitan Statistical Areas (MSAs) of 1 million or more people. Nearly one-third of all Canadians live in Toronto, Montreal, or Vancouver, which are the country's three largest metropolitan areas. Half of all Canadians live in one or another of their country's eight largest urban areas. Urban percentages in both countries have grown steadily, and no decline is foreseen.

In short, these are urban countries. Most people were born in urban areas; few of those presently in urban areas are likely to move to farm locations. The rural-farm component of the population, already small, continues to decline.

Several elements of urbanization will be emphasized in our discussion. Cities have a particular *form*, a particular layout. Most American cities have a rectangular-grid pattern, partly a result of cultural attitudes, partly a result of a desire for efficient transport before the automobile, and, to a degree, simply because that pattern is an easy way to survey the land. Modern shopping centers are usually in the suburbs near major highways. Within the city there is a collection of industrial and commercial centers, residential areas, warehouses, and so on.

Cities exist for many different reasons. They may have an important transportation role, such as New Orleans has as an ocean port or Omaha, Nebraska, has as a railroad center. They may provide an important administrative function, as the Canadian federal capital at Ottawa does or as New York City does as the administrative center for many corporations. Perhaps your city is a center of recreation, like Las Vegas or Atlantic

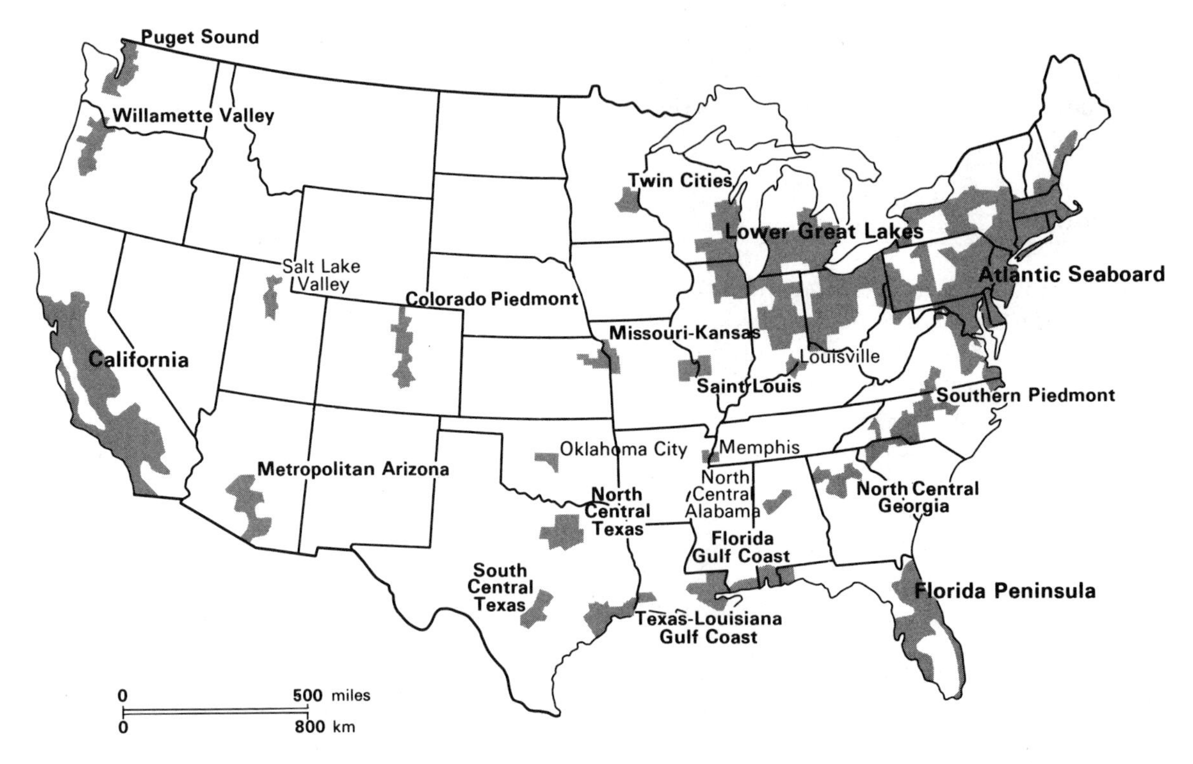

Figure 1-1 Predicted urban area expansion to the year 2000. By the year end of this century, the major urban regions in the United States are expected to be much more extensive than at present. The largest may have merged across vast areas as shown.

City. Others, such as Hamilton, Ontario, or Detroit, are usually thought of as manufacturing centers. Thousands of smaller towns and cities in both countries exist primarily to provide goods and services to residents of the town and a surrounding trade territory. Most cities, certainly all large ones, contain many different urban *functions*. Nevertheless, many are characterized by certain dominant functions that were the reason for their initial development or for much of their early growth and that today continue to give them their special character.

The pattern of continuing and often rapid urban growth in the two countries during the last 100 years is another key element in their geography. Coupled with the increasing mobility of the urban population, this growth has stimulated a great sprawling pattern of urbanization. The total urbanized area of Canada and the United States increases at a rate of about 800,000 hectares (2 million acres) each year. This loss of rural land has generated a growing concern for the future availability of adequate amounts of good agricultural land. In some areas, the result of urban spread is urban coalescence, with the edges of different urban areas meeting and blending. As independent cities grow together, the role and meaning of urban political organization changes, and attempts may be made to create new administrative organizations to cut across city boundaries (see the discussion on political complexity later in this chapter).

Industrialization

The United States is the world's largest manufacturing power. No other country in the world can come close to matching the volume of output of American manufacturing, nor can any country match the variety of U.S. products. Canada,

Urban housing built before the advent of the family automobile often consisted of closely-packed houses and walk-up apartments like these units in Boston. Afterward, housing spread and cities exploded outward as people commuted great distances by car. (Donald Dietz/Stock, Boston)

being much smaller than the United States in population, does not, of course, reach the magnitude of America's industrial output. Furthermore, with its much smaller population and less available financial capital, Canadian manufacturing lacks the diversity found in U.S. manufacturing; economies of scale similar to those of its southern neighbor are less possible with Canada's smaller market. Thus, Canada depends more on imported manufactured goods than does the United States, and many of the items it does manufacture are somewhat more expensive than their counterparts in the United States.

Still, both are highly industrialized countries. Each manufacturing job generates two or three other jobs. This *multiplier* effect emphasizes the importance of manufacturing as *basic* employment. Such employment is that which generates wealth in an area by serving the needs of people who live beyond the margins of the area. *Nonbasic* employment, on the other hand, is that which does not bring in additional wealth from outside because it is concerned primarily with serving the population of the area itself. Thus, it is an area's basic employment that is most important to long-term growth; only by bringing in wealth from outside can more activities, basic and nonbasic, be supported. Although manufacturing may be the most important of the basic industries in many cities, other activities can also be basic. For example, if a university attracts most of its student body from places beyond the bounds of the community in which it is located, the faculty have basic occupations. Police, retail sales personnel, public school teachers, and local bankers are among the many in the community who have nonbasic occupations.

A substantial part of the total employment in both countries, thus, is related to manufacturing either directly or indirectly. Furthermore, most cities were founded and experienced their major periods of growth when manufacturing was easily the primary factor in urban growth. The basic urban pattern, the pattern that reflects where most people in the two countries currently live, can be viewed as a direct result of industrialization.

Substantial regional specialization in manufacturing exists in both countries. This is partly the result of variations in the availability of industrial raw materials. No one should be surprised that North Carolina is a leader in cigarette manufacturing, for the state is a major producer of tobacco, or that Quebec, with vast forests in its interior, leads both countries in pulp and paper production. The industrial pattern is also a result of *industrial linkages:* manufacturing concerns that produce component parts of some final product are located near each other as well as near the final assembly site to minimize total movement costs. The many manufacturers of automobile parts in northern Ohio and southern Michigan is a classic example.

Other important sources of variation may be differences in labor availability or labor skills, the quality of transportation facilities, and local political attitudes. Regions tend to specialize in the production of whatever it is that they can best produce. And with this regional specialization

Sun City, located near Phoenix, Arizona, had a population of 11,000 in 1980 and may reach 50,000 by the end of this century. It is basically a retirement community. Recent urban expansion in both countries often ignores the grid street system that dominated American cities from the early nineteenth century to mid-twentieth centuries. (Dell E. Webb Development Co.)

has come regional interdependence; few sections of either country are truly self-sufficient in manufacturing, in spite of what local pride might lead us to believe.

High Mobility

America's extensive transportation network is an important element in its high level of economic interaction. Goods and people move freely within and between most sections of both countries. Regional interdependence is great; it is made possible by these interregional flows. Relative isolation is uncommon, but it does exist; in a few cases, it may be important in a region's economy (e.g., see Chapter 8).

The populations of both Canada and the United States have high residential mobility. Nearly all Americans move more than once during their lives. On the average, nearly 20 percent change their residence in any one year. Although much of this residential migration is local in nature, it does result in substantial interregional population movement. A decision to migrate is a function of the balance between negative characteristics at the origin (push factors) and positive characteristics at the destination (pull factors). The easier it is to move, the more likely this push-pull balance will be the primary influence on the migration decision.

Migration has traditionally been toward areas perceived to contain greater economic opportunity. Until the last decade of the nineteenth century in the United States and the 1910s in Canada, there was a strong westward population shift toward frontier agricultural lands. The focus of opportunity then changed and migration became oriented overwhelmingly toward urban areas. More recently, particularly in the United States, the national economy has entered what some call a postindustrial phase: employment growth is primarily in professions and services rather than *primary* (extractive) or *secondary* (manufacturing) sectors. Such employment is much more flexible in its location, and there has been a more rapid growth in such employment in areas that appear to contain greater amenities (Figure 1-2).

Resources: Abundance and Dependence

The United States and Canada are among the world's most important suppliers of raw materials of many types. Foodstuffs are the key raw material export of both countries. Almost half of all world food exports of the 1987–88 crop year were from the United States, including 44 million tons of corn, 43 million tons of wheat, and 22 million tons of soybeans. Total U.S. grain and soybean exports rose from 56 million tons in 1970–71 to an estimated 122 million tons in 1987–88. The United States exports a majority of its wheat crop, 40 percent of its soybeans, and one quarter of its corn. The country is the leading international exporter of most of the world's major grains, including wheat, corn, and rice. It also leads in the export of such products as soybeans, tobacco, and cotton.

In 1988 the United States had a total trade deficit of $133 billion. This resulted from a net deficit of $146 billion from the sale and purchase of manufactured goods and minerals (especially oil) and a surplus of $13 billion from the sale and purchase of agricultural products. (For comparison, the previous edition of this text used 1981 data. At that time the total trade deficit was $28 billion resulting from a surplus of $22 billion in the sale and purchase of agricultural products and a deficit of $50 billion for manufactured goods and minerals.) Both Canada and the United States grow enough food in a normal year, measured in terms of calories, to feed more than twice their combined populations. About 25 percent of the total land in row crops in the United States produces export crops; in Canada the percentage is even higher. The surplus moves onto the international market.

Canada in particular is also an important exporter of a variety of other raw or semimanufactured materials, such as petroleum, coal, pulp, lumber, iron ore, natural gas, copper, and nickel. The United States is easily the most important buyer for most of these raw materials, although other countries also purchase large amounts. For example, Japan buys much of the coal and forest products exported from the west

Figure 1-2 Relative population change, 1980–1990. The size of each circle is proportional to the degree of relative change for that state. Populations in most states in the South and West are expected to grow comparatively rapidly for the rest of this century, whereas that of much of the Northeast and Middle West will lag in population growth.

coast of Canada. The United States is able to satisfy much (but not all) of its gigantic demand for industrial raw materials with domestic sources, an indication of its large production of raw materials. The United States does have the potential to be a major supplier for a few non-agricultural raw materials internationally and is the world's leading exporter of coal.

Although the populations of both countries are predominantly urban, the extraction of natural resources from this abundant base does require a large nonurban labor force. Furthermore, particularly for agriculture, the development of these resources often involves a substantial land area. As a result, the relationship between the physical environment and human adaptations to that environment are clearly visible in many of the landscapes of both countries. Governments play an important role in this relationship by establishing controls on land use and agricultural production and by regulating the development of many resources. It is partly because processes inherent in urbanization and industrialization lead to high demand for raw materials that both countries, especially the United States, have become dependent on imported raw materials in spite of great natural resource abundance.

The massive Ford Motor Company assembly plant at River Rouge in Detroit employs many thousands of workers. The complexity and intensity of manufacturing is a basic element in the economic character of both countries. (Ford Motor Company)

High Income and High Consumption

Canada and the United States are rich countries. Their populations earn a great deal of money, and they spend it. The demands of the populations and of the economies themselves generate massive consumption.

There are substantial regional variations in the income within both countries (a topic that will be treated in several chapters), but as national units, both countries have per capita incomes that rank among the highest in the world. Income levels in the United States are somewhat higher than those in Canada; indeed, the average U.S. citizen does have a somewhat higher standard of living than the typical Canadian. Still, the difference is often between two or three cars per family or one or two televisions.

The economies of both countries have grown increasingly energy dependent. Their high national incomes are achieved through high levels of worker productivity. High productivity levels require a greater use of machines. And modern machines are fueled by inanimate energy sources. High mobility also implies a high use of energy resources. High income spread somewhat evenly among a large share of the popula-

The Chesapeake and Ohio railyards and associated coal docks at Newport News, Virginia. Coal is brought here by rail from the coalfields of the central Appalachians and then sent by ship to points along the east coast of the United States and throughout the world. (Daily Press-Times Herald Photo/Newport News, Virginia)

tion will generate high product demand. All this increases energy consumption. North Americans presently consume about 30 percent of the world's total energy production, and the absolute amount used continues to increase.

As mentioned, the United States also has grown increasingly dependent on imported resources, with Canada a principal supplier. The United States now imports half the petroleum it consumes, an increasing share of its iron ore and natural gas needs, nearly all of its tin and aluminum, and large quantities of many other mineral ores. There are few industrial raw materials in which the country can still claim self-sufficiency. Coal is the most notable exception. Although Canada is less dependent on foreign sources, it too purchases from other countries materials not available domestically plus a substantial minority of its needed petroleum resources. Canada will become self-sufficient in petroleum when its domestic resources are better developed and when it completes a pipeline to move surplus western supplies to the eastern part of the country.

High income also affects the diets of Canadians and Americans. They eat far more meat products and have a substantially more varied diet than most of the world's population. Beef and dairy production are, therefore, especially important in the agricultural economy. Although both countries are major food exporters, both must turn to foreign suppliers to obtain products not grown at home. This is especially true of Canada, whose climate does not allow production of the variety of foodstuffs that can be grown in the United States.

The high consumption that goes along with high incomes has a number of important manifestations. Travelers are bombarded with a constant series of ads, signs, carefully designed buildings, and the like, all of which urge them to part with at least a small portion of their money. The resources of the two countries are consumed at a rapid rate in the attempt to satisfy their populations' appetites for manufactured goods and specialized foods. The proven oil reserves in the United States are steadily shrinking as annual withdrawal outstrips discoveries. There recently has been an increased recognition and concern that continental reserves of many materials are definitely limited. The implications of high consumption in a finite environment and of high consumption's natural partnership with high income will be among the most important, and most difficult, issues facing the United States and Canada during the next few decades.

In spite of this high overall income, problems of poverty remain important to both countries (Figure 1-3). Many rural areas of French Canada and parts of the rural Atlantic Provinces have area income averages that are little more than half that found in Toronto or Vancouver. Much of the rural U.S. South, Mexican-American and American Indian areas of the Southwest, sections of Appalachia, and nonwhite ghettos in most American cities have populations with average incomes far below the national average. Current information suggests that this pattern of regional inequality intensified during the 1980s. The persistence of these poverty areas is a major problem for the two countries because the poverty represents both a real economic loss to the national economies and a large number of people who are not sharing in the abundance of their countries.

Figure 1-3 Median household income, 1979. While this map is a decade old (1990 data were not available at the time of publication), only the income levels themselves have shifted significantly. The basic geographic pattern changed very little during the 1980s. Nine of the 10 states with highest household income in 1990 were coastal, with the top 5 all in the Northeast.

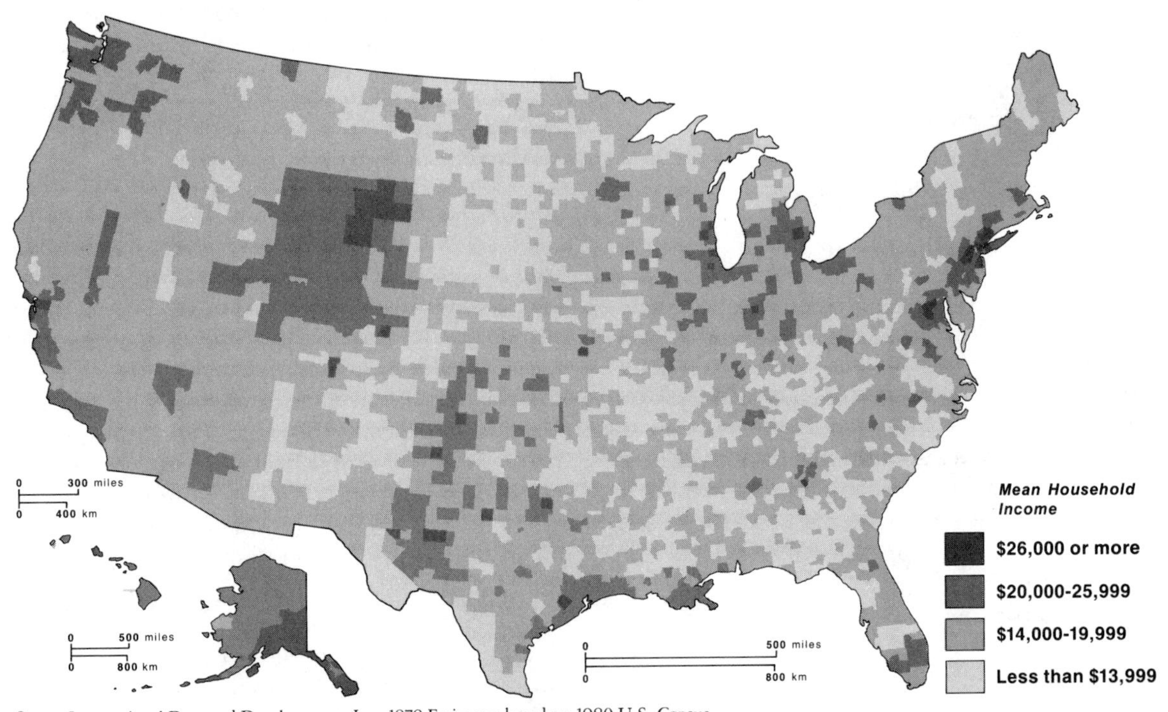

Source: International Data and Development, Inc. 1979 Estimates based on 1980 U.S. Census.

Environmental Impact

One consequence of the combination of high consumption with resource abundance and dependence has been strong disruption of the physical environment. Resources seldom can be removed from the natural landscape without some visible (or invisible but still important) impact, and the manufacture and use of these resources often harms our air and water. The increased severity of such environmental impacts has enlivened the age-old argument between development and conservation. The argument recently has stimulated greater governmental intervention in both processes as federal and state/provincial governments attempt to establish a middle ground between these two divergent viewpoints.

The governmental problem frequently is described as one of establishing regulations that provide some appropriate and meaningful level of environmental protection without raising extraction costs so much that the development of a necessary resource is impossible. As domestic resources become increasingly scarce and their costs of extraction and production increase, the importance of this conflict also will grow. The entire field of environmental impact promises to be one of the major regulatory battlegrounds during the next several decades.

Political Complexity

Both the United States and Canada have complex political structures, with jurisdiction over an activity or territory divided among many different decision-making bodies, some elected and some appointed. At the national level, Canada has a parliamentary government after the British model, whereas the United States separates its executive and legislative branches. The Canadian federal system provides for substantially more provincial independence than the U.S. Constitution allows individual states. The current demand by Quebec for greater autonomy within Canada would be impossible to accept in the United States, but it is not impossible in Canada. Canada has also begun to restructure the relationship between the provincial and federal governments in a way that presumably will add to provincial prerogatives.

Below the state and provincial level, the two countries have rather similar political structures, and their complexity presents a major problem in the effective and efficient distribution of governmental services. Counties, townships, cities, and towns are all governed by their own elected officials. Many special administrative units, some appointed and some elected, oversee the provision of specific services, such as education, public transportation, and water supply. The resulting administrative pattern is often nearly impossible to comprehend, because there may be many overlapping jurisdictions that provide one service or another in a given area.

One problem is that a jurisdictional unit, once created, almost never disappears and seldom is willing to surrender any of its powers. It is rare for political units to join their efforts and form administrative organizations under a single new structure as populations grow and merge spatially. Thus, certain service inefficiencies increase over time, but the spatial pattern of administration remains the same (see Figure 5-9).

Variety of Cultural Origins

Finally, both Canada and the United States have grown from diverse cultural backgrounds. Much of this diversity remains evident within the two countries. The most obvious example of the differences that can continue as a result of varied backgrounds is the distinction between British and French cultural areas in Canada. Canada is officially bilingual, and the government of the country follows a program that stresses this bicultural development.

There are a number of obvious cultural differences within the United States. One example is the important contribution of blacks to the national culture and particularly to that of the South. Another example is the distinctive culture region that has developed in the Southwest with an admixture of Hispanic Americans, American Indians, and Anglos. Another might

Americans value what they perceive as their rural or small town background. The essence of that perception is captured by small New England villages such as East Cornith, Vermont. (Vermont Travel Division)

be the contribution of Chinese in such cities as San Francisco and New York. Several other examples will be treated at some length in the text, for this cultural diversity is an important element in the distinctive character of many places in both countries.

REGIONS

One notable geography text, written decades ago, began with the observation that "Hell is hot." This was not a veiled threat to those students who might treat the material in the book too lightly. Nor was it the author's suggestion that the book would have a strong religious tone. Rather, he was making a significant, yet obvious, comment about one of his culture's more widely accepted regions. Although substantial differences of opinion existed concerning its exact location, and many argued that it did not exist at all, most could agree on its general position. There were many different evaluations of the nature of its environment, although nearly everyone accepted the idea that intense heat was a necessary element. All the available information about the region came not from individual experience but from written materials, the words of others, folktales, and beliefs. Even though they personally had not experienced the place, most people were willing to state firmly their evaluation of the general nature of that region and of the quality of the experience to be had there. In short, it was a region in much the same way that the sections of the United States and Canada that we are discussing are regions.

A region, as we use the term, is very much a mental construct. One person may choose to divide an area into one set of regions; another may regionalize the same area in a different way. In part, this may be due to different goals or different information. A district sales manager who wishes to determine the precise borders of the sales regions of the people working for him or her would probably create a pattern quite unlike that of a botanist interested in regional variations in tree types. Similarly, a regionalization based on agricultural patterns might bear little resemblance to one based on manufacturing activities. One set of geographers, when subdividing the United States and Canada into regions, may well create a map noticeably different from that formed by a second group of geographers simply because each group selected different elements as most influential to their patterns.

These various reasons for regionalizing and the alternative attitudes expressed by the regionalizers obviously can lead to the creation of many different maps of any given territory. No one regional scheme is necessarily "better" than another. If it meets the requirements of its creators, if those requirements were thought out reasonably, and if the map is executed correctly and factually, then the regionalization must be perfectly acceptable. Geographers use regions as a neat system of categorization, a way of organizing a complex, large set of facts about places into a more compact, meaningful set of areas. As with any categorization, the regions are satisfactory if they identify understandable patterns in the facts and if they help clarify the complex patterns.

To geographers, a region can be either *nodal* or *uniform, single featured* or *multifeatured.* A nodal region is characterized by a set of places connected to another place—a focus (or node)—by lines of communication or movement. The places that make up this region, therefore, are associated with each other because they share a common focus. The circulation area of a newspaper is an example of such a region, as is a "milkshed"—the set of places surrounding a city that send their raw milk production to the markets of that city. Each of these separate places in the region may be quite diverse in appearance, but because they are tied together, they constitute a single nodal region.

In comparison, a uniform region is a territory with one or more features present throughout the area and absent or unimportant (or much diminished) beyond the territory. The entire area where the cottonwood constitutes a significant part of the total tree cover may be such a single-featured, uniform region. Or a uniform region may represent some characterization of the total environment of a territory, including both its physical and cultural features. These quite general and frequently subjective regions are meant to reflect distinctive mixtures of all elements of the landscape. It is this last type of region that we use for the general structure of this book.

As we said earlier, there is an almost endless variety of elements that could be considered in the creation of these regions. In fact, however, our perception of the nature of a region, of the things that together shape its personality, is based on a relatively small group of criteria. In each major section of the United States and Canada, we have tried to identify the one or two underlying themes that reflect ways in which the human population has interacted (within itself or with the physical environment) to create distinctive regions. The most important identifying themes for a region may vary greatly from one region to another. It is impossible to speak of the Southwest of the United States without a focus on aridity and water erosion, of the North without its cold winters, or of the Northeast without cities and manufacturing. The key element that establishes a total uniform region, then, is not how that section compares with others on a

American landscapes are often a creation of the culture of the people who live there rather than the physical environment. The houses strung out along the main road, the large size of the church, and the long, narrow farm lots all help identify this as a village in Quebec. (George Hunter/Miller Services)

predetermined set of variables, but how a certain set of conditions blend there. It is not possible to create such regions by simply amassing a large set of statistics and then grouping areas and places on the basis of relatively similar numbers, proportions, or averages. To follow such an approach ignores the personality of a place. It is that personality that gives meaning to the regions we use in the text.

Students in a geography class at Ohio State University were asked to identify those parts of the United States that had meaning to them as regions and to locate these regions on a map (Figure 1-4). What emerged most clearly was that certain sections of the country were well defined in the minds of most of the students, with the result that these sections were named and located consistently by the students. Such sections as the Southwest, New England, and the South were almost always indicated, and their borders, broadly stated, were similar. Other parts of the country, such as the Middle Atlantic states, were often excluded from the regionalization entirely.

With minor variations, this pattern of identification would be repeated by individuals from

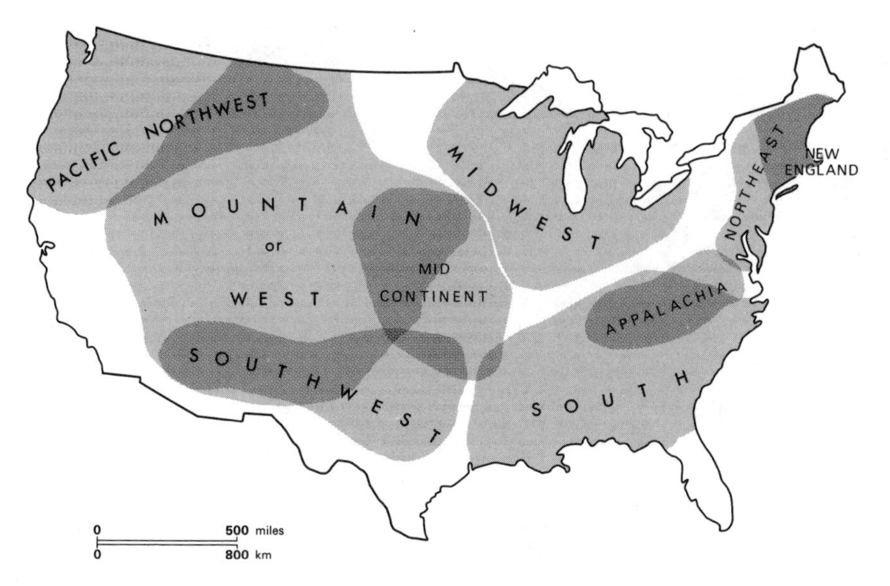

Figure 1-4 Perceived regions of the United States. Is this set of perceptions from students in Ohio likely to match patterns arising from groups in other parts of the country? How might they differ?

most areas of the country. Such places as New England and the South are regions with a firm personality, one that most Americans recognize and can at least loosely elaborate. As with Hell, the South exists as a region for almost all Americans, even though opinions about its character may vary widely. Similarly, to most Canadians, British Columbia and the Maritimes conjure up definite impressions, although residents of Nova Scotia and Ontario may differ outrageously in the specifics of these impressions.

An elaboration of the idea of a uniform region suggests that every such region is composed of several distinct zones. Somewhere within a region there is a core area in which the major identifying features are most obvious. This is usually a well-accepted area that is clearly a part of a region, even by the most stringent application of the selection criteria. Somewhere toward the margins of a region these identifying features grow fuzzy or less distinct; we grow less sure of the appropriateness of, or have greater disagreement over, the specifics of our regional assignment. For example, although nearly everyone would accept central Alabama and Mississippi as a part of the South, opinion varies as to what to do with southern Illinois or the Louisville area or much of northern Virginia. Boundaries between multifeatured regions are seldom sharp or well defined on the landscape. Regions blend into each other, and travelers are often well into one region before realizing that they have left another. The boundary between such regions is usually a transition zone, not a sharp line.

It is, therefore, often difficult to map the boundaries of a multifeature region. If we were to view regions simply as a form of categorization, we might convince ourselves that we must create regions that occupy mutually exclusive portions of the earth's surface and are represented equivalently by all of the data for that portion—that is, every place would be a part of one and only one region and would be characterized in the same way as all others. We feel that the arbitrary nature of such an approach is inappropriate and sometimes misleading.

Within this book, regions have been presented largely as though they are distinct territorially even though they are not. The "feeling" of a region we wish to present is a function of place, but it is also (and perhaps more important) a function of the subject theme chosen. Therefore, for example, the intense urban character of Megalopolis is discussed in Chapter 5, but the aspects of manufacturing that affect New York, Philadelphia, and Boston and other manufacturing core cites are presented in Chapter 6. There are two important aspects of regional "feeling" in the region usually called "the Midwest"—the urban-industrial and the rural-agricultural. Both are important enough for us to treat each separately in some detail. Therefore, students who wish to have a complete picture of North American geography should be aware of the two major themes of the Midwest, the two aspects of Megalopolis, the three or four pertinent phases of the West, and so on.

Rigid regional boundaries do not fit the landscape of the United States and Canada. A given portion of the continent's territory may be occupied by parts of two or more regions, but the boundaries of many regions may also be fairly broad transitional zones. These zones are areas that contain many of a region's characteristics even though their diminished strength indicates the area is beyond the edge of the region itself. At times, these zones mark an area where the mix of characteristics is so subtle or complex that it is difficult to assign the area to any one region. Parts of the margin between the Deep North and Great Plains are examples of this, as are sections of the transition between the Deep North and the Deep South. Therefore, students should be forewarned (1) that the regions we discuss in this text are complex, (2) that because of this complexity, the regions are not exclusive territories—they may overlap other regions, and (3) that the outer edges of the regions blend gradually into the neighboring areas and should not be viewed as hard-and-fast lines.

Regional boundaries and regions themselves are not static. Settlement patterns shift, society develops significant new technological abilities, and political patterns are altered; regions reflecting these patterns may expand, contract, appear, or disappear. A regionalization of North America for 1492 would have been quite different from one for 1776, 1865, or 1991. There is no reason to believe that the pattern for 2091 will be at all similar to that for 1991.

An examination of the regions that we have created for this text indicates a subdivision that should be generally recognizable (Figure 1-5). Again, this is because many of the regions of the two countries are widely accepted. Other regions represent combinations that are normally not expected. For example, consider the Bypassed East, a combination of the Atlantic Provinces of Canada, the Adirondacks of New York, and northern New England. Most casual observers firmly lump all of New England into one region. This is largely an historical carryover, reflecting the long-term identification of the states of New England as a separate region with strong cultural cohesion. But there have been great changes in southern New England in recent decades because of heavy immigration and urbanization. Still, the popular image of the New England Yankee and small towns with village greens has not changed.

Several of the regions closely follow political boundaries. The reason for this in Hawaii is obvious. California is separated from most of its adjacent landscape because of its leadership role in changing the culture of America and its statewide political "solutions" to its local resource problems. Megalopolis has been defined traditionally along county lines, and we choose to follow that approach. It does serve to make the collection of statistics about these regions much easier, for nearly all information bases follow standard political boundaries.

Finally, it is significant that only one of the regional boundaries follows the border between the two countries. Few international borders are as loose as this. Communication and transportation across the border is frequent, and economic and settlement changes influence both sides of the boundary. It should not be supposed,

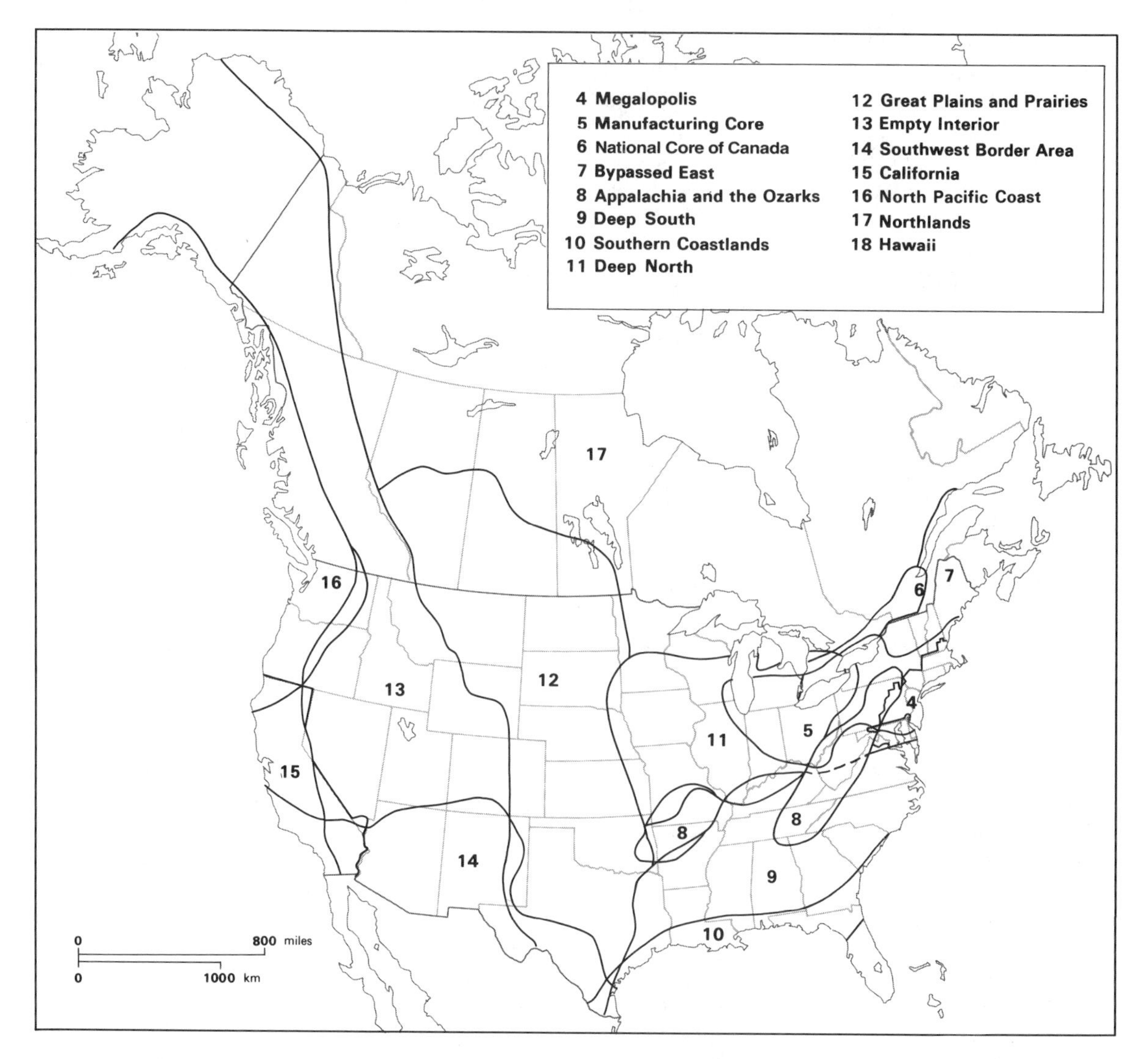

Figure 1-5 Regions of the United States and Canada. Separated by theme but overlapping territorially, the regions used in this book reflect the complex variations that characterize the geography of these two countries.

however, that the Canadian-American border has no meaning. Although the Great Plains region extends easily into both countries, differences in early settlement, governmental land policies, and general economic conditions result in noticeable differences in the plains landscape in the two countries. Canada and the United States are remarkably cooperative, and their settlement histories are similar, but there are many differences between them.

As we mentioned, each of the regional chapters is developed around one or a few basic themes (Table 1-1). Most of these regional themes are drawn at least indirectly from the

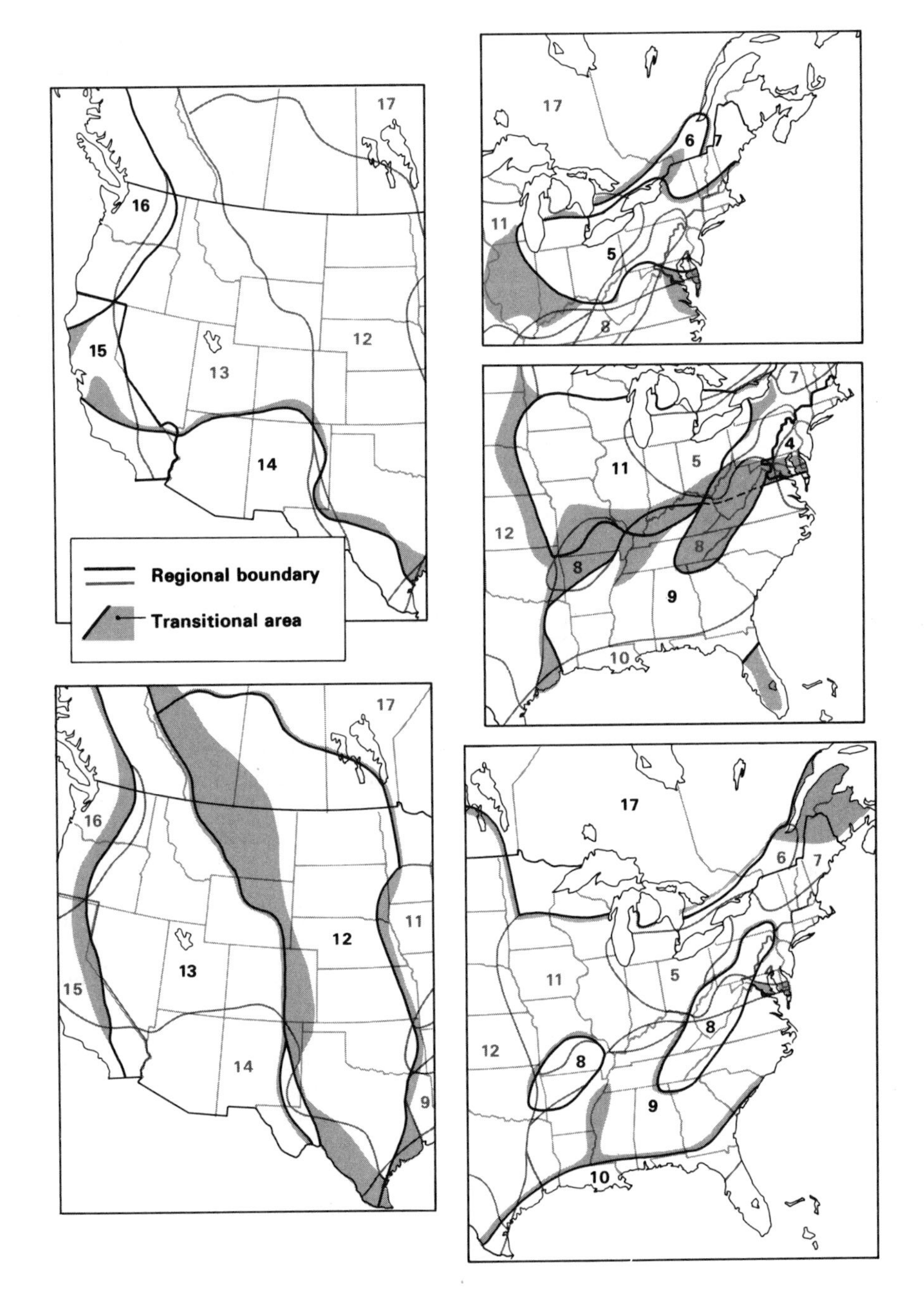

Figure 1-5 *(Continued)*

Table 1-1 Regional Themes

Chapter	Chapter Title	Themes
5	Megalopolis	Urban growth and sprawl, urban problems
6	North American Manufacturing Core	Industrialization, how regions change
7	The National Core Region of Canada	Federalism, cultural pluralism
8	The Bypassed East	Relative isolation, regional economic problems
9	Appalachia and the Ozarks	Resource dependence, governmental intervention
10	The Deep South	Culture
11	The Southern Coastlands	Amenities and economic resources, development conflict
12	The Deep North	Agriculture's cultural and economic foundations
13	The Great Plains and Prairies	Environmental perception, human/land interaction
14	The Empty Interior	Settlement in a difficult environment (environmental impact)
15	The Southwest Border Area	Ethnic diversity
16	California	Perception of amenities, population redistribution, environmental dependence
17	North Pacific Coast	Regional independence
18	Northlands	Settlement in a difficult environment, resource extraction, governmental intervention
19	Hawaii	Ethnic diversity, isolation, environmental impact

basic themes of the entire book. In certain regions, the expression of some themes is stronger or clearer than others. The themes are intended to provide an explicit basis for treatment of information about the region, although, in many chapters, it will not be difficult to identify elements of national or continental geography.

ADDITIONAL READINGS

Canada, Surveys and Mapping Branch, Geography Division. *National Atlas of Canada*, 4th ed. Toronto: Macmillan, 1974.

Doster, Alexis, et al. (eds.) *The American Land.* Washington, D. C.: Smithsonian Institution, 1979.

Hart, John Fraser (ed.). *Regions of the United States.* New York: Harper & Row, 1972.

Kazin, Alfred. *A Writer's America: Landscape in Literature.* New York: Alfred A. Knopf, 1988.

Jordan, Terry J. "Perceptual Regions of Texas." *Geographical Review*, 68 (1978), p. 293–307.

McCann, L. D. (ed.). *Heartland and Hinterland: A Geography of Canada.* Scarborough, Ont.: Prentice-Hall Canada, 1982.

Newmann, Peter C. *Sometimes a Great Nation: Will Canada Belong to the 21st Century?* Toronto: McCelland and Stewart, 1988.

Reed, John S. "The Heart of Dixie: An Essay in Folk Geography." *Social Forces*, 54 (1978), pp. 925–939.

Shortridge, James, R. "Changing Usage of Four American Regional Labels." *Annals of the Association of American Geographers*, 77 (1987), pp. 325–336.

Steiner, Michael, and Clarence Mondale. *Regions and Regionalism in the United States: A Source Book for the Humanities and Social Sciences.* New York: Garland, 1988.

U.S. Geological Survey. *National Atlas of the United States of America.* Washington, D.C.: U.S.G.S., 1970.

Watson, J. Wreford. *The United States: Habitation of Hope.* New York: Longman, 1982.

Zelinsky, Wilbur. *Nation into State: The Shifting Symbolic Foundations of American Nationalism.* Chapel Hill: University of North Carolina Press, 1988.

CHAPTER

2

Geographic Patterns of the Physical Environment

Some landscape characteristics are most easily discussed when they are used to highlight particular regional themes. Such characteristics are often reflected in the regional thematic organization of this text. For example, the basic pattern of manufacturing in the United States and Canada is discussed in Chapter 6 and that of religion is identified in Chapter 3. Some subjects are developed in several contexts. Ethnic diversity, for example, is a theme in Chapters 6, 7, 11, 15, and 19; agriculture is discussed in almost all of the chapters.

Nevertheless, several important geographic patterns are not a focus of any single region. They are so broad that they must be dealt with independently of the regional structure. Most elements of the physical environment—topography, climate, and vegetation—are important almost everywhere. Because our regions are expressions of human-landscape patterns, we consider the physical environment only one of several factors that may help explain the character of human geographic distributions in the United States and Canada. Even though the physical environment may be the most important element in some regions, we have chosen to discuss its broad patterns in this chapter.

The physical environment of the West of the U.S. and Canada, especially, is often characterized by extreme variation across very small distances. This eastern California view is from the Owens Valley, in the Intermontane Basins and Plateaus physiographic region, looking westward toward the massive east face of the Sierra Nevada, part of the Pacific Mountains and Valleys region. (Peter Menzel/Stock, Boston)

THE PHYSICAL ENVIRONMENT

Topography

Compared to a continent such as Europe, where a traveler can encounter very different landforms within a small area, the topographic regions of Canada and the United States tend to be large (Figure 2-1). If you select your route correctly, it is possible to drive for several days through areas with little variation in landforms.

The major topographic features of the continent tend to extend north-south across the two countries. The interior of the continent is a vast, sprawling lowland that stretches from the Gulf of Mexico to the Arctic. At times close to 5000 kilometers (3000 miles) wide, this lowland encompasses well over half the entire land area of the continent. Geomorphologists (scientists with an interest in landform development) place this expanse of flat land and gently rolling hills in three different physiographic regions—the Atlantic and Gulf coastal plains, the interior lowland, and the Canadian (or Laurentian) Shield. Some split the interior lowland into two parts, the Great Plains and the interior plains. These regions may be best differentiated by the processes that led to their appearance and by their subsurface (bedrock) materials. However, there are only moderate differences in the visual appearance of the landforms in each of these three major divisions. More significant in differentiating the regions, at least to the untrained eye, are variations in climate and in the cultural landscape.

The Atlantic and Gulf coastal plains reach north along the east coast of the United States as far as the southern margins of New England. Underlying this area are beds of young, soft,

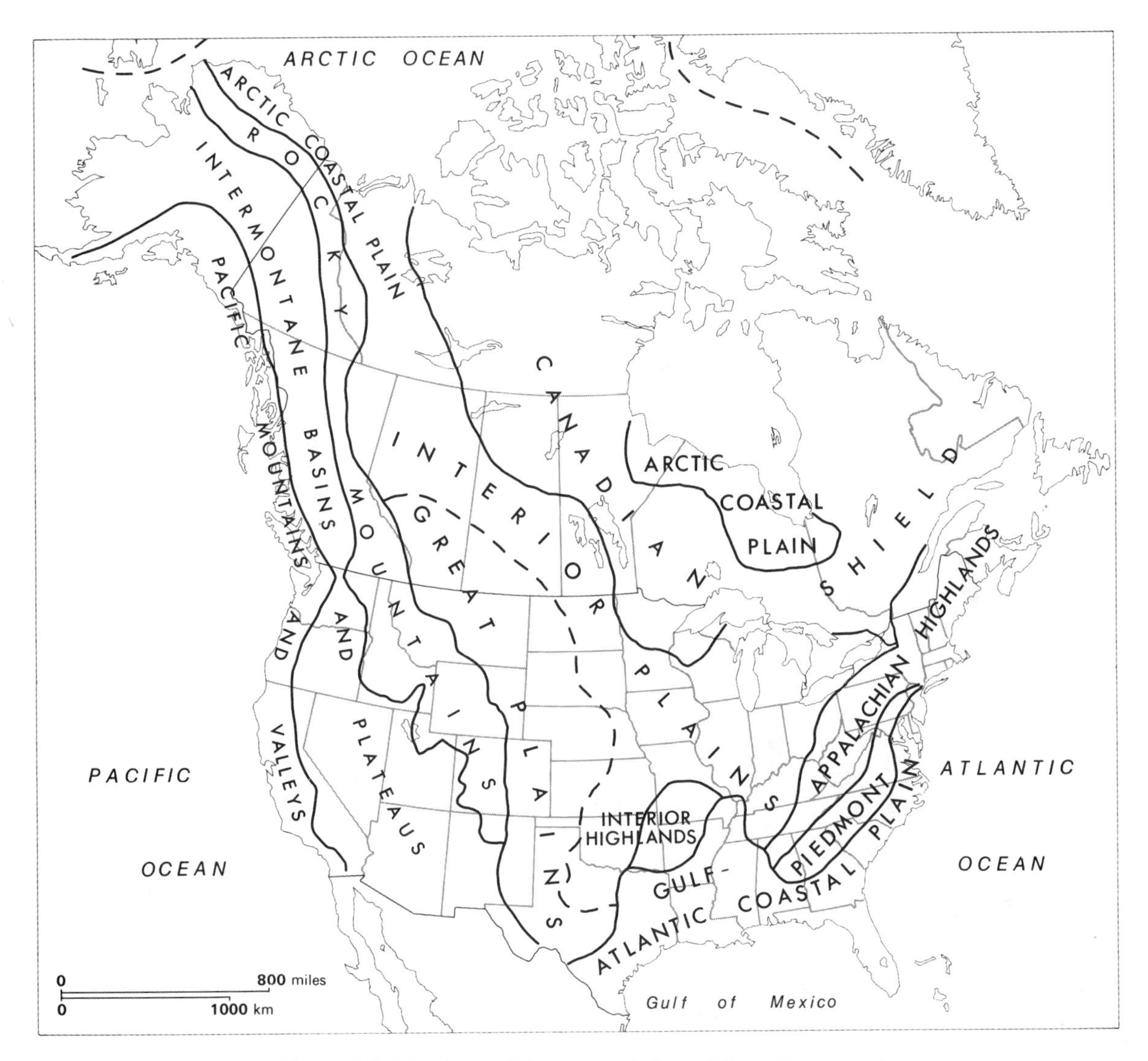

Figure 2-1 Physiographic regions (adapted from Fenneman).

easily eroded rock deposited in recent geologic time as shallow seas lapped back and forth across the land. These low plains actually extend well out under the current ocean surface to form a continental shelf, which in places extends as much as 400 kilometers (250 miles) beyond the shore.

The coastal plains are generally flat and often swampy. For example, a geomorphology field trip conducted annually near Baton Rouge, Louisiana, visits a hillside where the students can observe the boundary between one rock bed and the one above it, a difference in elevation of little more than 1 meter (3 feet). In many locations across the coastal plains, farm fields are bounded by drainage ditches 5 or more feet deep

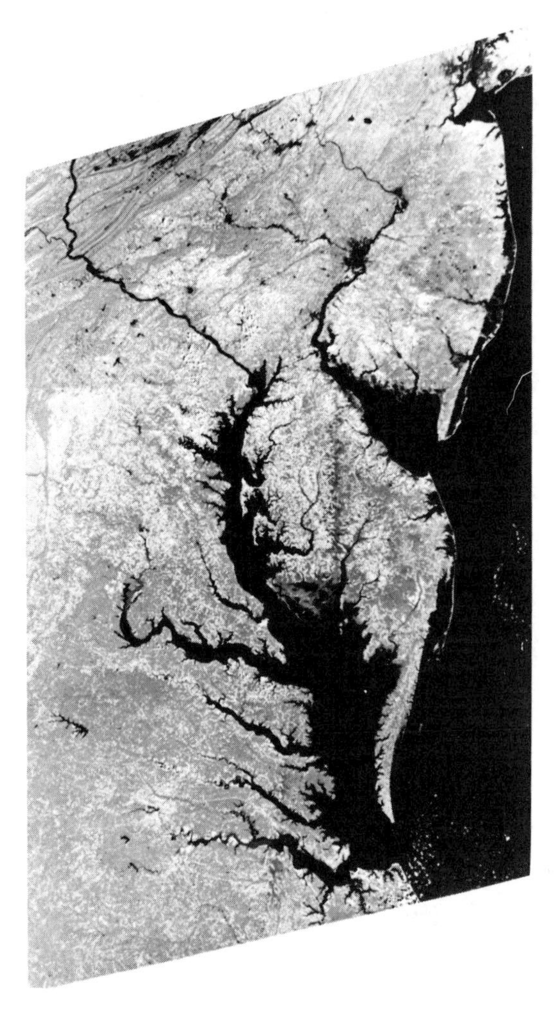

This is a composite satellite photograph of the U.S. coast from New York City to Norfolk, Virginia. The excellent harbor facilities, seen here in Chesapeake Bay, Delaware Bay, and New York harbor, have played an important role in the economic growth and urbanization of this region. (NASA)

to handle runoff from heavy rains and to lower the water table. These are examples of a very flat environment.

Northward, the interior of the continent is occupied by the central, or interior, lowland. Although noticeably hillier than the coastal plains, there is almost no rough terrain in the lowland. This region is like a saucer, turned up at the edges and covered with a deep series of sedimentary rocks. The rocks are older and more durable, can withstand erosion more effectively, and hence contribute to a more varied topography than do those of the coastal plains. These sedimentary beds are generally quite flat, and most topographic variation in the interior lowland is the result of local erosion or, in the north, of glacial debris deposited during the Ice Age. Still, to observers accustomed to a mountainous landscape, the overwhelming topographic impression of the lowlands is a bland, slightly modified flatness.

The geologic structure of the Great Plains differs little from that of the interior plains. The sedimentary beds dominate, although in the north they are broken by some eroded domes, most notably the Black Hills of western South Dakota. While nearly horizontal, the sedimentary beds do dip gently toward the west to a trough at the foot of the Rocky Mountains. The Colorado cities of Denver and Colorado Springs are both located in that trough. The boundary between the Great Plains and the interior plains is marked by a series of low escarpments that indicate the eastern edge of the mantle of loose sediments, eroded from the Rocky Mountains that covers the Plains. The sediment cover creates a gentle eastward slope from an elevation of 1750 meters (5500 feet) at the foot of the Rockies to 650 meters (2000 feet) at the section's eastern edge. One result of the regularity of this gradient in the U.S. Great Plains is a parallel west-to-east stream drainage pattern.

The character of this massive interior lowland area has had a number of important influences on the economic and settlement history of the United States. In addition to the vast agricultural potential it provides, fully half of the country can be crossed without encountering any significant topographic barriers to movement. This has facilitated the integration of both this region and the distant West into the economic fabric of the country. Nearly all of the interior lowland is drained by the Mississippi River or its tributaries. This drainage pattern also assisted regional integration by providing a transport and economic focus for the land west of the Appalachians. The Mississippi River system easily outstrips the Great Lakes in terms of the total volume of U.S. commerce it carries. To many of its residents, this region is the heart of America—vast, economically powerful, productive, separated from the outside influences felt to be so common along the coast. These residents may overstate their case, but it is hard to deny that this is the region that most clearly represents many of the popular images of America.

North and northeast of the central lowland lies the Canadian Shield. At over 2.8 million square kilometers (1 million square miles), it is the two countries' second largest physiographic region. Here, old, hard crystalline rocks lie at the surface. Farther south in the lowlands, similar rocks are covered by the sedimentary beds deposited under the sea that once filled the midsection of the continent. A long history of erosion has worn down the surface of the Shield into a lowland of small local relief.

The Shield, more than any other North American physiographic region, has had its landforms remolded and shaped by massive continental glaciers during the last 1 million years. Vast sheets of ice, probably originating on either side of Hudson Bay, have repeatedly crept outward before melting back. At their greatest extent the massive ice sheets covered nearly 13 million square kilometers (5 million square miles), two-thirds of the total land area of the continent. These continental glaciers covered most of Canada and Alaska east of the Rocky Mountains and the coast ranges, and they reached southward to approximately the present valleys of the Missouri and Ohio rivers. The massive weight of the ice, up to 1.6 kilometers (1 mile) thick in places, gradually depressed the land on which it rested. Much of the Shield is still "rebounding," slowly rising in response to the removal of the ice.

Major landscape changes were caused by this moving ice cover. The ice could pluck rocks weighing many tons off the surface and carry them great distances. Massive boulders called *erratics*, often many feet in diameter, are strewn across the landscape of the Shield, resting where they were dropped by the glaciers. Preexisting drainage patterns meant little to the glaciers, and old patterns were disrupted greatly by the passage of the ice. Ice melt along the peripheries of the glaciers created major rivers and cut broad new pathways to the sea.

Glaciation scoured much of the Shield's surface. Today the soil cover of the region remains thin or nonexistent. The heavily disrupted drainage pattern dammed many streams with debris and led others into the area's labyrinth of lakes and swamps rather than to the sea. Central and northern Minnesota, for example, called the "Land of 10,000 Lakes," is part of the southern lobe of the glaciated shield that extends into Minnesota, Michigan, and Wisconsin.

The landscape impact of continental glaciation is also found far south of the Shield. Southward, where the ice was not as thick and its force correspondingly less, the glaciers were diverted or channeled by higher elevations. For example, the ice was blocked in central New York by the highlands south of the Mohawk River. Narrow probes, however, did push up the valleys of streams tributary to the Mohawk, gradually broadening and deepening them. As the ice began to melt around the glaciers' peripheries faster than it was replaced by new ice, the glaciers "retreated" northward. They rested for a long period where they had piled up against the highlands before sufficient melt occurred to pull the ice away from the margins of the hills. During this period, as the ice melted, the debris that the glaciers had collected was deposited along the glacier's margins. These deposits created a series of natural dams across the tributary streams. Today the deep, narrow Finger Lakes of New York State fill these glacially enlarged valleys and form one of the country's truly beautiful landscapes.

All along and beyond the southern edges of the glaciers, deposition replaced erosion as the prime result of glaciation. Large areas of the interior lowland are covered by a mantle of glacial *till* (rocks and soil dropped by the glaciers). The till covers the land to depths varying from a meter or less to more than 100 meters (330 feet). Where the glaciers remained stationary for long periods of time, higher hills, called *moraines,* were created. In the east, Staten Island, Long Island, Martha's Vineyard, Nantucket, and Cape Cod are end moraines that mark the farthest major extension of glaciers toward the southeast. The landscape south of the Great Lakes is laced with long, low, semicircular moraine ridges and other glacial deposits.

One section of the generally glaciated portion of the interior lowland escaped glaciation. The southwestern quarter of Wisconsin and the adjoining 400-kilometer (250-mile) stretch of the Mississippi River valley was apparently spared by the barrier effect on the flowing ice of the Superior upland to the north and by the channeling of the ice by the deep valleys of Lake Michigan and Superior. The result is the Driftless Area (*drift* is another name for till), a local landscape that is more angular, with fragile rock formations like natural bridges and arches that were destroyed in glaciated areas. The more rugged topography, and poorer quality soil resulting from the lack of a till cover, makes the Driftless Area less useful agriculturally than the surrounding glaciated areas.

As the ice retreated, massive lakes were created along the glacial margins as the ice blocked normal drainage patterns. On the northern Great Plains two huge lakes, Agassiz and Regina, together covered an area larger than today's Great Lakes. With continued glacial retreat, these lakes mostly disappeared. Their existence is now marked by the former lake bed, a flat area covering parts of Manitoba, Saskatchewan, North Dakota, and Minnesota. The St. Lawrence River, the outlet for the Great Lakes, was dammed for a time by the ice; the flow from the lakes was diverted southward into the Mississippi and Susquehanna rivers.

Sea level was significantly lower during periods of widespread glaciation. This lowered the *base level* of many rivers and, thus, fostered increased erosion by those streams. Furthermore, many of these stream valleys extended well into what is today the ocean because of the lower sea level. Along with many others, the Susquehanna and Hudson rivers cut much deeper valleys during this period. As the ice retreated and sea level rose, the ocean filled these deepened valleys. Two of the world's finest harbor areas were formed in this way: New York Bay with the deep Hudson River and the protective barriers formed by Staten Island and Long Island, and Chesapeake Bay, the drowned valley of the Susquehanna River and some of its major former tributaries, such as the Potomac and the James rivers.

In the east, the coastal plains are gradually squeezed against the coast northward along the ocean by the Appalachian Highlands until the lowland disappears entirely at Cape Cod. From there northeastward to Newfoundland, the coastal landscape is a part of the northern extension of the Appalachian system. The Appalachians separate the seaboard from the interior lowlands along much of the eastern United States. These are old, eroded remnants of what were once much higher mountain ranges. Today they are less than half the height of their counterparts in the interior west. Mt. Mitchell, in North Carolina, the tallest peak in the east, is 2036 meters (6684 feet) high.

These mountains are important, despite their relatively low elevations. The two regions dominated by these peaks and ridges, the Bypassed East and Appalachia, have had a settlement and economic history profoundly affected by the mountains. Soils in most parts of the Appalachians and the Bypassed East are shallow, and as Robert Frost once noted, rocks are often a farmer's most dependable crop. The steep slopes, difficult to farm under any circumstances, are totally unsuited to modern agricultural practices that emphasize mechanization. Large-scale urban or industrial growth is cramped by the small local lowlands. Pittsburgh lays claim with considerable justification to the distinction of being the hilliest large city in North America. Early settlers found the Appalachians from the Mohawk River in New York southward to northern Alabama to be a surprisingly effective barrier to western movement; there are few breaks in the mountains' continuity. Inadequate transportation remains a major regional problem.

The Appalachians, although generally rugged throughout, do contain several noticeably different kinds of landforms. South of New England, the eastern Appalachians are an intricate pattern of folded linear ridges and valleys. In the west, the mountains offer a jumbled landscape that was created as many streams cut into the region's level-bedded sedimentary rocks. In Newfoundland and the Atlantic Provinces, massive bubbles of igneous rock created huge domes that have been heavily eroded by water and ice. These domes are most obvious in the highlands of central New Brunswick and on Cape Breton Island in Nova Scotia. New England's landscape appears to represent a little of both, with perhaps a dash more of the combination from the southern Appalachians. These variations will be discussed more thoroughly in the appropriate regional chapters.

If the physical landscape of North America from the Great Plains eastward can be characterized as a gently rolling terrain punctuated by mountains in a few places, the West is a land of mountains and of sudden, great changes in local elevation. Few parts of the West are beyond sight of mountains. Instead of punctuating flatness, the mountains form and mold the landscape, dominating all of its other elements. The physiography again is arranged in a series of three large north-south trending bands, with the Rocky Mountains on the east separated from the mountains and valleys of the Pacific coastlands by a series of high, heavily dissected plateaus. On the map, these landscape regions create the impression of great homogeneity. In fact, there is a great visual diversity in this section of the continent.

Starting in the east, the Rocky Mountains generally present a massive face to the Great Plains, with peaks occasionally rising a mile or more above the plains. The Front Range of central Colorado is a fine example of such a sudden change in elevation. Elsewhere, as in south-central Wyoming, the Rockies almost seem not to exist at all. Travelers passing through Wyoming on Interstate 80 must search to find the mountains through which they are passing. Really a western peninsula of the Great Plains, the Wyoming Basin has allowed millions of travelers to circumvent the more rugged sections of the Rockies. In the northern Rockies in Idaho, the north-south linearity of most of the region's mountains is replaced by massive igneous domes irregularly eroded into a rugged, extensive series of mountain ranges that contain the largest remaining area of wilderness in the United States outside Alaska. Farther north still, parts of the Canadian Rockies, carved by alpine glaciation during the Pleistocene, contain some of the most beautiful mountain scenery on the continent.

The high plateaus of the interior West are also varied in their origin and appearance. The southernmost subsection, the Colorado Plateau, is a series of thick beds of sedimentary rocks rising more than 1000 meters (3300 feet) above the lowlands' elevation and tilted upward toward the northeast. The Grand Canyon of the Colorado is most dramatic when viewed from the south rim because the higher north rim blocks the view of the relatively flat, less exciting land beyond. The Colorado Plateau may be the most geologically colorful part of the continent. It is a land of spectacular canyonlands, volcanic peaks, sandy deserts, and the Painted Desert. More than a dozen national parks and monuments help preserve this natural treasure trove. The basin and range subsection is characterized by interior drainage, with many of the area's usually small or intermittent streams ending their courses in the area's lakes and dry sinks. Much of the region has been covered with a thick mantle of alluvium washed down from the mountains. To the east-west traveler, the area appears to be occupied by a series of narrow mountains and low barren lowlands in boring repetition.

Farther north, the Columbia-Snake Basin has been filled by repeated lava flows to a depth of

The West in both countries is a land of dramatic landscapes. Few places from the Rocky Mountains westward are beyond the sight of mountains. This view is of Mt. Baker in the North Cascade Range of Washington rising above the early morning fog. (Roy Bishop/Stock, Boston)

more than 1000 meters (3300 feet). Rivers, both past and present, have eroded into the rock. The resultant landscape is similar to that of the Colorado Plateau, although the stepped appearance resulting from the variable resistance to weathering of the eroded sedimentary rocks of the Colorado Plateau is missing. Volcanic cones also dot portions of the region, especially across south-central Oregon and in the Snake River valley in Idaho.

After a brief disappearance along the international border, the plateaus reappear farther north. They gradually widen northward, encompassing the valley of the Yukon River in Alaska. There is a complex of lava plateaus, low mountains, and river plains in British Columbia and the Yukon Territory. In comparison, much of central Alaska is a broad, flat lowland that is poorly drained.

Along the Pacific Coast, the pattern of complexity within apparent simplicity is repeated. In the conterminous United States (i.e., excluding Alaska and Hawaii), the region seems to consist largely of two north-south tending mountain chains separated by a discontinuous lowland. In southern California, the Coast Range is fairly massive, with peaks reaching 3000 meters (10,000 feet). From there almost to the Oregon border, the mountains are low and linear, seldom rising above 1000 meters (3300 feet) in elevation. This also is the major fault zone of the state and a region of frequent earthquake activity. Along the California-Oregon border, the Klamath Mountains are higher, more extensive, and much more rugged and irregular—much like the Idaho Rockies. Except for the Olympic Mountains on the Olympic Peninsula in northwestern Washington State, the Coast Ranges in the rest of Oregon and Washington State are low and hilly rather than mountainous. In Canada, the Coast Ranges, again higher in elevation, are called the Insular Mountains as they extend along the coast of British Columbia.

The interior lowlands along the coast—the Central Valley of California, the Willamette Valley in Oregon, and the Puget Sound lowland in Washington and British Columbia—are the only extensive lowlands near the West Coast. Filled with relatively good soils, these lowlands have supported much of the Pacific Coast's agriculture. Most of the poplation of the North Pacific Coast lives in the Willamette Valley and Puget lowland. The Central Valley, especially, has some of the flattest land to be found anywhere in the United States.

East of the lowlands are the Sierra Nevada and the Cascade ranges. The Sierra Nevada appears as though a massive section of earth was tilted upward relative to the areas to the east and west in what is called a *fault block*, with the highest, sharpest exposed face toward the east. Although the western approaches into the Sierra Nevada are reasonably gentle, on the eastern side, the mountains rise in some places more than 3000 meters (10,000 feet) in just a few linear miles. Volcanic activity was important in the formation of the Cascades. Some of the continent's best known volcanic peaks, such as Mt. Rainier and Mt. St. Helens in Washington, are found there. As the Cascades continue into Canada, they are renamed the Coast Mountains.

Climate

Climate is the aggregate of day-to-day weather conditions over a period of many years. It is the result of the interaction of many different elements, with the most important being temperature and precipitation. Climatic patterns are a result of the interaction of three geographic controls. The first is latitude. The earth is tilted on its axis with reference to the plane of its orbit around the sun. As it makes its annual revolution around the sun, first the Northern Hemisphere, then the Southern Hemisphere, is exposed to the more direct rays of the sun. During the Northern Hemisphere's summer, higher latitude locations have longer days, with far northern points experiencing a period of continuous daylight. Conversely, daylight periods during the winter months are shorter at higher latitudes. Places north of the Arctic Circle receive no daylight at all at midwinter. Latitudinal variation in temperature is, thus, minimized during the summer when the heating of the more direct sun's rays at lower latitudes in the United States is partially balanced by the longer period of daylight in northern Canada or Alaska. During winter, more southerly locations have both longer days and exposure to more direct rays than northern places and latitudinal variations in temperature are much greater. Thus, Canadians and residents of the U.S. northeast can journey to Florida in January and find temperatures averaging 15°C or 20°C (roughly 30°F or 40°F) warmer than those they left behind. Floridians making the reverse trip in summer would find temperatures only slightly [perhaps 6°C (10°F)] lower than those in Florida. Thus, seasonality in temperature is greatest in the middle latitudes and the high latitudes, and warm to hot summers are widespread.

The second control is based on the relationship between land and water. Land tends to heat and cool more rapidly than water. In a tendency called *continentality*, places far from large bodies of water experience greater seasonal extremes of temperature than do coastal communities. Parts of the northern Great Plains and the Prairie provinces experience annual temperature ranges close to 65°C (150°F); annual differences of as much as 100°C (180°F) [from +50°C (+120°F) to −50°C (−60°F)] have been recorded in some locations. The larger the landmass, the greater is this annual range in temperature at continental locations. North America is a very large landmass.

The converse effect occurs at maritime locations, especially on the west coast of continents in the midlatitudes. These locations have much smaller temperature ranges as a result of what is called a *maritime influence.* Vancouver, British Columbia, for example, has an average annual temperature range of only 15°C (27°F). Summer and winter extremes are moderated by the movement onshore of prevailing westerly wind systems from the ocean. Horizontal and vertical ocean currents minimize seasonal variations in the surface temperature of the water. The moderated water temperature then serves to curb temperature extremes in the *air mass* above the surface.

Proximity to large water bodies also tends to have a positive influence on precipitation levels, with coastal locations receiving generally higher amounts. The reason for this should be obvious; large water bodies provide greater levels of evaporation and thus increase the amount of moisture in the atmosphere. That, in turn, increases the possibility of precipitation. There are, however, notable exceptions to this general rule. An example is the dry coast of southern California. Moisture-producing weather systems that develop over the north Pacific and then move southward are blocked during much of the year by a large stable air mass that maintains the hot, dry weather of that coastline. The Arctic coastline of Canada and Alaska is even drier. Here the long winters minimize atmospheric humidity. Very cold, stable air dominates the Arctic for long periods, reducing still further the likelihood of snowfall.

The third prime geographic influence on climate is topography. Most obvious is the relationship between elevation and temperature, with higher elevations cooler than lower elevations. The influence of topography can be broader, however, because of its effect on wind flow. If a major mountain chain lies astride a normal wind direction, the mountains force the air to rise and cool. As the air mass cools, the amount of moisture that it can hold is reduced. Precipitation results if the cooling causes the *relative humidity* to reach 100 percent. Moisture falls on the windward side, and the lee is dry. The wettest area in North America is along the Pacific Coast from Oregon to southern Alaska, where moisture-laden winds strike mountains along the shore. Average annual precipitation is more than 200 centimeters (80 inches) throughout the area and in some places exceeds 300 centimeters (120 inches) (Figure 2-2). Mountains also can reduce the moderating effects of maritime conditions on temperature, as happens in the interior of the Pacific Northwest in the United States. Within the United States and Canada, the Western Cordillera (mountain mass) confines West Coast maritime climatic conditions to that coast. Some of the greatest variations in both precipitation and temperature to be found across a small distance anywhere on the continent exists between the west and east sides of parts of the Coast Ranges. The aridity of the central and northern interior West is due in large part to the barrier effect of the north-south-trending mountain ranges of the West.

East of the Rockies, the topographic effect on precipitation eventually disappears, partly because the eastern mountains are much lower, and thus pose less of a barrier to moving air, and partly because much of the weather of the interior is a result of conflict between two huge air masses that are unimpeded, one flowing northward from the Gulf of Mexico and the other flowing southward out of Canada. The contact of these two different air masses creates what are often violent displays of weather in the region.

This illustrates a fourth major and complex influence on climate, the impact of air mass characteristics and wind systems. The weather of most of the two countries is affected markedly by the confrontation between polar continental air masses (usually cold, dry, stable) and tropical maritime air masses (warm, moist, unstable). The former push farthest south in winter, whereas the latter extend farthest north in summer. Most parts of North America are subject to a generally westerly wind flow that tends to move most weather systems eastward. The continental climate of the interior is, thus, pushed

The close relationship between climate and vegetation is well illustrated by the tree line, clearly visible in this photograph of mountains in Glacier National Park. Lower elevations are more darkly shaded because of the trees growing below the tree line, while the lighter shaded mountain slopes above the line experience climatic conditions too cold for tree growth. (Stephen S. Birdsall)

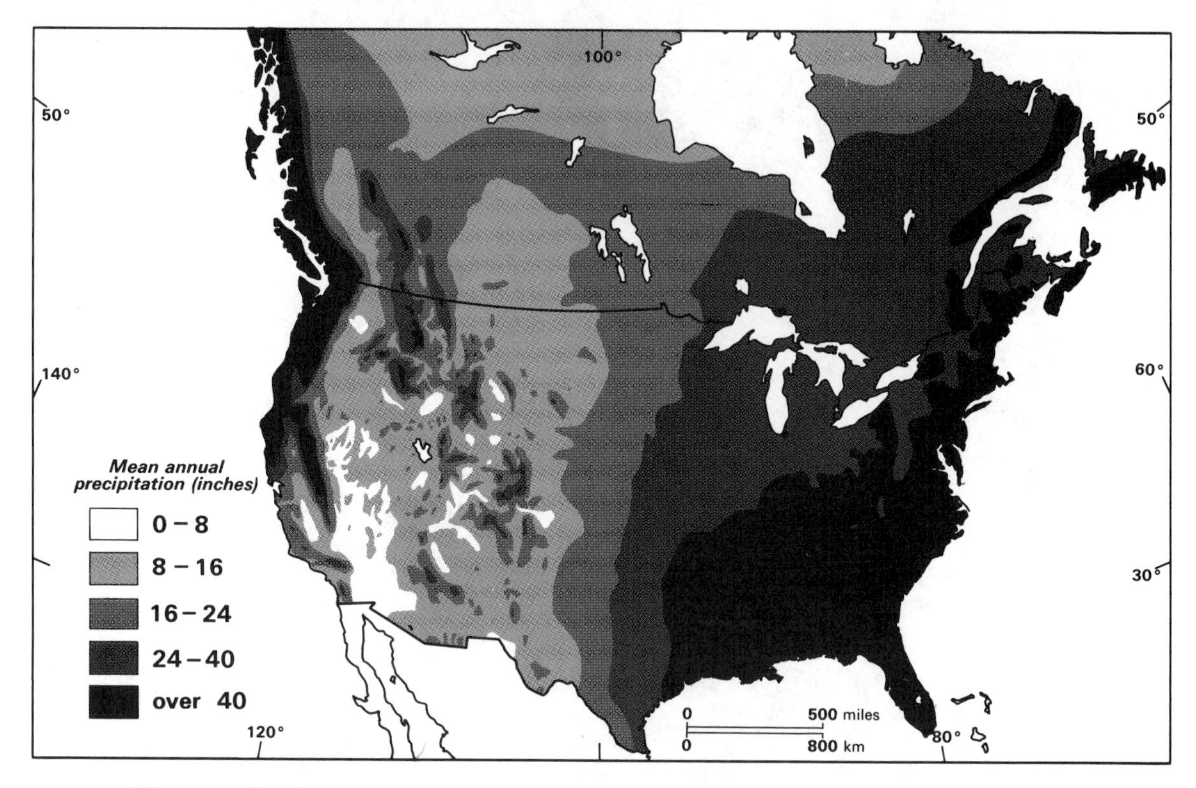

Figure 2-2 Precipitation. From the Rocky Mountains westward, precipitation is strongly influenced by topography, whereas to the east most precipitation is generated by storm systems. One result is a pattern of dramatic local variation in the West compared with broad areas of similar moisture conditions in the East.

onto the east coasts of the two countries. Coastal New England and the Atlantic Provinces, for example, have a climate with much greater average seasonal variations in temperature than do similar latitudinal locations along the west coast of the United States and Canada.

The interaction of all these climatic controls creates a pattern of climatic regionalization that is decidedly different between east and west. In the east, the principal element in climatic variation is temperature; in the west, it is precipitation. In the east, the divisions between the climate regions are based largely on the length of the growing season—the period from the average date of the last frost in spring to the first frost in fall—and on the average summer maximum temperature or winter minimum temperature. In the west, average annual precipitation is the key, although moderated temperatures are, as mentioned, an important aspect of the marine West Coast climate. To be sure, in the east the more northerly areas are generally drier; in the west they are colder. Still, in the east the key is temperature and in the west it is precipitation. In the east, the major influence on climatic variation is latitude; in the west, it is topography. Many climatic margins are indistinct. It is often difficult to tell exactly where one influence is less important and another becomes more important. In general, the western influences reach beyond the Rockies onto the higher, drier western portions of the interior grasslands.

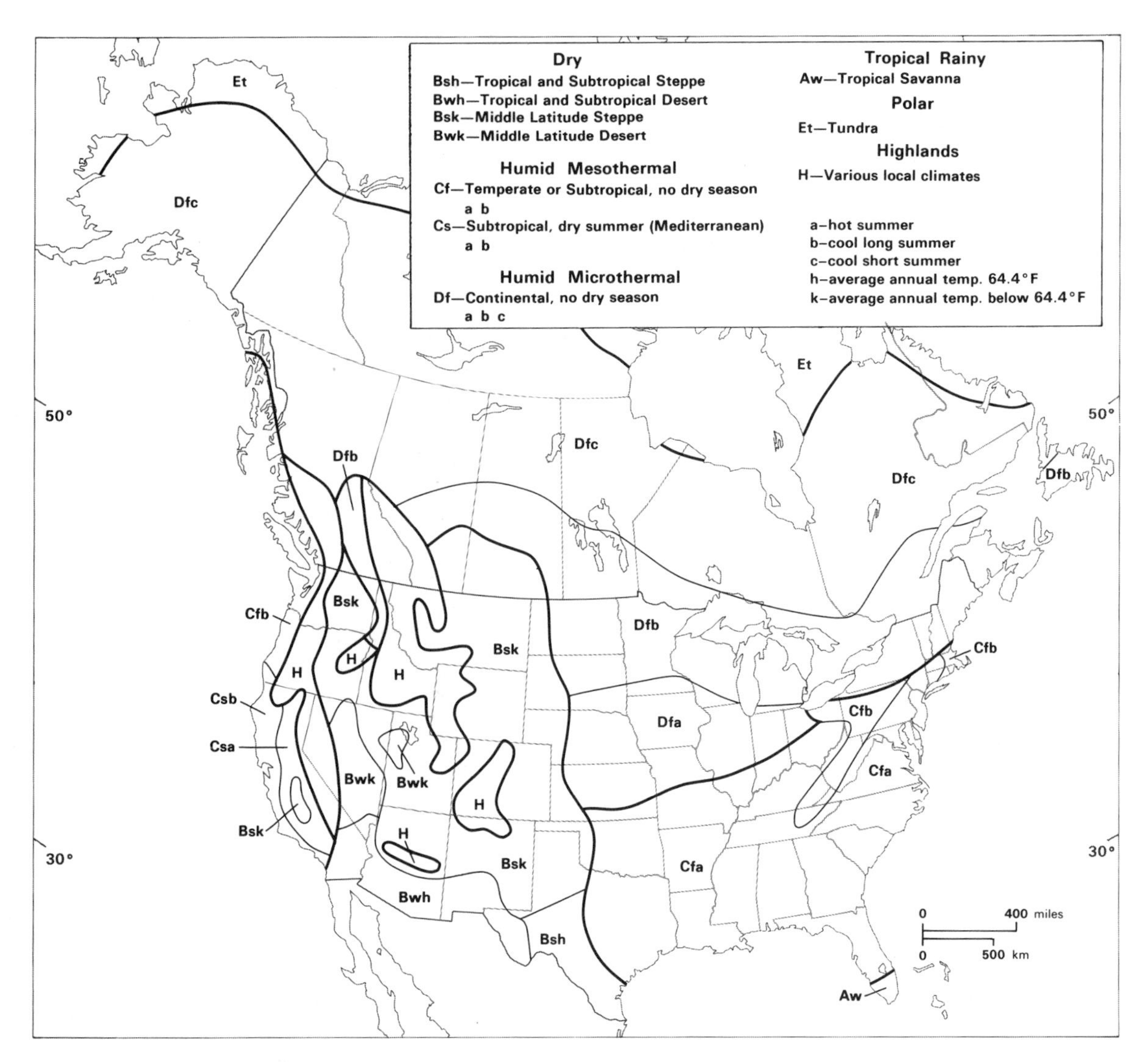

Figure 2-3 Climate regions (after Köppen). Climate results from the combination of the long-term pattern of moisture and temperature conditions.

Many different systems of climate classification have been developed. Greek scholars in the first or second century B.C. divided their known world into three climate zones: torrid, temperate, and frigid. The classification system that is by far the most widely used today was developed by Wladimir Köppen (1846–1940), a German climatologist and amateur botanist (Figure 2-3). In the Köppen system the determination of climate classes is based on the distribution of major plant associations. Thus, regional boundaries in the Köppen system represent the precipitation and temperature association with vegetation limits. Köppen used only mean annual and monthly temperature and precipitation data in the construction of his system. Compare

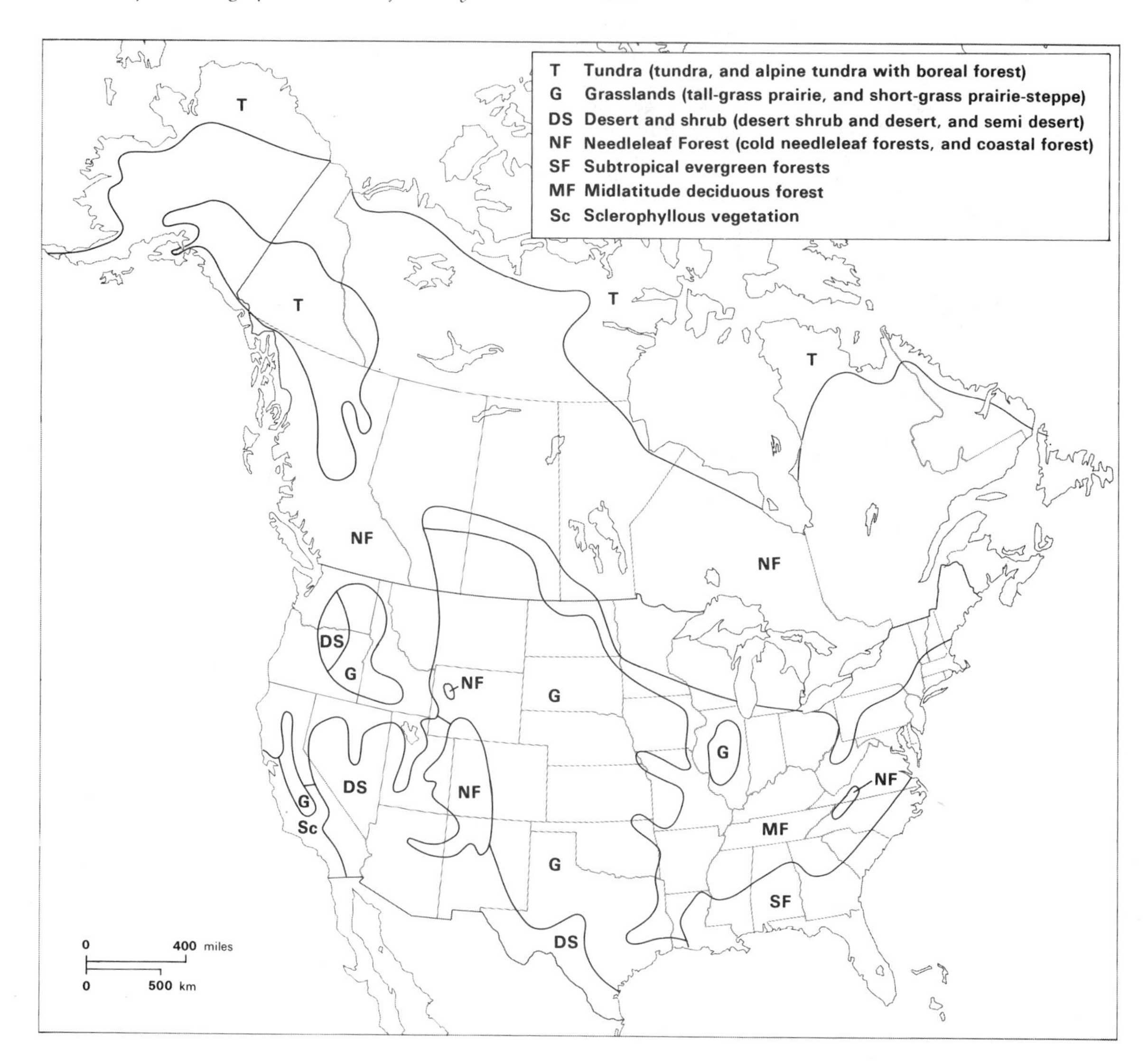

Figure 2-4 Vegetation regions. Notice the similarity between vegetation patterns and the map of climate regions.

Köppen's map for North America with the map of vegetation (Figure 2-4). The fit is certainly not perfect, but the association is clearly evident.

Vegetation

Botanists speak of something called *climax vegetation.* It is defined as the assemblage that would grow and reproduce indefinitely at a place given a stable climate and average conditions of soil and drainage. For most of the inhabited portions of North America today, that concept has little meaning. The "natural" vegetation, if it ever existed, has been so substantially removed, rearranged, and replaced that it seldom is found now. In the Southeast, for example, much of the land is presently covered by pine trees. The original mixed broadleaf and needleleaf forests were cut and replaced by the economically more important needleleaf forests.

The grasses of the plains and prairies of the interior are now mostly European imports. Their native American predecessors are gone either because they offered an inferior browse for farm animals or simply because they could not withstand the onslaught of modern humanity and its imported weeds. Most of what climax vegetation remains in the two countries is in the mountain west and the north. Logging and settlement expansion continue to reduce even that small remnant.

Still, vegetation regions are worth mentioning for at least two reasons. First, the area association between climatic and vegetative regions clearly indicates the central importance of climate on vegetation. Second, even with all the human changes, areas that once were forested continue to have substantial tree cover today, and areas that were grasslands still have few trees.

There are several ways of creating vegetation regions. Perhaps the simplest is to divide an area into three broad categories: forest, grasslands, and scrublands (see again Figure 2-4). Forests once covered most of the East, the central and northern Pacific Coast, the higher elevations of the West, and a broad band across the interior North. Forests of the Pacific Coast, the interior West, the North, and those of a narrow belt in the Deep South were all needleleaf and composed of many different trees. Much of the Ohio and lower Mississippi river valleys and the middle Great Lakes region was covered by a deciduous broadleaf forest. Between the broadleaf and needleleaf forests were belts of mixed needleleaf and broadleaf forests.

Grasslands covered much of the interior lowlands, including nearly all of the Great Plains from Texas and New Mexico to central Alberta and Saskatchewan. This is an area of generally subhumid climate where precipitation amounts are not adequate to support tree growth. An eastward extension of the grasslands, the Prairie Wedge, reached across Illinois to the western edge of Indiana. Precipitation in this area is clearly adequate to support tree growth. Although there is disagreement about why the Prairie Wedge existed, fires set by the American Indians to improve buffalo browse probably had a major impact.

Scrublands usually develop under dry conditions. They are concentrated in the lowlands of the interior West of the United States, with some minor extensions into Canada. Actual vegetation varies from the cacti of the Southwest to the dense, brushy chaparral of southern California and the mesquite of Texas. The scrubland has expanded into some parts of the drier southwestern sections of the grasslands as overgrazing destroyed the fragile grass cover (see Chapter 14).

The tundra of the Far North is the result of a climate that is too cold and too dry for the growth of other vegetation. The tundra vegetation does not fit well into our three-part classification. Grasses, lichens, and mosses dominated the region's vegetation. Tundra exists in small areas far southward into the United States, where climatic conditions at high elevations are inhospitable to tree growth. Northward, the altitudinal tree line is found at lower elevations until, eventually, the latitudinal tree line is reached.

Soils

Soil is a mixture of weathered rock material and organic matter that has been shaped and altered by the physical environment. The soil of a place owes its characteristics to such things as the parent rock material, climate, topography, and decaying plants and animals. Hundreds of different types of soil result from the interaction of these elements. Any particular soil is unique because of its mix of properties (such as color and texture) and composition (including such conditions as organic content and the action of soil colloids). Even so, general soil zones can be specified.

Colloids are small soil particles. Their properties and influences on soil are complex and often important. Soil acidity (or alkalinity), for example, is a result of the alteration and integration of soil colloids. Acid soils are characteristic of cold,

A desert landscape rich in a variety of cacti, such as the majestic saguaro and intricate cholla, and ropelike ocotillo in the scene from southern Arizona—so much a part of the image of the interior West after countless western movies and television shows—is, in fact, uncommon in the United States. (Alan Pitcairn/Grant Heilman)

moist climates; alkaline soils typically are found in dry areas. Most soils of the major agricultural zones of the eastern United States and Canada are moderately to strongly acidic. Lime must be added periodically to neutralize that acidity before these soils can be used to produce most row crops.

Color is perhaps the most obvious soil property. The red clays of Georgia, the black soils of the Red River Valley in Manitoba—these colors may be the first characteristic to catch an observer's attention. A dark color usually indicates an abundance of organic materials, and red, the presence of iron compounds. Generally, however, color is a result of the soil-forming processes. For example, the pale-gray soil of the northern needleleaf forest results from the *leaching* of organic matter and minerals from the soil's surface layer.

Soil texture refers to the proportion of particles of different size in the soil. Sand is the coarsest measure of soil texture, silt is intermediate, and clay is the finest. Soil texture depends largely on the composition of the parent material. Texture is important because it determines a soil's ability to retain and transmit water. Soils called "loams" contain substantial proportions of each of the three particle grades and are considered best. They are fine enough to hold moisture yet are not so fine that they cannot easily take up water.

Where soils have developed in place for a sufficiently long period, they take on a characteristic layered appearance. These layers are called *soil horizons.* The uppermost, the A-horizon, is the zone from which certain colloids, chemical compounds, and other matter has been removed; below that is the B-horizon, the zone of accumulation of the matter removed from the A-horizon. The red color of the clay of the Georgia Piedmont, for example, appears because the coloring materials other than iron have been removed from the A-horizon, a result of the substantial precipitation of the area. The lowest soil zone, the C-horizon, is the parent material. A thin O-horizon at the surface, the zone of active addition of organic matter, can also be identified in many soils.

There have been many attempts to classify soils. Soil scientists usually prefer a genetic classification, one that focuses on the nature of soil creation. At present the most widely accepted in the United States is the Department of Agriculture's soil classification. It is often called the Seventh Approximation because it is the Department's seventh classification in a gradually evolving process. Six earlier versions were given to soil scientists for evaluation during the 1960s before the final version, properly titled the U.S. Comprehensive Soil Classification System, was adopted. It is based on the identification of precise soil horizons that develop under particular environmental conditions. The map of soil regions based on the Seventh Approximation (Figure 2-5), indeed any regionalization of soils, must be viewed with some question, for it hides the tremendous complexity that can exist in small areas. For example the system is subdivided into suborders, with about 50 recognized in the United States; great groups, with 225 in the United States; and eventually soil series. About 12,000 soil series have been identified in the United States, and the number is increasing steadily. The regions simply indicate the most important soil type for the area.

Aridisols gain their name from *arid.* These soils of dry climates are low in organic content and have little agricultural value. *Spodosols* develop in cool, moist climates. In Canada and the United States, spodosols are largely the soils of the northern needleleaf forests. They are quite acidic and low in nutrients; they are of agricultural value only for acid-loving crops such as potatoes or blueberries. *Tundra soils* are associated with a cold, moist climate. The soil is shallow, frequently water saturated and with a subsurface of perennially frozen ground (see Chapter 18). Tundra soils also have little agricultural value. Highland soils are little developed and agriculturally worthless.

Mollisols are grassland soils of the semiarid and subhumid climates of the midlatitudes. They are characterized by a thick dark brown to black A- and B-horizon, loose texture, and high-nutrient content. They are among the most naturally fertile soils in the world and produce most of North America's cereals.

Alfisols are second only to mollisols in agricultural value. They are soils of the midlatitude forest and the forest-grassland boundaries. They are very much "middle" soils in a climatic sense. They are located in areas moist enough to allow for the accumulation of clay particles in the B-horizon (their characteristic identifying feature), but not in areas so moist as to create a heavily leached or weathered soil.

Alfisols are divided into four subcategories, each with its own characteristic climatic association. *Boralfs* are soils of the boreal needleleaf forests of central Canada. They are usually thin, acidic, and have little agricultural potential. *Udalfs* are soils of the deciduous forests of the American Middle West and southern Ontario. Somewhat acidic, they are, nevertheless, highly

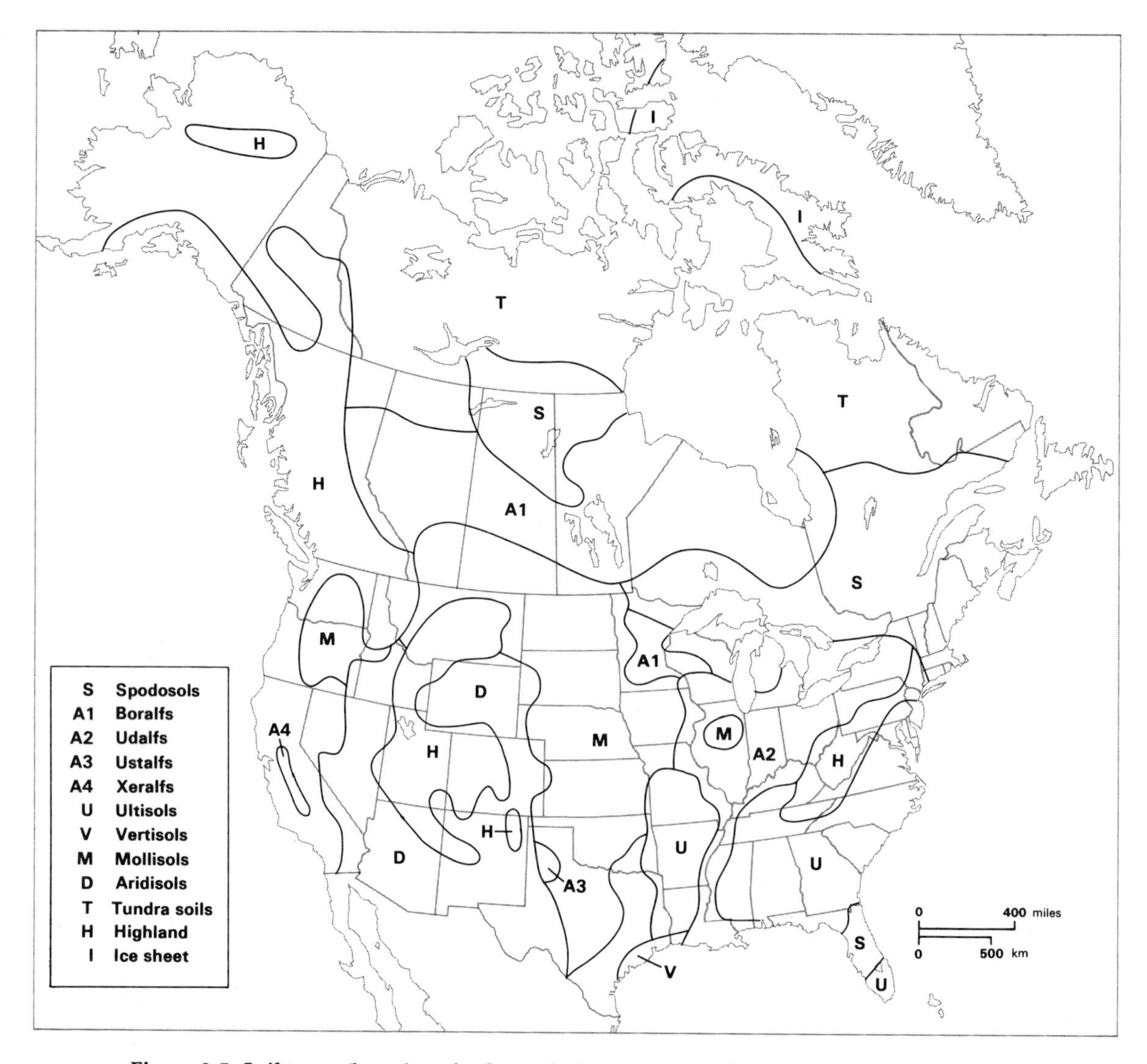

Figure 2-5 Soil types (based on the Seventh Approximation from the U.S. Department of Agriculture). This map is very generalized. The elements that together influence soil type are so complex and can be so localized that a single small field or city block often contains several different soils.

productive when lime is used to reduce the acidity. *Ustalfs* are found in warmer areas with a strong seasonal variation in precipitation. In the United States they are most common in Texas and Oklahoma. They are highly productive if irrigated. *Xeralfs* are soils of a Mediterranean climate of cool, moist winters and hot, dry summers. Found in central and southern California, they too are highly productive.

Ultisols represent the ultimate stage of weathering and soil formation in the United States. They develop in areas with abundant precipitation and a long frost-free period. Particle size is small, and much of the soluble material and clay has been carried downward away from the A-horizon. These soils can be productive but high acidity, leaching, and erosion are often problems.

Another soil class, *entisols,* is nowhere sufficiently predominant to be found on the soils map, but it is important locally. These are recent soils, too young to show the modifying effects of their surroundings. They are widely scattered and of many types, from the Sand Hills of Nebraska to the alluvial floodplains of the St. Lawrence and Mississippi river valleys. The agricultural potential of entisols varies, but the alluvial floodplain soils, drawn from the rich upper layers of upstream soils, are among the continent's most productive.

The soil of a place is not an unlimited resource. The quality of these different soil horizons differs substantially, with the greatest amount of nutrients commonly concentrated in the upper horizons. (USDA—Soil Conservation Service)

MINERAL RESOURCES OF NORTH AMERICA

There is a distinct association between the location of minerals that meet the needs of heavy manufacturing and the land's subsurface rock structure. Each of the three major forms of rock is capable of containing a type of mineral economically useful to humans, although this potential is not always realized. Sedimentary and metamorphic rocks are the most prevalent rock substructure and are more likely to contain minerals of broad utility than are *igneous* rocks, the third category.

Sedimentary rock is the result of the gradual settling of small solid particles in stationary water. For example, if a shallow sea was located adjacent to an arid landscape subject to occasional rainstorms, sand particles would be washed into the sea and spread across its bottom by water currents and the force of gravity. As this process continued—for a length of time perhaps 500 times that since Imperial Rome's legions marched or a length 5000 times that since the United States of America was conceived—each layer of sand would press down on the layers beneath it, squeezing and solidifying the sandy mass that had been deposited only a few thousand years before. When this seabed was raised and folded into mountains by shifts in the earth's crust, the method by which at least some of the rocks were formed was betrayed by the presence of layers of sandstone. A wide variety of other sedimentary rocks were also formed in this general fashion from various kinds of parent material and dependent on local conditions.

About 300 million years ago during what earth historians call the Carboniferous period of the Paleozoic era, conditions present in most existing land areas were such that unusual sedimentary sequences were created. Heavily vegetated and thick, swampy regions were drowned and covered with another layer of sediment. In some cases, the organic matter came to be represented in liquid form, trapped between folds of impermeable rock and eventually drawn off as petroleum. Most of these petroleum deposits are found in conjunction with another by-product of the period—natural gas. In other cases, the organic matter became solid layers of coal that were sometimes only inches thick but occasionally found dozens of feet thick.

In North America, vast regions are underlaid by sediments formed during the Carboniferous period. These areas where coal, oil, or natural gas might be found are located in the interior and Great Plains physiographic provinces (Figure 2-1), sections of the Gulf coastal plain, portions of the Pacific mountains and valleys, the Arctic rimland, and in folded and broken form along the western margins of the Appalachian Highlands and into the eastern Rockies.

Large deposits of mineral fuels have been identified across extensive portions of these sedimentary lowlands. The most important coal deposits on the continent have been mined in the more rugged Appalachian field. In spite of irregular surface topography, the coal seams are generally easy to work. Most seams are between 30 centimeters and 3 meters (1 and 9 feet) thick, and in many areas are close to the surface. Both factors facilitate the use of machinery for extraction. Mines throughout this nearly continuous field in eastern Kentucky, West Virginia, and western Pennsylvania were the earliest to be brought into production, and they continue to supply over half of the nation's coal needs. And the reserves that remain in this field still comprise about 20 percent of the total reserves in the United States (Figure 2-6).

Until recently much of the remaining coal mined in the United States has been obtained from the Eastern Interior Field. This large coalfield underlies most of Illinois and extends into western Indiana and western Kentucky. Although some of the Eastern Interior Field's coal has been used in iron and steel production, its higher sulfur content has restricted most use to heating and electric-power generation. Recently, this sulfur content drew the attention and ire of those concerned with air pollution. A by-product of burning this coal is the noxious gas, sulfur dioxide. The consequent acid rain has affected the quality of downwind lakes in southern Canada and the northeastern United States. The Western Interior Field is also large, located under Iowa and Missouri with a narrowing extension southward into eastern Oklahoma. The coal found in this field is of slightly poorer quality than that found in the eastern fields and has only recently begun to be mined. There are also many small and a few large bituminous deposits scattered through and along the eastern margins of the Rocky Mountains. Extensive deposits in Wyoming and Montana have come into production in the last two decades. Culminating two decades of rapid mining expanison, Wyoming now leads the United States in coal production (see Table 9-1). There are also several extensive fields of lignite (brown coal) in the northern Great Plains and the Canadian western prairies.

Although petroleum and natural gas are not formed in quite the same manner as coal, their genesis is similar (Figure 2-7). Scattered deposits of petroleum and natural gas are found through the Appalachian coalfield, including the first commercial well in the United States at Titusville, Pennsylvania. Southern Illinois and south-central Michigan produce some petroleum, as do scattered sites across the northern Great Plains and the northern Rockies. Easily the most important petroleum fields, however, have been those in the southern plains, along the Gulf Coast, and in southern California. One great arc of producing wells is located along the full length of the Texas and Louisiana coasts. Another slightly broken arc extends from central Kansas south through Oklahoma and westward across central Texas to New Mexico. Between and beyond these two large areas lie two more fields of great importance, the East Texas field and the Panhandle field in northwest Texas. Separate from these fields but also of major importance are those located in southern Califor-

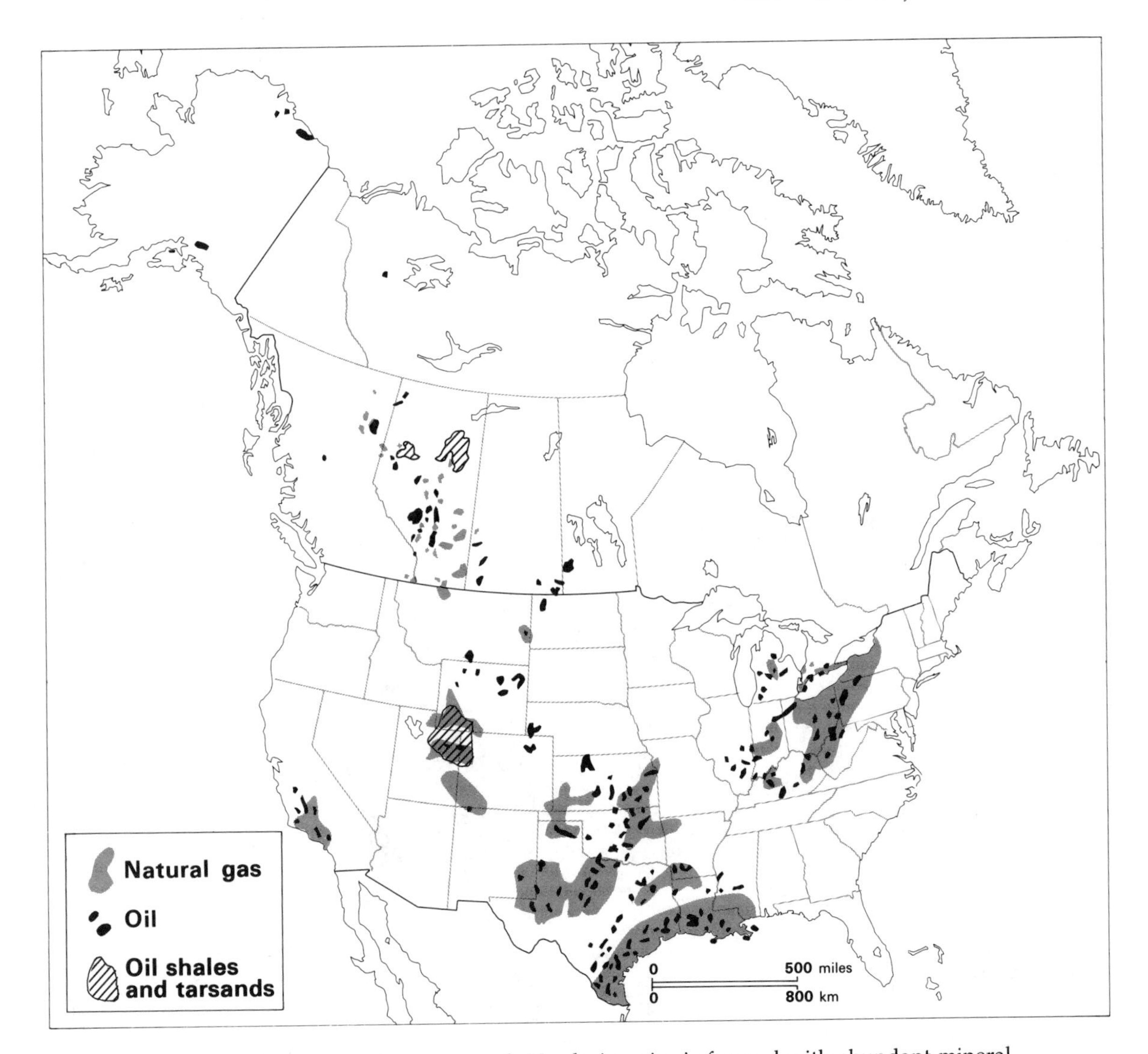

Figure 2-6 Mineral fuels except coal. North America is favored with abundant mineral fuels resources. The two countries also have a large demand for fuels, which generates a constant search for new sources. Major new petroleum and natural gas fields are being sought off the eastern seashore, in northern Canada and the Arctic Ocean, and in the interior West.

nia. In the mid-1960s, exploitation of deposits of petroleum and natural gas was begun along the north Alaska slope and in the Arctic fringes of Canada; the development of significant deposits in Alberta has increased the economic influence of that province within Canada.

The broad pattern of mineral fuel deposits in North America, then, is one of extensive deposits of coal south of the Great Lakes and west of the middle Appalachians and of very large petroleum and natural gas fields southwest of the major coal reserves, stretching from the mid-

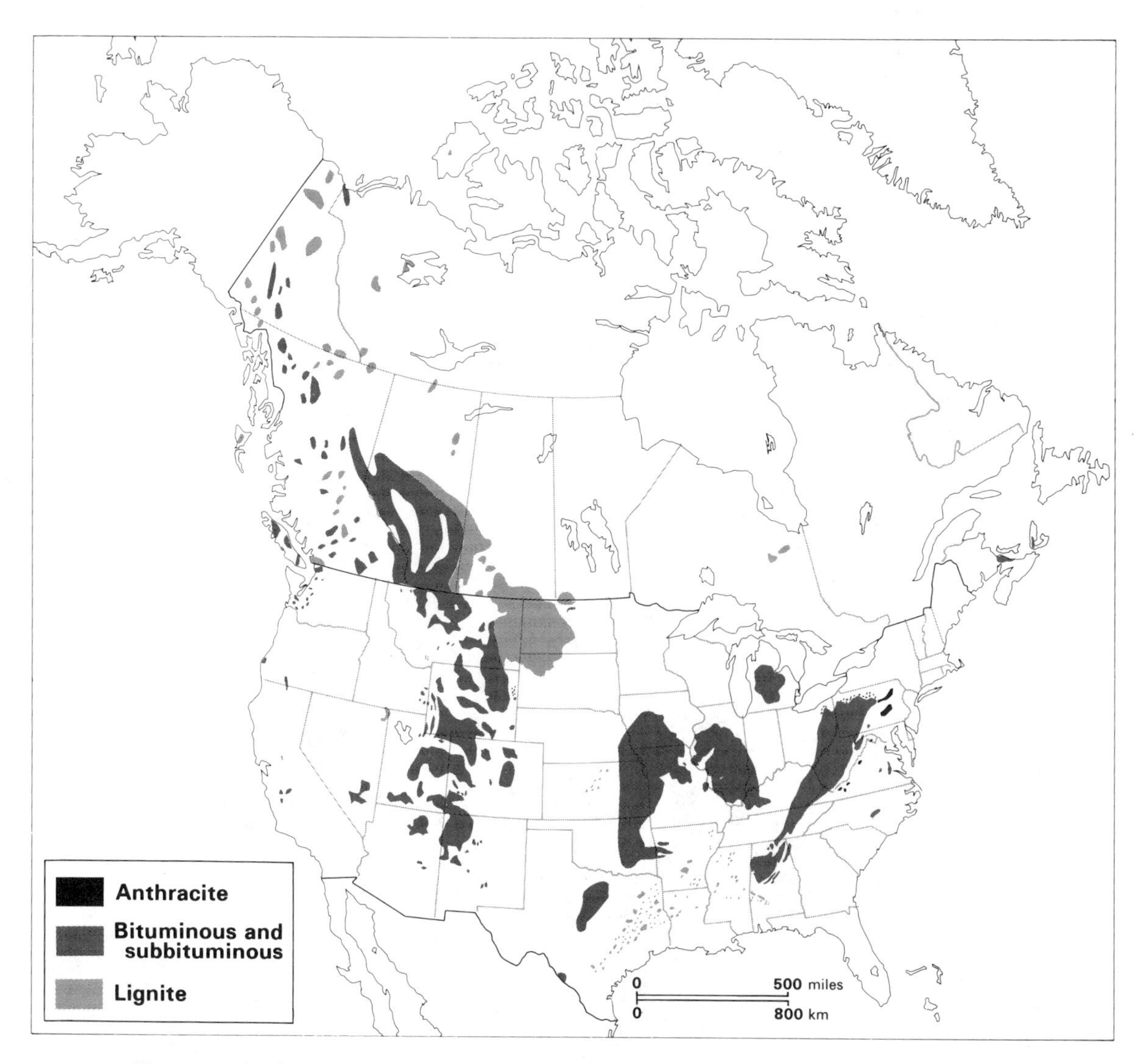

Figure 2-7 Major coal fields. Canada and the United States have a coal reserve that would last hundreds of years at present use levels. However, the increasing cost of alternative fuels plus continuing concern over the safety of nuclear power should result in an increase in coal production over the next several decades.

dle Great Plains south to the Gulf Coast. Scattered, and occasionally significant, deposits are located west and northwest of the major reserves.

Metamorphic rock is formed in a quite different manner than sedimentary rock. Under the tremendous pressure exerted through the gradual deformation of the earth's crust, the internal structure of previously formed rocks can be metamorphosed (changed). So great is the pressure exerted over thousands of years and so great is the heat generated, that the very molecular structure of the rock is altered. This transformation indicates why metallic minerals in economically extractable quantities are located most often in areas of metamorphic rock. In North America, therefore, the primary deposits of metallic ores are located in the three major zones of metamorphic rock: the Canadian Shield, portions of the Appalachian Highlands and their eastern Piedmont, and in sections of the western mountain ranges (Figure 2-8).

Many of the mining sites for early exploitation of the metallic minerals on the Canadian Shield were located near the margins of the Shield. The pattern of mineral production follows a long arc extending from the North Atlantic and St. Lawrence River estuary across the Great Lakes and northward to the Arctic Ocean. This arc defined the edge of the Shield itself in the broad horseshoe shape it exhibits around Hudson Bay. As the interior of the Shield began to be developed in recent decades, the arc has blurred. It includes the cluster of iron deposits on the Quebec-Labrador border at Burnt Creek and Schefferville; a tremendous concentration of minerals on the Quebec-Ontario border, including the nickel-rich region at Sudbury; and such self-explanatory places as Cobalt, Ontario, and Val d'Or (Valley of Gold), Quebec. The arc continues on both sides of Lake Superior: in northern Michigan, Wisconsin, and Minnesota with copper and iron and in Ontario with iron at Steep Rock. Then it swings north through Flin Flon (copper, lead, zinc) to such likely places as Uranium City, Yellowknife (gold), Port Radium, and Coppermine.

A second zone of metamorphic rock is located along the eastern Appalachian Mountains. Iron mines at Wabana, Newfoundland, supply the ore for a small steelworks. Copper and iron were important minerals found locally by New England colonists. The most important centers in this zone still producing are Ducktown, Tennessee (copper), and Birmingham, Alabama, where iron ore can be mined at the southern end of the Appalachia chain. This narrow, linear band of metamorphic rock remains only moderately important as a source of metallic minerals. Most of the small deposits were depleted as demand continued to grow, and transportation has eased shipment from more distant locations with more easily worked deposits. But these eastern minerals were significant during the early years of industrial growth in the United States.

A third and extensive region of metallic minerals is formed by the western mountains. Scattered deposits of gold and silver, a few of them rich, drew prospectors and mining companies to isolated locations scattered from south of the Mexican border to central Alaska—and provided Hollywood writers with a setting for countless (even if mythical) thrillers. Of great industrial importance are the large deposits of copper, zinc, lead, molybdenum, and recently, uranium found in this western region. When this list of major deposits is swelled by adding the minerals found in smaller deposits, such as tungsten, chromite, manganese, and many others, we can appreciate the great flexibility this diversity gave to the growing industrial economies in the United States and Canada.

It should not be assumed that the two countries' industrial requirements were met fully by the tremendous variety of minerals found in these three zones of metamorphic rock. A few minerals needed by modern industry have not been located in North America in sufficient quantities to satisfy domestic needs (e.g., tin, manganese, and high-grade bauxite for aluminum). In addition, the growth of industrial capacity has been matched by a growth in demand for many minerals. Demand now often exceeds the diminishing domestic supply of these nonrenewable resources. Few other countries, however, ever equaled, or even approached, the original quantity and diversity of metallic minerals and mineral fuels located in the United States, especially when complemented by the vast reserves in Canada. This abundance of minerals has been critical in assisting the development of the immense North American manufacturing-industrial complex (Chapter 6).

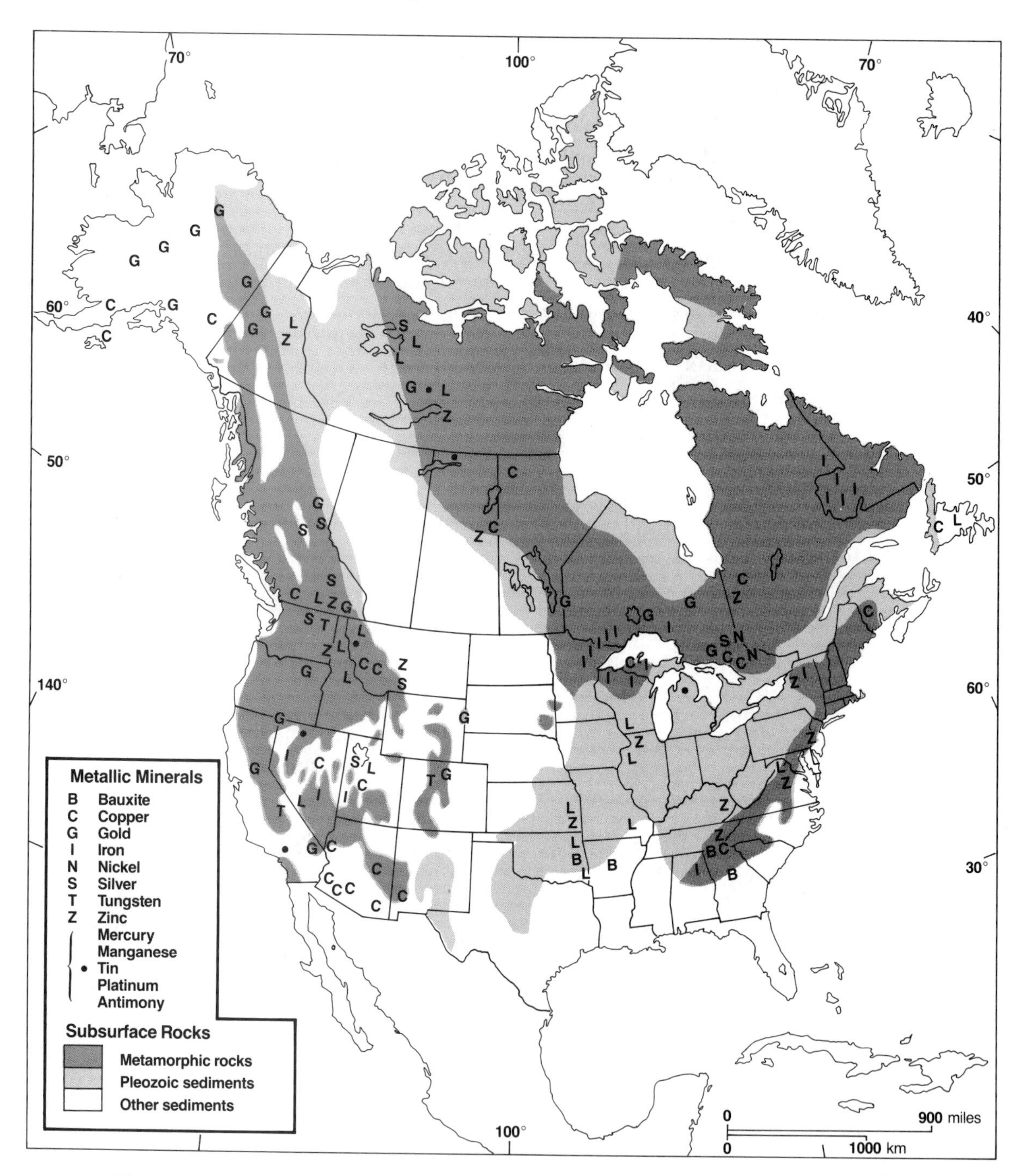

Figure 2-8 Metallic minerals. The metamorphic rocks containing metallic minerals in the two countries arc around the interior lowland and its fossil fuels.

GEOGRAPHIC INSTRUCTION CASE STUDY: WHICH SOIL CLASSIFICATION?

The Seventh Approximation is not the first soil classification system to be officially supported by the U.S. Department of Agriculture. It 1938 it adopted a *genetic* classification of soils, meaning that it was based on soil-forming conditions and processes. The relationship between soils and other elements of the environment was central to the system. Thus, the distribution of soils could be compared directly with such things as vegetation and climate. The genetic nature of the system, and the opportunities for comparison and association that it provided, made it very popular with geographers.

One could, for example, clearly see the distribution associations between Köppen's humid subtropical climate, the subtropical evergreen forests, and something called red and yellow podzols. Red and yellow simply referred to characteristic color. *Podzolization* is a soil-forming process found in areas with a distinct winter. Podzol soils are shallow and acidic. The point is, the geographic patterns of the three are clearly similar. One can describe and analyze those associations. It all has a clear geography.

Soil scientists found the 1938 system both imprecise and uncomprehensive. The Seventh Approximation allows for an unlimited number of soil series, thus solving both problems. However, it is based on the observable characteristics of the soil, and not on its genesis or association with other elements of the environment. The scientific logic of the system is clear. A soil type has properities that can be carefully identified and measured. Thus, one soil can be compared with another. Still, for us geographers it lacks that easy opportunity for spatial association. We have accepted it, but feel a certain nostalgia for the genetic classification.

ADDITIONAL READINGS

Atwood, Wallace W. *The Physiographic Provinces of North America.* Boston: Ginn, 1940.

Bennett, Charles F. *Conservation and Management of Natural Resources in the United States.* New York: John Wiley & Sons, 1983.

Bryson, Reid, and F. Kenneth Hare. *World Survey of Climatoloy,* Vol. 2, *Climates of North America.* New York: Elsevier, 1974.

Dixon, Colin J. *Atlas of Economic Mineral Deposits.* Ithaca, N.Y.: Cornell University Press, 1979.

Falconer, Allan, et al. *Physical Geography: The Canadian Context.* Toronto: McGraw-Hill Ryerson, 1974.

Farb, Peter. *Face of North America: The Natural History of a Continent.* New York: Harper & Row, 1963.

Gaff, William L. (ed.). *Geomorphic Systems of North America.* Boulder, Colo.: Geological Society of America, 1987.

Gersmehl, Philip J. "Soil Taxonomy and Mapping." *Annals of the Association of American Geographers,* 67 (1977), pp. 419–428.

Hare, F., Kenneth, and Morley R. Thomas. *Climate Canada.* Toronto: John Wiley & Sons Canada, 1974.

Hunt, Charles B. *Natural Regions of the United States and Canada.* San Francisco: W. H. Freeman, 1973.

Kuchler, A. W. "Potential Natural Vegetation of the

Coterminous United States'' (Special Publication no. 36). New York: American Geographial Society, 1964.

Steila, David. *The Geography of Soils*. Englewood Cliffs, N.J.: Prentice-Hall, 1976.

Thornbury, William D. *Regional Geomorphology of the United States*. New York: John Wiley & Sons, 1965.

Visher, Stephen S. *Climatic Atlas of the United States*. Cambridge, Mass.: Harvard University Press, 1954.

Williams, Michael. *Americans and Their Forests: A Historical Geography*. New York: Cambridge University Press, 1989.

CHAPTER 3

Foundations of Human Activity

POPULATION AND SETTLEMENT

Canada and the United States had a small, dispersed native population at the time of initial European discovery. American Indian and Eskimo populations totaled perhaps 1 million people, four-fifths of them in what came to be the United States. These peoples usually were organized in small tribal units. By the time of European arrival in North America, there was great diversity among American Indian cultures. Several hundred dialects were spoken along the coast of California alone. And the diversity was more than linguistic. The Pueblo, who lived in what is now New Mexico and who were probably influenced culturally by the Aztecs to the south, resided in permanent towns and constructed extensive irrigation systems. At the same time, the Piutes of the Great Basin lived in temporary thatch dwellings and pursued a seminomadic existence based on available wild edible vegetation and small game. The Inuit or Eskimos, who were the most recent of the pre-European arrivals, shared close cultural ties with Inuits in Greenland and Siberia.

Although American Indians did represent a real barrier to the expansion of European settlement at times, for the most part, their impact was minimal. Many died of imported European infectious diseases such as smallpox or measles before they experienced direct contact with the Europeans themselves. Occasionally, the Indians made important contributions to the arriving Europeans, especially during the first decades of settlement. Still, the cruel fact is that, in general, they simply were ignored as much as possible. Most often, they were killed or shunted off to reservations somewhere to the west, where it was hoped they would stay out of the way. As the settlement frontier moved westward, so did the Indians and their reservations. There was almost no integration of the Indians into the European settlement pattern, something that was common in portions of Latin America. The Canadian record on relations with the Indians was—and is—a little better than that of the United States, but neither country deserves any humanitarian awards for their treatment of the previous inhabitants of the two countries. About two-thirds of Canada's American Indians reside on some 2200, mostly small, reservations scattered throughout the country, whereas in the United States one-third live on 281 reservations with the largest on the dry western portion of the Great Plains and the Empty Interior. Most of the other two-thirds live in urban areas, with the largest number in the Los Angeles metropolitan area.

Johanna Gentsch-Kuschne, seen here on Ellis Island off the southern tip of Manhattan in New York harbor, was one of the many millions of late 19th and early 20th century European migrants whose introduction to the United States came at this immigration examination and acceptance facility. Ellis Island has been restored as a national historic site. (Andrew Holbrooke/Black Star)

European Settlement

What was easily the greatest long-distance migration in human history followed in the wake of the retreating American Indians. Although it is impossible to state precisely how many people entered what is now the United States and Canada from Europe and, to a much lesser extent, from Africa, a reasonable estimate would place the figure at somewhere close to 60 million.

The American Indians of pre-European North America were divided into many, many different tribal and cultural groups. Here, for example, is a scene from the northern Great Plains of a tribal group actually heavily influenced already by the introduction of the horse. (Smithsonian Institution National Anthropological Archives. Photo by William S. Soule)

The overall pattern of source areas of European immigrants to North America is complex, but a few generalizations can be made. Most early immigrants came from northwestern Europe, primarily the British Isles (Table 3-1). At the time of the first national census of the United States in 1790, more than two-thirds of the white population of the country had an ethnic origin on the British Isles, with Germans and Dutch next in importance. Nearly all of the French migration to Canada came early in the European settlement history, mostly in the seventeenth century. The total number of French who made the trip was small, fewer than 15,000. Their number has increased hundreds of times over in the ensuing 300 years, a striking example of human capacity to increase in numbers under reasonably good conditions.

Immigration into North America slowed during the period between 1760 and 1815. This was a time of intermittent warfare in Europe and North America as well as on the Atlantic Ocean. Between about 1815 and the start of World War I in 1914, the volume of immigration tended to increase with each passing decade. Periods of war or widespread economic depression were exceptions, marking times of reduced movement. At the beginning of this period, which

These Pueblo Indians, photographed late in the 19th century, are performing the Sun Dance. Many elements of American Indian culture have survived through centuries of European contact. (Library of Congress)

lasted a century, annual migration was perhaps 10,000. In 1913, nearly 1 million immigrants entered the United States and another 400,000 streamed into Canada.

The United States passed its first major legislation to restrict the volume of immigration in the 1920s. Canada followed suit a decade later. These enforced limitations, coupled with the depression of the 1930s and a war in the 1940s, cut immigration to a fraction of its annual high in 1913. Since 1945, the number of arrivals has increased somewhat. Today the United States typically receives roughly 550,000 legal immigrants annually; Canada perhaps 110,000. An even larger number of illegal aliens probably enter the United States each year, although the estimates of their numbers vary greatly. The impact of illegal aliens on the U.S. population is discussed in more detail in Chapter 15.

There has been a great change in the major sources of immigrants to the United States with the passage of time. During the period from 1815 to 1913, the home country locations of migrants from Europe gradually shifted southward and eastward. For the first half of the period, most migrants continued to come from northwestern Europe. There was a decline in the percentage from Britain, but an increased number arrived from Ireland and Germany. By late in the nineteenth century, these migration origins became relatively less important. In the 1880s, there was a major outflow from Scandinavia, followed in subsequent decades by streams of people from southern and eastern Europe. In 1913, well over four-fifths of all immigrants were from these areas of Europe, especially Italy, Austria-Hungary, and Russia.

The reasons for this southeastward shift are based on the impact within Europe of the diffusion of the Industrial Revolution. Beginning in the British Isles and the Low Countries across the English Channel in the eighteenth century, it spread southeastward across Europe during the following 150 years or so. With industrialization

Table 3-1 Immigration into the United States and Canada

	United States		*Canada*	
Decade	Number in Thousands	Principal Sources	Number in Thousands	Principal Sources
1820s	129	Ireland, Britain		
1830s	538	Ireland, Germany		
1840s	1427	Ireland, Germany		
1850s	2815	Ireland, Germany	253	
1860s	2081	Germany, Britain	156	
1870s	2742	Germany, Britain	329	
1880s	5249	Germany, Britain, Scandinavia	850	
1890s	3694	Eastern Europe, Italy, Germany	373	
1900s	8602	Eastern Europe, Italy, Russia	1401	Britain, United States
1910s	6347	Eastern Europe, Italy, Russia	1859	Britain, United States, Russia
1920s	4296	Canada, Latin America, Italy	1273	Britain, United States, Ireland
1930s	699	Canada, Germany, Italy	251	United States, Britain
1940s	857	Latin America, Canada, Britain	429	Britain, Eastern Europe
1950s	2300	Germany, Canada, Latin America	1540	Britain, Italy, United States
1960s	3212	Latin America, Canada, Asia	1375	Britain, Italy, United States
1970s	4493	Latin America, Asia	1589	Britain, United States, Hong Kong
1980–87	4600	Latin America, Asia	906	Asia, Europe

Source: M. C. Urquart, ed., 1990 *Historical Statistics of Canada* (Toronto: Macmillan Co. of Canada, 1965); Statistics Canada, *Canada Year Book,* 1965, 1981, 1990; W. Zelinsky, *Cultural Geography of the United States* (Englewood Cliffs, N.J.: Prentice-Hall, 1976); *Statistical Abstract of the United States* (Washington, D.C.: U.S. Government Printing Office, 1970–90).

came a rapid rise in population size as mortality declined. The economy shifted to manufacturing, urbanization increased, and there was a proportional decline in the agricultural population. The growth in the demand for urban labor did not match the increase in the potential labor force, and thus, there were many willing emigrants.

The destination of these immigrants within North America was largely the result of the distribution of existing economic opportunity. It has been suggested repeatedly that migrants chose areas that were environmentally similar to their European homes. The substantial Scandinavian settlement in Minnesota and the Dakotas is indicated as a case in point. There may be some small truth in this, but it was more important that those states happened to represent the principal settlement frontier at the time of major Scandinavian immigration. Did Germans move to the Texas Hill Country because it reminded them of home? Unlikely. Did southern Italian immigrants find Boston or Newark much like the farms and villages of their Mediterranean origin? Certainly not. For the most part, the mosaic of ethnic patterns in North America is the result of a movement toward opportunity—opportunity first found most often on the agricultural settlement frontier and then in the cities.

There are several notable exceptions to this pattern of essentially free choice of location. Restrictions of American Indians to reservations and the use of blacks as slave labor are the most obvious examples (see Chapters 10 and 15). Much of the early heavy Irish immigration in the late 1840s and 1850s was a panic move by a destitute population held in virtual serfdom by British landowners. The migrants were fleeing starvation that resulted from massive failures of the country's principal food crop, the potato. These Irish had no money to move beyond the cities of the East Coast and became the first major poor immigrant ethnic population in many of America's rapidly growing cities. French Canadians

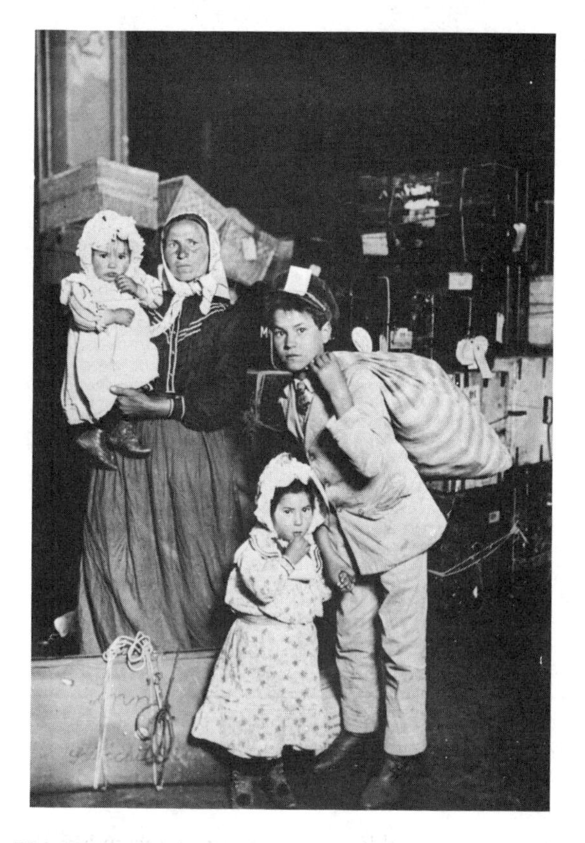

The immigrants who have come to North America over the past 150 years have added to the diversity of the culture of the two countries while at the same time integrating into that which was already there. This family is arriving in New York City around 1910. (Lewis W. Hine photo from the Library of Congress)

moving into the northeastern United States or Mexicans in the Southwest or, more recently, Cubans into the Miami area have all chosen locations close to their origins, although not necessarily because they lacked money to move elsewhere.

Non-European Settlement

Until recently, most immigrants into Canada and the United States were from Europe. The major exception was black settlement in the American South. Forced to move as slave labor for the region's plantations, this was a small part of the large movement of Africans to the Caribbean Basin, the northeast coast of South America, and the American Southeast. Next to the European exodus, this was probably the second largest long-distance movement in human history. Perhaps 20 million left Africa. It is believed that fewer than 500,000 blacks came into the United States, although accurate statistics are not available. Most who came probably arrived from the islands of the Caribbean rather than coming directly from Africa. At first, slaveholding was legal throughout the United States; in practice, however, the great majority went to the plantation South. The 1790 census indicated that fully 20 percent of the American population was of African origin. There was little African immigration after that date, and the percentage of the population that was black declined slowly but steadily until well into the twentieth century.

Immigration laws severely limited the possible source areas of immigrants after the 1920s. Essentially, except for a large quota for other Western Hemisphere countries, most migrants had to come from Europe. After 1881, Chinese immigration was excluded entirely from the United States, and the Chinese also were banned by Canada in 1923. Far more liberal immigration laws were passed by both countries in the 1960s. Coupled with a decline in interest on the part of most Europeans to move to North America, this has meant that most immigrants to the United States no longer come from Europe. Typically only about one quarter of the immigrants are from Europe, somewhat more from Asia, and the rest from the Americas. Mexico, the Philippines, and the West Indies provided the greatest number of migrants to the United States in the late 1980s, followed by Korea, China, the Dominican Republic, and Guatemala. Most Canadian immigrants are now from Asia, and the non-European component's share is increasing.

Most non-Europan migrants settle near their points of entry or in larger cities. Thus, Vancouver now has a very large ethnic Chinese population, while Miami is the home for migrants from throughout the Caribbean Basin and Middle America. Most of the larger cities in both countries now contain many different ethnic neighborhoods representative of the rich cultural diversity of these recent migrants.

Settlement Expansion

The European settlement of North America started in the east and moved westward (Figures 3-1 and 3-2). That is a simple, obvious statement. Yet it is the single most important element in understanding the European occupation of the two countries. Settlement did develop first in the east, and it did gradually extend westward. Although there are certainly some interesting variations in the trend—places that were settled earlier or later than might have been expected—in general, you can predict the date of earliest settlement at any given place merely by knowing how far it is from the East Coast.

The first settlements were small, clinging to the ocean and looking more toward Europe than toward the land that crowded in about them. When settlement pushed tentatively away from the oceans, it still followed the waterways, for they offered trade pathways to the coast and an important link to Europe as well. Thus, the British settled the indented coastline of the Chesapeake Bay and its tributaries, or they spread a thin band of settlement along the rugged coastline of New England, the Dutch moved up the Hudson River from New Amsterdam (New York), and the French gradually settled the

Figure 3-1 Expansion of the settlement frontier in the United States. For the first three-quarters of the 380 years since its initial settlement, the United States was in a period of active frontier expansion. Many have suggested that this long history was critically important in shaping American culture and ideals.

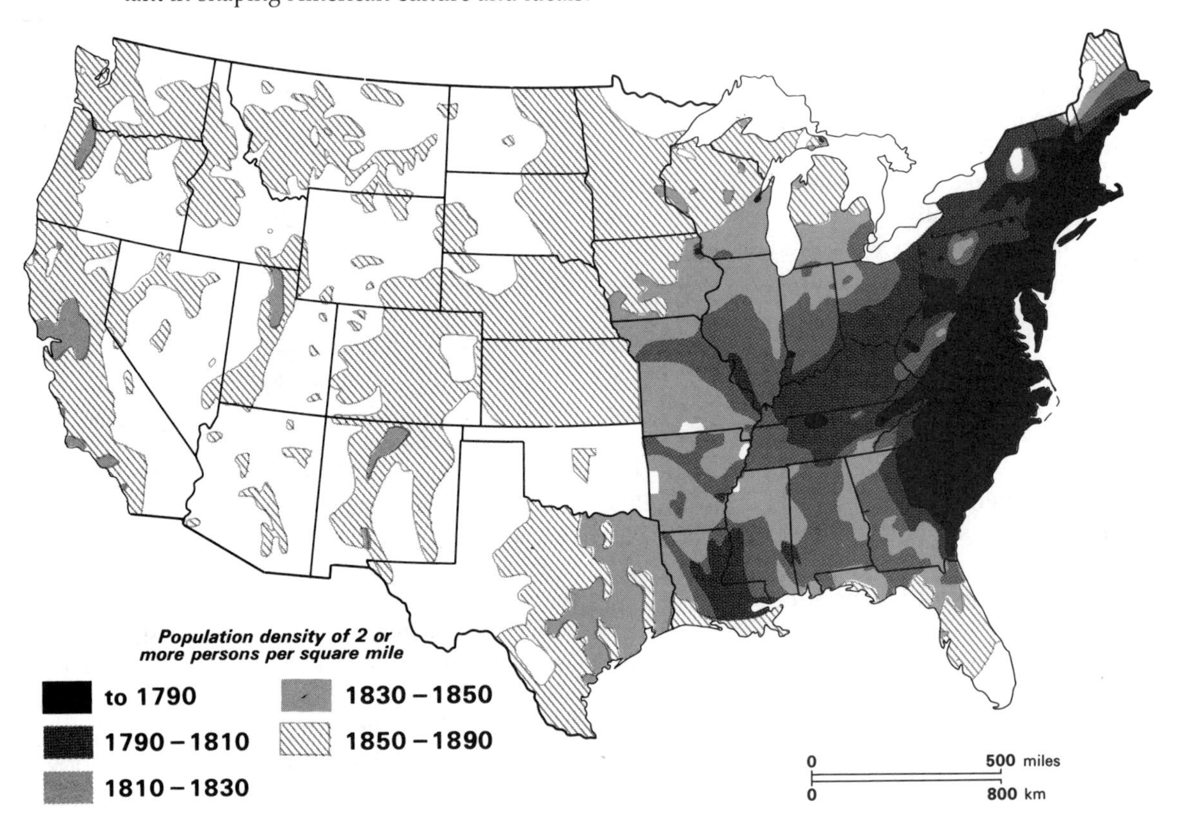

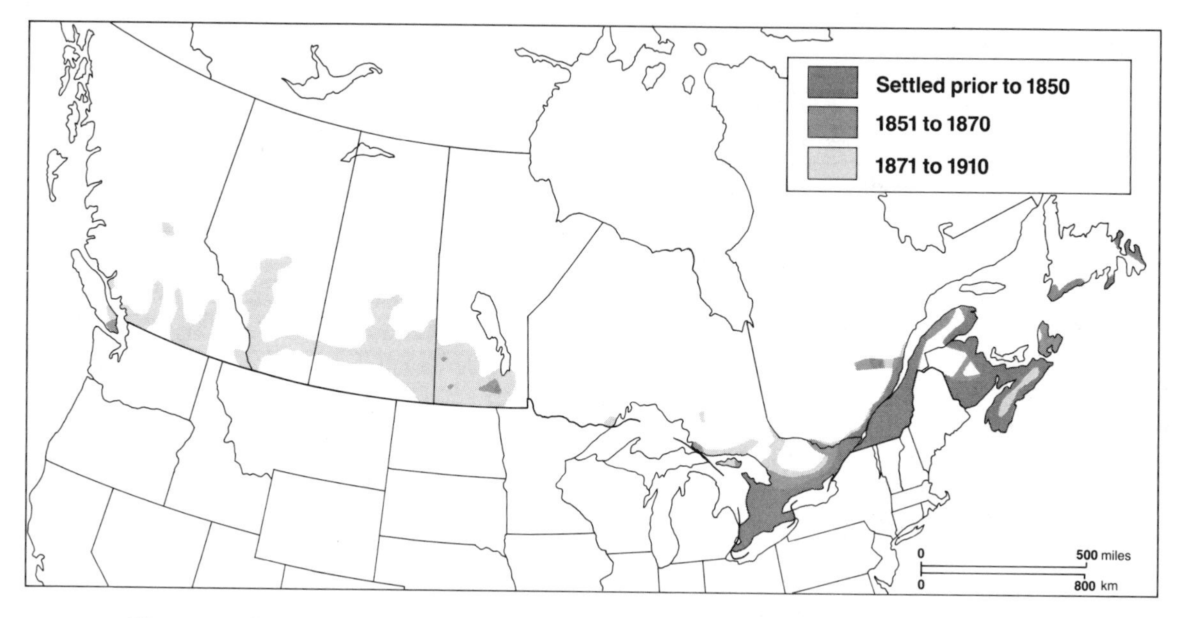

Figure 3-2 Expansion of the settlement frontier in Canada. The Shield was a major barrier to the north and to westward expansion in Canada. The St. Lawrence lowland and southern Ontario thus developed as the dominant Canadian settlement focus.

banks of the upper St. Lawrence. To break away from the water was in many ways a break from Europe. That step was taken slowly and with trepidation.

Settlement expansion in both countries began slowly, then gradually picked up speed as time passed. For example, in the United States during the first 150 years after the beginnings of permanent European settlement—until about 1765—Europeans had moved westward only as far as the eastern flanks of the Appalachians. Within a century after that, the frontier had reached the Pacific Ocean, and by 1890 the U.S. Bureau of the Census was able to announce that the American settlement frontier was gone entirely. This increasingly rapid settlement expansion resulted from a reorientation in attitude away from Europe toward North America. By the early nineteenth century, an increasing number of Americans viewed the occupation of the continent as their manifest destiny. The land laws of the country, culminating in the Homestead Act of 1862, became increasingly proexpansionist. This made it much easier to obtain land along the settlement frontier. Also, as the annual population increase grew—the result of both natural increase and immigration—there were more people who hoped to improve their lot by moving westward. The expansionist policies of both Canada and the United States lead to an almost constant state of underpopulation and surplus production. More people were needed for the opportunity that was available, but both countries already were producing more raw materials than they needed. To Europe, both countries were economic colonies, providing raw materials and buying finished products.

In the eastern half of the United States, about as far west as Kansas and Nebraska, settlement expanded westward in a generally orderly fashion. To be sure, advances were more rapid along certain transportation routeways, such as the Ohio River, and slower in other places. Some areas were bypassed by the early settlers in favor of more promising lands farther west. At times, this created a hollow frontier. For example, parts of the Adirondack Mountains of New York remained unoccupied while the principal line of settlement advance pushed across the Mississippi River. A somewhat similar pattern of even expansion developed in southern Canada along the St. Lawrence River and the lower Great Lakes.

Settlement moved westward onto the interior grasslands more rapidly in the United States than in Canada. Whereas the Mississippi River and its many tributaries offered easy routes in the interior of the United States, movement west beyond the Great Lakes in Canada was primarily overland. Furthermore, and more important, American settlers found a vast expanse of excellent agricultural land with a generally good climate for crop production that stretched from the western margins of the Appalachians well into the Great Plains. Canadian settlers, on the other hand, faced some 2500 kilometers (1500 miles) of land nearly worthless for agriculture that extended from about 150 kilometers (95 miles) north of Lake Ontario to just east of the site of Winnipeg, Manitoba. This is a portion of the glacially scoured Canadian Shield with its thin, rocky soils. Although most of the Great Plains of the United States were settled between 1860 and 1890, the prairies of Canada were not opened to settlement until near the end of the nineteenth century. Even then, many of the early settlers were Americans who had moved northward across the border.

Both countries adopted grid systems of land subdivision to facilitate the survey and settlement of their large interior lowlands. The result is the regular, rectangular pattern of fields and roads so familiar to residents of the territory. (George Hunter/National Film Board-Phototèque)

Elsewhere, in northern Canada, in the United States and Canada from the Rocky Mountains westward, and in Alaska, an even pattern of settlement expansion did not occur. These lands were characterized by generally poor agricultural potential. Much of this broad area was either too dry, too hot, or too cold for farming. Rugged topography hampered transportation and further limited agricultural development. Settlement congregated in areas that offered an identifiable economic potential. For example, agriculture developed in small areas of good soil and adequate precipitation, such as in the Willamette Valley of Oregon. Mineral resources offered another settlement impetus, as in the case of the mining centers of north-central Ontario. Other sites developed to service the transportation lines that extended across these empty areas. Whatever the specific reason for growth, the result was a pattern of point settlement scattered across an otherwise nearly unpopulated landscape.

Population Distribution Today

The 1986 census of population in Canada revealed a population of about 25.3 million. In 1980, the United States had a population of 226 million, and by 1990, its population approached 250 million. This 10:1 ratio has been maintained for about a century. Canada had a 1986 population density of about 2.5 people per square kilometer (6.5 per square mile), whereas the coterminous United States (i.e., excluding Alaska and Hawaii) had a 1990 density of roughly 235 people per square kilometer (85 people per square mile).

Neither country has anything approaching an even population distribution (Figure 3-3). Three principal zones of population can be identified. First, a primary zone fills a quadrant defined approximately by the cities of Montreal, Chicago, St. Louis, and Washington, D.C. This is the traditional population core of the United States and Canada. Most of the larger cities in the two countries are found in this zone: 60 percent of the Canadian population lives in southern Quebec and Ontario; 7 of the 12 most populous states in the United States are here. It is the area of earliest major population growth in both countries and long their most advanced sections economically. Fine natural routeways and many excellent harbors along the Atlantic shore have been augmented by one of the densest transportation nets in the world (Figure 3-4). Some of the continent's best agricultural lands plus rich mineral resources are either within the region or around its periphery. Now in decline relative to some other parts of the continent, this region is still North America's vital core.

Wrapping around the southern and western margins of the core and extending westard to the eastern sections of the Great Plains in the United States, there is a secondary zone of population. Much of the best agricultural land of the country is in this zone. With relatively minor exceptions, this area is amenable to agricultural development, the greatest part of its potential agricultural lands are farmed. Most of the area is populated although densities are generally much lower than are those found in the core. Cities are spaced more widely and more evenly in this zone than in the core, and they are primarily service and manufacturing centers for the regional economy. As a group, these cities have less significance nationally than the cities of the core.

Finally, a peripheral population zone fills the land from the central Great Plains westward. The pattern of population and economic growth at locations of special potential in an otherwise limited region continues to dominate. This is a pattern that developed early in the region's settlement history. Although some areas are now densely populated—notably California's San Francisco Bay area and Los Angeles Basin and the Fraser Delta-Puget Sound Lowland in British Columbia and Washington State—most of the land remains sparsely populated.

Canada, beyond the southern Ontario and Quebec core, does not fit easily into this scheme. The Canadian population is split into about four major populated areas, each separated from the

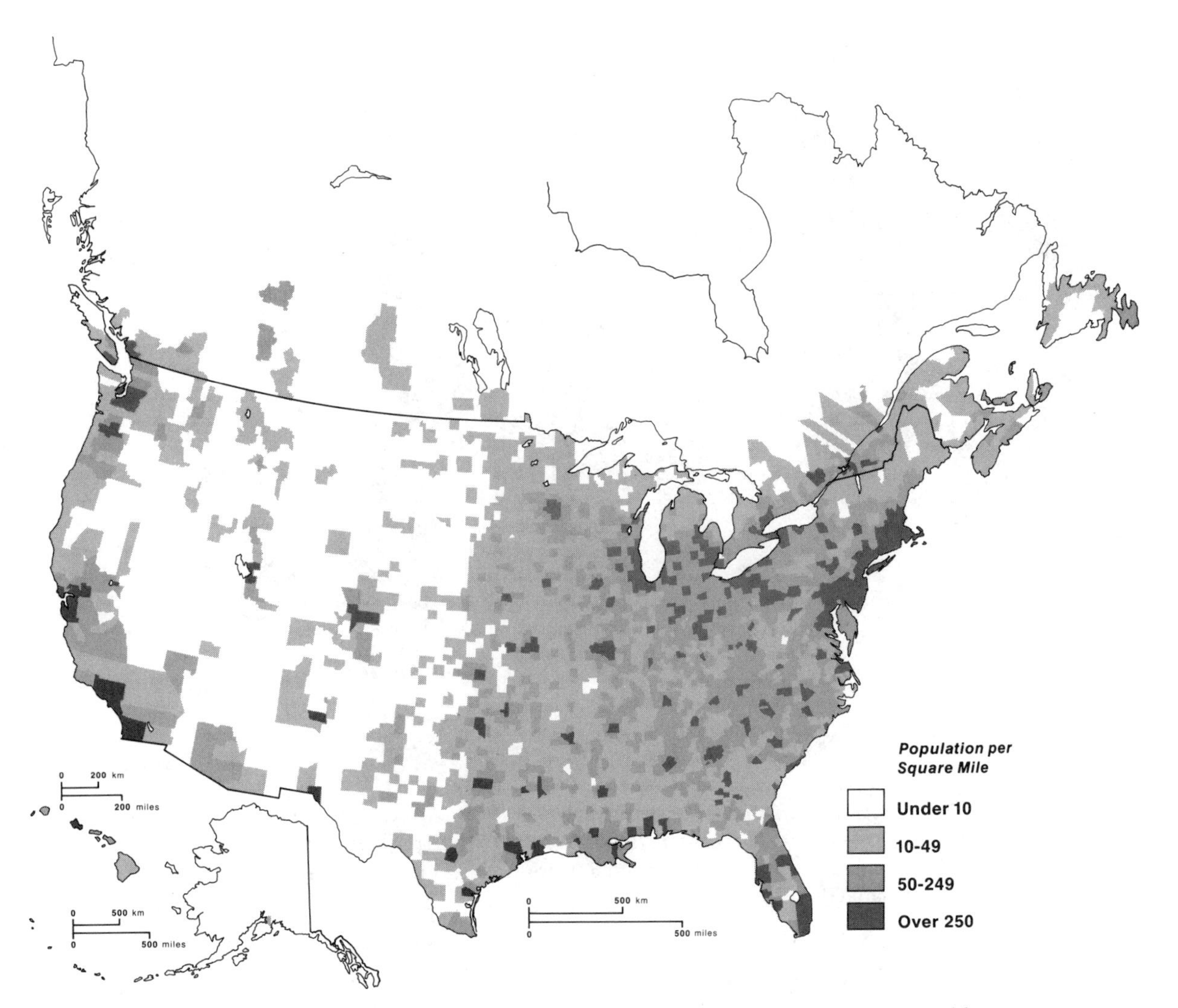

Figure 3-3 United States and Canada population density. Most geographers would argue that we have the technological ability to work within any physical environment, and humans can live anywhere on the earth's surface. Nevertheless, notice how areas of generally lower population density are associated with those of lower precipitation (Figure 2-2) and colder temperatures (Figure 2-3).

others by extensive empty areas. If we view the Quebec-Ontario core as a single unit, it is separated from the settlement of the Prairie Provinces by the Canadian Shield in western Ontario and eastern Manitoba. Otherwise, the farmlands of the Prairie Provinces might be viewed as a part of the secondary zone of population that we saw in the United States. The Prairie Provinces are themselves separated from the coastal population of British Columbia by the Coast Ranges and the Rocky Mountains. This far western region is similar to the peripheral zone in the United States. To the east, the population of the Atlantic Provinces is scattered widely along an

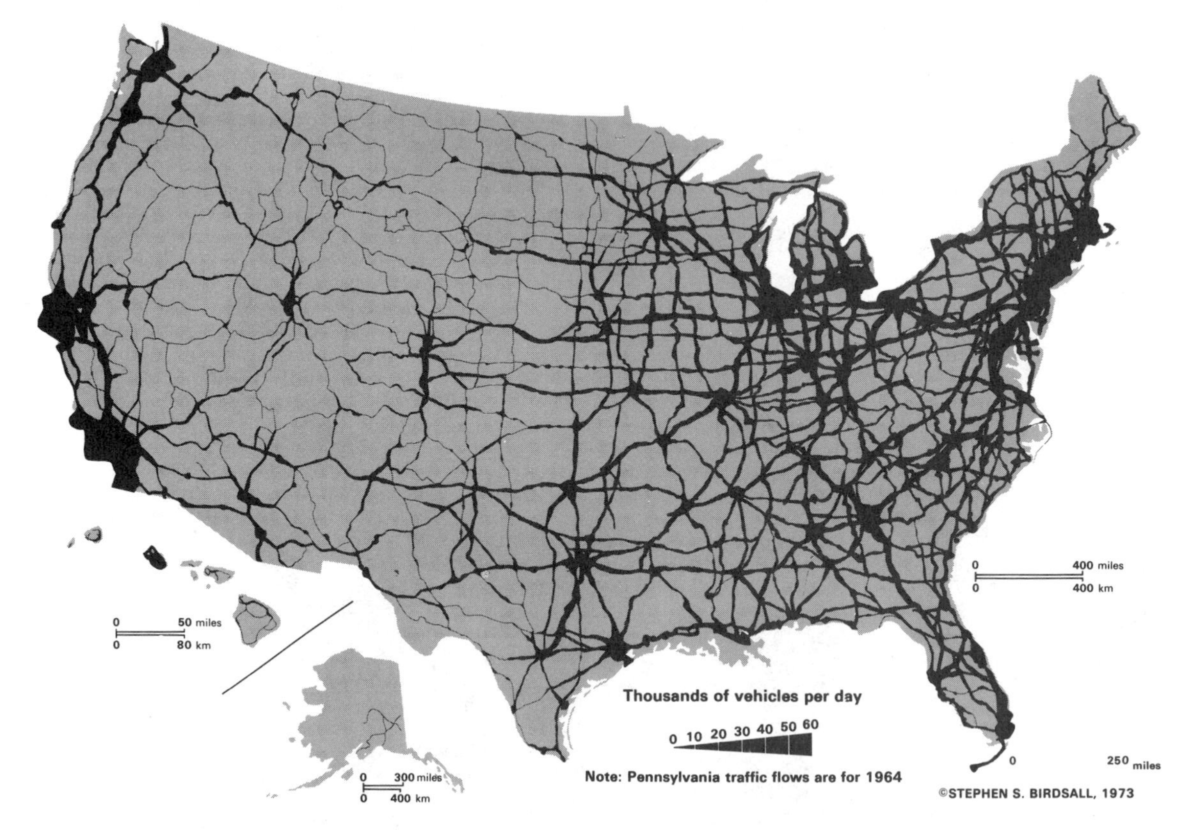

Figure 3-4 Highway traffic on primary roads, 1969–1970. Air, rail, and highway movement all focus upon the urban centers of the country. While the overall volume of traffic has increased substantially over the past 20 years, the geographic pattern remains substantially unchanged.

indented coastline and in a few interior lowlands. It is connected to the rest of the country by a thin, tenuous line of settlement along the southern shore of the St. Lawrence River and overland into New Brunswick. Taken together, these zones include well over 95 percent of the Canadian population.

Each of these four population clusters is pressed against the southern margins of the country. In fact, most Canadians live within 150 kilometers (95 miles) of the country's southern border. They have often had closer economic and cultural ties with the areas south of the border in adjacent parts of the United States than with other parts of Canada. This has been particularly true in the past. Rising Canadian nationalism plus an effort by the Canadian government to improve interregional communications has lessened this exterior association somewhat.

Population Redistribution–Mobility Patterns

Population redistribution is a pervasive aspect of the geography of the two countries (almost 1 American in 5 changes his or her residence in an average year). Substantial internal redistribution plus continued immigration has maintained a high potential for rapid regional change. The pace of population increase is used frequently as a measure of local or regional well-being. If a regional population is growing rapidly, the region's economy and stature are viewed as sound; if the population is declining, grave concern usually is expressed about the future.

"Go West, young man," that famous admonition attributed to Horace Greeley, has long characterized one element in the internal migration patterns of these two mobile populations. In fact, the center of population in the United States (defined as the intersection of two lines that equally divide the country's population north-south and east-west), located southeast of Baltimore at the time of the first national census in 1790, has moved westward during each ensuing decade. By 1980, it was found southwest of St. Louis, west of the Mississippi River for the first time. Until the mid-1960s the dominant reason for this westward shift was the movement of people within the country. Since then, immigration into the country has also been important. Most Asian and Latin American migrants into the United States enter through Southwest and West Coast cities, and many decide to remain in those areas.

It is possible to divide the mobility history of both countries into three periods. First came the period of east to west movement, then one from rural to urban areas, and, finally, the present period, when most long-distance movement is between metropolitan areas. If the U.S. population has moved westward with every decade, it has urbanized in an equally unvarying fashion. Whereas only about 5 percent of the population could be defined even loosely as urban in 1790, over three quarters of the country's population was urbanized by 1990. The 1986 Canadian census also classified over 75 percent of all Canadians as urban. These statistics reflect not only a relative decline in rural population, but also an absolute decline in farm population as well. For example, between 1960 and 1987 the farm population of the United States fell by over one-half; from more than 15 million to under 6 million. Although not as dramatic, Canada's farm population has also decreased substantially. The movements from east to west and from rural to urban America were both clearly in response to the perception of economic opportunity. First, more and more farmlands became available in the two countries as the settlement frontier pushed westward. Then there was a tremendous surge in urban employment generated by the technological advances of the Industrial Revolution. To reemphasize a point, both of these trends developed together throughout the nineteenth century and continued into the twentieth century in Canada. The westward movement was relatively more important at first, whereas the urbanward movement had gained ascendancy by the end of the period. And once North Americans were predominantly urbanites and economic opportunites were also urban based, variations in these opportunities ensured that most subsequent population migration would occur between metropolitan areas.

U.S. population statistics for the 1970s and 1980s suggest that a fourth major mobility period is at hand. Areas that had long experienced no change, or even declining population size are growing. Much of the South is a prime example. For more than a century, it was a region of out-migration as southerners were drawn to the opportunities located in the cities of the north and west. Recently, however, it has become a growth region; more people are entering the region than are leaving it. The Bureau of the Census predicts that this pattern of population change will continue through the remainder of the twentieth century.

Perhaps this shift should not come as a complete surprise. Many observers of the American economy have suggested that the United States has become the world's first postindustrial country. That is, the major growth areas in the national economy are not in manufacturing. Instead, they are in what are called *tertiary* and *quaternary* occupations. These are, respectively, occupations that provide services and those that manipulate and create information. The number of Americans employed in manufacturing has increased only slightly during the past two decades, whereas tertiary and quaternary employment has boomed. Much of what increase there has been in manufacturing employment has been in the production of high-value, lightweight products, such as electronic components.

It is suggested that this employment shift has freed many employers and their workers from traditional locational ties. In the past, the economic advantages offered by locations within the continent's core region were the most important considerations in attracting population. High-value light industry presumably can locate almost anywhere. High-speed, inexpensive communications and transportation allow many tertiary and quaternary employers to locate in any one of many possible locations. People, or at least more and more people, may live where they want, not where a specific economic function dictates they live. The Sunbelt states of the South and West provide the kind of residential environment that Americans desire. Outdoor recreation and mild winters—these evidently represent the goals of many Americans today. Now that more and more can live where they want and find employment as well, they are moving to the South and the West.

The socioeconomic nature of the migrants presently leaving and entering the South seems to support these comments. For both blacks and whites the average immigrant into the South is better educated and better paid than the average outmigrant. This suggests that the new arrivals are there in response to a growth in the number of high-quality jobs available to them and that fewer are seeking opportunity in the industrial centers of the North.

URBANIZATION

The United States and Canada are urban countries. This has been a fact of national life since World War I. The two countries' urbanizing trend is significant because it involves a spatial rearrangement of population, but it also represents a change in the type of space occupied.

By residing in urban locations, North Americans live in an environment quite different in character from that affecting most people several generations ago. The differences are more than technological. Rural and urban environments are different. The former is characterized by individual isolation created by distance; personal isolation in the latter is more likely a result of self-imposed or socially prescribed psychological distance. Economic activities in rural areas tend to be extractive and individual, whereas in cities most economic activities are joint efforts at manipulating materials, service opportunities, or ideas. And the rural environment offers a life-style that retains at least a degree of self-sufficiency, whereas the urban environment contains people entirely dependent on each other for virtually every aspect of their lives. The highly interdependent urban environment clearly has a spatial organization distinct from that typical of rural areas.

Spatial organization is the way a set of activities, each of which takes place at a separate location, is linked together by the interactions between them. Urbanized areas possess a more intense and complex spatial organization than rural areas. In large cities, the variety of activities is greater than in small towns, and the intensity of spatial organization varies accordingly.

Most of North America possesses urban areas which have grown in population and extent. In a few instances, the spatial growth was so great and the size of the core cities became so large that major urban areas merged and formed clusters of cities. Geographers and others began to treat these clusters as something quite different in the history of urbanization. The group of large cities extending from Boston, Massachusetts, to Washington, D.C., along the northeastern U.S. coast is the clearest example (see Figure 1-1). Another group of urban areas—more widely dispersed and containing central cities that are not as large as those in the Boston–Washington, D.C., complex—is found along the southern margin of the Great Lakes. Milwaukee and Chicago anchor this region in the west and Buffalo and Pittsburgh do the same in the east. Southern

California, from San Diego to San Francisco, is offered by some observers as yet another set of urban areas that will be merged by the end of this century. Others expect little spread in southern California beyond the current San Diego–Santa Barbara cluster considering the urban resource limitations of California (see Chapter 16). Much of east coastal and central Florida is expected to be urbanized by the year 2000, and numerous small clusters of cities repeat the pattern in other sections of the country. The one cluster of coalescing urban areas that overshadows all of the others, however, is the urbanized northeastern seaboard of the United States (Chapter 5).

Each major cluster of urban places and the isolated towns and cities, too, exist where they do for identifiable reasons. We might ask: Why do cities come to be located where they do? In a simplified and general way, we can say that most of the largest urban places have developed where transportation routes connect with each other. Quite often it is the land-water connection that is important. Other factors matter, of course: hinterland quality, proximity to alternative transportation, security, and even healthfulness of the local environment may be important. However, where goods and people must transfer from one form of transportation to another (called *break-in-bulk* locations), there are a great many opportunities to process, exchange, manufacture, repackage, sell, and buy goods. Cities, in other words, are usually located where they are because of certain economic advantages associated with movement to and from their location.

Think for a moment about the major urban centers of the United States and Canada (Figure 3-5). Many of the largest are located adjacent to navigable water. Some are on a seacoast or large estuary: Boston, New York City, Baltimore, Quebec, Miami, Tampa–St. Petersburg, Mobile, San Diego, Los Angeles, San Francisco, Seattle, Honolulu, and Vancouver, British Columbia, are examples. Others are on naturally navigable waterways, such as Chicago, Milwaukee, Minneapolis–St. Paul, Detroit, Cleveland, Buffalo, Hamilton, Toronto, Montreal, Philadelphia, New Orleans, Memphis, St. Louis, Kansas City, Cincinnati, Pittsburgh, Louisville, and Portland, Oregon. Still others are on rivers or channels that have been modified extensively just to give the cities water access: Houston and Knoxville are examples. There are exceptions to this water orientation, of course, such as Atlanta, Calgary, Denver, and Dallas–Ft. Worth, but these, too, were on early transport routes of some kind. Atlanta, for example, located at the southern tip of the Appalachian Mountains, had become a key inland center for railroad transportation in the South by the time of the Civil War.

The specific reasons for any particular city's growth and why it grew where it did are often divided into aspects of urban *site* and urban *situation*. The site of a city is its immediate physical environment—that is, the characteristics of the landscape in which the city is located. Pertinent information includes the topography across which the city spreads—Is it flat or rolling? Is it on a plateau or in a valley? its soil characteristics—Are large areas of poorly compacted river soils present? Are there good water-bearing sands? drainage—Are there swampy areas? Does it have a river for water supply or water transportation? the depth of the bedrock; and so on. Many times such site characteristics help account for the difference between rates of urban growth for competing cities.

Urban situation is the set of factors associated with the position of the city relative to other places. Situation is sometimes referred to as *relative location*. Several aspects of situation are proximity to other centers—short distances might make trade easier; position between two areas that produce different products—this may provide the opportunity to serve the intermediary function in exchange; and the productivity of land using the city as a regional focus. In urban geography, situation is used to describe the locational features of a city that link it to other places and form the relations the city has with other cities, other regions, or locations of potential and actual activity. In other words, a city's situation reflects the support it receives through the way it

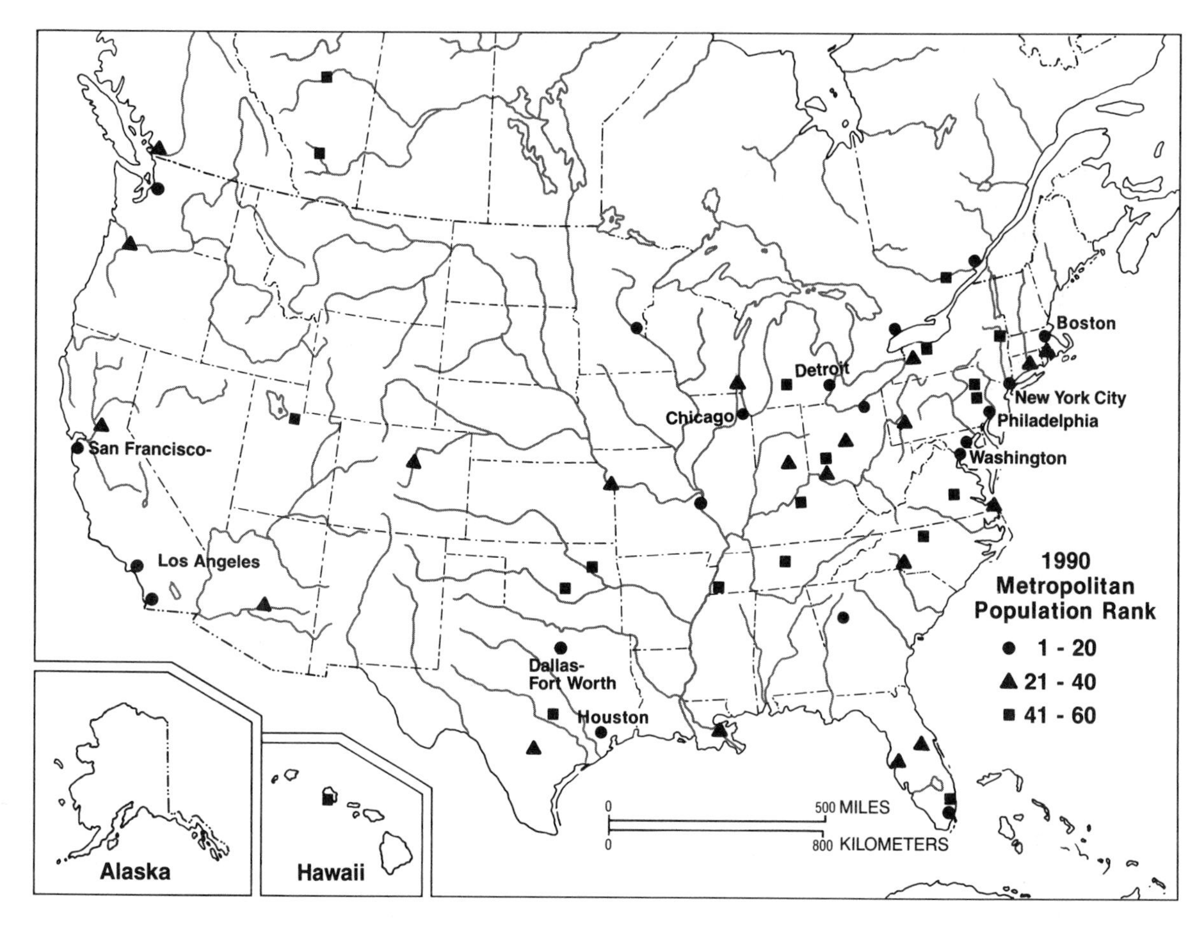

Figure 3-5 The largest urban areas in Canada and the United States. Most of these large urban areas in the two countries are at highly accessible ocean or inland waterway locations.

is associated with other places. In some ways, situation is defined by a city's accessibility to regions that contribute to the city's growth. Thus, situational factors are related much more exclusively to the process of urban growth, whereas site factors, as we have just seen, are related to both the initial appearance of settlement and its growth.

PATTERNS OF REGIONAL CULTURE

The territorial space of Canada and the United States has been neatly subdivided into political units—2 countries, 50 states, 10 provinces and 2 territories, over 3000 counties in the United States and hundreds in Canada, and many thousands of smaller units. These units mark and bound our political activity. They are far less reliable, however, as indicators of the spatial expression of culture. Although some of our political subdivisions do have substantial meaning as culture regions (Quebec is perhaps the best example), most do not.

The term "culture," and the concept behind it, have long been debated by anthropologists, geographers, and others. No universally acceptable definition is possible. Basically, culture is our assemblage of learned behavior. Its complexity and durability sets humans apart from other animals. A common culture often represents the strong bond that ties together the people of a country, an ethnic or social group, or a region. In turn, it is the group's heritage—its shared set of experiences—that is of greatest importance in the creation of that cultural togetherness. Culture is transmitted through symbolic means, such as language, not through biological means. Thus, a common language is perhaps the critical element in the maintenance of a culture or culture region. Some have argued that one of the greatest strengths of the United States is that it is the world's largest and most populous country joined geographically and socially by a common language.

Do you live in Providence, Rhode Island? Then you probably call a big bun sandwich a grinder, or perhaps a torpedo. New Orleans? To you, the same sandwich is a Poor Boy. Pittsburgh? You probably call it a hoagie. Southern Florida? Does the term Cuban Sandwich sound more familiar? Many of the rest of us refer to that same basic sandwich as a submarine. The name of this dining delight may not be especially important, but the title we choose for it is an element of our culture. The point is simply that there are clear regional differences in American culture. Indeed, most of the regions identified in this text are at least in part *culture regions*.

Regional variation in culture may be expressed in many different ways. Indiana, Kentucky, Ohio, and Illinois, often thought of as centers of college basketball fever, really do produce far more big-time college basketball players per capita than the national average. The vast majority of early country music singers were from the upper South, especially two core areas—the southern Appalachian Mountains and the Nashville–Blue Grass basins of Tennessee and Kentucky. While the South is now developing into a true two-party region, for decades many white Southerners thought of themselves as "yellow dog Democrats"—given the choice, they would rather vote for a yellow dog than for a Republican.

The landscape of an area blends the natural environment and a cultural imprint. The rectangular land survey system used widely in the United States during the nineteenth century created a striking regularity to the landscape of much of the Middle West. The German and English farmers of southeastern Pennsylvania built a large, even massive, cattle and hay storage barn with a second-story extension over the first on one side. While students of folk architecture may argue its origins, most agree that this "Pennsylvania barn" is a key identifying element in the landscape of the Pennsylvania culture area. Ethnic areas in many cities can be located simply by looking at the names on small neighborhood stores and restaurants.

While many aspects of culture are conservative and consistent, change is nevertheless a constant part of American culture. Many of the alterations result from changes in technology and economic conditions. Migration is another key ingredient. The arrival of millions of Latin American and Asian migrants in both countries in the last two decades is resulting in new patterns of cultural change. The Hispanic influx into southern Florida has dramatically altered the fabric of life and thinking in the southern part of the state. Vancouver's large Chinese population has added a new dimension to the city's culture. Nearly every large city, and a considerable share of smaller ones, is sharing in these changes.

American Religion

Of the individual elements of American culture, one of the most interesting and telling is religion (Figure 3-6). This is largely because it is an essentially conservative culture trait. If we were to ask most Americans why they are members of a particular denomination, the answer would either be because their parents were members or that they had always been members of that group.

A number of the larger Christian churches were brought to North America by European migrants. The distribution of these denominations closely matches the areas where those

There has been a significant Chinese presence in many large American cities for many decades. Recent large-scale Chinese immigration has often reenergized these "Chinatowns" and fueled their geographic expansion. This street is in New York City. (Thelma Shumsky/The Image Works)

various migrant groups and their descendants form a large part of the population. For example, German and especially Scandinavian settlers carried their Lutheran church to the northern Great Plains and northwest portion of the Deep North. Hispanics in the Southwest; Southern and Eastern Europeans in the Northeast, Middle West, and in most large cities everywhere outside the South; and French Acadians who were forced to leave eastern Canada—they migrated in substantial numbers to south Louisiana—all help explain the widespread distribution of Roman Catholicism in much of the country outside the South and Utah.

The United States has also been a place of active denominational creation. This is the result of a combination of a number of elements. One is the lack of an association between church and state and the relative wealth of American society (operating an independent denomination can be expensive) is another. A third element is the desire for national or regional identity. Several denominations, such as the Episcopalians (formerly part of the English Anglican church), were created at the end of the American Revolution in reaction against any continued dominance by Britain. The Civil War led to a similar north-south split in the country. Presbyterianism in the United States is divided into several denominations as a result of a post–Civil War split in which each subdenomination felt that the others did not adequately reflect its particular interests.

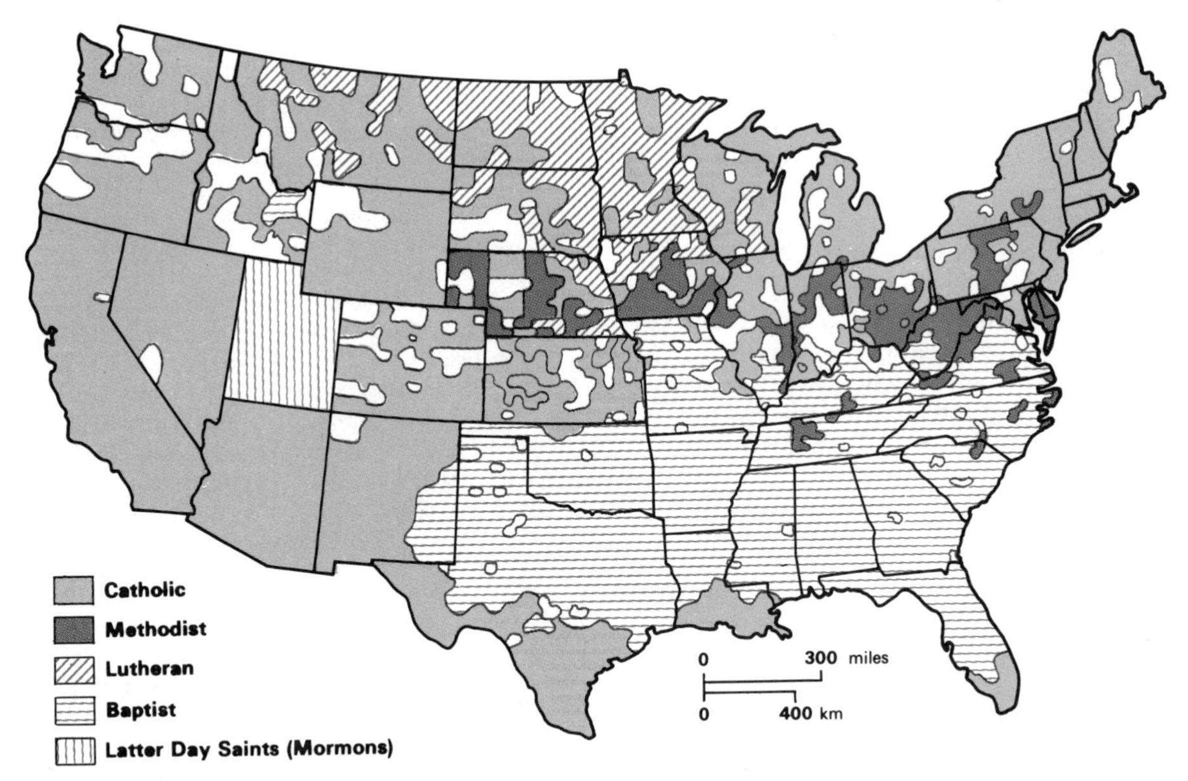

Figure 3-6 Predominant religious affiliations in the United States. (Adapted from *Churches and Church Membership in the United States, 1971* by Douglas W. Johnson, Paul R. Picard, and Bernard Quinn. Copyright National Council of Churches, 1974; published by Glenmary Research Center, Washington, D.C.) International migration into the United States (Catholics and Lutherans), internal migration (Mormons), and the development of regional identity (Baptists) are among the influences on the religious geography of the country.

Another explanation has been the creativity of American religion. Individuals would establish their own churches, or congregations or groups of congregations would leave a denomination to form a new one, because of disagreements over such questions as biblical interpretation or church administration. Many of these native American churches follow a conservative theology. Many were founded in the South. Although today often large in size, they are nowhere numerically dominant and hence are not on the map.

One native American church that is on the map is the Church of Jesus Christ of Latter-Day Saints, commonly known as the Mormon church. Founded in upstate New York in the midnineteenth century, it was gradually carried westward by its followers in search of an isolated place to settle and follow their beliefs (see Chapter 14). They eventually chose Utah. Today, most residents of Utah are Mormons. Mormons are procreative—they believe in large families. One result is that average completed family size (the total number of children ever born to an average family) in Utah is the highest in the United States, standing at about 3.2 children. The average for the country is near 1.8 children.

Southern Baptists (the territory on the map

dominated by Baptists is almost entirely Southern Baptist in denomination) are an interesting joining of several of the foregoing explanations. Baptism was brought to North America by early European migrants as a nonestablished church seeking freedom of worship. Although founded in the South, it did not become the dominant regional church until after the Civil War. During the last third of the nineteenth century, it was almost the religious expression of Southern culture and became easily the dominant regional church. One measure of whether a community is culturally part of the South surely must be the existence in it of at least one Southern Baptist church.

ADDITIONAL READINGS

Abler, Ronald, et al. *A Comparative Atlas of America's Great Cities: Twenty Metropolitan Regions*. Minneapolis: University of Minnesota Press, 1976.

Adams, John S. *Housing America in the 1980s*. New York: Russell Sage Foundation, 1987.

Allen, James P., and Eugene J. Turner. *We the People: An Atlas of America's Ethnic Diversity*. New York: Macmillan, 1988.

Bryan, Roderick, and Kevin McQuillan. *Growth and Dualism: The Demographic Development of Canadian Society*. Toronto: Gage, 1982.

Canada Yearbook. Ottawa: Statistics Canada (annual).

Gastil, Raymond. *Cultural Regions of the United States*. Seattle: University of Washington Press, 1979.

Greenlie, Barrie B. *Spaces: Dimensions of the Human Landscape*. New Haven, Conn.: Yale University Press, 1981.

Halvorsen, Peter L., and William M. Newman, *Atlas of Religious Change in America, 1952–1971*. Washington, D.C.: Glenmary Research Center, 1978.

Harris, R. Cole. *Historical Atlas of Canada. 1: From the Beginning to 1800*. Toronto: University of Toronto Press, 1987.

Koegler, John. *Canada's Changing Landscape: Air Photos Past and Present*. Toronto: Douglas Fisher, 1977.

Monkkonen, Eric H. *America Becomes Urban: The Development of U.S. Cities and Towns, 1780–1980*. Berkeley and Los Angeles: University of California Press, 1988.

Rooney, John F., et al. (eds). *This Remarkable Continent: An Atlas of United States and Canadian Societies and Culture*. College Station: Texas A & M University Press, 1982.

U.S. Bureau of the Census. *Statistical Abstract of the United States*. Washington, D.C.: U.S. Government Printing Office (annual).

Vale, Thomas R., and Geraldine Vale (eds.). *U.S. 40 Today: Thirty Years of Landscape Change in America*. Madison: University of Wisconsin Press, 1983.

Zelinsky, Wilbur. "North America's Vernacular Regions." *Annals of the Association of American Geographers*, 70 (1980), pp. 1–16.

CHAPTER

4

Key Geographic Issues in North America's Future

Is the future of North America bright with promise or fraught with problems? The answer, of course, is "yes." The future may be one or the other, or a mix of both. There is an inherent tension in attempts to look seriously into the future, a tension between imaginative fancy and skeptical, measured judgment.

Each of us holds a different attitude toward the future: It will be better. It will be worse. It won't matter because I won't be around. Whatever will be, will be. Problems that arise cannot be anticipated, but they can be fixed—by science, by education, by regulation, by knowing the past, by trusting in God.

Throughout this book, we take the position that current geographic patterns evolved from previous conditions. The present geography of the United States and Canada is a snapshot of processes playing themselves out across this portion of the earth's surface. The landscapes of the regions in these two countries have evolved from the interplay of many human and natural processes across earlier landscapes.

If the past is prelude to the present, then the present may also be prelude to the future. If we can know enough about the present geography of North America and how that geography came to be, then we may be able to sketch at least the outlines of the region's geography in the near future.

Rather than attempt a portrayal of broad geographic patterns in the United States and Canada as they might appear in the year 2010 or later, we describe a number of key issues that are apparent now and that will have an effect on the geography of the future. How these issues are addressed (or ignored) will help create the broad patterns in North America's next century.

There are many issues, and our selection is far from exclusive. Another pair of authors might select an entirely different set of issues as especially significant. An entire volume could be devoted to the topic. We have selected five with clear geographic foundations and clear implications for the United States and Canada during the next several decades. They are representative of the diverse and difficult problems with which these countries are faced. The five issues are multilingualism, competing uses of the land, adequate water supplies, the countries' changing age structure, and waste disposal. It bears repeating that these are not the only issues of a geographic nature to affect the two countries into the next century. As you read the rest of this text, you should be alert to other changes that will have equivalent impacts on the continent's future geography.

Each issue selected is broad, at least continental in scope and perhaps even global. In each case, however, local conditions are affected. Typically, when public reactions of sufficient intensity are expressed, policy is offered by government either to address the issue or to address the public reaction, or both. Geographers are especially interested when such policy affects places, specifically, and the people who live in them.

In the next decade we must address an issue which has proven to be difficult to grasp politically. How will we manage unwanted and difficult materials so that local environments remain healthy and attractive places in which to live? (Topham/The Image Works)

MULTILINGUALISM

Language is the means by which we most commonly share information and our ideas and beliefs about this information. Without this sharing, understanding is difficult or impossible, and a group of people remains a collection of individuals rather than a community or nation. Growth in the number of Spanish-speaking people in parts of the United States during the 1980s raised questions in the country about the function of English-language dominance. Canada's official bilingualism (English and French) remained in effect, but political arguments in 1990 over cultural and linguistic status threatened to lead to the country's dissolution. The geography of the United States and Canada could be very different in 20 years if disagreements over language and cultural dominance are not soon resolved.

The United States never specified a single language as the one to be used officially in the course of its governance or commerce. During the three centuries following Columbus's first voyage across the Atlantic, settlers from a number of European countries established zones of language concentration across the eastern and southern margins of North America. Some languages proved more persistent than others, just as did the settlements. But four European languages were important at the time of the American Revolution, and numerous others established themselves as locally important during the succeeding 200 years.

English was most extensively spoken in the United States at the end of the eighteenth century, and it continues to hold its position of linguistic preeminence. But other languages faded only slowly or not at all. German-speaking Quaker, Amish, and Mennonite settlers concentrated early in Pennsylvania. There were 15,000 in the territory by 1730, with more to come in succeeding decades. Dutch-speaking settlers along the Hudson River and near Philadelphia established areas in which Dutch was dominant until the early decades of the nineteenth century. Spain exercised control over much of the southern margin of North America, although it was in the lands from Mexico north that Spanish was most important. French-speaking Acadians, or "Cajuns," expelled from Canada in 1755, settled around the Atchafalaya swamps in southern Louisiana and remained linguistically distinct long after English-speaking settlers moved in. The French presence was so important in Louisiana, in fact, that when the state constitution was drafted, it included a statement that all official business must be conducted in the language "in which the Constitution of the United States is written."

Immigration to the United States from non–English-speaking countries contributed further to the country's cultural and linguistic diversity (see Chapter 3). In many cases, of course, a generation or two of residence meant linguistic absorption into the overwhelmingly dominant pool of English, but many pockets remain. Rural communities in Pennsylvania, Michigan, Texas, New Mexico, and other states are isolated enough to continue to be non–English-speaking for internal business and social affairs, even while their inhabitants know enough English to permit exchanges for commerce and governance with the larger population. Within urban areas, the cultural mosaic could also be relatively permanent with ethnic communities lasting generations and each community's residents more or less bilingual with English used in combination with the other language. Where Native American nations were relatively large and isolated, their languages have remained the most important means of intragroup communication.

Canada is bilingual by design and official decree. As early as 1763 and by Royal Proclamation, French and English were both given official status. This formal equivalence was reinforced by the Constitutional Act of 1791, in which these two languages were given equal status in both culture regions in Canada. Despite occasional proposals to eliminate the position of French and to assimilate French-speaking Canadians into the politically dominant English Canadian core, the Canadian Constitution of 1867 declared French and English legal and judicial languages in all federal activities and in Quebec. The constitution also declared protection of the right to denominational schooling, an indirect

The ideal that each group learn the languages of the others with whom they interact becomes more challenging as the number of local languages increases. (Terry McKoy/The Picture Cube)

protection of French speaking because the Catholic schools of French Canadians could thereby be taught in French.

The 1867 constitution did not prevent Canadian provinces from placing their own restrictions on French language interaction. French was restricted in some areas in New Brunswick in 1871 and in Prince Edward Island in 1877. Public support for French language instruction was ended in Ontario in 1913 and was not resumed until the 1960s. Manitoba banned bilingual teaching in 1915.

This drift away from bilingualism and its effect on the social and political fabric of Canada was addressed formally in the 1960s by a Royal Commission on Bilingualism and Biculturalism. The report of the Royal Commission recommended a series of steps whereby French and English were given more explicit practical equivalence. While the report recognizes the impossibility of requiring full bilingual facility, it sought to ensure that policies would not be established that would make Canada a unilingual country. Not all commission recommendations have been implemented, and this has encouraged some in Quebec to continue to argue that the only way to preserve French Canadian culture is to separate politically from the rest of the country.

Growth of the Spanish-speaking population and its relative concentration in the United States has triggered efforts there to give English an official status. Other languages are affected by efforts to assert English language dominance, as well, but local impacts from increases in the number of Spanish-speaking peoples have been considerable, especially in Texas, California, and Florida. The population growth rate among the United States' Spanish-speaking population was 57 percent between 1970 and 1980, for example, while the growth in the total population was only 11.4 percent in the same decade. The higher growth was the result of both higher fertility and immigration from various countries in Latin America.

At least as important as the growth in numbers was the uneven distribution of the Spanish-speaking populations. A significant Spanish-speaking population was present before the arrival of Anglo settlers in what came to be New Mexico, Texas, California, and Colorado. Further migration from Mexico into Texas and California followed World War II, and these states have been the primary destinations of migrants from Middle American countries during the 1970s and 1980s, although some moved on to the larger urban centers on the Great Lakes, especially Chicago. The Jones Act of 1917 confirmed U.S. citizenship status for Puerto Ricans, and in the 1940s and 1950s, especially, many moved to New York and New Jersey urban areas to seek jobs. And the influx of Cuban immigrants in the 1960s and early 1970s (see Chapter 11) led to a significant concentration of this latter group in urban southern Florida.

With the rapid growth in the number of

Spanish-speaking people in these areas, pressure accumulated to accept communication in Spanish as equivalent to communication in English. School systems were challenged to provide instruction in Spanish. Proponents argued that instruction in a student's native language would encourage greater understanding of the concepts taught where there was incomplete mastery of English. This idea was not new; 4.5 million students attended German-language schools in Pennsylvania and Ohio between 1830 and 1890, but this practice gradually died out, especially after each world war. As multilingualism resurfaced as an issue in the late 1960s and 1970s, other ethnic groups sought equivalent opportunities. Across the country in 1978, the U.S. government funded bilingual/bicultural projects in 68 languages, but in 1982 the responsibility for funding bilingual education was passed to the individual states. A majority of states now require bilingual education in public schools.

Officially recognized opportunities for bilingual communication increased during the 1970s and 1980s, but the growth in non-English communication also produced negative reactions. In many states experiencing rapid growth in the Spanish-speaking population, there were attempts to forbid public school instruction in Spanish. It was argued that English-language instruction was needed to speed assimilation of the new groups into the dominant culture and its language. By 1986, seven states had declared English to be their official state language, and bills for this purpose were introduced in 12 more. Although the market for Spanish-language television broadcasts passed $40 billion as early as 1980, referenda were passed in several cities (San Francisco and Miami) prohibiting the use of public funds for broadcast services in languages other than English.

A possible future for the United States and Canada is one where tensions over language and cultural supremacy escalate in some regions of each country, perhaps across one or both countries. National fragmentation because of these tensions is more possible in Canada, but this does not ensure that the United States will not experience equivalent problems as a result of disputes between sizable population groups. An alternative future is one in which the countries continue to contain a rich mix of languages and cultures. This is a future in which more primarily English-speaking people can also communicate easily in the language of whatever other ethnic group is regionally important. In this alternative future, a sharing of information, ideas, and beliefs would occur at several levels and in various languages.

COMPETING USES OF THE LAND

This is an old issue, but one that remains important. At the most basic level, when the way a parcel of land is used changes from one activity to another (say, from agriculture to urban, or from recreation to mining), something is gained by society and something is lost. The long-running argument, couched in many different ways, is over whether what is gained is greater than what is lost. This is not a simple issue and extends into such realms as land ethics and political values. Specific instances of land-use competition are found throughout the United States and Canada and will continue to press local and national populations for solution well into the next century.

One of the most persistent land use issues is that of conversion of land from agricultural use to nonagricultural use. When privately owned agricultural land is sold, it usually remains in private hands. For this reason, many people do not recognize the loss of agricultural land as a problem, observing that if farming were the best way to use a particular parcel of land, that parcel would remain in agriculture. Others take a longer view, suggesting that loss of land suited to an activity as important and land dependent as agriculture will have long-term undesirable consequences.

The United States has 1.3 billion acres of farmland, although less than 350 million acres are called "prime" agricultural land. Canada's more northerly latitudes limit its farmland acreage to about 125 million acres. This may still seem an inconceivably large amount of farmland, far too much to worry about the loss of a few hundred acres here and a few thousand acres there. However, the total amount of agricultural land lost to other uses each year is also very large. The U.S. Soil Conservation Service estimated that about 3 million acres of farmland are lost each year, with 70 percent of that loss to urbanization and the remaining 30 percent to water impoundment. Others say the loss in the United States exceeds 5.5 million acres if transportation, land speculation, recreation, and other development-related activities are included. Even in Canada, with less farmland to lose, approximately 250,000 acres of farmland were urbanized between 1976 and 1981.

Much more important than the mere numbers of acres are two geographic factors in the equation: the character (or quality) of the land and its location. Most of the best farmland is also near the countries' largest urban areas. High capability land is therefore lost at a disproportionate rate (Figure 4-1). This is especially true in Canada, where it has been said that on a clear day, more than one-third of the country's prime agricultural land is visible from the top of the CN Tower in Toronto. Or looking at the issue in other terms: in 1981, almost 47 percent of the value of Canadian agricultural production came from areas within 50 miles (80 kilometers) of the country's 23 largest cities.

More widely recognized as a conflict between competing land uses than the loss of agricultural land are those cases involving alteration in the way public land is used, especially when the alteration is for the private benefit of a relative few. This is a significant issue because of the amount of public land in both countries, that is, land owned by governments. Forty percent of the land in United States is government owned, with the federal government having responsibility for 85 percent of this total. Fully 92 percent of Canada is government owned, although only about 43 percent of public land in Canada is federally owned with other governments, primarily the provinces, retaining the remaining 57 percent. The distribution of federal ownership is also uneven in both countries, with most of the federally owned land in Canada north of the 60th parallel and most federally owned land in the United States lying west of the 100th meridian (see Figure 14-3).

One source of pressure on government lands, especially in the United States, is a growing demand for recreation in a natural setting. Attendance at U.S. national parks approached 300 million visitor-days by the end of the 1980s, with a consequent strain on the parks' facilities. Traffic jams and parking problems have affected the quality of visits to Yosemite and Yellowstone National Parks for decades. More than 1.6 million visitors per year at Mammoth Cave National Park created sewage problems. Where a national park is near other federally owned land, some of the demand has shifted to alternative public areas. Recreational use of National Forest land has increased even more rapidly than use of National Park land since the 1960s, and even Bureau of Land Management areas void of public facilities have experienced recreation growth since the 1970s.

Conflicts occur because of crowding. People have different ideas of what constitutes a satisfactory outdoors experience. Some want to see new places and travel about without sacrificing the comforts of home, and this is possible with large recreational vehicles. Others want to spend time in places so isolated that contact with another human is rare. Some seek excitement by traveling across country in an off-road vehicle where no roads exist, mindless of the fragility of the landscape traversed. Others gain their greatest satisfaction by separating themselves from vehicles of all kinds and going where only foot-traffic is possible.

There are three forms of pressure on public recreational land in the United States and Canada. One is the demand to increase opportuni-

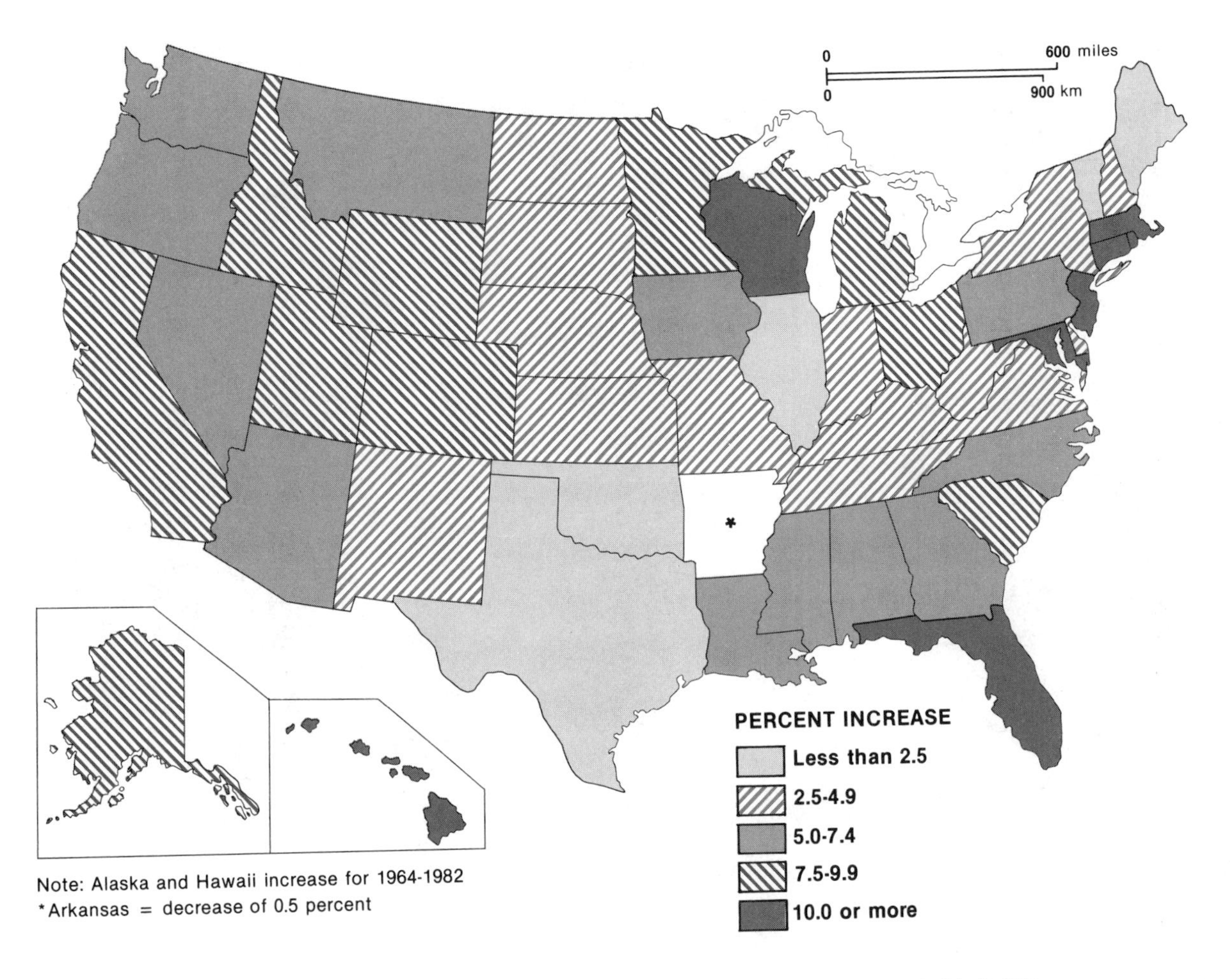

Figure 4-1 Increase in land use for purposes other than agriculture or forestry, as a percentage of total state land area, 1954–1982. Loss of agricultural land is especially pronounced in the heavily urbanized northeastern seaboard states and in Florida, Hawaii, and Wisconsin.

ties for recreation in existing lands set aside for that purpose. A second is a demand to make more exclusively recreational some of the lands that are now multipurpose. And the third pressure calls for resistance to the intrusion of nonrecreational activities into recreation areas. Many places in the western half of both countries are experiencing one or more of these pressures.

The national parks at Yellowstone and the Grand Tetons and their surrounding areas offer a good example of all three forms of pressure. Larger parking lots and wider roads in some areas within the parks have been built in re-

Wilderness is increasingly difficult to find for Yellowstone National Park visitors who come by the thousands each day to witness Old Faithful's eruptions. (Yellowstone National Park)

sponse to larger crowds of visitors. Jets, accompanied by the inevitable noise and smoke, land in Grand Teton National Park, interrupting the quiet. National Forest lands adjacent to park boundaries on the west are clear-cut by timber companies, and gold mining was begun recently within a mile of Yellowstone. Oil and geothermal drilling near Yellowstone was halted, at least temporarily, because of fears that subterranean pressures will be affected permanently, even to the point of eliminating geyser activity in Yellowstone, such as Old Faithful. In addition, these parks' popularity means millions of visitors annually, an attractive market to retailers of all types that have turned park entrances into urban strips, as well as ski resort and real estate developers who can alter much larger landscape swaths. The parks' popularity contributes to the pressures they are experiencing.

As demands for access to "natural" areas increase, it is the extractive industries that have drawn the attention (and ire) of those seeking to preserve tracts from the effects of those industries. In the United States, the federal government controls most of the western timberlands. Although logging is not permitted in National Parks, the National Forests are, by definition, open to logging. In Canada, provincial governments control most of the land but are often bound by law to provide a certain amount of

Some may see this photo as an uninteresting picture of an airplane, but others will note the hint of grandeur in the Teton Range beyond and be disturbed by the implications of the noisy and smelly machine's presence. (Wyoming Travel Commission)

timber to purchasers. In both countries, the issue of *sustainable yield* has drawn attention in the past decade. Trees grow more rapidly in some environments than in others. Sustainable yield refers to the amount of timber that can be cut out of a forest without diminishing the amount that will be available in the future. The sustainable yield of a forest answers the question: How much timber can be cut each year and still remain in environmental balance with new tree growth? In British Columbia in 1988, logging was conducted at a rate of 91 million cubic meters of timber per year, even though the sustainable yield was only 55 million cubic meters per year. Clearly, in addition to the impossibility of maintaining the rate of extraction, the competition with recreational and other uses in the area is also intensified when sustainable yields are exceeded.

A more complex argument about competing uses is based in the concept of *biological diversity*. Biological diversity refers to the variety of life on the earth, or more locally, in a region. Diversity is greatest where human impact on the environment is least. Biogeographers and ecologists have long recognized that environmental components are interconnected. If too great a shock is imposed on a local environmental system, by removal of all mature growth evergreen trees, for example, other plants and some forms of animal life may be lost, as well. If the system changes are maintained, it is possible that soil quality, water availability, and even the local climate could be changed permanently. Those groups who see recreational, economic, and moral value in maintaining biological diversity or slowing its decline favor land uses that minimize the shock to the local environment.

Few alternatives to direct public pressure on the governments responsible for public land are currently available, and this pressure has had mixed success. Several strategies, however, have been developed to reduce the competition between different land uses on private land. Zoning is the most widespread and best known method of influencing land use, but the issue of landowner rights is the usual contrary concern. There have been scattered attempts to preserve farmland by zoning minimum lot sizes at 40 acres (DeKalb County, Illinois), 100 acres (Napa County, California), and even larger. When this approach was attempted in New Jersey, however, opposition was based on the grounds that the poor and minorities were excluded unreasonably. A provincial zoning policy in Ontario was applied in an attempt to restrict developers' seeking to convert prime agricultural land in the Niagara fruit belt to urban uses, but legal disputes have kept the policy from being heralded a clear success.

A second strategy is for private landowners to enter into agreement with a government whereby control over future development of the land is at least partially in the hands of the government. The government purchases the development rights to the land, paying the landowner for the control it will have. In most cases, agricultural land continues to be farmed until the farm owner decides to develop the land for some other purpose or to sell the land for development. The government then exercises direct control over the type and degree of development, rather than acting indirectly through zoning. This strategy is not especially popular because it requires the expenditure of considerable amounts of public money for a benefit that will not be realized for an indefinite period into the future.

A third method for restricting private land use change is by placing the land into a trust. A land trust is similar to a development right purchase in that it is an agreement between the landowner and an organization about the future use of that land without changing how the land is currently used. A land trust differs in that the long-term goal of the trust is generally to preserve the existing category of use, not to have an influence on the form of development the land would experience in the future. The cost of trust programs has limited their widespread use, but the Maine Coastal Heritage Trust and the Lake Champlain Islands Trust are two examples of private trusts that have been successful. The Nature Conservancy in the United States is a private nonprofit organization that functions as a kind of transfer trust by purchasing land and then donating it as a restricted-use area, or by acting as a mediator when areas of privately held wilderness are transferred to government control.

Although there are many groups interested in reducing the conflict between alternative land uses, their effects have been extremely limited in both countries. Continuing population and economic growth and continuing differences in deeply held values pertaining to the land guarantee that this issue will remain important in the evolution of landscapes in the United States and Canada.

WATER SUPPLIES

Water is needed wherever people live. The water is expected to be clean and fresh—not just for drinking and washing, but also for agriculture, construction, industry, recreation, electric-power generation, and many more activities we take for granted. In the United States and Canada, if people choose to live where there is not enough water, they demand that more be brought in. If the number of people living in an area grows, the demand for water grows.

To the likely surprise of many, much of the United States and parts of Canada will experience periods of significant water shortage by the end of this century. These shortages will not be limited to the arid western third of the United States where the adequacy of supplies has challenged human ingenuity for as long as humans have been present in the region. The shortages will also occur in portions of the east in both countries, areas that receive enough precipitation annually to grow crops and other plants without irrigation. Although these shortages may be blamed on climatic change, a more complete explanation must include recognition of the very large amounts of water used by the average person in the United States and Canada.

The amount of water used is related to a society's technological level. The economically complex economies of Canada and the United States have very high levels of water consumption per capita, not because of personal consumption patterns alone but because of numerous agricultural, industrial, and municipal uses. In addition, many North Americans are accustomed to abundant natural water supplies so they are profligate in their use, even in areas where natural supplies are limited. Culture, too, has an effect on water demand and water use.

The migration trend in the United States during the past 50 years has been from the north and east to the south and west. Three of the attractions of the Southwest, its dry air, its abundance of sunshine, and the region's low population density, are directly related to the very low annual precipitation received in the region. The growing population has demanded more and more water from finite supplies. Tucson, Arizona, grew in population from 45,454 in 1950 to 330,537 in 1980 and then to 402,506 by 1990. In a region receiving only 11 inches of precipitation in an average year, the population is dependent on groundwater mined from deep aquifers in southern Arizona. Hundreds of millions of cubic meters of water are taken from the ground and not replenished by natural discharge. The U.S. Geological Survey reported that between 1952 and 1978, thousands of square miles of southern Arizona landscape sank as much as 7 feet because of groundwater withdrawal.

Agricultural production has also shifted westward. While half of the U.S. agricultural production in 1920 was located west of the 100th meridian, this proportion had increased to two-thirds by the 1980s. Irrigated agriculture is very productive in areas with abundant sunshine, so agricultural investment will seek out such areas as long as water remains inexpensive. This last point is critical, because it has long been state and federal policy to supply water inexpensively to agricultural consumers. One effect has been to encourage very high farm productivity in these areas, with the side effect of relatively low-priced produce for consumers. However, farmers are not encouraged to select crops on the basis of a limited resource (i.e., amount of water needed). Subsidized, low-cost water to farm consumers also ensures a greater agricultural concentration in the section of the country with the lowest natural supplies of water.

Simply in terms of economic return for the water used, agricultural irrigation often does not provide the greatest dollar benefit. A 1987 evaluation of alternative water uses in a section of New Mexico found that the economic returns from recreational use of a river system were 20 to almost 30 times the return from irrigation. Even if this study proves to be a special case, and it may not, how much water and the ways water is used for agriculture in the arid West is likely to change during the next several decades, perhaps before the end of the century.

Large and growing urban areas represent tremendous concentrations of water demand, and these concentrations now strain supplies even in the east where moisture is more abundant naturally. Water is used in New York City, for example, at a rate of about 190 gallons per person per day. With a central city population of more than 7 million, an average annual precipitation of 42 inches, and a realistic collection efficiency of one-third of the precipitation, all of the rain and snow that falls over land within a radius of 85 miles of New York's center would have to be reserved for the city population's daily consumption. Of course, there are many other large urban populations within 100 miles of New

Self-propelled center-pivot irrigation, using water pumped to the surface from aquifers deep underground, will permit a farm crop to be raised amidst the sagebrush in this large circular field in western Kansas. (Grant Heilman/Grant Heilman Photography)

York, each with their own fields of water need, so the reach for water into the catchment areas of the Northeast is long and competitive. Indeed, many smaller cities in the area argue that New York is taking what should be "their" water. Many have also found themselves in need of additional water supplies as their own populations have grown and nonresidential uses have increased average consumption.

One option available to many of the larger centers needing additional water is improvement in the efficiency of existing water supply systems. Water distribution systems in many of the older cities of the Northeast, for example, were built a century ago or more and are in constant need of repair. Boston has been estimated to be losing almost half of its water supply through broken and leaking pipes. Estimates of this loss for New York City range from 71 million to 200 million gallons per day. Urban water systems are generally newer and more efficient in the arid half of the country, so improved efficiency is not as frequently considered a viable solution to water shortages there as it is in the east.

A second option, more often discussed in terms of western water needs, is the transfer of water supplies from surplus regions to deficit regions. Large-scale water transfer has been a necessity in the arid west since early in the twentieth century. Los Angeles receives water from the Owens Valley (300 miles away) and the Colorado River (almost 200 miles away). San Francisco depends on the Hetch Hetchy Valley reservoir 150 miles to the east. Denver's water supply east of the Rocky Mountains is augmented by water pumped through a tunnel bored through the Rockies to the headwaters of the Colorado River west of the continental divide, but not without objections by states west of the mountains who were already drawing from the Colorado (see Chapter 14). As intrastate water sources in California are fully tapped, discussions of building aquaducts and pipelines to bring more water into the thirsty southern half of the state from the Columbia River have been rejected by Oregon and Washington, states from which the water would have to be taken (see Chapters 15 and 16).

At a mundane yet significant level, a third option is to reduce demand by educating people to be more frugal in their water consumption and by modifying water-using equipment and fixtures. Washing machines, dishwashers, toilets, faucets, and shower heads are among the items that could be made more efficient in their water use without affecting the way they are meant to function. A controversial approach to reducing demand is to change the cost of water to consumers generally, or to charge more for water destined for specific uses, such as irrigation.

Another most promising option is the recycling and reuse of wastewater. Residential water is of drinking quality, but not all home uses require water of this quality, and most nonresidential uses can use water that is not pure enough to drink. Technology is already in use in central Colorado to process and recycle home wastewater to sufficient quality that it could be drunk, but widespread implementation of this technology required education as well as money. Recycling is not limited to home consumption, however. Some newer industrial plants have incorporated water recycling technology with startling results. A steel mill in Kansas City, Missouri, for example, draws into the mill only 9 cubic meters of water for each ton of steel produced, in contrast to the 100–200 cubic meters per ton of steel in more traditional plants.

All of these options are expensive in one way or another, as is a fifth alternative: a gradual deterioration in water quality and availability. Arguments are beginning to be heard about how high the quality of our water must be. Even taking into account technological options (and their costs) and also accounting for periodic shortages due to drought, the distribution of water shortages in the near future will reflect where attitudinal changes about water use have and have not occurred.

CHANGING AGE STRUCTURE

The "baby boomers," those born between 1947 and 1964, surprised population experts by their large numbers when they came into the world; their appearance depended on millions of private and individual decisions unanticipated by population experts. There is much less surprise possible about how many of these "boomers" will pass their 65th birthdays and when that will be; it will occur between the years 2012 and 2029.

Aging is gradual enough that the drama of this experience is diffuse. For three reasons, however, increasing attention has been and will continue to be given to the share of North America's population who are 65 or older. Each reason has geographic implications. First, as those in the baby boom generation passed through their child-bearing years without producing the same number of children per family as their parents did, it suggested that the proportion of the U.S. and Canadian populations who were elderly would increase rapidly in the second and third decades of the next century. Second, our culture presently expects many in their sixties to change employment status, with all of the economic and social implications that such a change entails. And third, in a similar vein, the type and level of health care needs changes with age, especially beyond about age 75.

Not that 65 represents a sudden or universal physical transformation in all who pass this age; far from it. But 65 is the age used currently to distinguish between those who are elderly (65 or older) and those who are not, mostly because this became the traditional age of retirement. The elderly are, in fact, an extremely diverse group. All races, religions, and life-styles; all levels of income, employment experience, and health status; all personal philosophies of life, death, and aging are represented.

Just as there is diversity among its members, the older populations of Canada and the United States are not evenly distributed across these countries. In Canada, the populations of Saskatchewan and Prince Edward Island in 1986 had the highest percentage of their population 65 or older at 12.7 percent, with Manitoba close behind at 12.6 percent. The northern territories had the youngest populations with only 3.7 percent of the Yukon's population elderly and only 2.8 percent above this age in the Northwest Territories. In the United States, only 3.2 percent of the population of Alaska was 65 or older in 1985 while Florida was at the other end of the scale with 17.6 percent of its population elderly.

A geographic aspect to these population changes is tied to where the "boomers" will live as they age into their 60s and 70s. The distribution of elderly population can change in at least three ways, and each type of change can be caused by different factors. One basis for change in the pattern is through migration of the elderly. If people over 65 move in (or out) of a region in large enough numbers, the age structure of the region will reflect this. Florida's age structure is the best example. A second basis for change in the distribution of the elderly is also through migration, but in this case through migration of the younger population. If young people in their twenties and thirties move out of a region, perhaps because the economy is depressed and few new jobs are available, then the result is an increase in the share of the remaining population that is elderly. The number of people over age 65 may not have changed, but the proportion has. Some of the Great Plains states and the prairie provinces fall in this category. And the third basis for change in the pattern of older persons is differential mortality. If the health of the population, or a significant portion of it, experiences a short-term calamity or suffers some long-term deleterious effect so that fewer people survive into their sixties, the population will have a smaller percentage who are in this older age category. Alaska and the Northwest Territories may fit this description best, but strong inmigration of younger people clouds the simple explanation in such areas.

A much larger number of older people will be living in the United States and Canada in the first third of the twenty-first century. This has implications for regions in which they will be concentrated. Special health care needs of most people older than 75 years suggests one challenge for regions with large proportions of this age group. If health care is to be provided primarily in institutions, whether hospitals, nursing homes, convalescent centers, or some alternative, then a greater share of the regional economy for construction, operation, and maintenance will be obligated to health care. On the other hand, if home care becomes more available and used, the mix of incentives and recognition society gives to home care providers will have to be changed considerably.

Accompanying the higher demand for health care facilities in regions of greater elderly concentration may be a more subtle change in the social and political environment. The likes, dislikes, and preferences of people vary with age, even controlling for other factors like income and physical energy. Those who have lived through seven, eight, or more decades are likely to have a different view of an event or activity than they did when in their twenties or thirties. At risk of belaboring the obvious, we can say that experience affects outlook, and a region with a high proportion of people with many decades of experience is likely to have that experi-

For those elderly people in good health and with sufficient financial support, a move to a region with abundant recreational opportunities has considerable appeal. (Janice Fallman/The Picture Cube)

ence expressed in selective social and political outlooks at odds with younger populations in the region and elsewhere.

Geographers and sociologists have recognized the importance of age and experience in individuals' behavioral choices and opportunities by referring to a series of *life-cycle stages* that most everyone passes through in life. Of course, not everyone passes through each stage in lock-step, but the concept suggests different types of opportunity in each age range. The early years are ones of dependence on parents and other adults. Then comes a stage of unmarried independence, followed in most cases with a period of family responsibility, perhaps to a spouse and children. At some point, usually, the children leave home and the parents move into another stage, perhaps married independence. Then, retirement leads to another shift in life-cycle stage, still one of independence but with different responsibilities because of the change in employment status. For some, this stage lasts until life's end. For others, a final stage of dependence is entered if sufficient physical or mental capacity is lost.

The point of all this, once again, is that regions with very large numbers of people in the final two stages of the life cycle will be different from regions with most of the population in earlier independence and family responsibility stages.

Finally, the economic base of regions with many immigrant elderly will be different, as well. Significant numbers of elderly have very low incomes, and this creates distinctive demands on regions where they are concentrated—such as sections of large metropolitan areas and rural areas with overall declining populations. However, elderly who are able to move to favored destinations, such as parts of Florida, central Arizona, Texas, North Carolina, British Columbia, and other sections of the two countries, usually carry with them pensions and retirement incomes that will bring funds into the region for many years (Figure 4-2). This has spawned a "retirement industry" that responds to concentrations of these funds. Only recently has it begun to be recognized widely that this industry can be a significant stimulus in a local economy, providing benefits for the nonelderly population as well.

WASTE MANAGEMENT

Every activity that produces an item needed or wanted by people also yields something not sought or desired. For every good produced, "bads" are also produced as one form of waste or another, sometimes appearing as an unwanted by-product immediately and sometimes becoming waste only after the product is ready

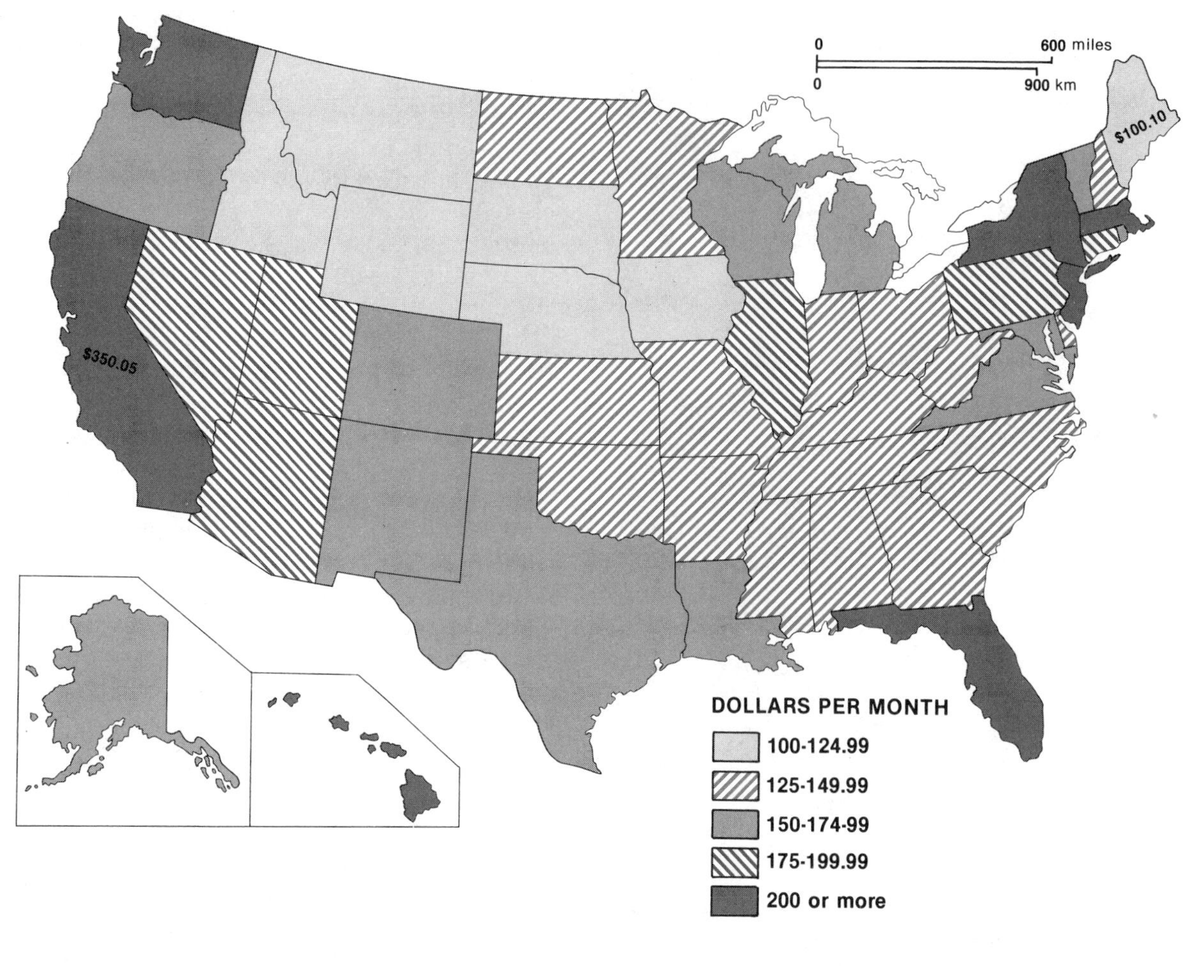

Figure 4-2 Average monthly federal Social Security Insurance payment per person qualified by age, September 1990. The range in average social security payment is considerable. Where large numbers of recipients are concentrated, the impact of these funds on the local economy can be great.

to be discarded. The challenges of managing the bewildering variety and overwhelming abundance of waste in Canada and the United States are now seen by many as challenges that cannot wait to be solved decades into the future.

There are many types of waste. Some come from production, some from consumption. Or we can think in terms of unwanted leftovers. We might consider pollution, for example, as waste that needs management much as does highly toxic chemical by-products or nuclear radiation. Some even argue that any corruption of the environment, any departure from a predetermined standard of environmental quality, is the result of poor waste management.

The geographic implications of waste management are imbedded in the uneven distribution of the impact of waste and in the ways that environmental quality contributes to the character of places. The unevenness of the distribution can be a result of factors of production (clusters of consumers, transportation routes, resource patterns), or it may reflect the relative ability of the environment to absorb or disperse the excess. The quality of a place can be damaged by too much waste or by waste that is too obvious. As our national economies have become more technologically complex and demanding (producing goods at higher levels of activity and intensity), unwanted by-products have become more pervasive.

Rubbish and garbage produced in thousands of homes and businesses are usually grouped together into what is called "municipal waste." Localities faced with the disposal of mounting volumes of municipal waste traditionally have dumped it into landfills or burned it in large incinerators, or both. So great are the volumes of rubbish, however, that landfills are filling up. Only about half the number of landfills in the United States in the late 1970s were still open a decade later. Some were closed because they were polluting the local environment. Some were closed because they were full and no obvious alternative site was available locally. From Passaic County, New Jersey, garbage was shipped half way across Pennsylvania while a new incinerator was constructed. A 3100-ton barge loaded with garbage from Islip, New York, received national attention in 1987 as it was pushed for weeks from port to port and country to country seeking a place to unload.

While there are many forms of waste, we have selected three—hazardous waste, radioactive waste, and acid rain—to provide examples of the geographic issues involved.

Hazardous Waste

At first glance, waste is easily defined as what is discarded after production or consumption is completed. However, is material "waste" if it is discarded today but reused tomorrow? And if waste material is collected or stored for recycling at some indefinite future date, when does it cease to be waste—when it is put into storage? when plans are begun for its later use? when it is physically removed from storage? when it is treated and recycled? To make answering these questions even more difficult, consider that discarded material stored for recycling may cause environmental damage while awaiting a decision about how or when to reprocess it into something useful.

According to the statute mandating control of hazardous waste by the U.S. Environmental Protection Agency, hazardous wastes are those that pose "substantial present or potential hazard to human health or the environment." But as with waste in general, this attempt to define hazardous waste in simple terms was inadequate; after all, all wastes are potentially hazardous. As a result, those engaged in waste management planning have participated in a long process whereby fine distinctions between different wastes are made on physical, chemical, and biological grounds, and these distinctions remain arguable. In the end, we might consider hazardous wastes as those that are present or will remain present in the environment in concentrations clearly toxic to human and other life during a relatively short period into the future. Using even this rough definition, two points become clear: hazardous wastes do not break down (or dissipate) easily in the environment, and hazardous wastes are produced almost every place where items are manufactured or manufactured items are discarded.

In the United States, approximately 80 percent of all hazardous wastes end up in landfills. Until the last several decades, a landfill was simply a place where waste was dumped until it was covered with dirt. Now, for hazardous wastes at least, a large depression or hole is selected. One or more liners are placed in the bottom of the depression below the wastes and a means of removing liquids that may drain from the landfill is constructed. Finally, a cover is added over the wastes.

Critics of hazardous waste landfills argue that, sooner or later, there will be a failure in one or more of the protections installed. Either the liner

Toxic waste can be at once a source of soil pollution, air pollution, water pollution, and even visual pollution. Sites such like this one are far more numerous and widespread than most people realize. (Paul Fortin/Stock, Boston)

or the cover will deteriorate and leak, or the drain will clog and fail. In both cases, hazardous waste will eventually find its way outside the landfill. The basic argument, given that a landfill is to be used to eliminate waste, is over how much management of the landfill is necessary after the waste is in place in order to postpone "indefinitely" the various potential failures.

There are alternatives to landfills—such as incineration—but like landfills, all are vulnerable to loss of some of the waste material into the local environment unless very expensive procedures are adopted. Once in the political arena, the issue becomes twofold: How much is society willing to spend relative to the increase in risk due to the hazardous waste? And will the costs be met by charging producers of the waste or by charging society as a whole through use of tax revenues. Meanwhile, hazardous waste continues to be produced faster than it is "managed," and only a portion of the discarded waste from previous years is brought under some type of management procedure. The extremely large cost of management, the incremental accumulation of hazardous wastes, and the delayed effects that many such wastes have on humans and other living organisms practically guarantees that this problem will be present and an even more serious issue to be addressed in the next century.

Radioactive Waste

There are some similarities between the issues of radioactive waste and hazardous waste management. Both involve materials known to be toxic, although there has also been much disagreement in both cases over the level of exposure that is toxic. Both types of waste are being produced faster than they are being managed. And both are generally seen to have been inadequately managed to date.

There are also some major differences, largely because radioactive material, while hazardous, has several very distinctive properties, and these properties require different kinds of attention. Radioactive materials, for example, lose their toxicity over very long periods—so long in some cases that for all intents and purposes, they remain toxic indefinitely. Radioactive materials are not easily incinerated, so there presently is no alternative to placing the material in some isolated and insulated site. And at least partly because radioactivity cannot be identified by people's natural sensory mechanisms—that is, because people cannot tell when a familiar material is radioactive—there is a great deal of suspicion and fear of anything having to do with the production, use, or disposal of radioactive wastes.

Prior to 1980, producers of low-level radioactive waste sent their waste to a limited number of burial grounds—the largest being in Washington, Nevada, and South Carolina. Low-level waste is generated by many industrial and health-related activities and, in general, the largest producers of low-level waste are in states other than those containing the dumps (Figure 4-3). Eventually, the main receiving states said that there was a limit to the amount of additional waste they would accept. In addition, there was increased public concern about the risks of transporting radioactive wastes long distances and through heavily populated areas. This led to the Low Level Radioactive Waste Policy Act of 1980, one of two radioactive waste management policies enacted by the U.S. government in the early 1980s.

The 1980 act asserted that each state would be responsible for the low-level waste it produced. To meet this responsibility, a state could set up its own low-level dump by January 1986 or by the same date it could join with other states in a regional compact. Each compact would then designate one of its member states to receive all the low-level waste from the other compact members. A number of compacts were formed (Figure 4-4), although some states chose to remain independent.

There are several problems yet to be resolved. If a state chosen by a compact to be the new

Figure 4-3 Major flows of low-level radioactive waste to the primary burial sites, 1979–1982. Cross-country transport of low-level radioactive waste was common and was not based on shortest distance to the burial site.

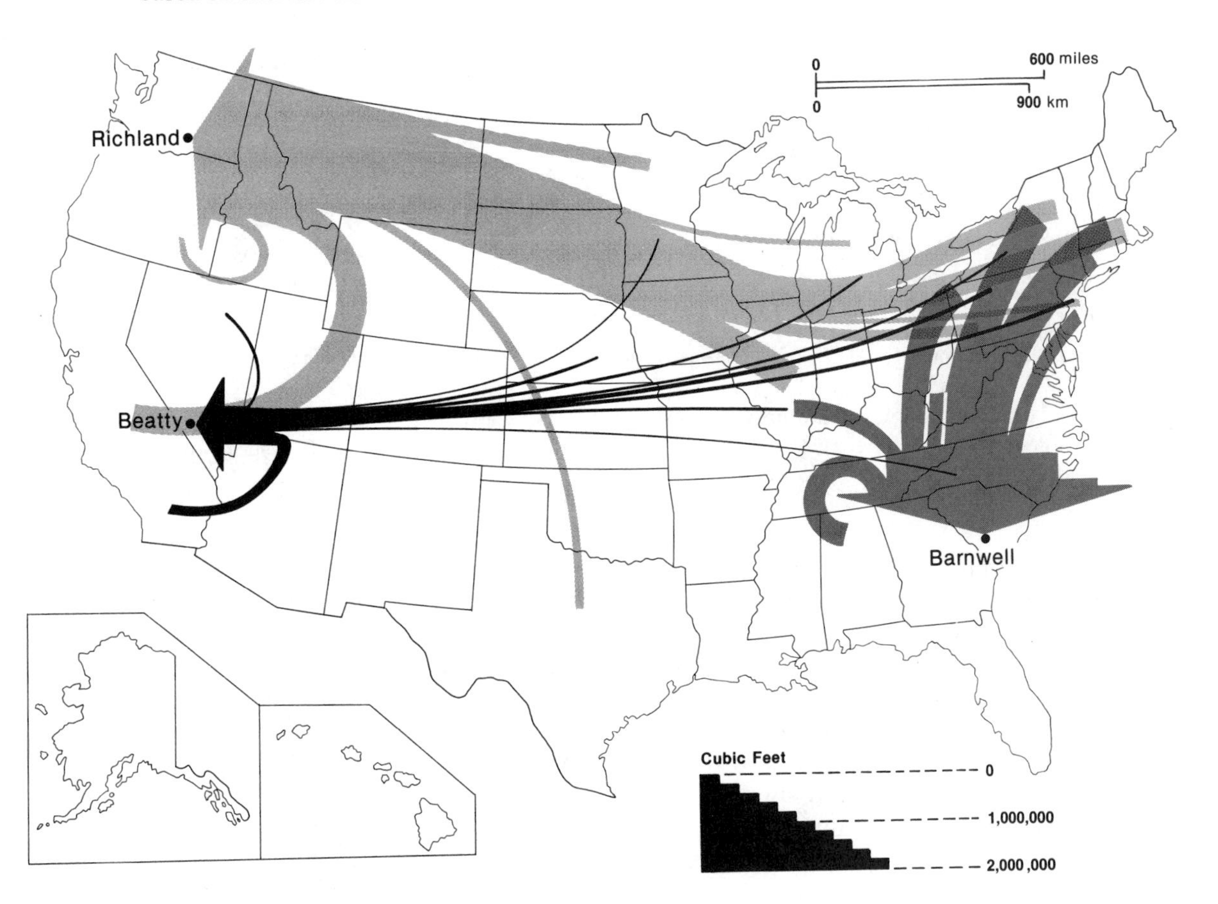

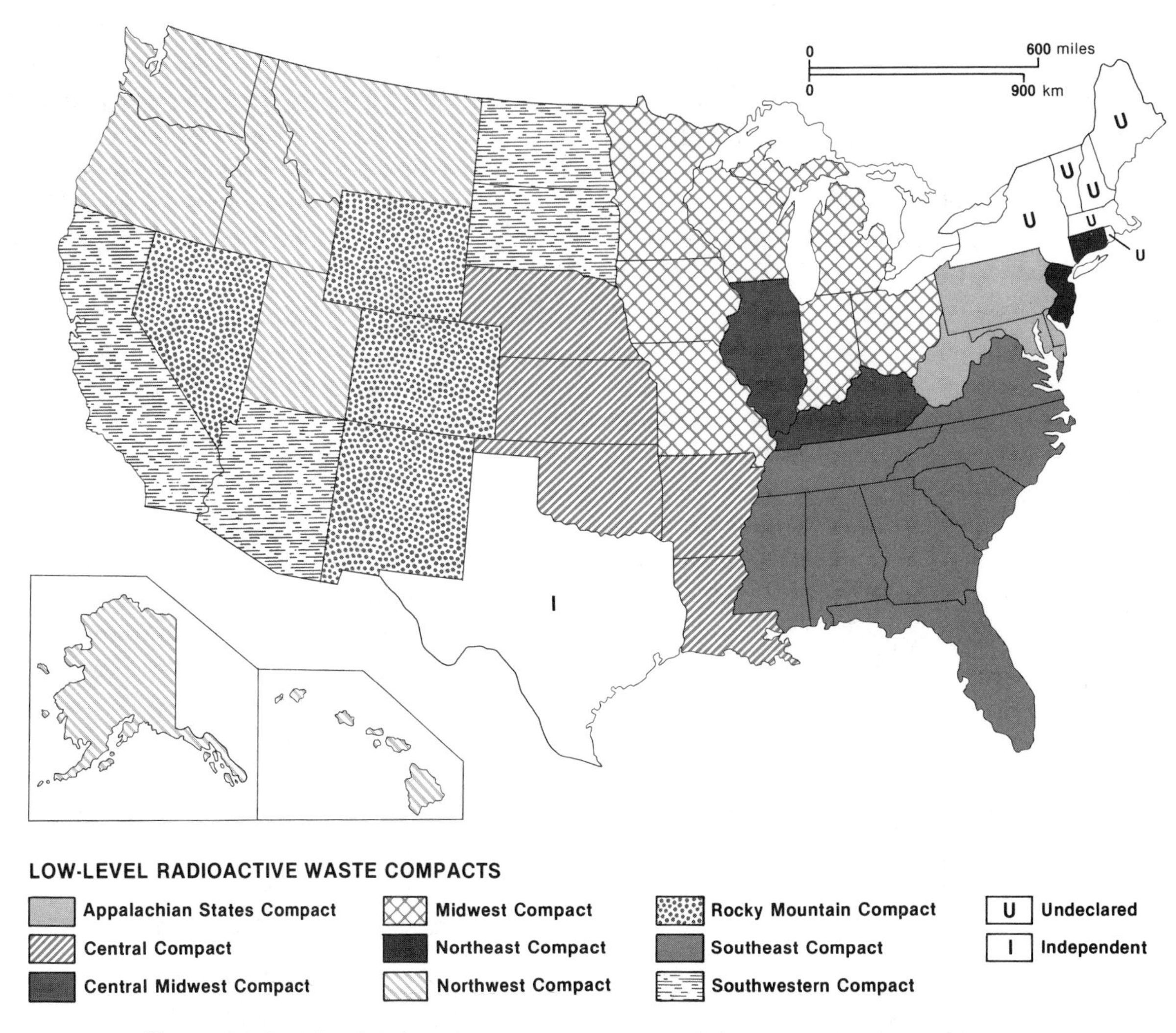

Figure 4-4 Low-level radioactive waste compacts. Many states chose to join others in agreement over where waste would be deposited rather than remain solely responsible for their own.

dump site rejects this designation, what are the consequences? Can low-level radioactive waste be stored in a manner adequate to protect nearby populations indefinitely? And given the shifting standard of "minimum acceptable exposure," what assurance can be given that present acceptability standards will not be revised downward once again, thus transforming what was thought to be a safe exposure into an unsafe exposure?

A second national policy in the United States, the Nuclear Waste Policy Act of 1982, was established to address high-level radioactive waste management. High-level radioactive waste is produced almost entirely by nuclear power generating plants in the form of spent fuel, although there is waste from nuclear weapons preparations, as well. Currently, high-level waste in increasing amounts is stored on site at the power generating plants or the weapons centers. The 1982 policy described a procedure whereby a single large site was to be selected, a site with natural geologic and environmental characteristics sufficient to store safely for 10,000 years all current high-level waste plus all to be produced well into the next century. In addition to the necessary shielding from radioactive leakage, the site must be dry. Following a lengthy search, Yucca Mountain, in southern Nevada, was selected as a satisfactory depository, construction contracts were awarded, and preparatory work was begun.

Most work on the Yucca Mountain site was stopped in 1988 by a lawsuit brought by the state of Nevada. At least part of the basis for the suit was a draft research report arguing that the depository site, although well above the current water table, was liable to flooding from below. Effectively, the report stated that the Yucca Mountain site was subject to earth fracturing, that this fracturing had in the past permitted the current water table to rise above the planned deposit site for the nuclear cannisters, and that there was no assurance such fracturing and water table rise would not occur in the next 10,000 years. Although a 1989 review of the draft report did not support its findings, the argument over the Yucca Mountain site—or any other—is likely to continue for decades, if only because conclusions about future events occurring in complex environmental settings are necessarily ambiguous.

Acid Rain

Essential to agriculture and as welcome as the rainbow following a spring shower, rain has long been seen as a refreshing, life-giving component of nature. But this image has changed recently. Acidic rain is now recognized as an immediate source of damage for thousands of lakes in which no fish live and for significant dieback in European and North American forests. Although many forms of precipitation can be acidic, it is common to refer to such precipitation generically as "acid rain."

Precipitation, whatever its form, can become more acid than normal when the atmospheric levels of certain oxides—primarily sulfur dioxide (SO_2) and the nitrogen oxides (NO_x)—are high. These oxides either combine with water in the air to increase the acidity of the precipitation, or they are deposited as dry particles. But whatever the form, acid rain may be absorbed by plants and accumulated in lakes and ponds at levels that damage sensitive ecosystems.

There are at least three geographic components to the acid rain issue. First, human sources of excess oxides are not evenly distributed, with fossil-fuel burning power plants and smelters the primary source of $S0_2$ and NO_x, with NO_x also from mobile sources. Although the distribution of sources for this airborne pollution is not even and the wind and storm tracks affecting the pollution sources are not obvious, the northeastern United States and the southeastern portion of Canada are the regions most affected. Unpolluted rain is generally accepted to be slightly acidic with a pH of 5.0–5.5 (pH of 7.0 is neutral), but an extensive area in eastern North America receives precipitation with a pH below 4.6 (Figure 4-5).

Second, the effects of acid rain can be changed by the rate at which it is introduced into the environment and by the nature of the land on which it falls. Locations where acidic snow accumulates throughout the winter are subject to a kind of "acid shock" when the snow melts. Conversely, adverse effects of acid rain are lessened somewhat if the soil on which the rain falls is alkaline. The acid-dead lakes in New England, Ontario, and Quebec, for example, are believed to be more the result of their surrounding nonalkaline soils' inability to neutralize the acid rain that runs off into the lakes than the small amount of precipitation that actually falls on the lakes themselves.

And, third, the political and economic geography of acid rain may be even more complex than its environmental issues. These complexities are independent of the scientific difficulties of finding agreement on details of emissions dispersal

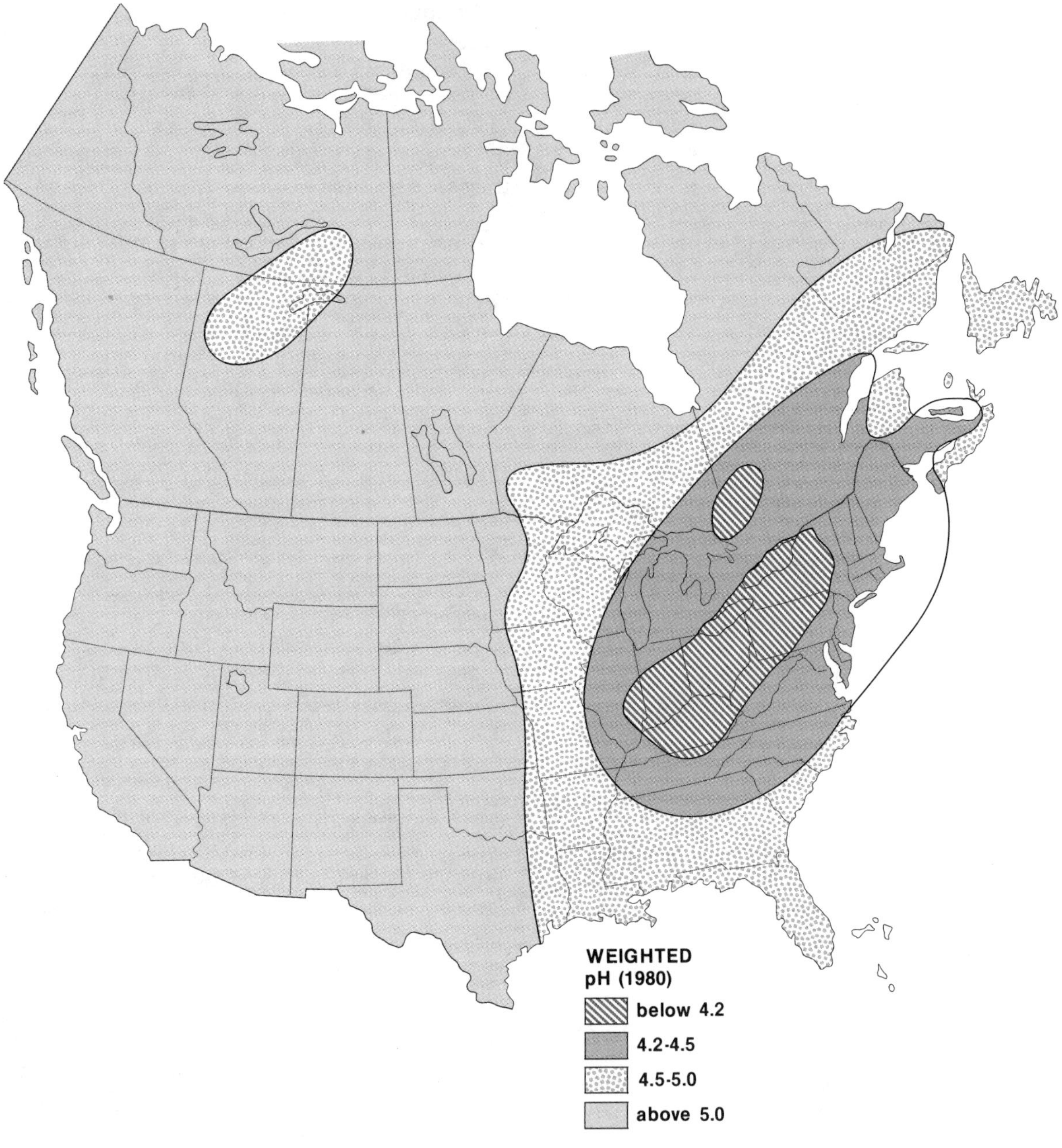

WEIGHTED
pH (1980)
below 4.2
4.2-4.5
4.5-5.0
above 5.0

Figure 4-5 Distribution of precipitation acidity. The locations of industries that put key pollutants into the atmosphere and the direction of prevailing winds go a long way toward explaining the distribution of acid rain.

and acid rain effects. The most severely affected portion of North America straddles the boundary between the United States and Canada, but no agreement on a solution, an approach, or even a definition of the problem was reached during the 1980s.

Canada saw the problem as transnational: both countries have sources of acid pollution emission, and both have areas affected by acid rain originating in each country. Canada argued that the problem must be addressed bilaterally. The United States, on the other hand, argued that the problem should be treated unilaterally. United States officials saw the problem as a domestic one. Each source of acid pollution is in one country or the other (not both), each affected area is similarly in one country or the other, and attempts to control emissions will have economic consequences in each country, not necessarily in both.

The different scales of SO_2 and NO_x emissions in the two countries also affects each country's view of the problem. With a population ten times the size of its northern neighbor, the United States production of SO_2 and NO_x is much larger than is that of Canada. Even accounting for wind patterns, the Canadian government estimates that the United States receives only one-fourth as much SO_2 from Canada as does Canada from the United States.

The different domestic sources of most oxide emissions also affects the way the problem is viewed in each country. In Canada, nonferrous ore smelters are the primary source of SO_2 emissions, while in the United States thermal power plants produce the bulk of the airborne SO_2. Lacking an agreement with the United States, Canada unilaterally commited itself to reducing its SO_2 emissions and wanted the United States to do the same. But electric utilities companies in the U.S. northeast have been resistant to implementing expensive emissions controls. Although additional costs could be passed on to consumers, the utilities are afraid of losing a share of the market to Canadian hydroelectric power companies. Hydroelectric power generation is already cheaper than fossil-fuel generated power, and it does not produce SO_2.

Throughout the 1980s, the official United States policy was that, given the cost of emissions controls, no action would be taken until there was stronger proof that specific acidic emissions and specific sources produced specific changes in acid precipitation. This position changed in 1991 with an initial agreement between the two countries to address the acid rain issue. However, it is likely that there will continue to be disagreements within and between Canada and the United States over the severity of impacts from acid precipitation and the responsibility of each of the parties in finding a solution.

The issues related to waste management are not simple and are only hinted at here. If we accumulate too much of what we want for the moment, it can be harmful. Too much of what we do not want—such as waste—can be disastrously harmful. Even so, agreement about how to address problems of waste management has proven difficult to achieve.

ADDITIONAL READINGS

Alvo, Robert. "Is the Laughter of the Loons to Be Stilled on Our Acid Lakes?" *Canadian Geographic,* 107 (June/July 1987), pp. 46–50.

Bartlett, Donald L., and James B. Steele. *Forevermore: Nuclear Waste in America.* New York: W. W. Norton, 1985.

Buchanan, Carrie. "Garbage Blues." *Canadian Geographic.* 108 (February/March 1988), pp. 30–39.

Davis, Charles E., and James P. Lester, eds. *Dimensions of Hazardous Waste Politics and Policy.* New York: Greenwood Press, 1988.

Elliott, Thomas C., and Robert G. Schwieger, eds. *The Acid Rain Sourcebook.* New York: McGraw-Hill, 1984.

Marston, Sallie A. "Adopted Citizens: Discourse and the Production of Meaning Among Nineteenth Century American Urban Immigrants." *Transactions of the Institute of British Geographers,* N.S., 14 (1989), pp. 435–445.

Murray, Raymond L. *Understanding Radioactive Waste,* 3rd ed. Columbus, Ohio: Battelle Press, 1989.

Panel of Social and Economic Aspects of Radioactive Waste Management, National Research Council. *Social and Economic Aspects of Radioactive Waste Disposal: Considerations for Institutional Management.* Washington, D.C.: National Academy Press, 1984.

Piasecki, Bruce, ed. *Beyond Dumping: New Strategies for Controlling Toxic Contamination.* Westport, Conn.: Quorum Books, 1984.

Schmandt, Jurgen, Judith Clarkson, and Hilliard Roderick, eds. *Acid Rain and Friendly Neighbors: The Policy Dispute Between Canada and the United States,* rev. ed. Durham, N.C.: Duke University Press, 1988.

Wallach, Bret. "Taking Heart from Upper East Tennessee: Politics in a National Forest." *Focus,* 38 (Winter 1988), pp. 22–27.

Wescoat, James L., Jr. "On Water Conservation and Reform of the Prior Appropriation Doctrine in Colorado." *Economic Geography,* 61 (1985), pp. 3–24.

CHAPTER

5

Megalopolis

In 1961 a French geographer published a monumental study of the highly urbanized region located in the northeastern United States.[1] Professor Jean Gottmann spent 20 years researching the area extending from southern New Hampshire and northern Massachusetts to Washington, D.C., and its northern Virginia suburbs (Figure 5-1). He argued that this was a "very special region" and named it Megalopolis. Gottmann argued that this was a type of urban region differing from anything that had existed before. This region is worthy of attention not only because of its importance in North America, but also because it provides an indication of a future for similarly evolving areas elsewhere in the world.

Megalopolis was formed along the northeastern coast of the United States by the gradual coalescence of large, independent metropolitan areas such as Boston, New York, Philadelphia, Baltimore, and Washington, D.C. As the population of such central cities grew, the effects of growth spilled over into the surrounding rings of smaller places. Larger suburbs in these rings made their own contributions to the total urban sprawl. The outer fringes of the resultant polynodal, or many-centered, metropolitan regions (called *conurbations*) eventually began to merge with each other to form the extensive urbanized region that is Megalopolis.

[1] Jean Gottmann, *Megalopolis: The Urbanized Northeastern Seaboard of the United States* (New York: Twentieth Century Fund, 1961).

Intense, complex urban activities define the landscape in which most North Americans live. Nowhere is this definition more powerfully present than in Megalopolis. (Fredrick D. Bodin/Stock, Boston)

The dominant theme of Megalopolis is "urbanness." Throughout the region, urban activities dominate the people's lives. In varying degrees, urban services provide for the millions who live in this region, and urban forms—dense patterns of streets and buildings, industrial centers, retail and wholesale clusters, governmental complexes—are never far away. Indeed, if one were to drive along Interstate Highway 95 between Boston and Washington, urban features assail the senses. There are office and apartment buildings, small shops and mammoth shopping centers, factories, refineries, residential areas, gas stations and hamburger stands by the thousands, interspersed with warehouses to temporarily store the goods brought by ship, rail, and truck. All this and more can be viewed along the 800-kilometer (500-mile) route through Megalopolis.

But Megalopolis is not just a collection of urban forms tied together by a web of transporta-

Major Metropolitan Area Populations in Megalopolis, 1990

New York, N.Y.–N.J.	8,546,846
Philadelphia, Pa.–N.J.	4,856,881
Washington, D.C.–Md.–Va.	3,923,574
Boston-Lawrence-Salem-Lowell-Brockton, Mass.	3,783,817
Nassau–Suffolk, N.Y.	2,609,212
Baltimore, Md.	2,382,172
Newark, N.J.	1,824,321
Hartford–New Britain–Middletown–Bristol, Conn.	1,123,678
Providence–Pawtucket-Woonsocket, R.I.	916,270
Allentown–Bethlehem–Easton, Pa.–N.J.	686,688
Jersey City, N.J.	553,099

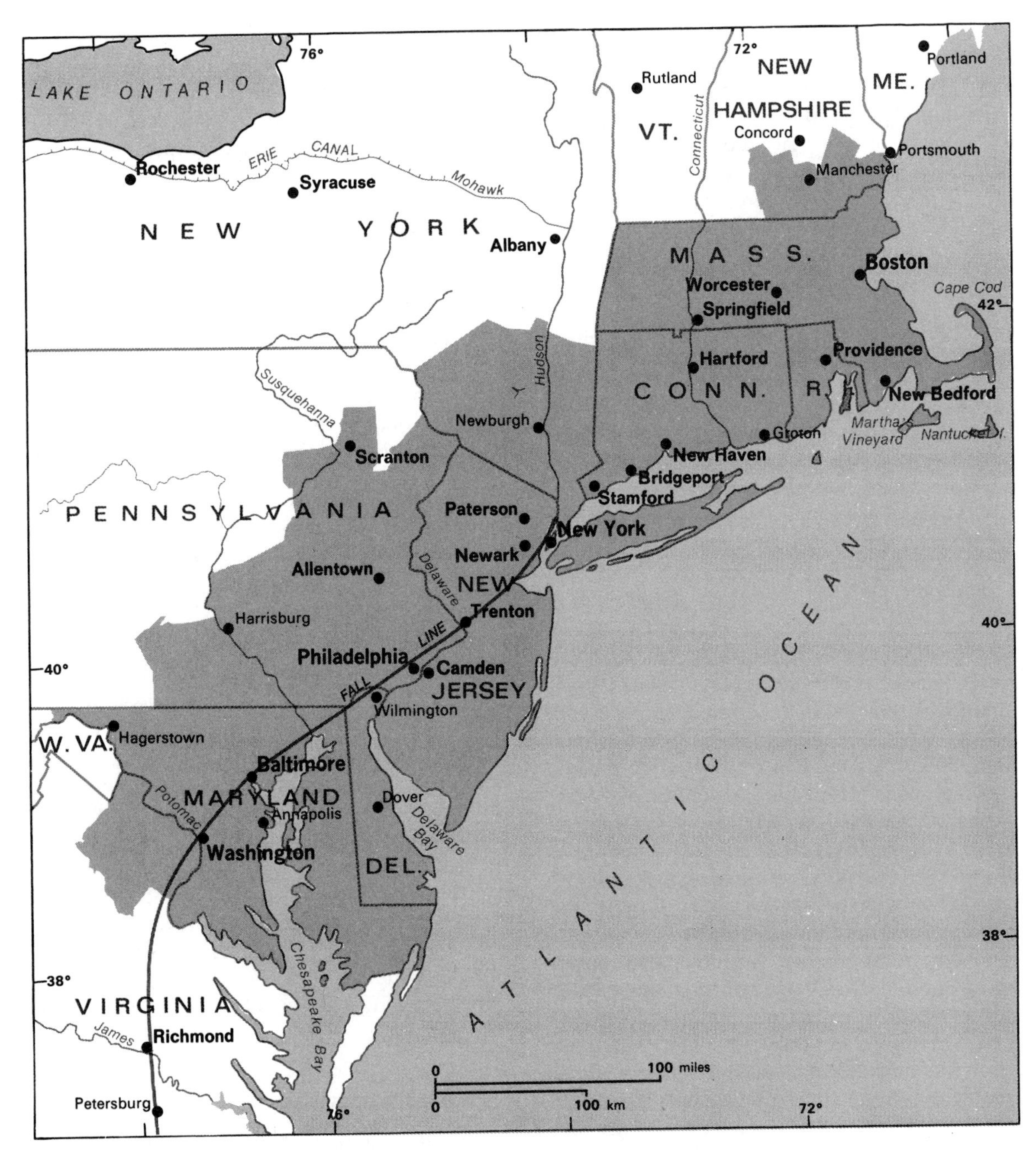

Figure 5-1 Megalopolis.

tion lines. It also contains many green spaces. Some are parks or other land available for recreation, but substantial areas remain devoted to agriculture. Farmland in Megalopolis comprised 37.7 percent of the total land area in 1959. Although this proportion had declined to 28.1 percent by 1969 and 23.9 percent by 1987, there remained almost 3.4 million hectares (8.3 million acres) devoted to farming within a very urban region.

Clearly, Megalopolis is a complex region. It is composed of at least five very large, but distinct, metropolitan areas with many smaller cities and rural pockets between and around these large urban zones. As Professor Gottmann put it,

> In this area . . . we must abandon the idea of the city as a tightly settled and organized unit in which people, activities, and riches are crowded into a very small area clearly separated from its nonurban surroundings. Every city in this region spreads out far and wide around its original nucleus.[2]

In spite of the mixed character of Megalopolis, it is its massive urban presence which makes this region so important in the United States. Ten of the country's 46 metropolitan areas exceeding 1 million population in 1990 were located in Megalopolis. The counties defined as Megalopolis by Gottmann contained an estimated 42,536,400 people in 1987, almost 36 million of whom lived in the seven largest metropolitan areas. The region held no less than 17 percent of the total U.S. population. You may begin to gain a sense of the continental importance of this region when reflecting on the implications of such a large population in an area comprising only 1.5 percent of the area of the United States. Relative economic, political, and cultural influence are all related to unusual concentrations of population, and Megalopolis is clearly such a concentration.

There is much more that makes Megalopolis important than the large number of people who live there. The region also possesses an unusual mix of economic qualities. Average per capita income is very high here. Of the almost 18 million persons employed in Megalopolis in 1986, a higher than average proportion worked in white-collar and professional occupations. Transportation and communication activities are very prominent in Megalopolis, partly a result of the region's coastal position. Approximately 40 percent of all commercial international air-passenger departures have Megalopolitan centers as their origin. And almost 30 percent of North American export trade passes through the six main ports of Megalopolis. The statistics could go on, but the point is clear. In spite of the attention given since 1980 to growth in other parts of North America, this has indeed been and continues to be a "very special region."

THE LOCATION OF MEGALOPOLIS

Megalopolis is unique in North America, but why has this particular portion of the continent developed as it has? Whenever a geographer is asked such a question—why a region developed the set of characteristics that distinguish it from other regions—the first aspect of the region considered is usually its location. In some cases, location is of secondary importance, but as we argued in Chapter 3, many times it is primary. What is it about this region's location that has helped generate the tremendous urban complex that is Megalopolis? Megalopolis is not a single city, but the *site* and the *situation* of this vast urban region hold clues to the region's origins and growth.

Site Characteristics in Megalopolis

The *site* of a place refers to the set of features making up the immediate environment of that place's location. Many of the site characteristics of Megalopolis are visible in the region's outline. Occupying a coastal position in North America, the eastern margin of Megalopolis is deeply con-

---

[2] Ibid.

voluted. Peninsulas jut into the Atlantic. Islands are scattered along the coast, some large enough to support communities. Bays and river estuaries penetrate the landmass throughout the region in a kind of mirror image of the land's penetration of the ocean. This interpenetrated coastline brings more land area close to the ocean and, in this way, provides greater opportunity for access to cheap water transportation than would a straighter coast. Access was one of the major supports of Megalopolitan growth.

Proximity to water, however, does not always equal accessibility. Quality harbors must also be present and Megalopolis straddles some of the best natural harbors in North America. To be sure, such cities as San Francisco, Seattle, Vancouver, and Mobile are built around excellent natural harbors. But the harbors at Boston, New York, and Providence, Rhode Island, and—to only slightly less degree—Baltimore are excellent *and* close to one another. Philadelphia, on the Delaware River, does not strictly have an ocean harbor, but dredging maintains that port's ocean access. Numerous smaller harbors, such as those at New Haven and Groton, Connecticut, New Bedford, Massachusetts, and Portsmouth, New Hampshire, also are located along the coast of Megalopolis. In this region, advantages of proximity to water are accompanied by many reasonably close harbors providing the ability to transfer goods and people across the land-water boundary.

Why so many good harbors located close to one another? Is it coincidence, or is there an explanation? Since Megalopolis is not a small region, anything that affected the whole region must have been of considerable magnitude.

As discussed in Chapter 2, the northern half of Megalopolis was covered by ice during the most recent of the series of continental glaciers. This glaciation contributed to the excellence of Megalopolis's harbors in several ways. Sea level was much lower during the Ice Age because a tremendous amount of the earth's water was held in the massive ice sheets. As the ice cover began to melt, large river spillways were formed. The erosive power of the rivers cut deeply into the flat coastal plain across which the silt and gravel-laden water rushed. As sea level rose, the lower river valleys were "drowned" to form estuaries and the ocean margin was shifted toward the land's interior. Chesapeake Bay and Delaware Bay were formed in this manner as the drowned lower courses of the Susquehanna and Delaware rivers, respectively. The navigability of the lower Hudson River was enhanced as well. These glacial river valleys, cut deeply by the meltwater, formed some of the harbors that were later to prove useful in the development of Megalopolis.

The other major contribution of the glacial period is more specific to one or two locations. Also as mentioned in Chapter 2, the large amounts of soil, stone, and other debris scraped up by the expanding glaciers were deposited as moraines when the ice front retreated. One such terminal moraine is the southern arm of Cape Cod. The rise in sea level and deposition of sandy soil by ocean currents sweeping north along the East Coast formed the distinctive north-reaching "forearm" of Cape Cod. An even longer series of ridges was left by retreating glaciers just south of what is now the coast of Connecticut. These moraines developed as an island when the seas rose, and it too was widened by deposition from the ocean. The island was not made so wide, however, that it could be called anything except Long Island.

Long Island has enhanced the quality of New York's harbor in two major ways. First, from the point of view of the ocean, one end of Long Island is tucked into the harbor mouth formed by the Hudson River without actually blocking it. The length of coastline available for port facilities, already significant along the Hudson, is increased considerably. Second, from the viewpoint of the land, when an urban area grows around a large, fully developed harbor, the growth creates a demand for more space. Good land to accommodate the New York area's urban growth was restricted to the west of the Hudson in New Jersey by tidal marshes and the erosion-

There are many fine harbors in Megalopolis, but none can match New York's. With the Hudson River on the west and the East River on the east of Manhattan, and the confluence of these rivers in the foreground also "protected" from ocean currents and storms, the city is a focus of interaction at the highest level of intensity. (Skyviews)

resistant ridge of the Palisades. To the east of the Hudson lies only a narrow finger of land, Manhattan Island, bounded on the north by the small Harlem River and on the east by Long Island Sound and its extension, New York's East River. But beyond the East River is Long Island, a flat to slightly rolling land without the barrier marshes of New Jersey. New York's boroughs of Brooklyn and Queens developed early at the western tip of Long Island, and the island offered a great deal of room for further urban expansion toward the east. Both the harbor and the port city which grew around it benefited from the presence and orientation of Long Island, a landform that owes its existence to glacial deposition.

Although Megalopolis possesses many high-quality harbors, there are few other site characteristics that contributed so positively to the region's urban economic development. The climate is not exceptionally mild, although the summers are generally of sufficient length and wetness to support farming. Soils are variable, with the good soils inland from Baltimore and Philadelphia better than most found closer to New York. From New York to Boston and beyond, the thin and stony soils are generally only fair, with the exception of the notably superior soils in the Connecticut River Valley. The agricultural resources of Megalopolis, while adequate, are hardly exceptional.

The general topographic features of Megalopolis south of New York do provide additional urban site benefits. If you were to travel inland from the Atlantic Coast, a sequence of distinct physiographic regions would be crossed (see Figure 2-1). A very flat coastal plain is succeeded by a rolling, frequently hilly landscape called the Piedmont. Further westward, you would encounter the more mountainous terrain of the Appalachians. The mountains do not parallel the coast but approach and eventually merge with the coast in New England.

Although the differences between the Piedmont and the mountains are more obvious topographically, the break between the Piedmont and the coastal plain may be more significant to Megalopolis. The irregularly rolling relief of the Piedmont is underlain by very old, very hard rocks. This surface is resistant to erosion, and the level of the Piedmont is maintained above that of the coastal plain. Wherever rivers flow off the Piedmont, therefore, a series of rapids and small waterfalls are formed. Because these falls are found along a line tracing the physiographic boundary, this boundary has been named the *fall line*.

Early European settlers found the fall line to be a hindrance to water navigation but an obvious source of water power. Settlements developed along the fall line, locations as far inland as possible but still possessing access to ocean shipping. In addition, because the fall line was often the head of navigation, goods brought inland or those to be exported had to be unloaded at fall line locations for transfer to another mode of transport. Because the goods were handled anyway during the transshipment, it was clearly advantageous to break the large, bulky shipments of ocean transport into smaller loads. These sites also benefited from goods moving for export from the interior to the head of river navigation. In many cases, manufacturing processes could be applied here as well. Thus, fall line sites operated as *break-in-bulk* points, and the characteristics of these sites provided early support for urban growth. Within Megalopolis, fall-line cities such as Washington, Baltimore, Wilmington, Philadelphia, and Trenton trace the Piedmont–coastal plain boundary. Only New York among Megalopolitan cities has water access beyond the fall line via the glacially eroded Hudson River Valley.

Situation Characteristics of Megalopolis

A broader view of Megalopolis's location is used when considering aspects of the region's *situation*, or location relative to other places, than that used when looking at site characteristics. One of the most important aspects of Megalopolis's situation is the region's location relative to Europe. The coastal portions of the New England and Middle colonies were readily accessible to trading ships sailing between Europe and the New World. It was not coastal position that was critical here—that is a site characteristic. This portion of North America was also on, or very near, the most direct sea route between Europe and the productive plantations of the Caribbean colonies and southern North America, at least on the homeward voyage. If we view a map with the North Atlantic as its focus, we can see the nearly straight line (or *great circle route*) between the Caribbean Sea and the British Isles (Figure 5-2). The dominant wind patterns and ocean currents also supported this shipping route. The ports around which Megalopolis later grew, therefore, were convenient stopping places and contributed actively to the transoceanic trade that expanded rapidly during the eighteenth and nineteenth centuries.

Ships stopped at New York, Boston, and the other nearby ports because there was business to be transacted. There were goods to be purchased for local consumption and manufacture and goods to be sold that could be found in few other places. Sugar and molasses were brought from the Caribbean to be distilled into rum. Cotton was transported from South Carolina to be woven into textiles. The obvious profitability of sea trade (or sea piracy) encouraged the growth of shipbuilding. The nearby forest resources were

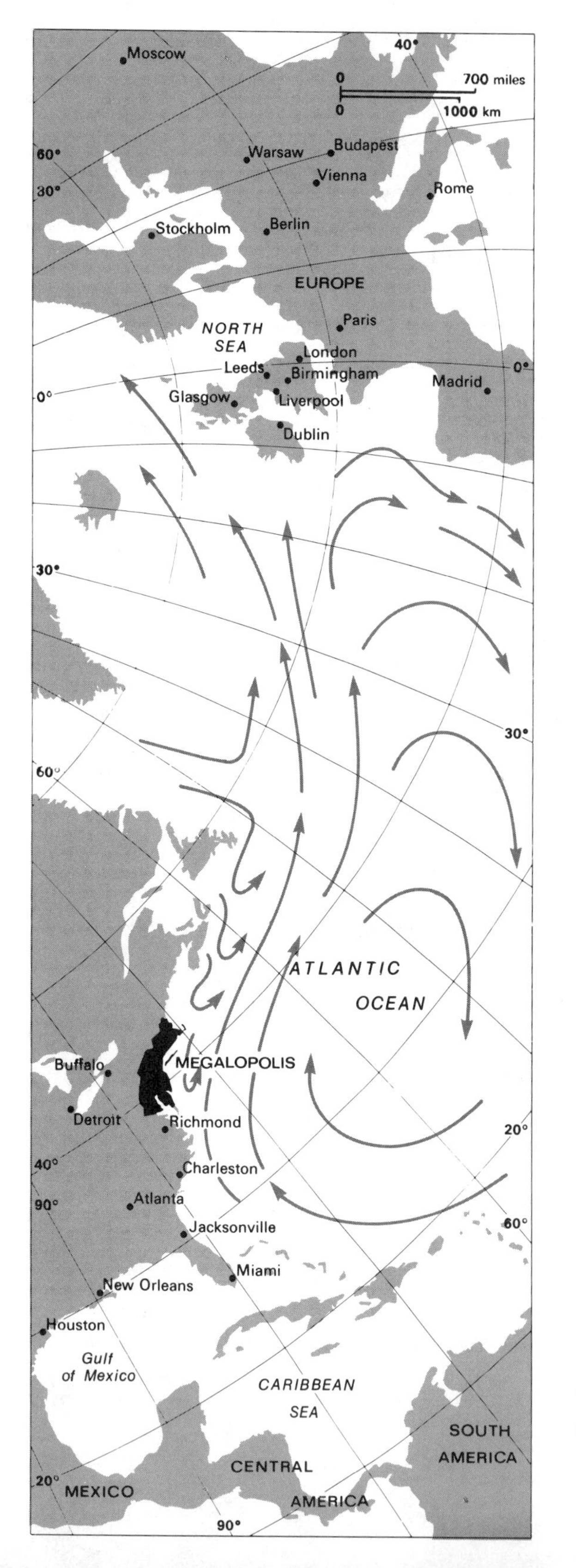

part of this region's situation and contributed to its early growth.

Also critical to this growth was the location of the core cities relative to the interior of the continent. Philadelphia and Baltimore grew rapidly because each was the focus of a relatively good-size and good-quality agricultural region. Southwestern New Jersey, southeastern Pennsylvania, northern Maryland, and much of the Delmarva (*Del*aware-*Mar*yland-*V*irgini*a*) Peninsula possessed good soils, among the best in Megalopolis. Access routes to the interior were constructed early and helped support the growth of the trading functions of these cities. Inland from Boston, in New England generally, the soils are too shallow and rocky and the terrain too rolling for this land to be considered good farming land. Agriculture was not as important a contributor to Boston's growth as it was to Philadelphia's and Baltimore's, although Boston did have the resources nearby to support the fishing and shipbuilding industries. The New England hills were covered with hardwood and pine forests nearly ideal for ship construction. Also accessible were the very productive fishing banks off the New England, Nova Scotia, and Newfoundland coasts and farther south in the rich Chesapeake Bay area.

The importance of accessibility in evaluating a city's situation, however, is most apparent in the case of New York. New York's local agricultural resources are better than those of Boston but not as fine as those of Philadelphia or Baltimore. Long Island had some good land for farming but little of exceptional quality was found nearby on the mainland. New York's chief advantage lay in its position at the head of the best natural route through the Appalachian Mountains. The Hudson–Mohawk River system, later amplified by the Erie Canal, railroads, and highways, pro-

Figure 5-2 Great circle route between the Caribbean and Europe. The high quality harbors of what came to be Megalopolis were located close to the most direct sea route from the Caribbean to northwestern Europe.

vides access to the Great Lakes. The Great Lakes, in turn, provide access to the broad interior of the continent. As settlement density and economic activity on the interior plains increased, some of the goods produced were shipped downriver to St. Louis or New Orleans, but large amounts were carried to the urban cores of Megalopolis. The city that benefited most from this growing trade was the one with the greatest natural access to the interior: New York.

Megalopolis as the Continental Hinge

It should be clear by this point that Megalopolis and the city cores that dominate it possess very real advantages for growth. Although none of these cities has great quantities of what are traditionally thought of as essential resources, such as high-quality agricultural land or mineral deposits, they do share a set of what might be called *accessibility resources.* Such resources are naturally occurring features of a place that facilitate movement into or out of that place. Good harbors, good routes to the interior, a location between regions conducting trade: all are accessibility resources, and all are definitive features of Megalopolitan centers.

During the colonial period, trade grew between Europe, the Caribbean area, and the North American mainland, and cities formed around the best harbors along the East Coast. Small-scale manufacturing began to appear in the larger port cities from Baltimore northward. The mutually reinforcing relationship between manufacturing and trade stimulated growth in both activities. As urban industry grew, the demand for labor increased, drawing immigrants from northwestern Europe or diverting large numbers of workers from farming swelling the populations of these cities as well. Banks and other financial institutions underwrote investments in local manufacturing and shipping. Service activities, wholesale and retail businesses, centers of information and control, all grew and supported further urban expansion.

The greatest growth occurred within the four large port cities in Megalopolis, New York, Philadelphia, Boston, and Baltimore. Each was the focus for an expanding *hinterland.* The richness of each hinterland and the city's accessibility to it were critical to the way growth proceeded. Boston was the largest city among Megalopolitan core cities through 1750. Its advantages were those that could dominate early development—proximity to the densely settled colonies of southern New England, easy access to the rich fishing banks nearby, resources for a shipbuilding industry, and the trade that accompanied growth in shipping. Philadelphia and, to a much lesser extent, Baltimore, enjoyed growth stimulated by early development of the rich agricultural possibilities in their general hinterlands and by the cultural and political activities focused on Philadelphia during the last half of the eighteenth century. By 1760, Philadelphia had reached a population of 18,756, exceeding that of Boston (15,631). As the land west of the Appalachians began to be developed, however, New York's preeminent site and situation began to be felt. The 13,040 people in New York in 1756 had increased to 98,373 by 1810, surpassing the population of every other city in the young nation (Figure 5-3).

The most unusual feature of this region is not the fact that these cities grew—the site and situation advantages were there—but that four such large cities (later to become five with the addition of Washington, D.C.) continued to grow in close proximity to one another. Washington is unique, of course, because even though it too is located on the fall line, its growth has followed directly from expansion of the national governmental structure. The other four cities, along with many smaller ones along the Megalopolitan axis, depended largely on economic stimulation. So great was national growth during the nineteenth century, and so strong were the linkages between the interior and these four ports, that none of the four was able to wholly absorb the flow of goods to any of their neighbors and competitors. By the turn of the present century, the combined economic resources of these cities'

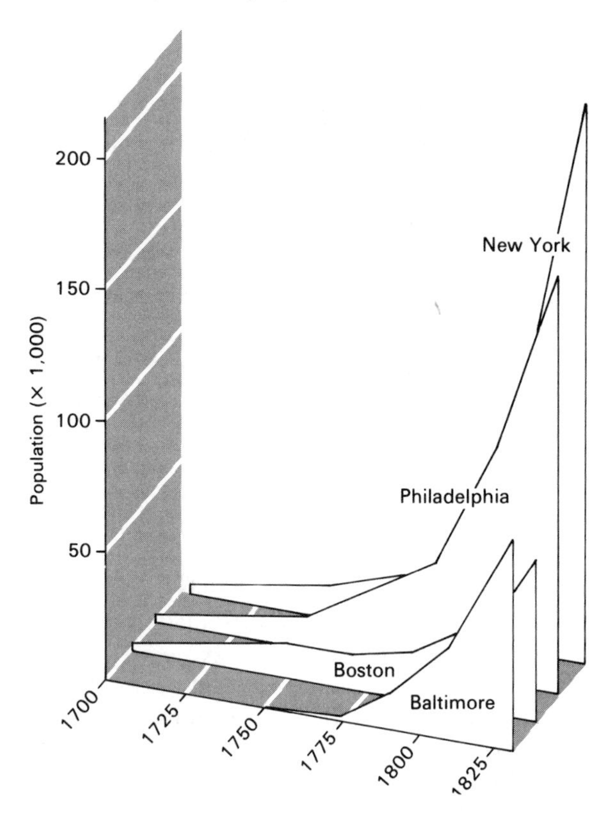

Figure 5-3 Population growth of selected U.S. cities, 1700–1830. The early growth of Boston and Philadelphia indicated their importance before the American Revolution, but New York surpassed both in size by the beginning of the nineteenth century.

hinterlands had reached the continental proportions described earlier.

Remember that by this time, initially important local resources were no longer significant in urban growth. It was the set of accessibility resources that came to be the most important. Flows of goods, people, and ideas entered Megalopolis from overseas and were absorbed, changed, and sent on toward the interior. Other goods, people, and ideas were transmitted through Megalopolis in the opposite direction. It was this dual set of flows that led Professor Gottmann to describe Megalopolis as the "hinge" of the continent, for this axis of urban activities functions much like a long pivot for North America's exchange with Europe, Africa, and western Asia.

THE URBAN ENVIRONMENT

There is a unity to the many functions and forms of the urbanized territory identifying Megalopolis as a distinct and coherent region. In a general way, the commercial "hinge" function indicates a similarity among sections of Megalopolis even if the various cities within the region do not coordinate their commercial efforts. The major parts of this region have all grown in spite of their relative proximity partly because the tremendous volume of goods transferred across the "hinge" has been much greater than the capacity of any single port. These port cities also have shared in the region's growth because each has sought competitively a larger share of national export and import trade, while no one city has an absolute advantage of position over all the others.

Throughout Megalopolis, however, it is the urban forms and urban functions that provide the most significant regional unity to the territory. It is the almost continuous—occasionally overwhelming—urbanness of Megalopolis that is its primary characteristic. Tall buildings, busy streets, crowded housing, and industrial plants, accompany an array of cultural opportunities—theaters, symphony orchestras, art museums, and large libraries. Also apparent are impressions of deterioration—dilapidated structures, traffic congestion, and air pollution. All of these, and more, are present in the metropolitan cores of Megalopolis.

These characteristics of urbanness are also found in any major city in the United States and Canada and in most large cities around the world. What distinguishes Megalopolis is that the urban characteristics in this region have spread out so far from their core cities that these urban regions have begun to merge with one another in a process of metropolitan coalescence. The process proceeded in Megalopolis earlier than elsewhere, partly because the degree of merging is more extensive, but primarily because there are so many large cities within a relatively small total region. Megalopolis became in this way a kind of gigantic laboratory in which intensively urban patterns and peculiarly urban problems could be observed developing at a very large scale. It is the most urban region on a continent with a largely urbanized population. Examination of its components helps set the tone for study of the other regions of Canada and the United States either through comparison or contrast.

Population densities in Megalopolis remain high, averaging about 305 persons per square kilometer (786 per square mile) in 1987. This average density continues well above Megalopolis's 1960 density of 266 per square kilometer (688 per square mile) in spite of national population redistribution that led to greater than average growth in other parts of the country.

Of course, population densities within Megalopolis are not the same from place to place. Some peripheral counties have populations that are only 10 to 20 percent of the region's overall average density. Settlement density increases as a city is approached, until very high densities are reached near the city core. In New York City, for example, 1987 densities exceeded an average of 226 persons per hectare (91 persons per acre), which amounts to more than 22,660 persons per square kilometer (58,700 per square mile). This general pattern occurs with considerable spatial regularity. There are variations from city to city in the rate of density increase as one approaches each metropolitan center and in the height of the density peak, but the pattern repeats itself again and again. What other features of city organization are associated with this consistent density pattern? What do these associations tell us about Megalopolis? How is the pattern changing, and what are the implications of these geographic changes for urban development elsewhere in North America?

Most geographers would agree that modern cities result basically from the locational consequences of economic activities. Of course, cities do have political foundations. They also constitute a form of social clustering. Some cities have been founded or have grown because of religious factors. However, when someone decides to move or to relocate his or her business into a city rather than a nonurban setting, it traditionally has been the fundamental economic advantages of such a choice that dominate the decision. So pervasive are these advantages that large numbers of people live very close to one another, often closer than they would prefer, and tolerate many negative consequences of this dense clustering in order to participate in the benefits of city organization. As Jane Jacobs stated recently, "There are advantages to density. Small things can be near each other. This is what happens in a lively, healthy downtown. The great advantage of cities is the choice it gives you."[3]

An increasing number of urbanites have minimized some disadvantages of city living by moving their residences to suburban locations. Others have moved even farther, well beyond the suburban fringe, into a zone referred to as *exurbia*. From small towns, areas of converted vacation homes, and rural estates, exurban commuters travel each day (or each week) to distant workplaces. But this spread of population has not eliminated the disadvantages of clustering, only shifted the population facing them. It has also spread the workplaces for those living in the region; a smaller proportion of a sprawling metropolitan population than previously still find it necessary to enter the central city for work. The spread has also created a wide variety of new problems, many of which are geographic. Before examining these new problems and their implications, the essential features of the urban landscape of Megalopolis should be made clearer.

---

[3] *Washington Post,* March, 8, 1987, p. A26.

The very great demand for land is dramatically portrayed in the abrupt contrast between New York's Central Park, preserved for recreation, and the tall, densely packed buildings crowding the park's borders. (Allen Green/Contact)

Components of the Urban Landscape

Interaction. In general, the cost of moving something is directly proportional to the distance it must be moved (i.e., the farther the item must be moved, the more it will cost to move it). Activities cluster in cities so that movement costs are minimized. *Spatial interaction,* the movement that occurs between places, is basic to the existence of a city, so the imprint of this interaction should be apparent. The visibility of interaction depends on what is interacting. The lines of interaction for human movement are visible in streets, subways, bridges, tunnels, sidewalks, and parking lots. In older cities, such as those in Megalopolis, the downtowns were laid out when travel was by foot and horse-drawn carriage; therefore, these cities have only about 35 percent of their total central areas devoted to visible landscape features that support human interaction. In newer cities, developed largely after the rise of the automobile, the proportion is much higher. In Los Angeles, for example, as much as 62 percent of the city's downtown is devoted to automobile uses. The critical importance of the ability to move from one place to another in urban regions is shown by these high proportions of urban land devoted exclusively to facilitate interaction.

Other forms of interaction may be equally important but less visible. The easy movement of information and ideas is supported in cities by the intensively developed telephone, telegraph, and teletype systems and other communications systems of this nature. Ninety years ago, telephone lines were not wound into cables and placed underground. The central business districts of all large cities were crowded with crisscrossing spiderweb networks of telephone lines testifying to the intensity of communication demands. The lines are no longer visible, and it is just as well because intraurban information flow has multiplied many times during the last century.

The movement of people and the movement of ideas and information are two obvious examples of interaction within cities, but many more things exhibit the same general pattern of most intense movement near the city core. Also hidden, and taken for granted unless something goes wrong, are such items as electricity, a fresh water supply, and sewage and garbage disposal. Until the volume of garbage reached high levels because of urban residential densities, it could be fed to animals, buried, or discarded in nearby dumps without becoming a general health hazard. In areas of rural density, individual septic tanks are often sufficient to handle a single family's sewage safely, and each dwelling can have its own well without overdrawing the underground water supply. Whereas some of these typically "urban" services have been extended into nonurban areas, the extension is relatively recent. Even electricity was not extended into most of rural America until well after the federal Rural Electrification Act was enacted in the 1930s.

As cities grew during the late nineteenth century, communication became intense. Telecommunication technology was still crude enough to lead to scenes like this one on Broadway in New York City in 1889. The landscape gradually changed after 1900 as wires began to be laid underground. (Culver)

An example of the tremendous intensity of urban interaction was provided by a story in *The New York Times* in 1973. In the process of constructing a new subway line, work was slowed by the intricate networks of utility lines buried beneath the pavement of Second Avenue in New York City. In one block, the crews found

> one 30-inch and two 10-inch gas mains; a four-foot water main and a 12-inch water main; a box sewer three and one-half by two and one-third feet; 16 three and one-half inch ducts for electrical wires along with 43 three-inch ducts; 64 three-inch ducts for telephones and 30 four-inch ones. In addition, they found footings for the old El; house connections for sewers, water and gas; cables for fire alarm-system boxes, cables for police call boxes and Transit Authority cables.[4]

Remember, all this was in only *one block* and found in the process of constructing a new interaction line!

[4] *The New York Times*, February 18, 1973, p. 69.

The intensity of intraurban interaction reflects the diversity, as well as the density, of activities. Movement between spatially distinct locations does not just happen. A cost must be borne because of the movement, and, by implication, a return must be expected to make the movement worthwhile. Perhaps the prime motivation for interaction is the geographic separation of supply and demand. Economically, this means that an item or a service is needed or desired at a location that cannot meet that need; the good or service must be obtained at another location. Interaction may occur as a result. When two or more places are deficient in items that can be obtained at the other places, they are said to be complementary, or to have *spatial complementarity*. When locations are complementary, interaction will occur unless they are too far apart, there is some closer place to obtain the good or service, or there is some other barrier to movement.

Functional Complexity. The implications of all this for city landscapes should be clear. A city is a region of considerable functional complexity. There are a great many different kinds of activities in a relatively small area. Some functions tend to cluster together while others are scattered more generally across the urban region. Since a wide variety of activities are carried on in a city, interaction is stimulated between functions or within zones of the same function. When the activities are mapped, the result is a greatly mixed pattern of land use. The generalized pattern of planned land uses in Philadelphia provides an example of the complexity that is presently found there and in other major cities (Figure 5-4).

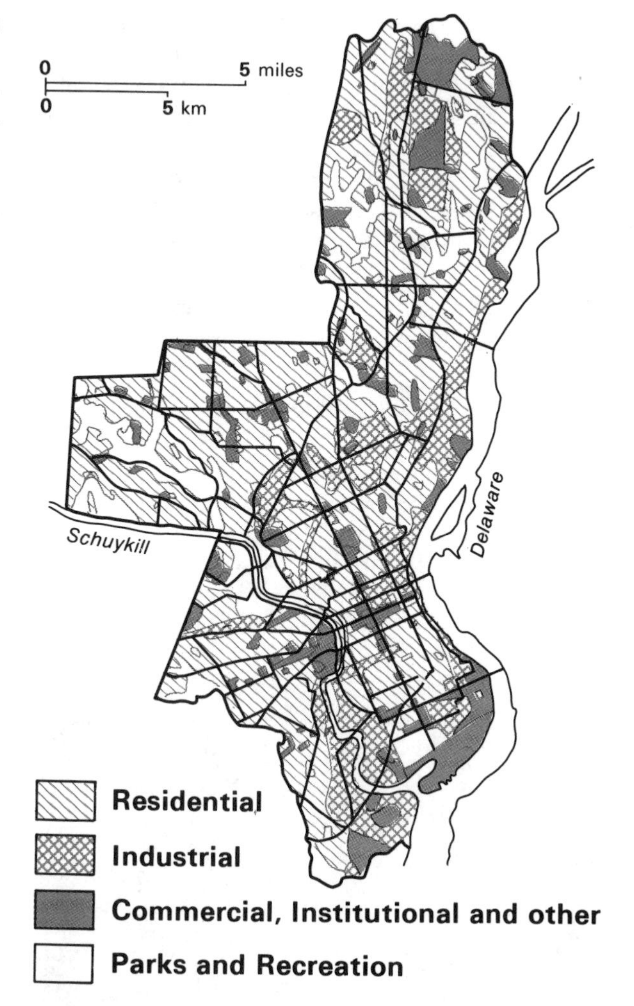

Figure 5-4 Land use pattern in Philadelphia. Even in this somewhat simplified map, the complex intermingling of urban functions is apparent.

Philadelphia's land area is dominated by residential and industrial uses. The waterfront is the primary industrial region. Low-cost water transportation and Philadelphia's role as a hub of international oceanic trade combine to generate the location of industry. This land-use pattern has also had an impact on the rest of the city. Almost the entire waterfront along the Delaware River is planned for industrial uses, as is a large area on the lower Schuylkill River. When we add the two or three linear industrial areas extending across the city, much of Philadelphia is seen to be close to some form of industry. Although industrial activity within the city has declined during the last decade or more as firms move elsewhere or go out of business, and although it is difficult to anticipate how the land uses will change, the complex pattern remains.

A tremendous diversity of function within each land-use type is masked by the generalized nature of the categories on the map. The densely packed and intensively interacting central business district of Philadelphia is not differentiated on the map from the retail and wholesale clusters located to the north and west of the city. Large shopping centers are not indicated in a manner different from that of the headquarters of a large insurance company, whereas the functions themselves are clearly complementary and produce interaction flows. Other interaction arises from institutional activities such as the city government, various centers of higher education, the state hospital and the adjacent county prison farm, and the Philadelphia naval shipyard. This wide diversity of functions is surrounded by, interspersed with, and often interacts with the resident population of ethnically and socioeconomically diverse communities. This complex pattern is repeated in other major cities of the region.

Public Services. A third aspect of the urban landscape is implied by the dense population, the complex functional mix, and the high level of interaction stimulated by these diverse activities. The concentration of population and urban activities requires supporting functions as well. Traditionally organized and operated as various branches of city government, these public-service functions are only indirectly productive,

The major port cities of Megalopolis continue to be important shipping and industrial centers. This view of south Philadelphia across the Delaware River demonstrates the impact of these functions. (Courtesy Delaware River Port Authority)

in an economic sense, but they are necessary for commerce and industry.

As the settled urban environment becomes more concentrated, by definition the services provided by the urban administrative structure also become more intense geographically. In addition to the services mentioned earlier—water, electricity, sewage, and garbage collection—cities also operate services such as police and fire protection, construction and maintenance of public movement facilities, health care, documentation of vital population statistics, and educational facilities.

There is much variation in the quality and quantity of these services from city to city—some cities, for example, extend support to cultural activities, as well. A few cities have chosen recently to turn operation of some public services over to private firms, but the large number of people and functions clustered in urban areas require that such services be present in some form.

As the United States and Canada have become increasingly urbanized—that is, as a larger proportion of the total population has come to live in urban areas—many public services have been extended far beyond metropolitan limits. In nonurban areas, nonurban governments (county, state, provincial, and occasionally federal) have also chosen to provide support for public services, although to a less intensive degree. Such services are often demanded by urbanites who have moved to nearby suburbs or semirural communities but who also consider these services essential. Thus, small communities experiencing a significant influx of new residents may be forced to change their tax structure to institute services more typically found in larger urban places.

So pervasive did these services become in large cities that an immense governmental structure developed to administer them. In the New York metropolitan region, to take the extreme case, more than 1550 distinct administrative agencies were in operation in 1982. More on this will be discussed later, but it should be clear by now that an extensive network of services is an important corollary to functional diversity and the interaction flows typical of urban areas, such as those within Megalopolis.

Accessibility. A fourth major component of the urban landscape is the high level of accessibility found there. This, too, is related to the other components. Accessibility is created and maintained as a public service, but one that is demanded by the great amount of interaction and shaped by the spatial arrangement of land uses. The degree of accessibility found at a given location also may affect the decision to locate an activity at that site. And of course, limited access between two places is a constraint on the amount of interaction possible.

Easy access between most sections of an urban area has not always been as dominant a consideration in city organization and structure as it has become in recent decades. The original street plans of most cities in Megalopolis, for example, followed the "rational" and simple grid pattern popular in the seventeenth and eighteenth centuries. Square or rectangular grids were laid down as fully as local terrain allowed in New York, Philadelphia, Baltimore, Washington, D.C. (with modifications), and most of the smaller cities and towns along the eastern seaboard. Even in Boston, where the configuration of the harbor and the winding rivers that enter the harbor make the square grid awkward, portions of the city were laid out in this fashion.

During the early years of development, when the cities' functional mix and the interaction it stimulated were less intensive than today, the grid pattern was satisfactory. As these cities grew, however, the inadequate access provided by this street system began to become apparent. A square grid, for example, possesses very frequent, right-angle intersections. Because the traffic flow is interrupted at each intersection, a greater traffic volume leads to longer pauses at each intersection. By 1900, Baltimore and Boston had each exceeded populations of 500,000, Philadelphia had reached nearly 1.3 million, and

New York was approaching a population of 3.5 million. The main impact of the automobile was then still in the future, but serious congestion was already present in these cities' centers. As long as all sections of the city were subjected to the same conditions of limited accessibility, pressures to change the form of access were not great enough to have much effect. When all sections of the population had to live within walking distance of work or within reach of the trolley tracks, none suffered significantly poorer access than any other.

Following World War II, a series of changes occurred that led to the rapid areal expansion of cities and increased use of automobiles for travel. An increasing proportion of the work force in the cities began to live at distances and in residential densities that made it uneconomic for mass transit to reach them. Economically, the speed and flexibility of truck transport accelerated the diversion of short-haul freight from rail to road carriers. Traffic planners responded by recommending construction of peripheral circumferential, or ring, highways and high-volume, limited access expressways. The goal was to separate local movement (that remained on the grid) from cross-town and through traffic. Partly successful, these changes, plus many others, also increased demand for access within the city center, between the center and the periphery, and eventually between sections of the periphery. The entire pattern of accessibility became more complex and difficult to manage.

Intensity of Change. All of this accentuates another component of the urban landscape. This is a landscape of change. All landscapes change, but urban patterns shift and are rearranged in ways made more vivid by the intensity and complexity of urban functions. Tens of thousands of new residents enter a large city like Philadelphia or New York each year. Even greater numbers have been leaving, some to distant cities and some only into the metropolitan fringes. Structures are destroyed, new ones built. Street patterns are altered, the pattern of functions is changed, and flows of people, goods, and ideas are shifted to fit these new patterns. The changes may be observed in any major region in North America, but the changes, in some ways, actually created Megalopolis. After 1940, this region became less and less a compact series of large, essentially separate cities performing somewhat similar functions for the continent. Increasingly, the sprawling metropolitan areas merged to form a single urbanized region. The tranformation to one "super city" is far from complete and will probably never occur because of yet other changes, but the process is more advanced here than anywhere else on the continent. The impact of these changes and the problems they have raised may well provide insight into some of the patterns developing elsewhere.

CHANGING PATTERNS IN MEGALOPOLIS

Perhaps the most fundamental and far-reaching change in Megalopolis during the last 40 years has been the great areal expansion of each of the major metropolitan areas (Figure 5-5). Greater New York clearly has extended its population the farthest, but the Boston, Philadelphia, Baltimore, and Washington urban areas have also grown greatly. New York had both the largest initial population and the most intensive economic concentration, but the other three port cities possessed firm foundations for growth as well (see Chapter 6). At the same time, the federal government rapidly expanded its operations. The District of Columbia was no longer large enough to absorb the burgeoning civil-service population and the growing number of people needed to feed, clothe, and otherwise serve them. Urban development spilled into Virginia and Maryland.

Changes on the Land

The spread of urban population far beyond city limits also had a strong impact on rural ac-

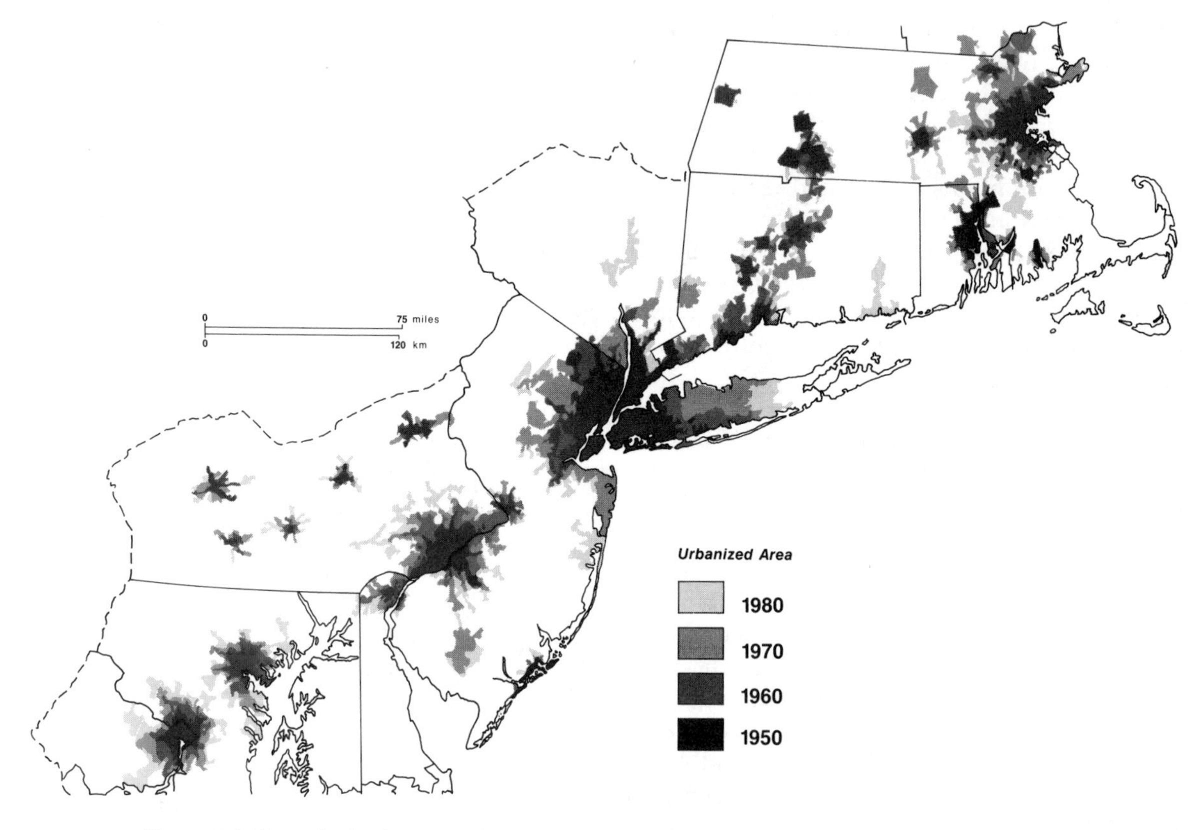

Figure 5-5 Spread of urban population in Megalopolis, 1950–1980. (After C. E. Browning.) The sprawling metropolitan regions in Megalopolis have grown at different rates, but the tendrils of urban densities from each center reach far from the original cores.

tivities in Megalopolis. As pointed out earlier, substantial areas within Megalopolis have yet to become urbanized, but the cities have had a profound influence nevertheless. There are two types of such influence: the effect on land use and the effect on land value.

As city populations grew, a greater number of people not directly engaged in food production had to be fed by foodstuffs shipped in from rural areas. The tens of millions of people in urban Megalopolis consume agricultural products from all across America and beyond. Many of those farming the agricultural land close to the cities, however, chose to specialize in higher-priced foods and in products with high perishability. The distance to market was short, and the price of the final product did not have to be raised further to cover high transport costs. Dairy products, tomatoes, lettuce, apples, and a wide variety of other intensively produced "table crops" came to dominate farm production in sections of rural Megalopolis.

Accompanying this shift in agricultural emphasis was the second influence of expanding urbanization. As land on the margins of urban areas was approached by dense settlement and intense economic activities, the price of land was driven up. Good farmland purchased at a few hundred dollars per acre in the 1920s or 1930s might be sought for residential subdivision at many times the price. For example, a 125-acre farm purchased decades ago for $20,000 as agricultural land might eventually be sold to a real estate developer for $1 million. The developer, in turn, subdivides the land into 250 home lots of 0.5 acres each and then, after putting in utilities and streets, sells each lot for $25,000, or a total $6.25 million. In many areas, a half-acre suburban lot is currently priced at two to three times, or more, this price.

Even if a farm family were able to resist the profitability of such a sale, the taxes on the land will rise sharply toward urban levels as nearby areas begin to be used for urban activities. Until land use controls began to be put in place to keep land in agriculture (see Chapter 4), the only way for a family to remain in farming was to pursue intensive agricultural production devoted to high-value products. As a result, there is a pattern of high-value farm-product operations clustered in between the expanding urban regions of Megalopolis (Figure 5-6).

Urban sprawl and the corresponding shifts in agricultural practices in Megalopolis were pulled along the main lines of interurban access between the major urban nodes. Strong flows of traffic developed early between the cities of Megalopolis. Each center stimulated growth in the others through complementarity as well as competition; transportation and communication lines were needed to carry the interaction. When people who continued to work within the major cities relocated their residences, it was only natural that a high proportion chose sites that allowed them easy access to the main cluster of workplaces. The arterial highways and, to a lesser extent, the interurban railroads and main feeder roads became the seams along which metropolitan populations spread first and farthest. As a consequence, the urbanized areas merged first along these main interurban connections. And the demand for increased accessibility generated the construction of even better interurban movement facilities.

Figure 5-6 Average value per farm of farm products sold, 1987. Agricultural production in Megalopolis is generally land intensive and high-value, with the greatest returns coming from the best quality land remaining available for agriculture.

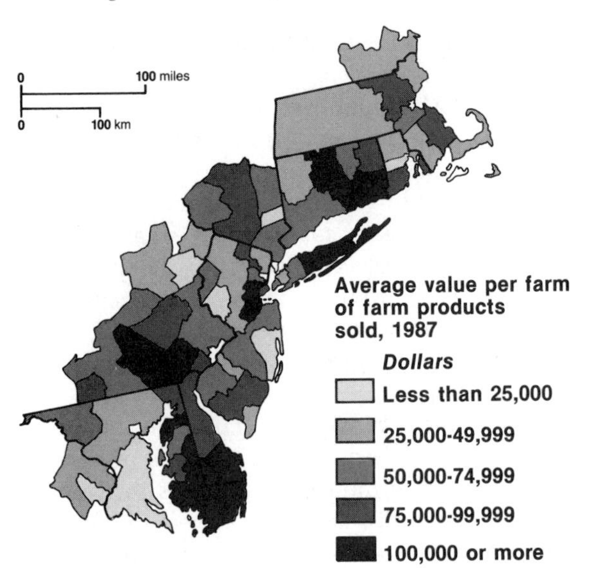

Changes in Population Composition

As the populations of the separate urban areas grew, the composition of the populations also changed. Prior to 1910, the cities of Megalopolis absorbed large numbers of immigrants from Europe. During the 1840s and 1850s, millions of immigrants arrived in the new world from northern and western Europe, with 70 percent of these from Ireland and Germany (Chapter 3). Throughout the latter half of the nineteenth century, large numbers of Italians, Poles, Russians, and Austro-Hungarians and others in small groups and families arrived from Europe. These migrants passed through one or another of the large Megalopolitan ports, usually New York. Those who did not continue westward into the farming areas or urban centers of the Midwest and Great Plains settled in dense clusters within Megalopolis's cities, usually forming communities of each nationality. Communities within the very large urban areas offered cultural and linguistic support for the newer arrivals. Many such communities have had enough social cohesion to remain spatially indentifiable (see also pp. 154–156).

When World War I broke out in Europe, the

flow of emigrants stopped suddenly and a new set of migration flows began to filter into Megalopolis. The new stream of migrants was in fact an expansion of an already established population movement. After 1910, the slow trickle of black migration from Southern states began to grow. Black migrants and groups or rural whites from the region repeated the patterns of settlement used by migrant groups from Europe. Most blacks settled within the cities in areas that were already occupied by small black populations.

As the black migration continued through midcentury, population densities increased and generated outward expansion from the original core areas. Often after many decades of population increase within a city, black densities began to increase in outlying regions of black settlement as well. In New York, for example, the black settlement area in the Queens borough on Long Island did not show a census tract over 90 percent black until 1930, long after much of Harlem, on Manhattan, had become an entirely black area.

This fragmented expansion of black settlement also illustrates the increased social and economic differentiation within ethnic and racial groups in these cities. As portions of each group found better-paying jobs and developed "mainstream" attitudes that helped broaden these opportunities, families moved toward the city's edge or beyond. Faced with individual and institutional discrimination, blacks and some other non-European minorities did not move as early or as far or in as large numbers as other urban groups, but they too fit the trend (Figures 5-7 and 5-8). In this way, the urban pattern of intense core and diffuse periphery is reinforced again. The most intense ethnic and racial separation near the city center gradually changes to loss of ethnic identity as one travels away from the center.

Central-city populations have declined since the mid-1950s, at least, but much of the population leaving the large core cities relocated to one of the nearby suburbs. Thus, the total metropolitan population continued to grow. Much has been written about this "flight to the suburbs" by those who could afford to move out and were not barred by racial discrimination from suburban entry.

Changes in Population Redistribution

Two entirely new aspects of urban change appeared during the 1970s and 1980s in the metropolitan regions of the United States. There are some indications that this change is national in scope, but it is the largest and oldest cities such as Megalopolis's ports that demonstrate the change most dramatically.

First, although almost all central-city populations continued to decrease and, in general, the suburban rings continued to increase, the 1980 population census indicated that suburban populations are no longer growing as rapidly as earlier. Sometime during the late 1960s, for the first time in U.S. history, people began to leave the largest metropolitan regions—considering both central city and suburbs together—in greater numbers than those who moved in. Suburbs still grew, but their growth no longer exceeded central-city decreases. Smaller cities and towns and the rural areas between them in general have been the primary recipients of these population shifts. This pattern is especially prevalent in communities and areas near metropolitan clusters but too distant to be considered part of them.

Much of Megalopolis is composed of medium-sized cities and nearby rural areas located between the largest urban cores. In spite of the significant shift in national population away from the northeast, many of the smaller places within Megalopolis continue to grow. One trend in population redistribution is reducing the number who live in the region as a whole, while another is rearranging the pattern of population within the region.

The second recent aspect of urban change is an eruption of high-rise office clusters at various locations in the metropolitan areas for reasons not yet made fully clear. The appearance of new metal and glass office skyscrapers has transformed the downtown skyline of many North American cities since the mid-1970s, but this feature has not been limited to the old city cores. Even more dramatic has been the emergence of massive office clusters within the suburban rings surrounding the central cities. Not only are these clusters new, they are very large—with many exceeding the square footage of office space of a major city downtown. For example, there is more office space in Rockville-Gaithersburg, a

Figure 5-7 Ethnic diversity in Boston, 1910. Areas with greater than 10 percent of their population native to the ethnicity or race are indicated.

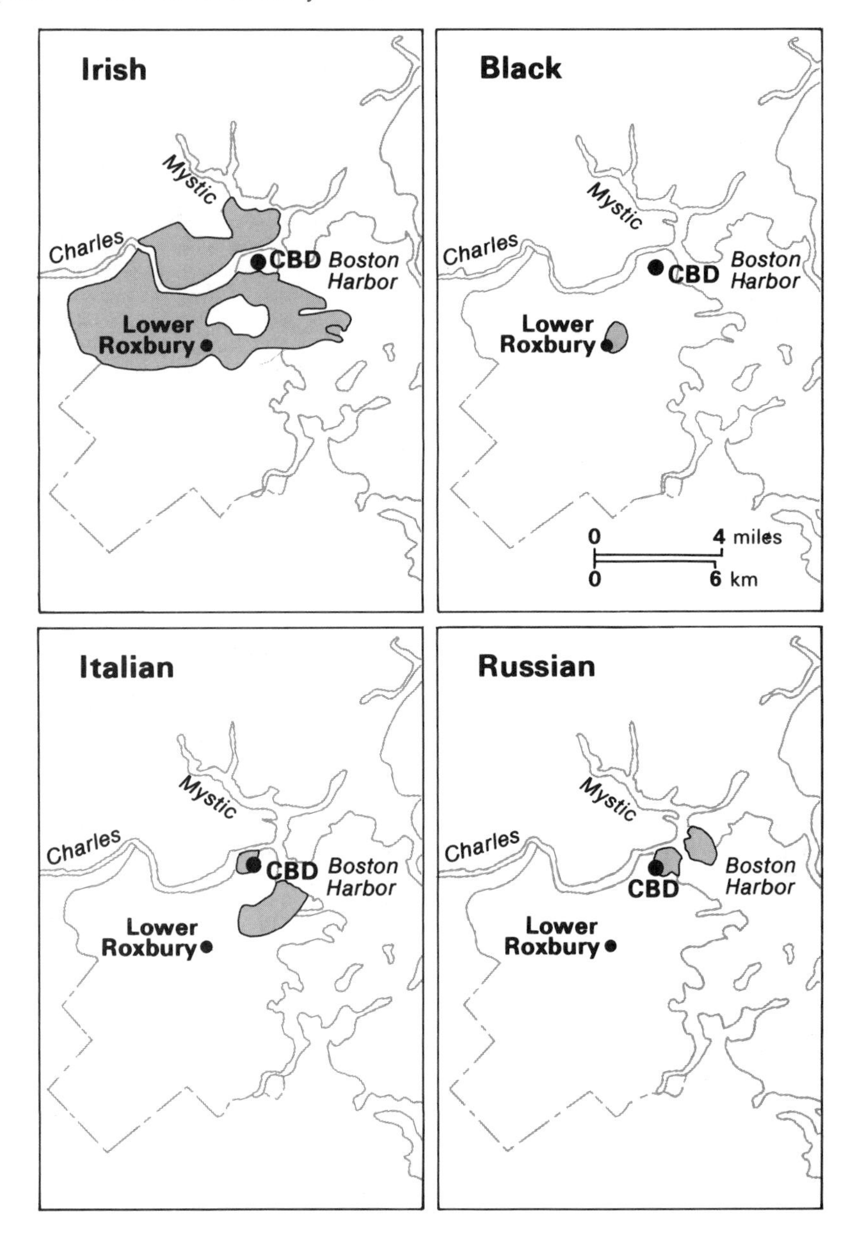

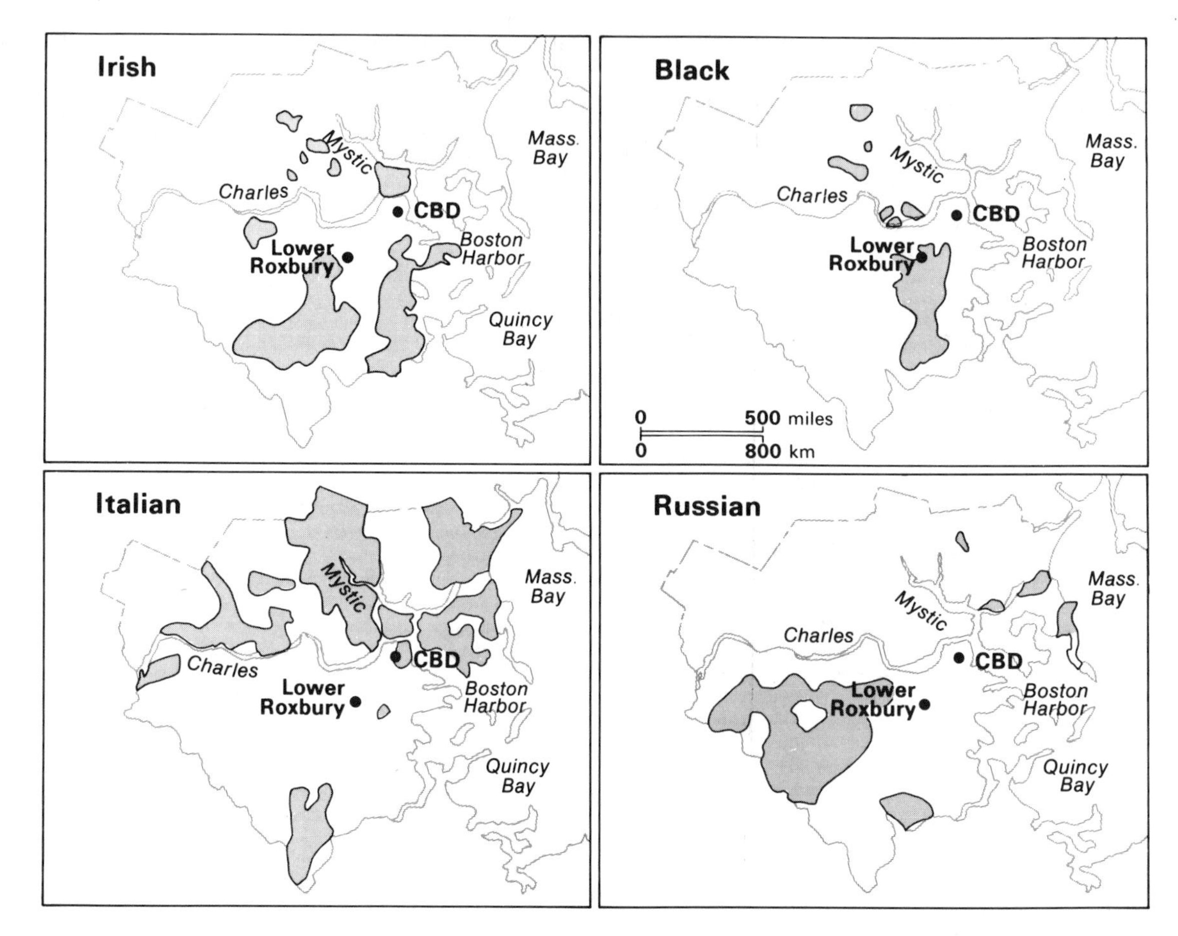

Figure 5-8 Ethnic diversity in Boston, 1970. Areas with greater than 10 percent of their population native to the ethnicity or race are indicated. Between 1910 and 1970, the city's black neighborhoods spread south between areas that were primarily Irish.

Maryland suburb of Washington, D.C., than there is in downtown Baltimore. And this cluster is only one of more than a dozen such new office zones around the District of Columbia. One of these, Tysons, is even larger than Rockville-Gaithersburg and exceeds Miami in leasable office space. The phenomenon is repeated throughout Megalopolis and across the country, at varying scales. This change is one that appears to affect most significantly the location of jobs and the pattern of travel to them rather than the location of residences.

Since Megalopolis comprises the greatest single collection of old, large metropolitan areas but is also a region of tremendous economic energy and potential, it is difficult to foresee the impact of national population redistribution on this region as a whole. The core-city populations have shared in the trend (Table 5-1) and many suburbs showed decline as well when 1980 census data were released, but the geographic advantages that stimulated commercial, financial, industrial, and cultural activities during two centuries of urban growth are still present and are indicated by the growth of office clusters. This is shown by the lower rates of population decline

Table 5-1 Megalopolis Central City Population Change, 1970–1990

City	1990 Population	*Percentage Change* 1970–1980	1980–1990
New York	7,033,179	−10.4%	− 0.5%
Philadelphia	1,543,313	−13.4	− 8.6
Baltimore	720,100	−13.1	− 8.5
Washington, D.C.	574,844	−15.6	−10.0
Boston	553,712	−12.2	− 1.6

Source: U.S. Census figures.

in the central cities during the 1980–1990 period than in the previous decade.

It should be clear by now that urban areas are landscapes of change and that the changes in Megalopolis suit the unusual character of the region. The changes here have been continuous, they have been drastic, and they have occurred on a scale unmatched anywhere else in the world.

Of course, there are clear reasons for the region's dynamic nature. Each aspect of change is a response to demands and pressures that could not be resolved otherwise. Some changes came about because of conditions hundreds or thousands of miles from Megalopolis, conditions that led to immigration, for example, or conditions that altered the cost of oceanic shipping. Some changes occurred in response to developments inside the region. In each case, new problems were created as previous pressures were resolved. The new problems often appeared more complex and more difficult than those faced earlier, although this may be largely because they were new. In a general sense, however, there are categories of urban geographic problems suggested by the experiences of Megalopolis.

PROBLEMS IN MEGALOPOLIS

Beyond other considerations, remember that the problems faced by the residents of Megalopolis are not merely the problems of any large city multiplied four or five times. Megalopolis is *not* a super city with super problems. While many of the difficulties of urban organization and growth are those found in most large metropolitan areas, this region also contains numerous large urban places competing economically while merging their territories in a complex, uncoordinated manner. Problems have arisen that are different from those of large cities—in kind as well as degree. In addition, the cities of Megalopolis are among the oldest of the presently large cities in the United States.

Most of the problems of Megalopolis can be treated in the context of three of the characteristics discussed earlier as representative of urbanness in general: density, accessibility, and spread. The first two include problems that are fundamentally urban, occurring in the cities of Megalopolis on a somewhat larger scale than elsewhere. The third characteristic, however, also includes a set of problems that are megalopolitan, or multiurban, in nature.

Disadvantages of Density

High population density and spatially intense economic activity are typically urban patterns, and both have problems associated with them. When cities experience a more rapid influx of migrants than can be accommodated by housing openings, residential areas become overcrowded. The core cities of Megalopolis contained overcrowded residential areas for a great many years as migrants from Europe poured in. Not only were new housing structures constructed more slowly than the increase in popu-

Abandoned buildings, evidence of inadequate public services, and limited local recreational opportunities are representative landscape characteristics in many sections of older Megalopolitan core cities. (Mark Antman/The Image Works)

lation, the new building sites were usually far away from where immigrants could afford to settle. The magnet of social or cultural cohesion and the constrictions of low migrant incomes and restrictive property tax structures exaggerated the extreme overcrowding in sections of the city with some of the oldest buildings and reduced owner incentives to maintain those buildings. Meanwhile, as outlined earlier, there were great opportunities for profit from land development around the expanding urban fringe.

Intense concentrations of economic activity also produced problem conditions exaggerated by high urban population densities. Many of these difficulties are related to the issue of waste management (see Chapter 4). The problem here is not that waste is produced in large quantities but that it is produced faster than its negative aspects can be neutralized. When this happens, the excess spills over into the environment, the environmental balance deteriorates, and the result is recognized as pollution.

Industrial activity frequently is singled out as the primary polluting activity in urban areas, and in many cities industries do contribute greatly to environmental deterioration. It would be a mistake, however, to lay the chief blame for urban pollution on industry. In general, cities in Megalopolis possess much lower proportions of heavy industrial activity than do many cities along the Great Lakes, yet their air- and water-pollution problems are severe. Washington, D.C., for example, has almost no heavy industry, but its population will testify to frequent smoggy days every summer.

More important than industry are what are euphemistically called "municipal sources." These sources are actually the urban dwellers who generate increasing levels of waste in such high densities that the waste cannot be absorbed locally. Hundreds of thousands of vehicles are driven in and out of each core city every day, many carrying only the driver. Thousands more travel between suburban locations. Millions of homes and millions of acres of office space in each metropolitan area are heated every winter and air conditioned every summer with public utility–produced electricity or public utility–supplied natural gas. Millions of gallons of waste water are expended from washing machines, dishwashers, garbage disposals, and flush toilets every day. And all of this is concentrated within a few hundred square miles. In the New York metropolitan region, for example, 5 million cubic yards of treated sewage residue (or "sludge") is produced each year. The accumulation of this sludge spread evenly over Manhattan would bury the entire island to a depth greater than 3 feet in just 12 years. And in Megalopolis, the pattern of urban concentration repeats the problem again and again within a small portion of the continent.

Two approaches are offered to the problems of pollution when the problems are addressed at all. One is a technological approach—develop and apply better ways of neutralizing the negative aspects of waste. The other is an ecological approach—find a better balance with nature's capacity to absorb waste by producing and consuming less, thereby reducing the output of waste. A third choice, a spatial approach—reducing waste density by spreading population and economic activity over a greater area—might be more often discussed, but there are already well-recognized problems associated with an increasingly dispersed population.

Accessibility and Density

High population densities are strongly associated with difficulties in getting from one place to another. Congestion occurs when the activity sites become so densely packed and the level of interaction between them becomes so intense that the routes for interaction become overcrowded. Real access is reduced while demand for access continues to increase. The phenomenon is familiar as a daily cycle to all who have lived in, or visited, the core areas of major cities. Morning and evening rush hours find movement so difficult that "rush" is no longer an appropriate description of these periods of the workday.

The problems of intracity accessibility are exaggerated in Megalopolis by two features. One is the obviously large number of very densely populated urban areas, including the continent's extreme case in New York City. In a 1989 study of the 31 largest metropolitan areas, the average time to commute downtown in New York City during a workday was 50–65 minutes, easily the highest figure among the places included in the study.[5] The other factor is related to these cities' sites. All were founded and grew because of their easy access to ocean shipping and the excellent harbor facilities provided by the sites. The coastal configurations, however, create severe problems for land movement. Boston and Baltimore have no land approach to their cores from the east and southeast, respectively. New York and Philadelphia cannot be reached directly by land from the southwest and the south, and Manhattan is an island. The rivers within these cities and in the District of Columbia congest movement by funneling it across a limited number of bridges and through tunnels. When they are all open to traffic, there are 14 bridges (including 3 for railroads) across two rivers in the District of Columbia, 22 bridges into or within Philadelphia, and 20 major routes of access for

---

[5] *Fortune*, October 23, 1989, pp. 79–93.

land movement onto Manhattan Island in New York. Baltimore, Boston, and many smaller cities in Megalopolis—New York's New Jersey suburbs; Wilmington, Delaware; Providence, Rhode Island; and Connecticut's Bridgeport, Hartford, and New Haven—also possess transportation networks that are constricted and limited by the expensive need to bridge or tunnel under water barriers.

The cities of Megalopolis have approached this problem in several ways. The first method used when existing facilities become overcrowded is almost always to increase the capacity of the facilities. Four-lane roads are widened to 8 lanes, and 8 lanes to 12 lanes—or even 16 lanes. Limited access expressways are constructed parallel to other limited access expressways, and additional telephone, water, sewage, and electricity lines are installed. Sooner or later—often by the time they are opened—these expanded routes are carrying traffic up to, or in excess of, their designed capacities. Clearly, this repeated experience shows that traffic congestion will not be alleviated solely by building more and wider roads. Some argue that traffic actually is encouraged to increase when new roads are built.

A second method for alleviating congestion is proposed frequently but, for a variety of reasons, is seldom successful. Rather than increasing movement capacity by building parallel routes, each existing line can be used more efficiently. An average automobile at rest occupies almost 20 linear feet of road and can contain up to four or six adults in moderate comfort, although only through carpooling are such high passenger levels reached. A bus, on the other hand, may occupy 50 linear feet of road but can carry six to eight times as many persons as a car. The ability of mass transit to move much larger numbers of passengers per unit of road space than the private car appears to offer a clear opportunity to increase movement capacity by making more efficient use of existing lines of access.

In fact, however, this apparently greater efficiency is not so obvious a solution. First, mass transit systems already carry a large proportion of commuter traffic into some Megalopolitan cores but have been in serious financial difficulties for decades. More than 73 percent of all commuters entering Manhattan in 1970 traveled by bus, ferry, railroad, or subway. Nevertheless, several commuter rail lines went bankrupt, subway fares continued to rise, and bus facilities struggled to keep pace with the demand.

All large cities in Megalopolis have responded to the demand for interaction by building numerous limited access highways. By eliminating intersections where such routes cross each other, the capacity of both is increased. At times, as in this case on Baltimore's 29th Street Expressway, the result appears more complex than the problem. (M. E. Warren/Photo Researchers)

Second, and strongly related to the financial problems of urban mass transit, is a long-term decline in transit ridership. Movement by private automobile gives at least the impression of greater comfort, convenience, and flexibility of

Almost everyone wants to travel someplace else within cities, at least several times each day. With so many origins and destinations and with so many people moving within a small area, trips are slowed to a crawl during the busiest times of the day. (Spencer Grant/The Picture Cube)

travel and greater personal safety to the passengers as well as the illusion of lower costs per trip. Only when fuel is less available (as briefly in 1973 and again in mid-1979) has public transit ridership increased significantly. Employer and municipal attempts to encourage carpooling have met with modest success.

Federal, state, and local governments have responded to private transit demands by constructing hundreds of miles of new expressways, building new bridges or increasing the capacity of existing ones, and in other ways expanding capacity for private movement in and around Megalopolitan centers. Meanwhile, public transportation has declined in quality and capability with a few notable exceptions such as the Metro subway system in Washington, D.C.

One aspect of urban landscape change well underway in Megalopolis is operating to alleviate the congestion resulting from centralized activities. As suburban populations have grown, many stores and services dependent on proximity to large numbers of potential customers have relocated out of the central city. As discussed earlier, new urban workplaces have also located increasingly in suburban and exurban locations rather than in metropolitan core locations. With more activities available outside the central city, fewer trips are taken into what had formerly been the main focus of metropolitan congestion:

the city's central business district (CBD). A 1978 *New York Times* survey[6] of suburban residents, for example, found that half made fewer than five nonbusiness trips per year into New York City and 25 percent said that, except for business, they never went into the city at all. These proportions have undoubtedly declined further since the survey was taken.

For many years, it was believed that the trend toward urban decentralization would have less effect on total traffic congestion in this larger region than in other urban areas. Occupations emphasized in Megalopolis's distinctive employment mix, that is, in white-collar occupations, require daily decision making or are activities that support those making the decisions. The advantages of face-to-face interaction in the decision-making process help to concentrate white-collar jobs. Finance and insurance (as activities controlling investment), publishing firms, major magazines and newspapers (indicating the flow of information), and cultural leadership for much of the nation (through the theater, opera, symphonic orchestras, and such activities as clothing fashions) are all much more important components of Megalopolis than comparable areas elsewhere in the United States and Canada. These activities employ, or are a response to demand from, white-collar workers.

However, even Megalopolis's central cities have recently experienced an outflow of decision-making jobs in unprecedented numbers. Among the nation's top 1000 corporations, for example, there are now more corporate headquarters in New York's suburbs than in the city. Improvements in computer technology and telecommunications have helped information be exchanged rapidly without requiring the former level of centralization, and such improvements promise to continue. But to some degree, this relocation may also have been stimulated by an accumulation of the disadvantages associated with central-city location.

The overall occupational structure of the central cities' employment has changed dramatically during the last 25 years, and especially since the early 1970s. For example, New York City's decline in manufacturing employment has been great (Table 5-2). Losses in some sectors through 1988 were not balanced by growth in finance, service, and government jobs. The result was a total decline in employment of more than 97,000 jobs during the previous 20 years while national employment grew by more than 50 percent. The continued decline in manufacturing lay at the root of the city's overall decrease in jobs. Computer applications reduced the

[6] *The New York Times*, November 12, 1978, p. B1.

Table 5-2 Distribution of Jobs by Industry, New York City, 1968–1988 (in thousands)

Industry	1968	1978	1988	Net Change 1968–1988
Manufacturing	840.0	538.9	369.4	−470.6
Wholesale and retail	745.7	620.2	641.6	−104.1
Transportation and public utilities	320.3	259.8	219.3	−101.0
Construction	102.5	69.2	124.6	+22.1
Finance, insurance, and real estate	437.1	418.6	539.5	+102.4
Services	747.9	818.1	1136.7	+338.8
Government	526.0	500.7	592.2	+66.2
Total	3721.7	3226.9	3623.9	−97.8

Sources: Employment and Earnings, States and Areas, 1939–1974, Bulletin 1370-11, U.S. Bureau of Labor Statistics. Washington, D.C.: 1975; and *Employment and Earnings,* Monthly Reports, U.S. Bureau of Labor Statistics. Washington, D.C.: November 1979; December 1989.

The resilience of the human spirit is demonstrated in a small but significant way by this community garden carved out of the difficult environment of the South Bronx. (Hazel Hankin/Stock, Boston)

labor requirements of data management and information retrieval just as these tasks were expected to more than make up for the loss of manufacturing jobs. And finally, the locus of national decision making shifted during the period from the private sector (and New York) to the federal government (and Washington, D.C.). Just as this region's cities led North America during a period of centralization, they may be leading as strongly the early stages of a long trend toward decentralization.

Some Repercussions of Sprawl

As urbanized areas expand, many of the problems found in the metropolitan cores are spread toward the peripheries, initially in the older, interior suburbs and later in what earlier were entirely separate cities and towns. More than one-third of the suburbanites interviewed in the 1978 *New York Times* survey[7] believed their communities were becoming "more like a big city" with all its attendant problems.

Two somewhat contradictory trends follow from the slow aging of the interior suburbs. One is additional urban sprawl as inner suburban residents relocate farther out toward the metropolitan periphery. An opposite trend is reflected in the still-modest inflow of population as central-city residences are rehabilitated in ways that attract middle- and upper-income families and individuals. In spite of the disadvantages, the large cities of Megalopolis possess great economic and social possibilities. These "new" in-

---

[7] Ibid.

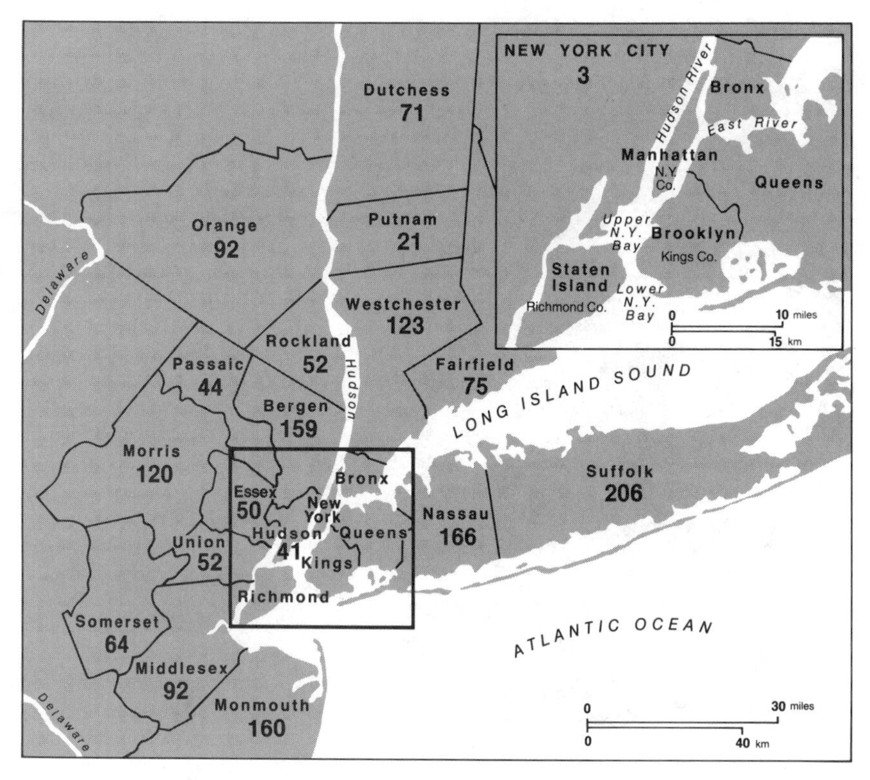

Figure 5-9 Number of jurisdictional units in New York metropolitan area, 1987. Governmental operations are made chaotic in a metropolitan region with so many separate jurisdictions when the urban problems faced are rarely contained within formal administrative units.

ner city locations are attractive to a small but significant number of America's metropolitan population.

An equally serious consequence of Spread City—as the sprawling metropolitan regions have been inaccurately called—is that they contain not one city but a great many independent municipal areas. Everyone familiar with the great cities of the United States has an image of a large city surrounded by a multiplicity of smaller suburbs. In fact, the process of suburbanization is now so advanced that many of the smaller cities near the major center are hardly *sub*urban. The urban character that typifies the original core city now extends uninterrupted far beyond its legal boundaries and includes hundreds of independent urban territories. The extreme case, undoubtedly, is the New York Metropolitan region, which contained some 1591 distinct political units in 1982 (Figure 5-9), a situation one observer described almost three decades ago as "a governmental arrangement perhaps more complicated than any other that mankind has yet contrived or allowed to happen."[8] These jurisdictions include municipal and township governments, school districts, and a variety of special districts, some with taxing power and some without. This pattern is repeated at varying scales throughout Megalopolis and much of the rest of urban North America.

[8] Robert C. Wood, *1400 Governments* (Cambridge, Mass.: Harvard University Press, 1961), p. 1.

The implications of this fragmented pattern of jurisdictions are staggering. The economic geography of the metropolitan region does not coincide with its political geography. Most people live in (and pay the bulk of their local taxes to) a city different from the one in which they work. Neighboring cities compete for the "right kind" of business or industry. The desirable industry is nonpolluting, pays a high average wage, and can contribute significantly to the municipality's tax base. The desirable resident is middle or upper income so that the city can accumulate tax revenue to pay for the public services it is expected to provide, and it can remain independent of the "less desirable" central city. The populations and industries most able to support these public services through taxes are also the ones most able to move to a new location if local conditions deteriorate. As the movement out of the core city continues, the burden increases for those who remain. New York's brush with bankruptcy in 1976, Cleveland's in 1979, and Philadelphia's in 1991 are reflections of these trends.

The central cities' problems are not isolated difficulties peculiar to the older sections of large metropolitan regions. These cities are economically and socially woven into the spatial fabric of their entire urbanized areas. What has been lacking is coordination and cooperation between the hundreds of politically independent municipalities in each region. Attempts have been made to develop regionwide coordination by the establishment of metropolitan study and planning regions that cross county and state boundaries. Boston, New York, the Delaware Valley area (Philadelphia), the Chesapeake Bay area (Baltimore), and the national capital region (Washington), all established the beginnings of planning commissions whose jurisdictions far exceed the geographic limits of each large central city. The real impact of each commission, however, is variable and certainly has never been very strong. Furthermore, the trend within Megalopolis and other large metropolitan areas in the United States lately has been toward a decrease of influence by these multimunicipal agencies in regional coordination and planning.

A logical question at this point would be: Has there ever been any equivalent regional planning agency for all of Megalopolis? No, there has not, and it is unlikely that one would be effective even if in existence. Megalopolis is a massive region of coalescing metropolitan territories with each territory containing a distinct urban core. But Megalopolis is *not* a "super city." The volume of interaction between the metropolitan regions is still much less than within urban areas.

Although not a super city and with its preeminence challenged during the last two decades, Megalopolis remains a very complex urbanized region of tremendous importance to the rest of the United States and Canada.

URBAN FRINGE CASE STUDY: QUEEN ANNE'S COUNTY, MARYLAND*

Ask farmers or fishermen with livelihoods rooted in tradition what they think about changes to Queen Anne's County on Maryland's Eastern Shore since the Chesapeake Bay Bridge gave easy access to the area and most say they don't like it. As Boyd Gibbons observed in *Wye Island,* "With little difficulty you can find people on the Eastern Shore who will tell you, without a trace of humor in their voices, that they would gladly blow up the Chesapeake Bay Bridge, if that

* Written by Margo L. Price.

would return the Shore to its former tranquility."[9]

When the Bay Bridge opened in 1952, Maryland's Eastern Shore was one of the rural oases in the creeping urbanization flowing outward from major urban centers between Boston and Washington, D.C. Queen Anne's County, like the rest of the Eastern Shore, was sparsely populated with rolling fields, scattered forests, waterfronts, and tranquil rivers. Because it was relatively isolated, strong traditions in farming and fishing methods, architectural styles, cooking, and even social relations evolved. It was, as journalist and politician Frederic Emory wrote, "at once a delightful place of residence and an exceptionally favorable location for farmers, stockraisers and 'truckers.' "[10]

Indeed, until the Chesapeake Bay Bridge was completed, the expanse of water between Annapolis and Kent Island literally kept the world at bay. Ferries running regular routes between the western and eastern shores brought summer vacationers and hunting and fishing enthusiasts from the growing cities of Baltimore and Washington, but this tidewater region's traditional lifeways of farming and fishing endured. The influx of outsiders was small and transient. Some stayed in resorts near the ferry docks and some built luxurious shoreside houses as weekend retreats. The more adventuresome, wealthy, or less pressed for time headed to the ocean beaches lining the Delmarva coast. These "outsiders" had scant effect on the self-sufficient and independent natives.

Certainly, Queen Anne's County's rural character was unaffected for most of the nation's history. It, like much of the Delmarva Peninsula of which it is a part, grew grains and produce in its fertile, easily worked soil to help stock the foodshelves of nearby cities. It, like much of Delmarva, was psychologically and physically isolated (Figure 5-10). Bounded by the Chesapeake Bay on the west and the Atlantic Ocean on the south and east, the world went around Delmarva. Ships chugged up the Chesapeake to unload goods and people at the port of Baltimore or destined for the politically important cities of Philadelphia and Washington. Others headed for New York, Boston, or Richmond. Overland transportation routes skirted the Delmarva Peninsula to avoid the natural barrier posed by the Chesapeake.

There was a gloomy aspect to life on the Eastern Shore, however. By the 1950s, the economy had suffered for decades—particularly in Queen Anne's County. Property values were declining. So were the number of farming and fishing jobs. Unemployment was a chronic problem. Those lucky enough to find jobs scraping boats or pumping gas barely made livings at subsistence levels. Shanty towns dotted the region, squalid homes to the laborers at canning and seafood processing houses in places like Kent Narrows. In search of better opportunities, many young people drifted away from the area to search for jobs. So did their parents.

To begin to address the economic problems of the Eastern Shore, a 1950 study by the Maryland State Planning Commission focused on the economic consequences of building a bridge spanning the Chesapeake.[11] The commission forecast an upswing in the Eastern Shore's economy based on the increased vacation trade a trans-Chesapeake bridge would create. They estimated that maybe 2 million vehicles would cross the bridge the first year it was open. With a regular increase

---

[9] Boyd Gibbons, *Wye Island* (Washington, D.C.: Resources for the Future, 1987), p. 113.

[10] Frederic Emory, *Queen Anne's County, Maryland* (Baltimore: The Maryland Historical Society, 1950),p.4.

[11] Maryland State Planning Commission, *Probable Economic Effects of the Chesapeake Bay Bridge on the Eastern Shore Counties of Maryland*, Publication No. 62, April 1950.

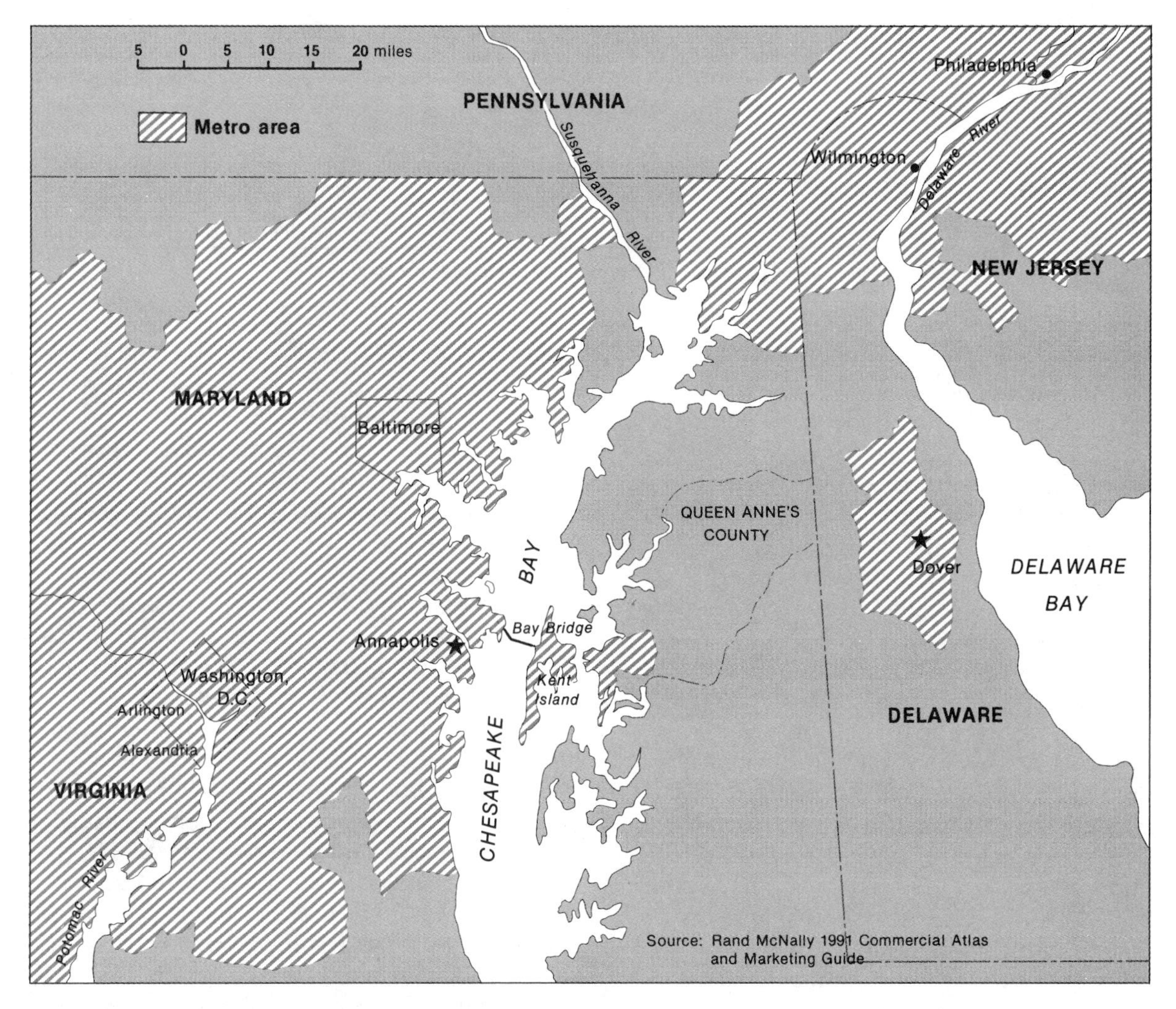

Figure 5-10 Queen Anne's county and urbanization on Maryland's Eastern Shore. The spread of urbanization onto the Eastern Shore was slowed for decades by Chesapeake Bay. But with the Bay Bridge providing easy access, commuters to the Baltimore and Washington areas now find residence in Queen Anne's County feasible.

in traffic each year, the money the tourists would drop in Eastern Shore gas stations, restaurants, and motels would boost the area's failing economy.

The commission's expectations have been met and exceeded. By improving access to nearby urban populations, economic improvements accompanied by population growth began slowly and became obvious even to casual observers by 1980 (Table 5-3). As a result, Queen Anne's County today is caught up in a dilemma typical of areas on the margins of Megalopolis's urban sprawl.

The county's rural appeal is waning as more and more settlers are drawn from the Baltimore and Washington metropolitan areas.

Table 5-3 Population of Queen Anne's County and Maryland's Eastern Shore, 1950–1990*

Year	Queen Anne's County	Eastern Shore	Maryland Total
1950	14,579	210,623	2,343,001
1960	16,569	243,570	3,100,689
1970	18,422	258,329	3,923,897
1980	25,508	296,620	4,216,975
1990*	33,300	337,900	4,666,200

* *Sources:* U.S. Bureau of the Census and (*estimate) Maryland Department of State Planning.

Based on commuting patterns, Queen Anne's County was added recently to the other six jurisdictions constituting the Baltimore metro area. Each morning, thousands of people travel U.S. Route 50 across the Chesapeake Bay Bridge to their jobs in the city and retrace the route back to the Eastern Shore at night. The irony, of course, is that the rural tranquility these commuters seek is destroyed by their very presence.

On Kent Island, Queen Anne's County's access point for the Chesapeake Bay Bridge, this slow destruction of rural character is particularly apparent. Although it still has the aura of countryside in many places, the island is under siege from urbanization. Housing developments are replacing farm fields. Shopping centers, not combines, cluster along the highway's edge. Interspersed between and among the shopping centers are the usual array of fast-food eateries and gas stations. Traffic jams on the island can be gargantuan, particularly on summer weekends when vacationers headed for Delmarva beaches compete for access with commuters returning home. Amid this, county officials grapple with how to pay for larger schools and increased services for a ballooning population. No longer a sleepy place of farmers and fishermen, Kent Island is becoming, in the words of Queen Anne's County planning director Joseph Stevens, a "bedroom community" for the Western Shore's metropolitan areas.

Robert Walters, a 63-year-old life-long county resident who has spent most of his life plying the Chesapeake's waters for clams or oysters, gave his impression of the change. "The way it is here now, if you haven't been some place in three months, you'll see something else," he said. "It used to be you could ride up and down the roads and just see fields and farms. Now there's just one shop right after another. It's getting like Glen Burnie," referring to a Baltimore suburb with a 1990 population of 37,305.

Unlike many other Kent Island natives, Walters said he isn't bothered by the influx of people. "Guess I'm different, but different people don't bother me," he said. "The way I see it, I couldn't live over there either. I don't blame them for wanting to come here." He admitted that because he lives in a tranquil part of Kent Island, he doesn't feel closed in by the growing congestion. In fact, because he runs a framing shop to supplement his income in the winter, Walters said he finds the influx of people has helped him. "There's new people over here, building homes, and they need what I have to offer."

But other locals aren't as charitable. Most give the short answer an 85-year-old waterman crisply offered when asked what he thought of the changes on Kent Island. "Not much," he said. Typically taciturn among strangers, locals—like the elderly waterman—keep to themselves, lament, and watch the changes unfold.

ADDITIONAL READINGS

Clay, Grady. *Close-up: How to Read the American City.* New York: Praeger, 1974.

Conzen, Michael P., and George K. Lewis. *Boston: A Geographical Portrait.* Cambridge, Mass.: Ballinger, 1976.

Cybriwsky, Roman A., and Thomas A. Reiner. "Philadelphia in Transition." *Focus.* 33, no.1 (September/October, 1982), 16 pp.

Domosh, Mona. "Shaping the Commercial City: Retail Districts in Nineteenth-Century New York and Boston." *Annals of the Association of American Geographers,* 80 (1990), pp. 268–284.

Hovinen, Gary R. "Suburbanization in Greater Philadelphia, 1880–1941." *Journal of Historical Geography,* 11 (1985), pp. 174–195.

Meyer, David R. *From Farm to Factory to Urban Pastoralism: Urban Change in Central Connecticut.* Cambridge, Mass.: Ballinger, 1976.

Muller, Peter O. "Transportation and Urban Growth: The Shaping of the American Metropolis." *Focus.* 36 (Summer 1986), pp. 8–17.

O'Brien, Raymond J. *American Sublime: Landscape and Scenery of the Lower Hudson Valley.* New York: Columbia University Press, 1981.

Olson, Sherry H. *Baltimore.* Cambridge, Mass.: Ballinger, 1976.

Schaffer, Richard, and Neil Smith. "The Gentrification of Harlem?" *Annals of the Association of American Geographers.* 76 (1986), pp. 347–365.

Warf, Barney. "The Port Authority of New York–New Jersey." *Professional Geographer.* 40 (1988), pp. 288–297.

CHAPTER 6

The North American Manufacturing Core

The United States and Canada are countries in which manufacturing is a very important economic activity. The evidence of this is everywhere, in the articles of clothing, the items of preserved food, the residential structures, the means of transport and communication, and in many other ways. In spite of the clear and growing presence of items manufactured in Europe or Asia, especially motor vehicles, small appliances, electronics, and some textiles, domestic industry remains paramount in the aggregate in both countries. Similarly, manufacturing plants are very numerous, especially in the United States. It is rare for any medium-sized town in these countries to be without at least some local employment in manufacturing.

Manufacturing is found throughout the more populated regions of North America, but it is not distributed evenly across the continent. The northeastern United States, excluding northern New England, and southern Ontario still comprise the single most significant region of manufacturing on the continent (Figure 6-1). The international boundary straddled by the region has little obvious impact on the region's shape. This territory of greater manufacturing concentration is loosely defined on three sides by the Ohio River Valley, Megalopolis, and the southern Great Lakes. The western margin of the region is less clear; it blends gradually with agriculture-dominant landscapes across southern Indiana, Illinois, and beyond. Manufacturing may be found west of the Mississippi River in many important clusters, but this activity's impact is definitely less concentrated than within the region defined as the continent's manufacturing core.

In spite of the region's moderate extent and the growth of manufacturing elsewhere, the *manufacturing core* continues to be of tremendous economic significance in North American geography. The region indicated in Figure 6-1 occupies about 3 percent of the total land area of the United States and Canada. However, this region contains factories that produce 80 percent of these countries' steel and nearly the same proportion of their motor vehicles and motor-vehicle parts. And, ranked in terms of 1985 sales volume, all 10 of the top 10 and 21 of the top 30 industrial corporations in North America had their headquarters located within the region. In addition, 11 of the United States' 24 largest metropolitan areas in 1986 and 20 of Canada's 52 cities over 50,000 population in 1981 were located in the region.

Because of its clear significance, the region frequently is described by geographers as the North American *core area.* A core is generally thought of as the center, as containing the heart, as comprising the essential characteristics of a thing or place. The idea of a core area is not easy to define precisely because it contains different elements of human activity. The political core of a country may not extend beyond the metropolitan region of the nation's capital city, or it may be the region from which the national government receives its major support. The economic core might contain many cities, extensive agricultural lands, and numerous industrial complexes all linked together by a dense network of transportation and communications. A cultural core may be the origin and continuing home area for a country's dominant people, or it may be defined

Industrial landscapes can have a stark beauty heightened by their contrast with the natural setting. Regardless of the season or the time of day, these places are busily meeting immediate goals in direct contradiction to the slow, evolutionary changes of nature. (Michael Hayman/Stock, Boston)

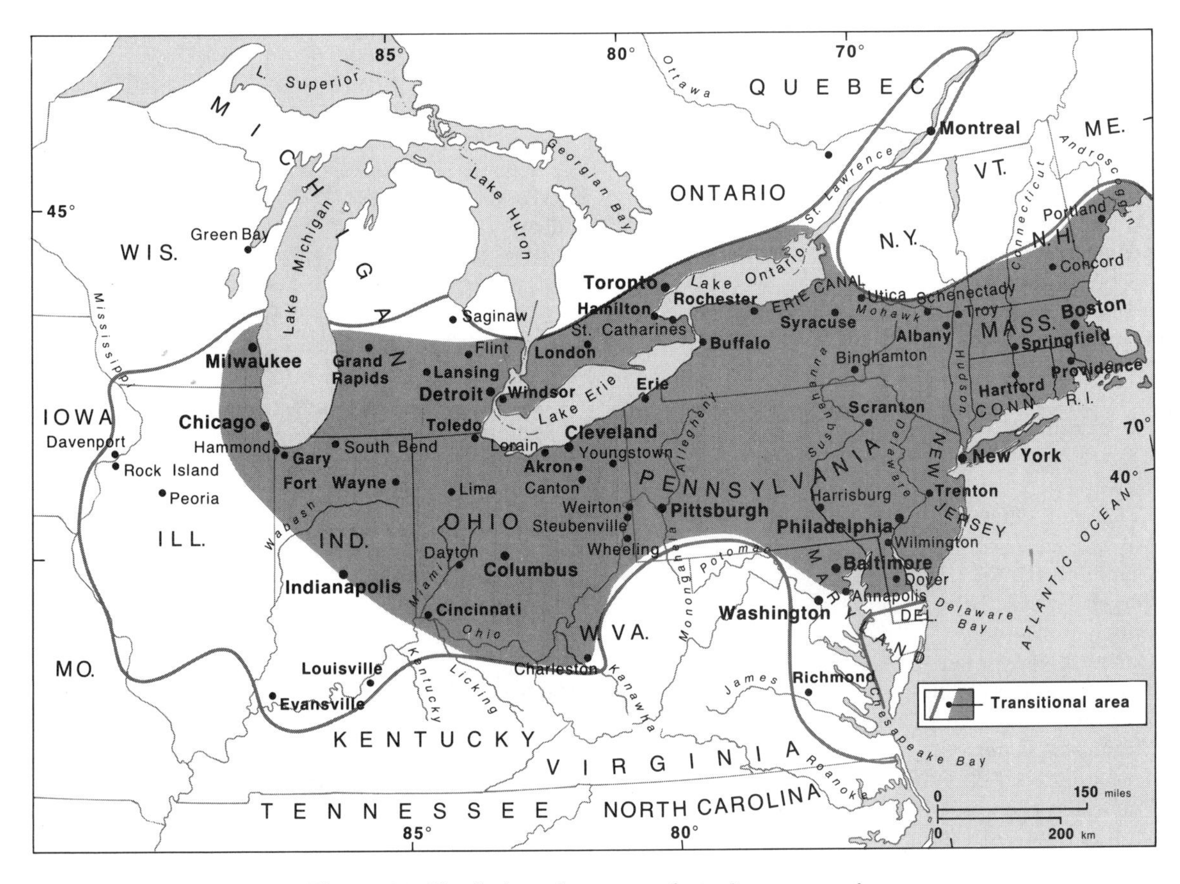

Figure 6-1 North American manufacturing core region.

only arbitrarily because important components of a nation's culture may originate in separate locations. The key difficulty lies in the identification of locations or areas that contain the bulk of the most essential components of each aspect of a country's human organization. By outlining the region of greatest overlap, the core area is defined.

This region of North America contains much of what would allow it to be considered the continent's core area. Most of both countries' industrial capacity is within the region. Most of the important ports, the main centers of communication, and the primary financial centers for both countries are within or very near the region. Both countries' political capitals, although outside the boundaries drawn here, are on the immediate margins of the region. The intensity of urbanization is much higher in the region than in any other territory of comparable size on the continent, and the region includes the two largest clusters of coalescing metropolitan areas: Megalopolis (Chapter 5) and the group of large urban regions between Milwaukee and Chicago on the west and Cleveland and Pittsburgh on the east. Much of the region contains land yielding great quantities of essential agricultural products. Manufacturing, however, is one of the two underlying features that make this territory distinctive and that have provided the economic support for much of what gives the core its character. The second theme, agriculture, is treated separately in Chapter 12. To be more accurate, then, the region described in this chapter can be

Major Metropolitan Area Populations in the Manufacturing Core, 1990

Metropolitan Area	Population	Metropolitan Area	Population
New York, N.Y.	8,546,846	Dayton-Springfield, Ohio	951,270
Chicago, Ill.	6,069,974	Providence–Pawtucket-Woonsocket, R.I.	916,270
Philadelphia, Pa.–N.J.	4,856,881	Albany–Schenectady–Troy, N.Y.	874,304
Detroit, Mich.	4,382,299	New Haven-Waterbury-Meriden, Conn.	804,219
Boston-Lawrence-Salem-Lowell-Brockton, Mass.	3,787,817	Scranton-Wilkes-Barre, Pa.	734,175
Toronto, Ont.	3,427,168[a]	Grand Rapids, Mich.	688,399
Nassau–Suffolk, N.Y.	2,609,212	Allentown–Bethlehem–Easton, Pa.–N.J.	686,688
Baltimore, Md.	2,382,172	Syracuse, N.Y.	659,864
Pittsburgh, Pa.	2,056,705	Akron, Ohio	657,575
Cleveland, Ohio	1,831,122	Toledo, Ohio	614,128
Newark, N.J.	1,824,321	Gary–Hammond, Ind.	604,526
Cincinnati, Ohio	1,452,645	Springfield, Mass.	602,878
Milwaukee, Wisc.	1,432,149	Harrisburg-Lebanon-Carlisle, Pa.	587,968
Columbus, Ohio	1,377,419	Wilmington, Del.–N.J.–Md.	578,587
Bergen-Passaic, N.J.	1,278,440	Hamilton, Ont.	557,029[a]
Indianapolis, Ind.	1,249,822	Jersey City, N.J.	553,099
Hartford-New Britain-Middletown-Bristol, Conn.	1,123,678	Youngstown–Warren, Ohio	492,619
Middlesex-Somerset-Hunterdon, N.J.	1,019,835	Lansing–East Lansing, Mich.	432,674
Rochester, N.Y.	1,002,410	Flint, Mich.	430,459
Monmouth-Ocean, NJ	986,327	Canton, Ohio	394,106
Buffalo, N.Y.	968,534		

Note: [a] Indicates 1986 population.

called the North American manufacturing core.

The manufacturing core area is given its highly complex but distinctive flavor by a set of landscape features. This is a region of many large cities and even more cities of medium size. It is a region that experienced growth of numerous industrial concentrations and the transport networks demanded by them. It is a region of smokestacks and factories, some of which may appear massively drab on the outside but can be vibrant and energetically alive inside. It is a region of great ethnic diversity, for the industrial cities of the manufacturing core grew just as waves of migrants left central, eastern, and southern Europe for new lives in the United States and Canada, and they have since been recipients of large numbers of black, Hispanic, and Asian migrants, as well.

A comprehensive understanding of the continental core is made even more difficult by its strongly dual character. Much of the territory occupied by the manufacturing core is also occupied by a major portion of the continent's *agricultural core.* In many respects, it was the vitality and productivity of the territory's farm population that created the resources and the demand for industrial production. Success in agriculture supported the region's early market centers, and it was the gradual mechanization of agriculture that demanded diversified manufacturing support. Mechanical reapers, winnowing machines, and cultivating implements by the tens of thousands were required during the later 1800s. Tractors, hay balers, pumps, other supporting equipment, and increasingly specialized farm machinery continued to be important local sources of industrial demand during the first half of the twentieth century. And transportation lines were improved and expanded to carry the tremendous volume of agricultural products produced on the region's farms. Thus, agriculture stimulated early urban and industrial growth, and successful manufacturing and rapidly growing urban centers continued to support intensification and improvement in agriculture.

Therefore, we encounter here a single portion of North America that must somehow be treated as two thematic regions without losing sight of the interdependence between the two aspects of the same territory. One theme, the urban and industrial nature of the region's manufacturing centers, is discussed in this chapter. The other theme, the rural and agricultural character of the territory's small towns and countryside, is presented later (Chapter 12), where it offers a contrast with another strong and agrarian culture, that of the Deep South (Chapter 10) and where it accompanies our discussion of the western reaches of the continent's agricultural heartland (the Great Plains and Prairies, Chapter 13).

The territories occupied by the two thematic regions of the continental core are not identical; the manufacturing core extends farther east to include Megalopolis, whereas the agricultural core reaches westward at least to the edge of the Great Plains. The regional label, the Midwest, is often applied loosely to the territory of overlap located in the United States, but a portion of the continent's core region is located in Canada. Furthermore, the landscapes that have evolved as reflections of each theme—the urban-industrial and the rural-agricultural—are too distinctive and too important to be presented without specific and separate consideration of each.

Keep in mind that manufacturing is carried on in many centers beyond the immediate boundaries used here to define the North American manufacturing core region. No major populated portion of the continent is without some light manufacturing. Even the more intense pattern of heavy industrial activity based on steel and metals fabrication typical of the core region itself declines only gradually outward from it, diminishing westward from Chicago across northern Illinois and from Cincinnati along the Ohio River, and pockets of heavy manufacturing can be found in almost every region. Should the manufacturing core be extended to include Peoria and Rock Island, Illinois, Davenport, Iowa, and even St. Louis, Missouri? Should the region reach downriver from Cincinnati to include Louisville, Kentucky, Evansville, Indiana, and the lower Tennessee River industries? A case could be made for, and just as easily against, such extensions. Any argument of this type, however, only serves to reinforce the concept of regional boundaries as transition zones (see Chapter 1). It also begs questions concerning the regional outline's permanence and, indeed, about the impact of changing patterns of economic activity on the permanence of any regional outline defined for the manufacturing core.

Turning to the manufacturing theme, we know where the core is, but why is it located there? What set of conditions or circumstances led to the development of so complex a mix of economic interrelationships on this portion of the continent? What is it about this region that encouraged the growth of heavy manufacturing industries and all of the related human activities that have come to dominate here? And what can be said about how pertinent these earlier conditions remain today as the national economies of both countries continue to evolve? General answers can be suggested by reviewing the features of North America that are essential to manufacturing: raw materials, transportation, and the population characteristics of labor and market. More specific answers can be approached by considering the interplay between the several manufacturing components and the variations within the manufacturing core.

MINERAL RESOURCES AND RELATIVE LOCATION

As discussed earlier (Chapter 2), the metallic mineral resources of the United States and Canada are found in three broad zones of metamorphic rock: the Appalachians, the western mountains, and the Canadian Shield. The mineral fuels, on the other hand, are found under much of the continent, especially beneath the sedimentary lowlands west of the Appalachians and between the Gulf of Mexico and the Arctic Ocean. The quality, quantity, and variety of minerals were extremely good. And in many cases, the deposits could be mined with no more than average difficulty. The United States and Canada, in other words, are large countries, and they have been well blessed with industrial resources. It is not surprising that manufacturing in the United States and Canada came to be significant.

Two additional geographic factors are critical in defining the size, character, and location of manufacturing clusters in North America—relative location and accessibility. Consider first the patterns of metallic mineral concentration in North America and their placement on the landmass: a long northerly arc covering the northeastern quarter of the continent (the Shield) and two linear areas, one extending northeast-southwest (the Appalachians) and one extending northwest-southeast (the Rockies). Next, consider these patterns with respect to each other. The eastern and western mineral zones form a broad V cut off at the base, and the Canadian Shield is almost fitted into the northern open end of the V. As a consequence, the broad interior plains are nearly enclosed by these zones of metallic minerals. Furthermore, much of these same interior plains are underlain by large deposits of high-quality mineral fuel, especially in the eastern section where the Shield and the Appalachians are only narrowly separated by the St. Lawrence lowland. In terms of the mineral requirements of heavy industry, then, the relatively small triangular region bounded by the Appalachians, the Canadian Shield, and the Mississippi River contains much of what is needed.

These three factors—resource quality, resource quantity, and close relative location—would have led to the development of a concentrated manufacturing core in this general region even without additional geographic advantages. But these other advantages do exist in the form of what might be called natural *accessibility resources.* Accessibility is a critical concept in human geography because it defines the ability to interact between locations and the relative ease and costs of movement. Accessibility can be created or improved through human effort. Roads can be built or widened. Railroads can be constructed and maintained. Airports, canals, and pipelines are all means that improve the accessibility of a place. Investment in transportation facilities, however, involves clear risk of financial loss unless the need for these facilities is overwhelmingly obvious. Navigable waterways, on the other hand, are natural routes of access that do not require much initial investment or contain the same possibility of financial loss. The presence of such waterways, therefore, is a very real resource during the early stages of economic growth when inexpensive movement facilities are needed.

The interior portion of the manufacturing core of North America possesses great accessibility resources. Connecting the mineral-rich Canadian Shield and the fuel-rich interior plains, the Great Lakes represent an internal waterway unlike any other in the world. The five Great Lakes are interconnected with only two significant changes of elevation. A small drop of about 6.7 meters (22 feet) between Lake Superior and Lakes Huron and Michigan was overcome by locks at Sault Sainte Marie, first opened in 1855. The much greater change of elevation between Lakes Erie and Ontario might have been a more serious barrier to water transport, but the Welland Canal (first opened in 1829) was built in Ontario to skirt Niagara Falls, and the Erie Canal was constructed (by 1825) in New York to permit

River barges on the Ohio are usually pushed by a single boat in groups two or three abreast and as many as six units long. This group is moving slowly past one of the coal-powered electric generating plants along the river. Note the variety of freight carried in this single barge train. (Charles E. Rotkin/PFI)

some freight to avoid Lake Ontario altogether. With these exceptions, the lakes offered well over 800 kilometers (500 miles) of inexpensive transportation to early European developers of North America. Later, in the nineteenth and early twentieth centuries, the same cheap transportation was of critical importance to those moving the Shield's iron ore to the coal in Illinois, Indiana, Ohio, West Virginia, or Pennsylvania. Much of the basis for the location of the industrial capacity that developed along the southern margins of the Great Lakes—Chicago, Gary, Detroit, Cleveland, Hamilton, Toronto, and many others—can be attributed to this natural accessibility resource. Although the Great Lakes are closed to traffic during the winter months, this has not eliminated their transportation importance.

Within the interior core, additional natural accessibility resources are also available. Flowing westward from deep within the coal-rich Appalachian region, the Ohio River crosses the interior plains for hundreds of miles, dividing the Eastern Interior coalfield before joining the Mississippi River. Dozens of tributaries supply the Ohio with its water and also provide further accessibility, either directly because they are navigable, too, or less directly because they offer easier routes of land movement through their valleys. Along the western margin of the core region, the Mississippi River and its tributaries provide access from the south and west. It is also a means of natural access between the industrial capacity of the core and the tremendous agricultural potential of the interior plains.

So unique is this combination of spatial and

mineral resources between the Appalachians and the Mississippi that the manufacturing core in the United States is often thought of as this interior territory alone. References to "the industrial Midwest" or more dramatically to "America's industrial heartland" may seek to fire the imagination, but they are geographically incomplete. The interior manufacturing core is a direct western extension of the large ports of Megalopolis. The Appalachians form a barrier to easy east-west transportation, except along the Hudson-Mohawk valley that connects New York City to the interior, but land connections were constructed to neutralize this barrier. A consequence is that the North American manufacturing core includes both the interior core and Megalopolis, the urban region through which the interior has its primary linkage to international commerce. The essence of Megalopolis is urbanism and exchange, but Megalopolis also

The valley of the Mohawk River in central New York is an excellent passage through the Appalachians. In addition to the river itself, a railway, two highways, and the limited access New York State Thruway provide direct surface transport connections between Megalopolis and the inland cities of the continent's manufacturing core region. (New York State Thruway Authority)

contains massive industrial development and is fundamentally a part of the continental manufacturing core region. Given the sequence of regional treatment used in this text, this overlapping of Megalopolis and a portion of the manufacturing core is a first example of how regions of North America are treated as thematically complex, not territorially separate regions.

Growth of the Region

The manufacturing core region is a development of the late nineteenth century. In fact, it was not until the 1870s that manufacturing industries began to appear in intensities that encourages extension of the tentative label of "manufacturing core" to the area west of the Appalachians. Manufacturing was not a major concern during the early years of European expansion and settlement. The initial paths of settlement penetration were along the natural lines of access, the waterways. Most families moved into this region via the Ohio River, along its many tributaries, and from the margins of the southern Great Lakes, although some came overland. By 1880, average population densities of 18 persons per square kilometer (46 per square mile) were reached almost everywhere in what was to become the manufacturing core region, but the overwhelmingly dominant economic activity was agriculture.

Prior to 1830, urban and industrial development of the region was limited almost entirely to the Atlantic Coast in the ports' immediate hinterlands. European settlement of the trans-Appalachian area was still limited to scattered subsistence agriculture and a few urban outposts. Between 1830 and the outbreak of the Civil War, population density in the interior increased, and agriculture intensified and began to produce a regular surplus, and demand grew for functionally efficient centers of exchange.

The foundations for growth of the region are reflected in the gradual shift of transportation concentration as railroad lines began to be spread across the interior plains (Figure 6-2). This was the period of rapid increase in water freight transport on the Ohio, on the Mississippi, and on the Great Lakes. The Erie Canal had been constructed earlier to provide a direct link between the interior and the rapidly growing port of New York. The other great ports of Megalopolis could not immediately match the support capabilities of the Erie Canal, but they could attempt to gain their portion of the productive interior by land transportation. Baltimore, with the Baltimore and Ohio Railroad, and Philadelphia, with the Pennsylvania Railroad, drove for the interior plains of the Midwest through gaps in the Appalachians. Not to be outdone, New York was also connected via the Mohawk Valley with the construction of the New York Central Railroad.[1]

Boston, north and east of the other ports, found its main lines of land access to the interior United States preempted by the other cities' connections and, instead, intensified the railnet in its more immediate New England hinterland much earlier than the other cities did. Canadian interest spurred construction and consolidation of rail lines between Montreal and Toronto and from there westward across Ontario to Michigan, as well as southeastward to Portland, Maine. The last decade of this period, 1850 to 1860, foreshadowed the tremendous intensification of interaction that was to follow the Civil War interruption, as an interwoven web of rail lines was flung across Ohio, Indiana, and Illinois.

After a decade of slowed growth, the 1870s produced several changes that allowed the core region's full range of advantages to be realized. The development of steel led to replacement of the original iron rails. This, in turn, allowed heavier railroad equipment, higher speeds, and longer hauls than had been possible before the 1870s. Similarly, the rail gauges and rolling stock

---

[1] Students who might puzzle over the railroad's name, New York *Central*, should note the location of the Mohawk Valley within the state.

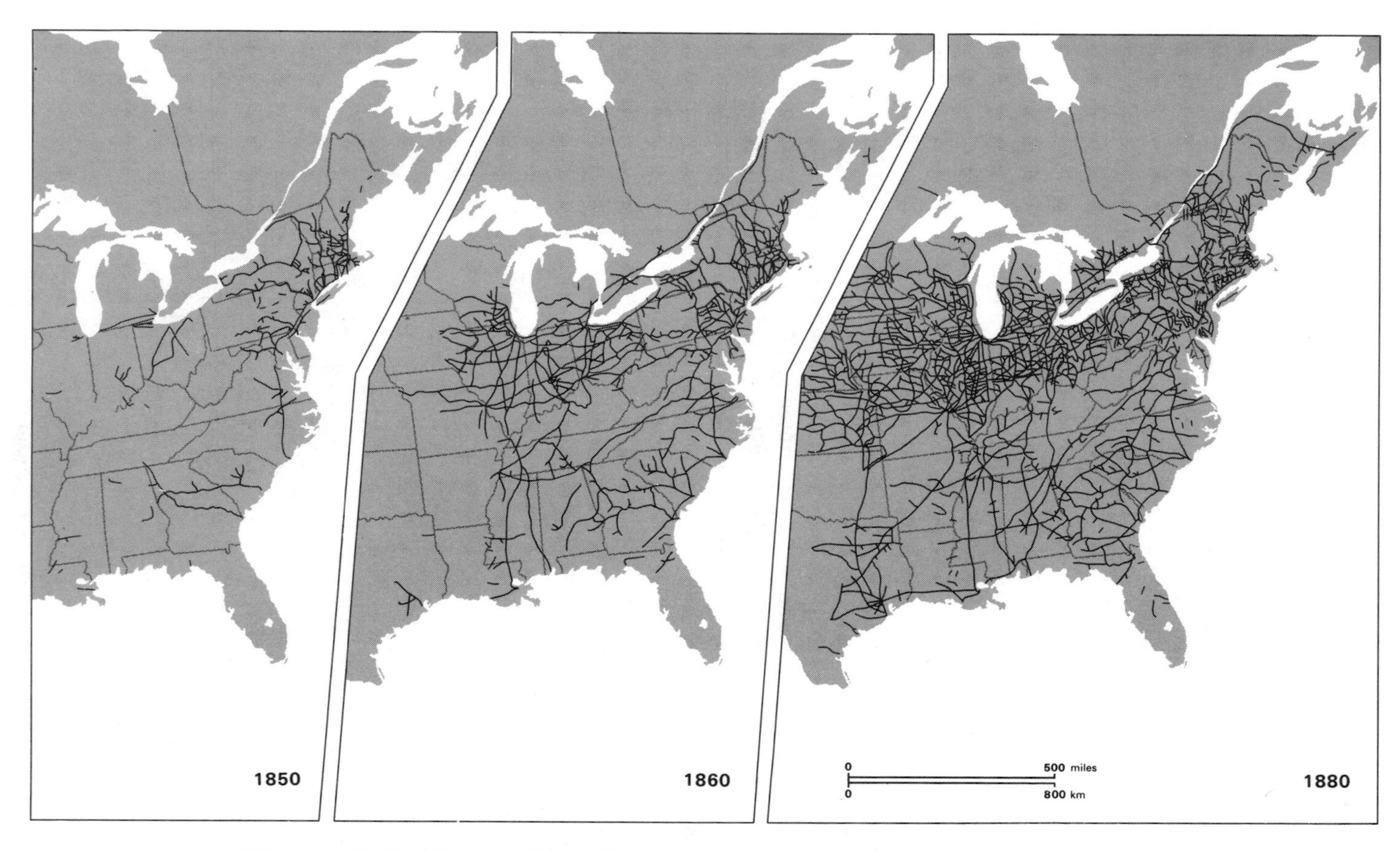

Figure 6-2 United States and Canadian railroads: 1850, 1860, and 1880. (With permission, Association of American Railroads.) The tremendous increase in developed land mobility across the northern half of the United States is in sharp contrast to the more limited rail construction in the South and in Canada.

were standardized to allow interline exchange and more efficient movement. The discovery of extensive iron ore deposits in the Lake Superior region, rapidly increasing demand for steel and the new ability to move the desired bituminous coal economically in large volume over longer distances and with more speed than earlier all combined to increase exploitation of the accessible Appalachian coalfields.

Cities in the coal region grew rapidly with the expansion of steel production. Pittsburgh is a prime example. Located well within the coal-producing region, it also enjoyed natural accessibility to the west by occupying a site at the head of the Ohio River and commanded good land connections to the east and west because it had been an early center of growth during the immediate postcolonial period. Other cities less closely tied to coal or steel production began to grow as coal and steam provided alternative sources of industrial energy to water power. Manufacturing was no longer tied to waterfront location, although if the waterway were navigable all of the nonpower advantages remained.

These changes and further intensification of agriculture are reflected in the rapid multiplication of rail lines by 1880 (Figure 6-2). With the spread of coal-powered electric generators during the 1880s, greater movement of coal to the ports along the southern margins of the Great

Looking eastward across downtown Pittsburgh, its strategic location at the confluence of the Allegheny River (to the north) and the Monongahela River (to the south) can be easily understood. The outline of Fort Pitt is visible in the park below the tall office buildings. Note also the conflict between the demands of water movement and land movement as demonstrated by the many highway and railroad bridges. (Courtesy Gulf Oil Corporation)

Lakes, and a continuation of improvements in transportation, a self-reinforcing cycle of growth was well underway before the end of the century, and the manufacturing core of North America was clearly outlined.

The technological changes that directly affected the manufacturing geography of the United States have been grouped by one geographer into a framework that is useful here. Professor John Borchert suggested that the pattern of metropolitan growth could be related to conditions that existed during four periods, or historical epochs as he called them.[2] Five categories of city population size were established, and the relative growth or decline of a city during each "epoch" was indicated by its change from one size category to another.

The earliest period, 1790–1830, was call the *Sail-Wagon Epoch* by Professor Borchert. During this 40-year period, almost all cities and towns were associated with water transportation (Figure 6-3). The Atlantic ports and towns that had their beginnings along some of the coastal rivers were the major urban centers. The greatest inland urban growth during this period occurred along the main inland waterways—the Mohawk River, the Great Lakes, and the Ohio River. Land transportation, primarily by wagon, was still too rudimentary to support equivalent city growth.

The second period, 1830–1870, was triggered by development of the railway, a radical innovation in land movement. The *Iron Horse Epoch* at first stimulated further growth in the already established port locations, both coastal and on the inland waterways. The new railway networks were constructed to focus on the port cities. Aside from additional growth of the larger port cities in what was soon to become Megalopolis, the greatest growth occurred in such cities as Pittsburgh, Cincinnati, and Louisville (all on the Ohio River); Buffalo, Erie, Cleveland, Detroit, Chicago, and Milwaukee (all on the lower Great Lakes); and St. Louis, Memphis, and New Orleans (all on the Mississippi River). Many other cities grew and a few declined in population, but the key to growth during this period was expansion of the hinterlands of already existing port cities by construction of railroad lines plus expansion behind the frontier of settlement. Because of the continuing importance of the Ohio, the Mississippi, and the Great Lakes, most of the rapid urban growth during this period was within what came to be the manufacturing core region.

As was discussed earlier, the *Steel-Rail Epoch*, 1870–1920, was stimulated by the development of steel, the replacement of iron rails with stronger and heavier steel rails, increased demand for bituminous coal, and the spread of electric-power generation. The greatest growth in national urban areas occurred in cities only peripheral to the manufacturing core, but there were several notable exceptions. The exceptions were the numerous smaller cities near the coalfields, near the Great Lakes, or on one of the major rail connections between larger cities. These were the cities able to establish themselves because the interconnecting rail network crisscrossed the region so densely between the Ohio and the Great Lakes. Akron, Canton, and Youngstown, Ohio, are clear examples, because they are located between the coal-and-steel city of Pittsburgh and the iron-ore port and steel city of Cleveland. The greatest decline during this period occurred among those cities most tightly tied to the navigable waterways. Within the broader core region, Erie Canal cities (Syracuse, Utica–Rome, and Albany–Schenectady–Troy), Louisville on the Ohio River, and St. Louis on the Mississippi were the most important losers west of Megalopolis.

Perhaps even more important for the formation of the manufacturing core region, the cost importance of two factors in the manufacturing process changed during this period. Before 1870, the greatest costs borne by manufacturers were shipment costs and power costs, but by the end

---

[2] John R. Borchert, "American Metropolitan Evolution," *Geographical Review*, Vol. 57, no. 3, (July 1967), pp. 301–332.

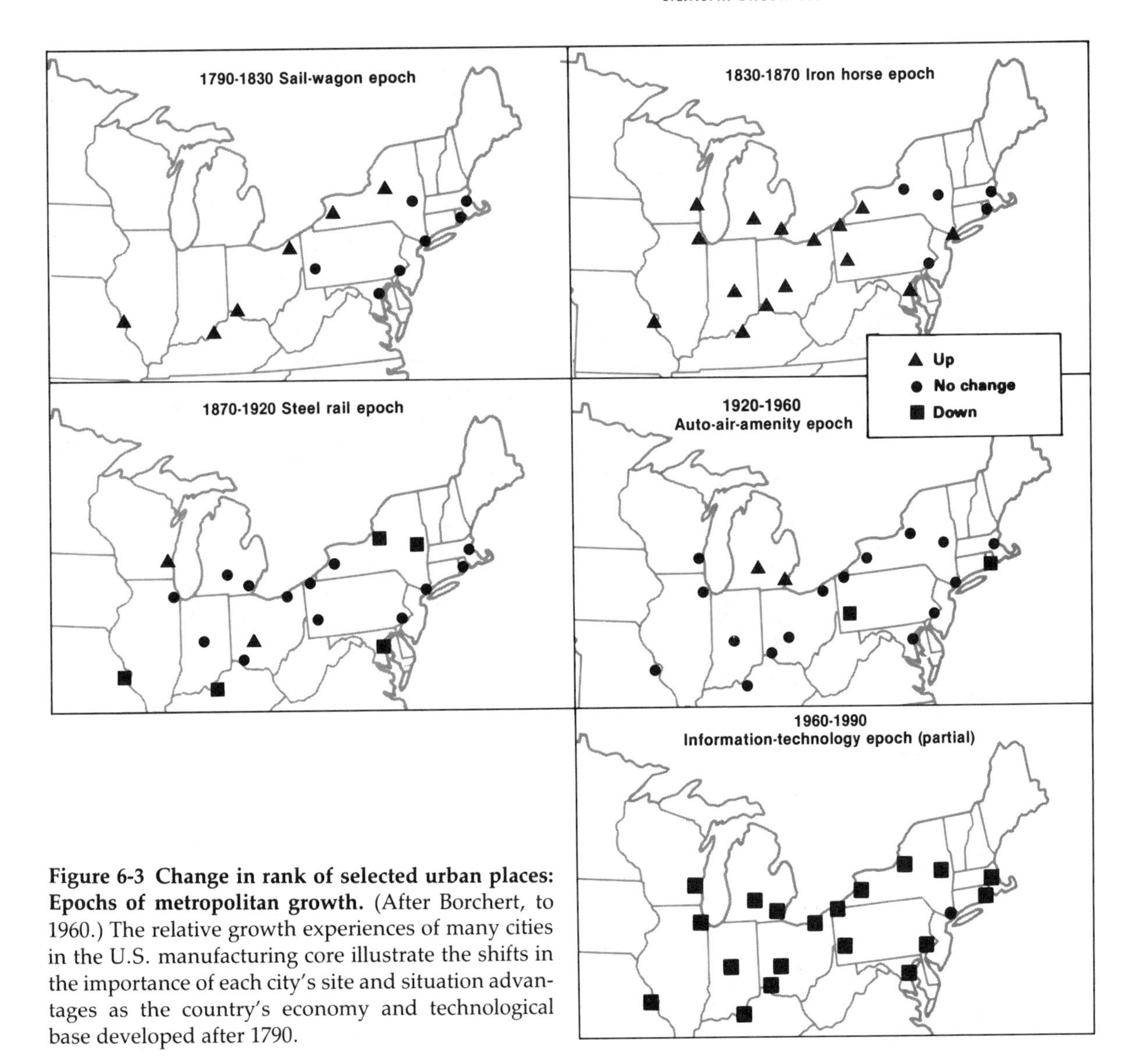

Figure 6-3 Change in rank of selected urban places: Epochs of metropolitan growth. (After Borchert, to 1960.) The relative growth experiences of many cities in the U.S. manufacturing core illustrate the shifts in the importance of each city's site and situation advantages as the country's economy and technological base developed after 1790.

of the epoch both began to be less dominant in location decisions. The costs of shipping raw materials was an important share of total costs then because a portion of the raw material would be used up or discarded during the process of manufacture. Therefore, at the beginning of the period, industry tried to minimize these costs by locating near the raw materials or where they might be shipped cheaply. That is, industry tended to locate in the port cities on navigable waterways. This locational tendency was reinforced at some river locations by the continuing dominance of water power as an energy source.

Toward the end of the century, however, cheap and efficient railroad transportation had removed some of the advantages of waterway

Large steel mills dot the southern margins of the Great Lakes where rail-carried northbound coal meets southbound iron ore transported by Great Lakes ore carriers. This Inland Steel mill on the shore of Lake Michigan near Chicago is a good example. (G. R. Roberts)

location for industry. The spread of electric power removed additional dependence on port cities. Although advantages remained for some port cities and for some industries, the general effect was to make it more worthwhile for many manufacturers to locate near their markets. In the Canadian section of the manufacturing core, southern Ontario developed more rapidly than did southern Quebec. Ontario shared the important Great Lakes waterway and it also lay much closer to the growing markets of the interior core in the United States. Toronto, Hamilton, and Windsor all have the advantages of both accessibility and market proximity.

Professor Borchert indentifies a fourth epoch, the period 1920–1960, as the *Auto-Air-Amenity Epoch*. The main effects of transport innovations such as the automobile and the airplane were to increase individual mobility and to minimize the impact of shipment costs in the production process. Migration within the United States was directed at those places perceived as having amenities (California, Florida, Arizona). The market and labor orientation of industry begun in the previous epoch was strengthened further as shipment and communication costs declined. Industry was drawn to areas of greatest population growth; these were primarily the amenity areas outside the traditional manufacturing core.

If we use hindsight to look at the manufacturing core region during Borchert's fourth epoch, stable growth until the 1960s with clear prospects of future decline might be the best description. Large markets and labor pools existed; manufacturing and the linkages between industries were well established. The main exceptions to the general stability were slower than average growth in the steel city of Pittsburgh and more rapid than average growth until the 1960s in the automobile production centers of Detroit, Flint, and Lansing, Michigan.

We argue that the two countries entered yet another period after 1960, one that will extend to the next century and which might be called the *Information Technology Epoch.* As the economies of the United States and Canada become more dependent on the production and exchange of information, the means of processing and transmitting this information will encourage the growth of industries which do not need cheap bulk transportation or even large population clusters. This suggests that those factors that supported growth in the large manufacturing core cities during the first two-thirds of the twentieth century will no longer provide those cities with special development advantages during future decades, even though these cities' skilled labor force, large markets, and established air transportation patterns will make some of them strong competitors for growth with the noncore cities.

THE ECONOMIC CHARACTER OF CITIES IN THE REGION

In discussing the patterns of growth within the North American manufacturing core, we have dealt almost entirely with metropolitan centers. In fact, until recent decades, little manufacturing in this part of North America was to be found located beyond the immediate vicinity of urban areas. The striking parallel between growth patterns of urbanization and manufacturing industrialization in much of the United States and Canada confirms the view of cities' growth as dependent on the economic advantages of locating these activities near each other. However, what stimulates growth in one city may be very different from forces stimulating growth in another. Places that accumulate urban intensities of people and economic activities can have very different character. As a result, the visible economic landscape of the cities of the North American core also varies.

The economic character of North America's manufacturing core cities falls into two broad regional categories: (1) that associated with eastern cities of the core, that is, the Atlantic coastal cities and the narrow region east of the Appalachian Highlands, and (2) that associated with the cities of the interior core between the Ohio River and the Great Lakes. Aspects of the former group of metropolitan areas have been described, for this is Megalopolis and its periphery.

Eastern Cities of the Core

With Boston, New York, Philadelphia, and Baltimore based early and firmly on commerce and the financial exchanges it stimulated, these ports and their satellites began to accumulate population long before manufacturing became dominant in the national U.S. economy. Although manufacturing was drawn to the eastern coast by the promise of matchless local markets, tremendous labor supplies, and easy access to water transportation, the economies of most of Megalopolis's cities maintained a distinctly professional character.

New England was exceptional within the eastern core by developing manufacturing at the same time that its ports were growing. So early was New England's industrial development that this region can be called North America's manufacturing *hearth.* More densely populated as a region than other portions of British colonial America, agricultural opportunities were relatively poor and nonagricultural labor became available earlier here than elsewhere. Shipbuilding industries thrived along the coast and generated countless subsidiary manufacturing operations needed to supply such a complex industrial undertaking. When factory industry began to grow in importance elsewhere in North America, New England had several advantages that kept manufacturing significant in this region. Experience and continued immigration from Europe were important, but the ready availability of power in New England's small but abundant rivers was primary. Although many of the textile industries that laid the foundations of the region's manufacturing development eventually shifted their sites to the South Atlantic states (Chapter 10), the highly skilled residual population drew in enough replacement industry to maintain a slow regional population growth and a high proportion of that population in manufacturing (Table 6-1).

Boston, as the regional capital of New England, characterizes many of the changes in this portion of the continental core. Boston's apparel and leather industries as well as shipbuilding in nearby Connecticut are remnants of an earlier period, but most growth since World War II has occurred in electronic components and machinery (very loosely called "high-tech" industry). Many of these newer industries located initially along Route 128, a limited access circumferential expressway 16 to 19 kilometers (10 to 12 miles) from the city center. There they took advantage of the high accessibility provided by such a location. Much recent expansion has occurred along more distant circumferentials. The growth of new industry on the periphery of the established city also emphasizes the decline of Boston's port. The harbor and the facilities found there remain excellent, but industry in New England (and Boston) now ships most of its products by land, either to markets in the rest of the United States or south to New York for export through Megalopolis's primary port.

Table 6-1 Percentage of 1985 Labor Force[a] in Manufacturing by State in Core Region

Region and State	Percentage by Region	Percentage by State
New England	24.2	
Maine		23.1
New Hampshire		26.4
Vermont		22.3
Massachusetts		22.6
Rhode Island		28.2
Connecticut		26.2
Middle Atlantic	18.8	
New York		16.7
New Jersey		21.0
Pennsylvania		23.0
Maryland		11.5
Delaware		24.6
Interior	25.1	
Ohio		25.6
Michigan		28.1
Indiana		28.0
Illinois		20.6
Wisconsin		26.0

Source: Calculated from *Statistical Abstract of the United States* (Washington, D.C.: U.S. Government Printing Office, 1987).
Note: [a] U.S. national average: 19.8.

New York's primacy among North American harbors has been discussed. As might be expected, manufacturing industries found proximity to this node of international commerce and the population cluster around it to be very advantageous. So strong was this pull that New York's industrial mix became extremely diversified. Many industries were located on Manhattan until after the beginning of the twentieth century. The increasing demand for space by the even more space-intensive office businesses, shown in the rapid construction of skyscrapers, gradually pushed the heavier industry to the very margins of lower Manhattan or beyond the island's confines into the New Jersey tidal marshes across the Hudson River. More recently (see Chapter 5), a great many manufacturing jobs have left the New York region altogether and have been replaced only barely by nonmanufacturing employment.

The New York metropolitan economy has been dominated for some time by office industries. These are the headquarters for activities of dozens of companies and corporations, the banking and insurance cluster, the publishing houses, and all the other service and control centers that require a worldwide information network and the facilities to transmit their responses rapidly. New York is industrial, by virtue of the large number of people employed in office industries; even with some recent losses, it probably remains the economic decision-making capital of the United States.

Philadelphia and Baltimore, so different in industrial inheritance and urban character, have shown indications during recent years that they may be becoming more alike. Philadelphia's manufacturing base is almost as diversified as New York's, although there is a greater emphasis on food-processing industries (from the truck farms of southern New Jersey, the lower Delaware River Valley and southeastern Pennsylvania, and even Eastern Shore Maryland) and on shipbuilding and ship repair than in New York. The growth of Philadelphia's industrial base suffered somewhat from the presence of New York's better harbor and superior access to the interior only 120 kilometers (75 miles) to the north, but Philadelphia's better access to the coal and steel regions of western Pennsylvania, its respectable port facilities, and its heritage as an early political and cultural center in the United States have maintained the Philadelphia metropolitan region's growth within Megalopolis. Baltimore, on the other hand, has always been on the periphery of the manufacturing core region. Like Philadelphia, its port possessed good rail connections with the coal and steel regions of the interior, and Baltimore's industrial mix reflects this. The manufacture of transport machinery is also important in Baltimore.

Two additional industrial sectors are well represented in Philadelphia and Baltimore. Metals fabrication and chemical industries are significant in both metropolitan economies. Copper refineries and steelworks at Sparrows Point east of Baltimore, similar steelworks near Trenton north of Philadelphia, and the chemical industries in Philadelphia and around Wilmington, Delaware, between the two larger cities, emphasize the coastal connections of these regions with the heavy industrial interior. It is because of this situational factor of location that the two are likely to become even more similar in the future. Increasing national dependence on imports of iron ore from Labrador, Venezuela, and Africa and continuing importation of petroleum from Venezuela, Nigeria, and the Persian Gulf states may increase the importance of this region's coastal access to the inland demand centers.

Cities of the Interior Core

The major cities of the other, larger portion of North America's manufacturing core region, the industrial Midwest and southern Ontario, have derived their primary character from their location relative to the rich mineral and agricultural resources of the continent's interior. Almost all of the large cities in the western portion of the manufacturing region are located along the Ohio River (or one of its tributaries) or along the shores of one of the Great Lakes (Figure 6-1). These comprise two of the three zones of manufacturing in the interior core. The third zone of generally smaller centers is that between the Great Lakes and the Ohio River Valley.

Most important in the development of urban centers in the interior portion of the manufacturing core has been the movement of metallic mineral ores from the margins of the Canadian Shield to the coalfields of western Pennsylvania and West Virginia, and the smaller movement of coal in the reverse direction (Figure 6-4). Iron ore is mined at the Mesabi range of northern Minnesota, at the Steep Rock deposit in western Ontario, and at the Gogebic, Marquette, and Menominee ranges in northern Michigan and Wisconsin. Mesabi ore is now processed into pellets at the deposit site to increase the iron content of the material shipped, but for decades unprocessed ore was carried to the southern shores of Lakes Michigan and Erie in large ships designed specially for Great Lakes travel. Pellets and ore are carried to the southern shore of Lake Michigan, to Hammond and Gary, Indiana,

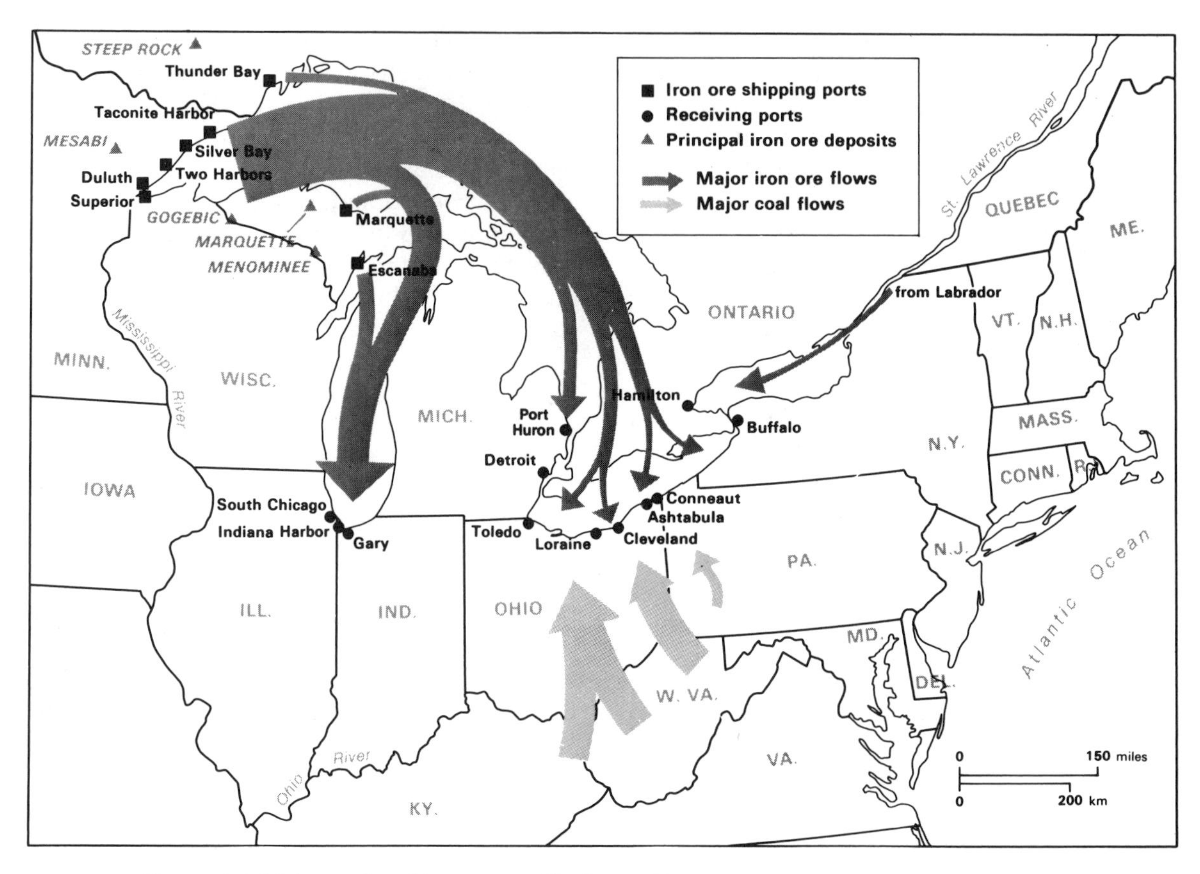

Figure 6-4 Major coal and iron ore traffic flows. Shipments of ore across the Great Lakes and both coal and iron ore between the Ohio River and the southern shores of the lakes have been fundamental to the development of urban industrial concentrations in the interior section of North America's manufacturing core region.

where these shipments are met by coal transported north by rail from the large Illinois coalfields. Most of the ore, however, is shipped to Lake Erie ports. From there, it is either carried south, primarily to the steel cities of the Ohio River, or converted to steel in the lakeside cities using coal carried north on the return rail trip from the Appalachian fields. Ore is also shipped beyond Lake Erie to the Canadian port of Hamilton on Lake Ontario supplementing iron ore brought up the St. Lawrence from Labrador deposits.

Of the cities of the interior core, Pittsburgh is the one whose name became synonymous with steel. Located at the junction of the Allegheny and Monongahela Rivers where they join to form the Ohio River, Pittsburgh was in an excellent position to take advantage of access to both raw materials and downriver markets. The Allegheny and the Monongahela drain the coal-rich margins of the Appalachians, and the Ohio flows along the southern margins of the agricultural core and into the Mississippi. The city's development as a steel-producing center was aided by the activities of Andrew Carnegie, the power behind what eventually became the United States Steel Corporation. As Pittsburgh grew, industries dependent on steel crowded onto the narrow river bottoms to take advantage of proximity to low-cost water transportation. Metal-fabricating industries, machine parts, and other industrial consumers of large quantities of steel located their plants in and around Pittsburgh. Nearby cities also benefited from the powerful pull of steel at Pittsburgh. Youngstown, Canton, and Steubenville in Ohio, Wheeling and Weirton in West Virginia, and New Castle and Johnstown in Pennsylvania shared to some degree in the industrial growth of this region and obtained steel and steel products industries (Figure 6-5).

Urban-industrial growth diu .not occur solely at the source region of coal. The iron ore shipped across the lake system had to be transferred to railroads at points along the Lake Erie shore for final movement to the Pittsburgh region, and railroad cars returning to the lake from the Pittsburgh region carried coal to the lake's southern shore. Metal products industries and other industry wishing to take advantage of the transportation focus along the lake shore and the *break-in-bulk* of materials transfer chose to locate at each of the port cities where iron ore was unloaded.

Cleveland was the largest of these Lake Erie port cities. Cleveland's initial growth was stimulated by a canal connecting the narrow and winding Cuyahoga River with a tributary of the Ohio. Although the city quickly outgrew this small initial advantage, it was enough to give Cleveland a head start on its nearby urban competitors. The diverse industrial base that resulted took advantage of the accessibility offered by the lakes and by the major east-west railroads in the United States connecting New York with Chicago and the agricultural core to the west. Cleveland's growth also spilled over into adjacent ports, such as Lorain, Ashtabula, and Conneaut, Ohio, and perhaps as far east as Erie, Pennsylvania, and as far west as Toledo, Ohio, as well as to complementary growth centers inland, such as the rubber-producing city of Akron.

Buffalo is located at the eastern end of Lake Erie, and this location has had three major effects on its industrial character. First, Buffalo was the premier flour-milling center on the continent until after World War II. While there has been a decline in the volume milled in Buffalo, wheat from the Plains states is still brought to the western Great Lakes and then carried in bulk to Buffalo for refining before sale in the large eastern markets in Megalopolis. In addition, Buffalo's locational advantages as a transfer site have been weakened further with the opening of the St. Lawrence Seaway. Second, the same factors that generated steel and metals manufacturing elsewhere along the lakeshore helped ensure that a significant portion of Buffalo's manufacturing would be connected to this type of industry. Third, and a more recent advantage, was the harnessing of Niagara Falls for hydroelectric power. This drew chemical and aluminum industries, among others, to the Buffalo region because of these industries' high electric power requirements.

Of the two major Ontario port cities in this section of the manufacturing core, it is Hamilton that shows the strong mix of steel-making, metals-fabricating, and automobile industries typical of the eastern Great Lakes. Much larger than Hamilton is Toronto, with a more diverse industrial base. The reasons for Toronto's size, a city having little obvious location advantage over Hamilton, are complex (see Chapter 7), and the result is a full mix of *light industry* rather than specialization in one or two heavy industrial sectors.

The cities on the narrow water passage between Lakes Huron and Erie, Detroit and Windsor, grew initially because this site controlled traffic between these lakes and was a natural crossing point between Ontario and southern Michigan. These apparent locational advantages were not significant during the early period of manufacturing in North America. Detroit did not grow rapidly until early in the twentieth century because it is located more than 80 kilometers (50 miles) north of the primary New York–Chicago rail connection. The more direct, but

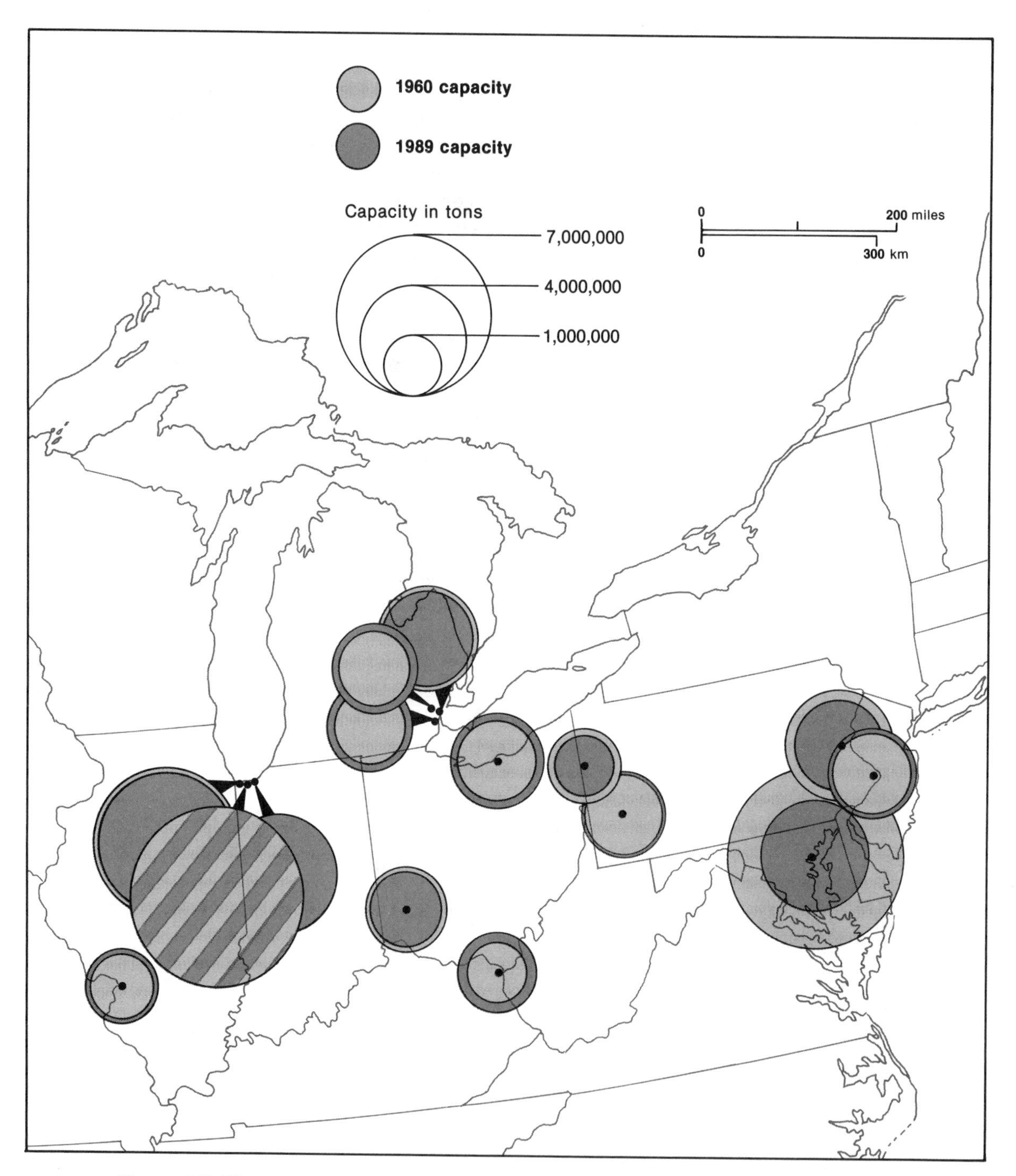

Figure 6-5 Changes in raw steelmaking capacity, 1960 and 1989. With several notable exceptions, most U.S. steelmaking capacity remains in the interior portion of the manufacturing core region. Furthermore, capacity at most plants has not changed greatly during the three decades shown.

Manufacturing core cities along the Great Lakes contain many industrial clusters, but The Flats along the Cuyahoga River in Cleveland are one of the most intense concentrations of heavy industry in the region. (Fred McConnaughey/Photo Researchers)

politically more difficult, rail route across southern Ontario via Buffalo helped stimulate some growth, but it was not until the rise of the industry that fostered the railroads' chief land transport competitor, motor vehicles, that Detroit developed the urban character for which it is best known.

The motor vehicle industry is very complex in the sense that an extremely large number of components are required to make up a single automobile or a single truck or bus. As the most successful automobile manufacturers concentrated in Detroit and nearby cities and the demand for automobiles skyrocketed, a wide variety of suppliers were drawn to southeastern Michigan. Detroit became the fifth largest city in the United States, a rank it held until the 1970s. Although in some ways Detroit was as specialized in its industrial mix as Pittsburgh and as susceptible to similar fluctuations and trends in demand and competition, the range of complementary industries was greater. This lent Detroit a somewhat greater buffer against decline in employment than Pittsburgh has had, even though Pittsburgh's economy has begun to diversify. For example, while metropolitan Detroit's nonfarm payroll employment increased by 10.6 percent compared to the national increase of 12.9 percent between 1980 and 1986, metropolitan Pittsburgh's equivalent employment decreased by 8 percent in this same period.

The smaller of the two remaining metropolitan centers located along the southern Great Lakes margin is Milwaukee. Milwaukee's industrial character is typical of Great Lakes port cities but with more agriculture-related industry, as well. In addition to the usual industrial mix of heavy industry and motor-vehicle manufacturing, Milwaukee is one of the leading centers of the continent's brewing industry, a result of the large number of German immigrants who settled in Wisconsin during the late nineteenth century. A significant food processing industry is also present in Milwaukee, as it is a major focus for the middle Dairy Belt of the state.

Chicago is easily the dominant city of the interior manufacturing core. So important did this city become that for many years it was called the

"Second City," recognizing Chicago's millions (2,725,979 in 1990) as second only to those of New York in number. Although Los Angeles grew rapidly enough to pass Chicago in population at the beginning of the 1980s (3,420,235 in the city in 1990 and with a much larger metropolitan area than Chicago), the informal "capital" of the Midwest persists as the strongest urban focus of the United States' interior.

Chicago had to overcome a series of early site disadvantages in order to grow to its present size and economic importance. The city occupied an undesirable swampy lake margin. Although ideal for mosquitoes and similar pests, it was viewed with little favor by early human inhabitants. Chicago's drinking water also had a distinctive flavor that was often unappreciated. The Chicago River was too small before improvements were made to provide access to the interior agricultural region. And, finally, after the city began to establish itself as a regional center, it was nearly consumed in the great 1871 Chicago fire.

Given these site disadvantages, it is testimony to the array of situational advantages that the city exists. Located along the southwestern shore of Lake Michigan, Chicago occupies the optimum location for people and goods transfer between lake transportation and the rich agricultural region to the west and southwest. The Illinois and Michigan Canal, located in part through the heart of the city, was completed in 1848 to link the Great Lakes and the Mississippi River system, the two greatest waterways of the continent. This focus of water transportation was supplemented four years later as Chicago was connected to New York by rail and became the primary regional rail focus in the Midwest.

Chicago drew on the support provided it as a focus of transportation and grew rapidly. The city absorbed many thousands of European immigrants throughout the later nineteenth century and spawned a radial network of rail lines into Illinois, Wisconsin, and the agricultural states beyond. Meatpacking developed around the city's large stockyards. Other industries such as furniture and clothing manufacturers located there to take advantage of the growing local market and good access to markets farther west. After the turn of the century, the steel industry was introduced to the Chicago region, south of the city itself but still along the lakeshore at East Chicago, Gary and Hammond, Indiana, and easily accessible to the city's unparalleled railroad network. Already a city of 1 million by 1890, Chicago doubled in size before 1910 and exceeded 3 million by the mid-1920s. The self-reinforcing cycle of population increase (labor and local markets), economic growth, more population growth, and so on clearly was operating before the turn of the century. The volume of Chicago's manufacturing activities (more than 540,000 employed in the city in 1987) is matched only by the immense diversity of products manufactured. In this volume and diversity, Chicago is at least partly an effective regional counterbalance to the set of intense economic nodes in Megalopolis. It is also appropriate that Chicago's industrial base was formed on both the typical Great Lakes port pattern of steel and metal products manufacturing and the agricultural orientation provided by the lands further west. Although changes in meat production techniques and shipping have closed the city's stockyards, the basis for Chicago's earlier vitality and its situational consequences were caught effectively in Carl Sandburg's description of the city as "Hog Butcher, Tool Maker, Stacker of Wheat, Player with Railroads and Freight Handler to the Nation."

The remaining manufacturing clusters and major urban centers within the continent's manufacturing core region fall into one of the other two general zones. One is the Ohio River complex that includes Pittsburgh and its immediate neighbors, the industrialized Kanawha River Valley, the older port city of Cincinnati, and many smaller centers in between. Almost all of the Ohio River clusters are related to the accessibility the river provides in moving coal and other Appalachian raw materials at low cost or in the return flow of materials from the Mississippi and

This small portion of Chicago's downtown illustrates the intense demand for space and the accommodation made for water traffic on the Chicago River and its access to Lake Michigan at the top of the photo. (Courtesy Kee Chang/Chicago Association of Commerce and Industry.)

beyond. Small, but representative, cities such as Ironton, Ohio, and Ashland, Kentucky, are sprinkled along the eastern half of the Ohio River, most with small steel plants or metals-fabricating factories. When the plants close in these small one-industry towns, whether temporarily or permanently, the effects can be devastating and require major local efforts to attract alternative employers. The Kanawha Valley industries in such places as Nitro and Belle, West Virginia, on the other hand, have used coal as a raw material as well as a fuel and specialized in chemicals, synthetic textiles and artificial rubber, plastics, and glass. Cincinnati also has gone through a change in industrial mix based on a shift in transport emphasis from a port city for agricultural product transfer to a more balanced manufacturing city, but one that bears the typical Ohio River stamp of metals industries.

Defining the other zone of manufacturing in the region are the diverse and scattered centers located between the Great Lakes and the Ohio River and between the Appalachian coalfields and the gradually dominant agricultural economy toward the Mississippi River. These cities—Indianapolis, Fort Wayne, and South Bend, Indiana; Dayton, Lima, and Columbus, Ohio; Battle Creek, Jackson, and Flint, Michigan; and many, many others—contain manufacturing activities that are partly a spillover from the larger centers surrounding the core's interior, partly a response to the intensely successful agriculture of the region (see Chapter 12), and partly a consequence of the network of interaction set up across the land area between the larger centers. Cities such as Fort Wayne, Lima, and Flint, with economies more strongly tied to heavy industries in the region have suffered along with these industries, but other cities, such as Indianapolis and Columbus, have continued to develop rapidly as service and market-oriented industries have grown.

THE DYNAMICS OF NORTH AMERICAN MANUFACTURING

Long-term trends in the U.S. and Canadian economies have been consistent with general patterns elsewhere. Agriculture's proportion of the total national economic effort has declined gradually during the last century, while manufacturing maintained its proportion until the last several decades. The drop in labor force engaged in agriculture from nearly 50 percent in 1880 to less than 3 percent in 1980 has been matched by an increase in what are called service, or *tertiary* activities. The tertiary sectors of both economies, comprised of activities that could loosely be described as serving the population (retail and wholesale, transport, recreation, finance, and many others), have grown as income became more generally distributed among the countries' inhabitants. As the *primary,* or extractive, sector (agriculture, mining, lumbering) and the *secondary,* or manufacturing, sector (steel, chemicals, consumer products, and so forth) became more efficient, fewer people were required to produce the same amount of goods. More of the labor force could be engaged in servicing people in the primary and secondary sectors (and each other).[3]

While the balance among the several economic sectors gradually shifted in both North American countries, the geography of these sectors changed as well. The application of new technology to agriculture permitted increases in total production even as the farm population declined. The tremendous growth in manufacturing in the continental core region created a demand for labor that attracted migrants from Europe and from rural America, especially the South. The sprawling industrial agglomerations of the region grew to city size and filled with multiethnic populations heavily dependent on the secondary sector. During the last several decades, forces encouraging industrial relocation have shifted growth to centers outside the core in spite of other forces resisting such a change.

---

[3] Some geographers have argued that there are what can be called *quaternary* and *quinary* sectors of the economy, as well. Under this approach, activities in the quaternary sector are those involved in information creation and transfer, while control functions make up the quinary sector. Examples of the former are education, portions of the communications industry, and research and development components of government and private industry. Quinary activities can be represented by decision makers, such as those in business headquarters and government policy forums.

More recently, others have argued that this sectoral analysis of national economies is useful only in a general, mostly conceptual manner. Activities do not fall neatly into one sector or another; most jobs involve more than one "sector" in their day-to-day activities.

The stability of a manufacturer's location is primarily an extension of the factors that led the firm to establish a plant at its specific site in the first place, although some factors are new. Where are the firm's raw materials or its suppliers of component parts? Where are the most important markets for the item produced? How good (that is, how reliable and how expensive) are the transportation facilities upon which the firm depends? Where is the labor supply? How important are local and regional amenities for maintenance of satisfactory life-styles by the firm's employees? How important are each of these (and similar locational factors) in relation to each other? Since the patterns of these factors are subject to changes beyond the immediate control of any single firm, the set of conditions that made a manufacturing site desirable at one time could make that same site relatively undesirable some years later. When a firm is unable to enjoy the best location for its activity for long, it must pay higher costs due to shifts in the factors affecting its production location. The costs incurred because of these shifts have been called *inertia costs of location*.[4]

The costs of locational inertia, then, are the costs each firm must bear if it is no longer situated in the optimum location. They are called inertia costs because it is the locational stability of a plant that creates the costs. The stability is necessary because a fixed investment in some location is required. The inefficiencies of a location may be slight initially, but changes in technology, population distribution, the locations of suppliers or competitors, and many other factors could raise this inefficiency considerably. If the costs of remaining in a location despite its locational inefficiency (i.e., the inertia costs) are small, the firm will not move. If, however, these costs grow until they exceed the costs of abandoning the existing facilities, moving to a new site, and constructing a new production center, then the firm is likely to move, especially if the firm is under competitive pressure.

Consider the implications of this for individual firms within the North American manufacturing core and for the shape of the core itself. Manufacturers engaged in production that is primarily supplied by, or consumed by, other firms within the core are not likely to suffer a rapid rise in inertia costs. So interdependent and interconnected are the suppliers, producers, and consumers of manufactured products that each supports the others' locational stability by their own. In a study that examined the changing factors affecting the location of one automobile manufacturer, the best site for that firm was estimated to have shifted only 16 kilometers (10 miles) in the 28-year period between 1935 and 1963.[5]

In the 1980s, motor vehicle production locations were affected by two innovations. In technology, the introduction of robotics changed the importance of the labor supply in locational decisions. And with the development of highly coordinated "on-time" component delivery systems, inventory costs were reduced. "On-time" systems also meant that final assembly sites no longer needed to be located very near the places where components were manufactured. These innovations have helped disperse the industry in the last decade.

Much has been made recently of higher relative economic growth in the Southeast and the Southwest United States, but the total locational inertia of the manufacturing core region is immense. Changes are occurring because national population and per capita income patterns are shifting, and because of changes in competition from foreign industries. The demand for some products is rising more rapidly than others. Whenever energy costs rise rapidly as they did

[4] Stephen S. Birdsall, "The Development and Cost of Location Stability: The Case of Oldsmobile in Lansing, Michigan," *Tijdschrift voor Economische en Sociale Geografie*, Vol. 62 (January–February 1971), pp. 35–44.

[5] Ibid., p. 43.

for the decade after 1973, the relative importance of transportation, fuel availability, and other factors of production and distribution are affected. But again, these major interregional shifts have been slow; they are very long term and important primarily because the core has been predominant economically for more than a century.

The persistence of manufacturing employment in core region states can be appreciated by examining those patterns over a long period (Table 6-2). If we consider the percentage of employment in manufacturing for the entire United States each year as our standard and compare state percentages to those standards, then changes in the national emphasis on manufacturing do not affect our state-by-state comparison.

Table 6-2 Percentage of State Employment in Manufacturing Relative to National Average, 1920–1985 (U.S. Average = 100)

State	1920	1950	1970	1985
Massachusetts	167	144	104	114
Rhode Island	191	170	129	142
Connecticut	175	164	135	132
New York	127	115	90	84
New Jersey	156	146	122	106
Pennsylvania	135	137	128	116
Ohio	135	141	132	129
Indiana	110	134	141	141
Illinois	108	124	113	104
Michigan	136	158	132	142
Wisconsin	111	118	119	131

Source: Calculated from *U.S. Census of Population (1920)* and *Statistical Abstracts of the U.S.*, (Washington, D.C.: U.S. Government Printing Office, 1951, 1971, and 1987).

Within the manufacturing core, changes in relative state employment concentration in manufacturing reflect the two broadest zones within the region. Generally, states east of the Appalachians were engaged more heavily in manufacturing in 1920 than were states of the interior core. But the eastern portion has experienced a decline in manufacturing employment since 1920 relative to the trend for the interior states. While these figures are rather general, together with those in Table 6-1, they suggest that the enormous economic mass of the interior manufacturing core region will resist rapid change even though specific industries within the core have experienced difficulty. The figures also suggest that absolute shifts in manufacturing employment will continue to alter the economic composition of other regions in North America.

URBAN CASE STUDY: NEIGHBORHOOD ETHNICITY

In addition to their industrial mix, the character of the major urban centers in the manufacturing core were also affected by their population compositions. As discussed earlier, most of these cities grew during the period of heavy European immigration before World War I. In coming, migrants sought employment opportunities where they were most clearly available—in the growth centers along the Atlantic Coast and across the interior manufacturing core. The large number of new arrivals and their ethnic diversity gave a distinctive complexion to this region's cities.

When looking for a place to live within one of the core's cities, neighborhoods chosen by more recent immigrants were strongly influenced by the ethnic residential patterns already in existence. Large ethnic neighborhoods formed and persisted through several generations. Although suburbanization has clouded the resulting mosaic to some degree and most of the more recent immigrants are not of European heritage, many of the neighborhood influences remain and continue to affect the urban cultural geography.

In his book about former Mayor Richard Daley of Chicago,[6] Mike Royko wryly catches the flavor and significance of the city's ethnic patterns. Although he is describing a Chicago of more than a generation ago, the scenario could as easily be for any of the large northeastern and midwestern cities in the United States during the period.

> Chicago, until as late as the 1950s, was a place where people stayed put for a while, creating tightly knit neighborhoods, as small-townish as any village in the wheat fields.
>
> The neighborhood-towns were part of larger ethnic states. To the north of the Loop was Germany. To the northwest Poland. To the west were Italy and Israel. To the southwest were Bohemia and Lithuania. And to the south was Ireland.
>
> It wasn't perfectly defined because the borders shifted as newcomers moved in on the old settlers, sending them fleeing in terror and disgust. Here and there were outlying colonies, with Poles also on the South Side, and Irish up north.
>
> But you could always tell, even with your eyes closed, which state you were in by the odors of the food stores and the open kitchen windows, the sound of the foreign or familiar language, and by whether a stranger hit you in the head with a rock.
>
> In every neighborhood could be found all the ingredients of the small town: the local tavern, the funeral parlor, the bakery, the vegetable store, the butcher shop, the drugstore, the neighborhood drunk, the neighborhood trollop, the neighborhood idiot, the neighborhood war hero, the neighborhood police station, the neighborhood team, the neighborhood sports star, the ball field, the barber shop, the pool hall, the clubs, and the main street.
>
> Every neighborhood had a main street for shopping and public transportation. The city is laid out with a main street every half mile, residential streets between. But even better than in a small town, a neighborhood person didn't have to go over to the main street to get essentials, such as food and drink. On the side streets were taverns and little grocery stores. To buy new underwear, though, you had to go to Main street.
>
> With everything right there, why go anywhere else? If you went somewhere else, you couldn't get credit, you'd have to waste a nickel on the streetcar, and when you finally got there, they might not speak the language.
>
> Some people had to leave the neighborhood to work, but many didn't because the houses were interlaced with industry.
>
> On Sunday, people might ride a streetcar to visit a relative, but they usually remained within the ethnic state, unless there had been an unfortunate marriage in the family.
>
> The borders of neighborhoods were the main streets, railroad tracks, branches of the Chicago River, branches of the branches, strips of industry, parks, and anything else that could be glared across.
>
> The ethnic states got along just about as pleasantly as did the nations of Europe. With their tote bags, the immigrants brought along all their old prejudices, and immediately picked up some new ones. An Irishman who came here hating only the Englishmen and Irish Protestants soon hated Poles, Italians, and blacks. A Pole who was free arrived hating only Jews and Russians, but soon learned to hate the Irish, the Italians, and the blacks.
>
> That was another good reason to stay close to home and in your own neighborhood-town and ethnic state. Go that way, past the viaduct, and the wops will jump you, or chase you into Jew town. Go the other way, beyond the parks, and the Polacks would stomp on you. Cross those streetcar tracks, and the Micks will shower you with Irish

[6] Mike Royko, *Boss: Richard J. Daley of Chicago* (New York: New American Library, 1971), pp. 30–32.

confetti from the brickyards. And who can tell what the niggers might do?

But in the neighborhood, you were safe. At least if you did not cross beyond, say, to the other side of the school. While it might be part of your ethnic state, it was still the edge of another neighborhood, and their gang was just as mean as your gang.

So, for a variety of reasons, ranging from convenience to fear to economics, people stayed in their own neighborhood, loving it, enjoying the closeness, the friendliness, the familarity, and trying to save enough money to move out.

ADDITIONAL READINGS

Benhart, John E., and Marjorie E. Dunlop. "The Iron and Steel Industry of Pennsylvania: Spatial Change and Economic Evolution." *Journal of Geography*, 88 (1989), pp. 173–183.

Berry, Brian J. L., et al. *Chicago: Transformation of an Urban System.* Cambridge, Mass.: Ballinger, 1976.

Borchert, John R. "Major Control Points in American Economic Geography." *Annals of the Association of American Geographers.* 68 (1978), pp. 214–232.

Cutler, Irving. *Chicago: Metropolis of the Mid-continent*, 3rd ed. Dubuque, Iowa: Kendall/Hunt, 1982.

Haynes, Kingsley E., and Zachary B. Machunda. "Spatial Restructuring of Manufacturing and Employment Growth in the Rural Midwest: An Analysis for Indiana." *Economic Geography*, 63 (1987), pp. 319–333.

Jones, Peter d'A., and Melvin G. Holli, eds. *Ethnic Chicago.* Grand Rapids, Mich.: William B. Eerdmans, 1981.

Lewis, Peirce. "Small Town in Pennsylvania." *Annals of the Association of American Geographers*, 62 (1972), pp. 323–351.

Proctor, Mary, and Bill Matuszeski. *Gritty Cities.* Philadelphia: Temple University Press, 1978.

Schenker, Eric, Harold Meyer, and Harry C. Brockel. *The Great Lakes Transportation System.* Madison: University of Wisconsin, Sea Grant College Program, 1976.

Smith, Neil, and Ward Dennis. "The Restructuring of Geographical Scale: Coalescence and Fragmentation of the Northern Core Region," *Economic Geography*, 63 (1987), pp. 160–183.

Warren, Kenneth. *The American Steel Industry, 1850–1970: A Geographical Interpretation.* Oxford: Clarendon Press, 1973.

CHAPTER

7

The National Core Region of Canada

Canada is a larger country than the United States [9.8+ million square kilometers (3.8 million square miles) to the United States' 9.3+ million square kilometers (3.6+ million square miles)], but it contains a much smaller share of the continental core region. The portion of the North American manufacturing core that is in Canada is nearly coincidental with that country's national core region. Canada's national core is entirely within the provinces of Quebec and Ontario lying along the northern shores of Lake Erie and Lake Ontario in a band that also reaches northeastward along the banks of the St. Lawrence River to Quebec City (Figure 7-1). While similar in extent to the continent's manufacturing core region, the Canadian national core extends farther to includes cities, such as Ottawa, Montreal, and Quebec, that are of strong cultural and political importance to that country but that are not really part of the continent's manufacturing core.

As a national core should, this Canadian region contains a large share of the nation's population and economic activity and was also the historic hearth of the country's development. More than 55 percent of the Canadian population lives within southern Ontario and southern Quebec, a regional proportion expected to reach almost 60 percent by the end of the century. Disproportionate shares of the urban population and national industrial capacity are also found here, the industry associated with the advantages of accessibility provided by the Great Lakes and the St. Lawrence or with the mineral-resource regions nearby.

It was into the riverine lowlands that French settlers spread in the century and a half after 1608; English settlement was most extensive inland from the lake shores in the late eighteenth and early nineteenth centuries. By 1867 when the new Dominion of Canada was formed, more than 90 percent of the population of this country resided in the southern portions of Ontario and Quebec and around the margins of the Maritimes to the northeast. Thus, it has been only during the most recent century of Canada's existence that permanent settlements have been established in much of the vast interior, but the core remains dominant.

In considering the essential character of this region, three themes can be suggested. The first deals with the relatively small size of the region. What factors have restricted growth within the core to the limited territory it occupies? Ordinarily, one might expect a region that is the development hearth of a rapidly growing and prosperous country such as Canada to have increased its area along with the growth in its economy, but this territorial expansion of the core did not occur. A second theme is the great diversity found within the region. Although in geography the concept of a core region is subject to differing definitions, a certain intensity and functional uniformity is often an underlying assumption to the delimitation of such a region. Appropriately, the cultural diversity that has threatened the very existence of the country as a single political unit finds its clearest expression within the core region. And, finally, the character of the national core is affected by its location within the larger core region for the entire continent. Thus, the third theme is the relationship between the United States and Canada as seen

Will Canada remain whole through the end of this century? Or will recent moves toward Quebecois independence finally split the world's second largest country into two or more parts? (Paul Chiasson/Canapress Photo Service)

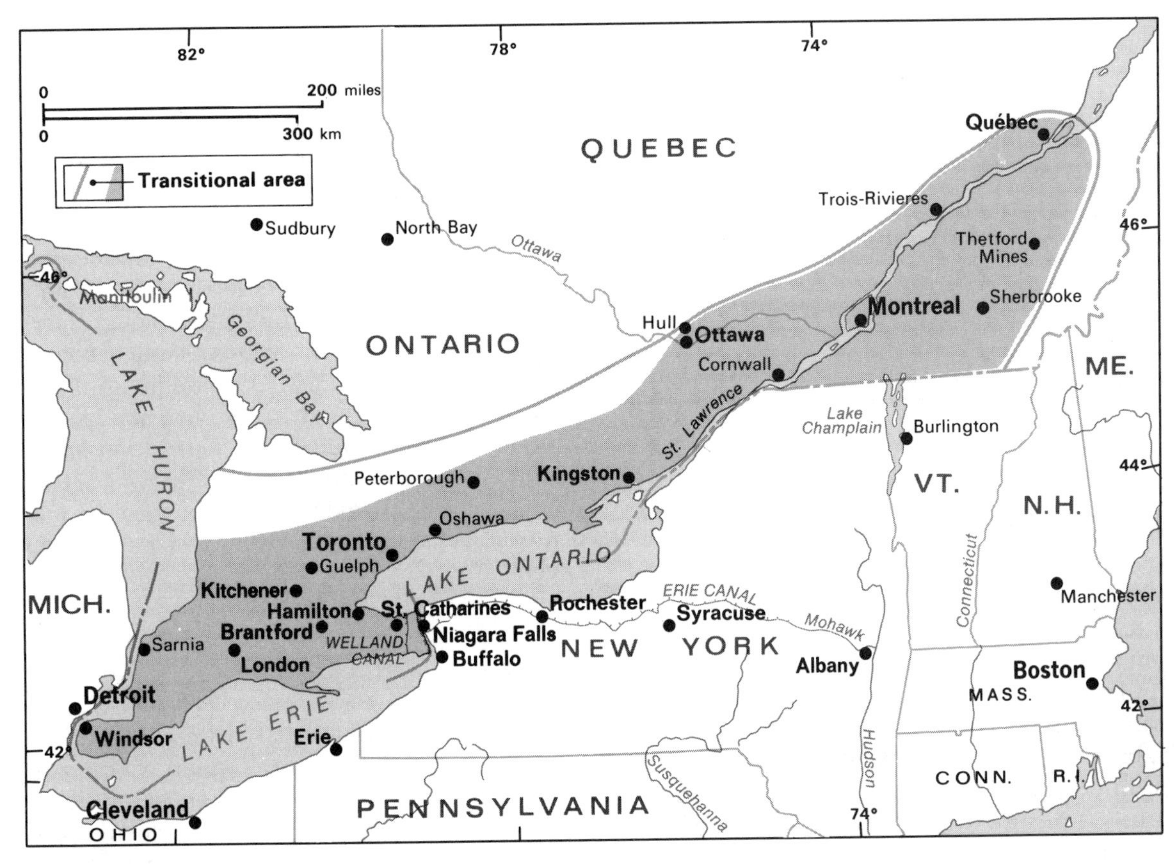

Figure 7-1 The national core region of Canada.

through the integration of the national core into the continental core.

The core region of Canada is located far from the center of the country, a consequence of the historical sequence of settlement from Europe and environmental patterns. The initial European entry into this portion of North America was from the northeast via the St. Lawrence River. Jacques Cartier reached Montreal Island as early as 1535; further extension of French interest along the Great Lakes followed the turn of the seventeenth century. Following the conclusion of the French and Indian War in 1763, the British encouraged settlement to balance the French and Catholic population with (mostly) English Protestants. After the American Revolution, there was additional settlement by Loyalists along the northern shores of Lakes Ontario and Erie as population spilled across the United States border. Because of the importance of the St. Lawrence as an *entry* route, the lands reached by this passage up the river were given the administrative labels of Lower Canada and Upper Canada during the first half of the nineteenth century. Although these regional names are no longer in use, the land in southern Quebec, occupied by the French Canadian population along the St. Lawrence River, was Lower Canada and the most southerly lands of all, that area bordering Lakes Ontario and Erie, occupied by the English Canadian population, were Upper Canada.

The pattern of settlement in Canada was for a long time keyed to production for export Europe

Major Metropolitan Area Populations in the Canadian Core, 1986	
Toronto, Ont.	3,427,168
Montreal, Que.	2,921,357
Ottawa–Hull, Ont.–Que.	819,263
Quebec, Que.	603,267
Hamilton, Ont.	557,029
St. Catharines–Niagara Falls, Ont.	343,258
London, Ont.	342,302
Kitchener, Ont.	311,195
Windsor, Ont.	253,988
Oshawa, Ont.	203,543

of what are called *staple products.* Staples are products that are basic economic goods and, therefore, are in steady demand. Early dependence on cod fishing off Newfoundland and fur trapping in the St. Lawrence Valley provided settlers with the means to purchase imports from Europe. As fur trappers exhausted the local supply and fur trapping shifted westward, Canadians turned to alternative staple products as the bases for their export economy: timber, metals, and agriculture. From the late eighteenth century onward, trade remained oriented toward Great Britain. Population pressure increased during the nineteenth century within the St. Lawrence Valley and southern Ontario, but the traditional economic and trade ties with Britain maintained a dependence on exploitation of primary products. The Canadian economy did evolve beyond the early, nearly absolute dependence on the export of staples from the core region—by expansion of influence from the United States, by growth in domestic manufacturing industries, and by the spread of settlement beyond the core—but these factors did not fully transform the national economy, as we shall see.

As a northern country, Canada has severe environmental conditions in much of its territory. Canada's most desirable early agricultural resources were located on the more southerly Ontario Peninsula and to a lesser degree along the narrow St. Lawrence lowland. With population growth in Canada prior to confederation (1867) supported by the production of staples for export, and with the rapid and leading growth of the nearby United States core, the pressure on land resources in the Canadian core became considerable. Dense settlement, however, was not pushed much beyond the present limits of the core. By the time population expansion became truly continent-wide, southern Ontario and southern Quebec had accumulated enough inertia--economic, political, cultural--and enough focus from transportation developments to remain the leading growth region in Canada. But why did the developing core remain so small during its period of formation? Why was settlement largely restricted to the region until the last century? What factors restrained expansion so severely?

LIMITS OF THE CORE

One of the most substantial factors limiting the size of the country's core region is physiographic. As discussed in Chapter 1 (and again later in Chapter 17), much of the continent north of the core is underlain by the Canadian Shield. The margins of this vast zone of igneous and metamorphic rock describes a long arc from the Arctic coast to the Great Lakes and the St. Lawrence Valley. Within the core, the Shield is present under a thin layer of soil from the mouth of the St. Lawrence estuary southwestward to about the present capital, Ottawa, south again almost to Lake Ontario, and then nearly due west to Georgian Bay (Figure 7-2).

Keeping this outline in mind, it is clear that early settlement in Quebec was very much restricted to land south of the thin and poorly drained soils of the Shield. The agricultural and commercial populations hugged the narrow band of land along the St. Lawrence, in most cases within a river valley only a few dozen miles wide. And while the Shield is not as obviously present in southern Ontario, it helped restrict the agricultural opportunities of the expanding

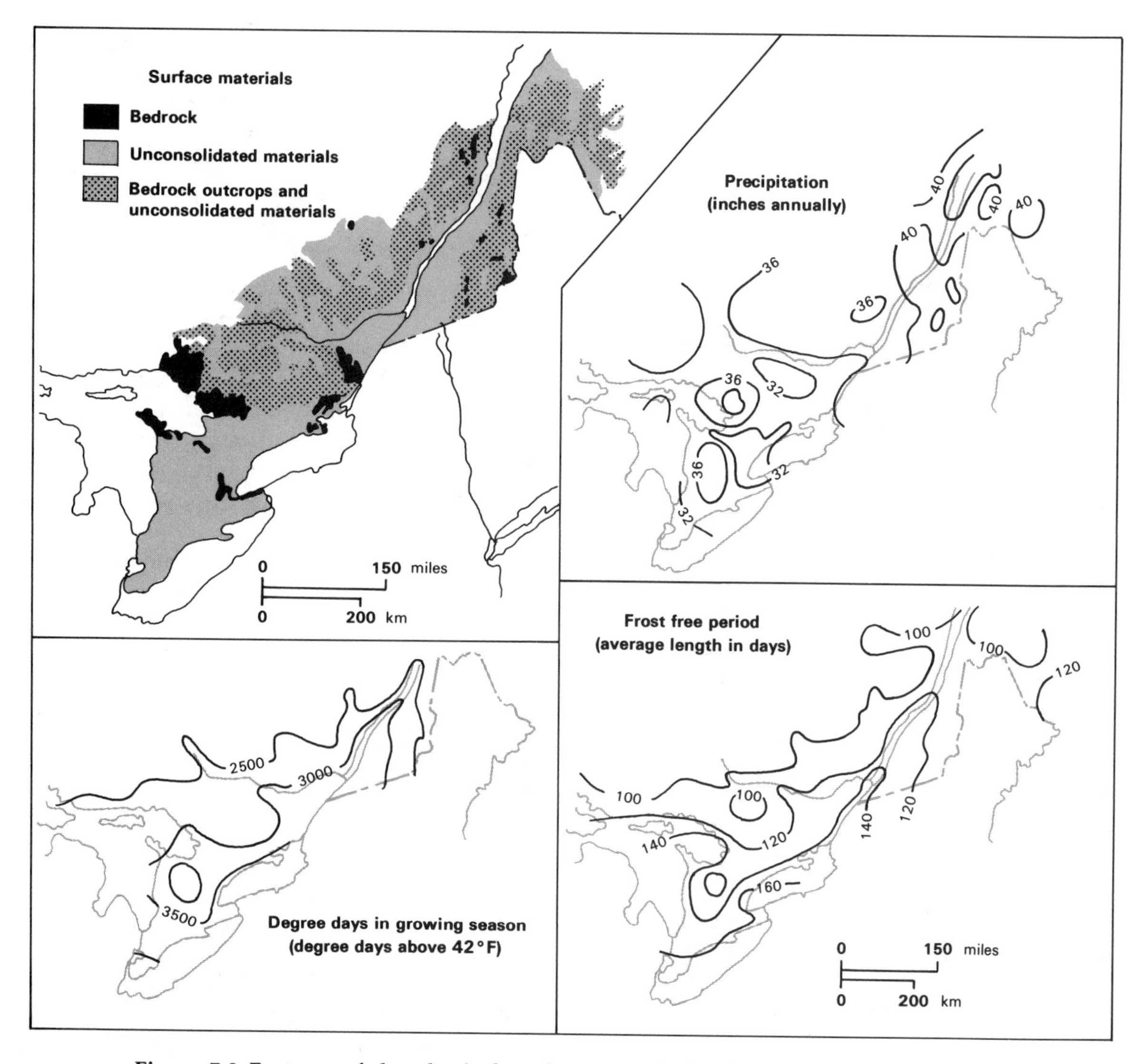

Figure 7-2 Features of the physical environment. Each of these four indicators of the environmental support for farming reflects the importance of southern Ontario and the St. Lawrence valley in the Canadian core's agriculture.

Canadian settlement during the nineteenth century. It was not the topography of the Shield that was the essential barrier to farming. Much of the Shield has the appearance of a gently rolling low plateau dotted with small lakes and marshes and covered with a mixed forest. Forests were removed in other parts of North America when there was a demand for farmland. In this case, the land was too poor and the climate was too harsh for successful farming.

The land of the Shield is poorly suited for agriculture because it is not well drained and the soils are too thin and acidic. As discussed earlier, the glacial ice cap crept outward during the Ice

Age over the Shield and beyond. As the glaciers moved, loose material was scraped from the underlying rock surface and deposited far to the south. Then, as temperatures gradually rose and the ice melted back, very little cover remained on the crystalline bedrock of the Shield. In the 8000 to 10,000 years since the glaciers retreated, too little time has passed for the slow process of soil formation in a cold climate to have developed anything more than a relatively thin, acidic layer of soil in shallow pockets on the Shield.

The combined forces of a scouring glacial sheet of ice and a basically flat, resistant bedrock also resulted in a heritage of poor drainage for this region. While most streams and rivers flow off the Shield to the Great Lakes, the St. Lawrence, or north into Hudson Bay, their paths are rarely direct or uninterrupted. Most surface water on the Shield moves leisurely from lake to lake in the general direction of the Shield's margins but with little apparent local organization. The result is a *hydrography* that discourages agricultural use because of an abundance of small swampy pockets and poor drainage. This also discouraged commercial development because only a few rivers, such as the Ottawa, could be used for the movement of goods.

South of the Shield in Ontario and in the St. Lawrence Valley in Quebec, the pattern of soils also was influenced strongly by the long period of glacial activity. Most of the better agricultural soils in the Canadian core region are derived directly from the material deposited by the glaciers as they melted (*glacial till*) or from the outwash material carried by the tremendous volumes of water leaving the melting glacier. The low relief of much of this land masks a complex pattern of alfisols (see Chapter 2), some of which are deep and moderately fertile, and substantial pockets of clay and sandy soils that are less suited to agricultural use. A recent estimate of soil capabilities in southern Ontario places 60 percent of the land in categories of soil that could support field crops with proper management, but two additional factors must be kept in mind. First, the technology for "proper management" was not available to Canadian farmers until well into this century, and, second, southern Ontario's arable 60 percent constitutes less than two-thirds of a very small portion of Canada.

The other environmental feature that contributed to the early limits of Canadian core development was climate. Canada is a high-latitude country. The short growing seasons and cool summers would have made the agricultural efforts of its European settlers difficult north of the core region even if the Shield's soil and drainage were less formidable. Lakes Huron, Erie, and Ontario bound the Ontario peninsula, however, and their proximity does hasten the spring thaw and postpone the autumn freeze to a small degree. The valley of the St. Lawrence and the lowland between the St. Lawrence and the Ottawa rivers also benefit from an extended growing season, almost all sections averaging greater than 120 frost-free days annually (Figure 7-2). But as Canadian settlers moved inland, farther from the moderating influence of the Great Lakes and farther north, they experienced growing seasons that were appreciably shorter. More important, they found that summer days did not accumulate enough heat energy for good crop growth. Measured as the number of degrees over 42°F (over 5.5°C) for the number of days this temperature was exceeded, the better agricultural areas in Canada are those exceeding 3000 annual *degree-days*. The boundary of the national core in Quebec coincides closely with the line indicating this level of mean annual heating, and all of the Ontario portion of the core receives an average of more than 3000 degree-days.

Moisture availability is generally not a problem in the Canadian core. While mean annual precipitation is less than 100 centimeters (40 inches) in most areas, the summer temperatures are cool enough that *evapotranspiration* rates are not high. Only along the Lake Erie and Lake Ontario margins are temperatures warm enough to produce a small water deficiency in the soil, but it is not sufficiently severe to hamper agricultural production unless the local soil is sandy and less able to hold moisture.

The location of the core in Canada is partly a result of the environmental limits on agriculture, partly a result of major routes of continental entry, and partly a result of the inertia provided by strong transport ties to Europe and later to western portions of the country. With limited technology, individual farmers could not establish settlement north and west of the Great Lakes across the Shield. With a diffused political organization, arguments for connecting the national core region with the western three-fourths of the country for any purpose beyond the export of staple products were also unconvincing until late in the nineteenth century. By this time, the agricultural base was strong, and southern Quebec and southern Ontario were well settled with two cities rapidly approaching populations of 100,000 (Montreal and Toronto) and another five exceeding 10,000 (Quebec, Ottawa, Kingston, Hamilton, and London). Patterns of transport and communication were well established. The economic and political foci for the country had already developed here. The inertia of national dominance was clearly present in this region.

DIVERSITY WITHIN THE CORE

The southern margins of Quebec and Ontario share many characteristics. Both are located along the main route of access between Europe and the Canadian interior. Both developed urban commercial centers that came to dominate the economic organization of each province's portion of the core and to compete with the other center for ultimate dominance of all of Canada. Both experienced the gradual population shift from rural to urban residences common to most economies undergoing industrialization. Both possess the moderately restrictive environmental base just discussed. And the economic organization of both areas is tied through investments and subsidiary ownerships to the economic structure of the United States. In spite of similarities such as these, striking diversity exists within the Canadian core.

Cultural Landscapes

Perhaps the most pervasive contrasts within the core are those related to Canada's cultural dualism. A large majority of the population of Quebec can trace its ancestry to French origins, while most of those living in southern Ontario are of British heritage. Although Canada has received large numbers of migrants from other countries of Europe and more recently Asia, and some population movement has occurred between these two provinces, the cultural boundary zone continues to approximate the provincial boundary (Figure 7-3).

The French Canadian cultural base is reflected in linguistic distinctiveness, religion, historical identity, individual attitudes, settlement patterns, and frequently in a political posture defensive of French Canadian culture. This last feature is a function of the minority position of French culture in Canada. Nine of 10 people whose mother tongue was French lived in Quebec in 1986, and although one-third of New Brunswick's population identified French as their first language and almost 500,000 in Ontario did the same, only one-fourth of the total Canadian population reported French as their first language that year. In addition, almost 250 million additional English-speaking people live immediately to the south.

French Canada has remained culturally distinct within British North America largely because it has maintained a defensive posture with regard to any natural or planned assimilation trends. When the British conquered the French portion of North America (1763), most of the land in the St. Lawrence lowland was occupied by French-speaking farmers. Quebec and Montreal were rapidly growing commercial centers, and French influence had spread beyond the Great Lakes to the Ohio and the Mississippi Rivers along routes frequently traveled by traders and fur trappers. The British seizure of governmental control from France in this part of North America effectively ended additional cultural infusions and most migration from France, but it

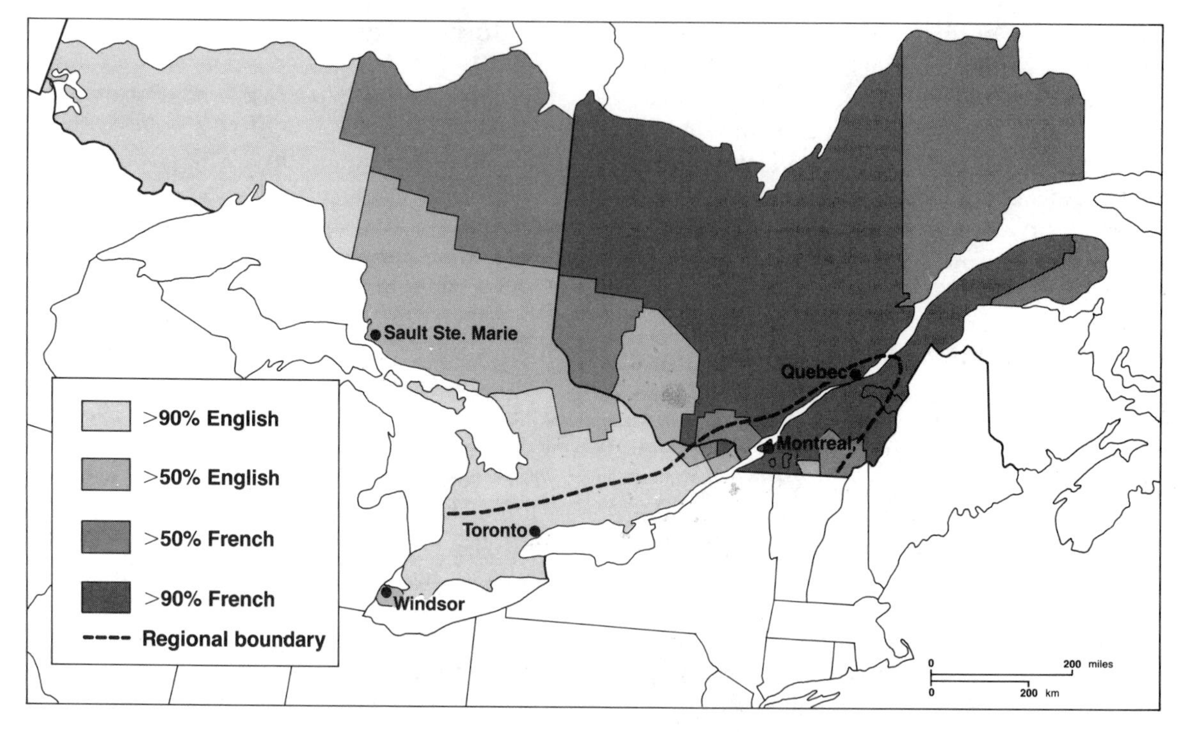

Figure 7-3 Linguistic dominance in Ontario and Quebec. The political boundary between these provinces does not match the linguistic boundary perfectly, but relatively few people in southern Ontario use French as their first language and relatively few in Quebec consider English their mother tongue.

did not mean that the cultural base in Quebec was destroyed. Quebec's economy simply became redirected toward British goals.

Following the American Revolution, British Loyalists moved into areas less well settled and possessing greater agricultural potential, so that there was little immediate dilution of the French Canadian culture within Quebec's lowlands. Even during the nineteenth century when the national economy was British controlled, Montreal maintained a French cultural dominance in this portion of the core region as urban commercial and industrial growth stimulated the growth of that city. The cultural distinctiveness persisted even though the city attracted many non-French Canadians as well as many from rural Quebec. The rural French Canadian population, more isolated from Anglo cultural influences, also grew rapidly, and migration to smaller cities and towns in the province helped maintain a French Canadian presence as the population gradually became more urban overall. A highly urban province in 1981 with 80 percent of its population in urban places, Quebec's nonurban population is also larger today than it was a century ago.

In order to accommodate the cultural and regional diversity of this large country, efforts toward unification—formally completed in 1867—required the governmental organization to be in the form of a *federation*. A federation is a form of government in which a group of states or provinces are united under a single, mutually acceptable, sovereign central authority, even while the states or provinces retain other governmental powers. Ideally, a federation (or a confederation as it was called in Canada) permits many of the advantages of a larger unified organization while leaving room to accommodate local regional diversity. If the differences are persistent and deep enough, however, there may be times when the required accommodations conflict with each other or with the demands of the central authority. In the United States—also formally a federation—such a conflict led to civil war.

In Quebec, the tension between the central government's powers and those of the regional government has revolved around issues of French Canadian cultural independence and equality within a largely non-French country. The country is officially bilingual. This ostensibly means that the French and English languages are of equal importance in the governmental operation of the country. In theory, a member of the federal civil service in British Columbia, for example, could address others in that province in French and expect to be understood. In fact, with extremely few in the western provinces speaking French as their first language, these areas do not participate in this bilingualism in spite of its official status. In recent decades, questions of cultural dualism, cultural integrity, and equality focused on the validity of bilingualism. Outside French-speaking portions of the country, this element of Canadian federal governmental policy accumulated considerable opposition. Either as cause or effect, there has been a correspondent heightening of Quebecois nationalism within *La Belle Province.*

Much more is involved than language differences, but contention over linguistic parity between English and French has long been used as a political issue in Canada. For example, when a study by the Canadian census in the late 1960s projected that a majority of Montreal's population would be non-French-speaking by the year 2000, French Canadian nationalism was aroused further. A provincial law was subsequently passed requiring that all signs in Quebec be in

Both French and English are used in Montreal and other large cities in Quebec, but French is the preferred language throughout much of the province. (Mark Antman/The Image Works)

French, and primarily in French alone. Although this was ruled unconstitutional by the Canadian Supreme Court, a new law was passed that clearly affects potential English-language migrants to the province. It requires that all children permanently residing in Quebec attend French-language schools unless the parents themselves were educated in Quebec in English-language schools.

The British impact on southern Quebec was not trivial, although it was southern Ontario that received the most broadly based cultural impression. Aside from a mild French imprint, the European sequence of entry into, and the taming of, the Ontario wilderness really began with the influx of Loyalists forced out of the new United States. They were followed by large numbers of Americans and British. Settlement proceeded slowly enough that the land was surveyed and *platted* before the population arrived. Roads were planned and settlement was directed inland from the main urban centers. The population that migrated to southern Ontario was drawn by governmental policy from the British Isles, partly to offset American invasions but also to balance the large and growing French Canadian population that had been farming along the lower St. Lawrence for more than a century. Many of the British settlers were Scots and Irish, although most were probably English. By the time of confederation, southern Ontario provided a definitive Anglo-Canadian culture hearth for new population diffusion around the Great Lakes and onto the Canadian prairies.

The long-term significance of the larger issue of separatism, or independence for French Canada, lies in the location of Canada's population and productivity (about 25 percent of both are in Quebec) and the location of the French-speaking population within the provinces (about 85 percent are in southern Quebec). As cultural differences are translated into strong political feelings, the Canadian core becomes threatened with division. The effect on the core should Quebec ever separate politically from the rest of Canada is difficult to predict. The region would hardly be a national core any longer, but the economic connections between Quebec and Ontario have been built over decades, even centuries. It is speculation, of course, but only if there is enough bitterness between the two populations would it be likely that the economic benefits of close association be surrendered with political union.

Rural Landscapes

Environmental conditions and the interplay of cultural histories have affected the agricultural emphasis of each part of the core, although similarities appeared during the twentieth century. For many decades, grain and animal production dominated the farmers' efforts in both regions. In the less productive interior, near-subsistence farming was often supplemented by maple sugar manufacture and lumbering during the long nonfarming season.

The imprint of each culture on the landscape has created differences in its appearance. Land was assigned to early French settlers within seigneuries, and, in these land grants, the *rang* system of settlement was used. Under the *rang* system, each farmstead was located on the river or lake shore, or on a road constructed parallel to the shore, at the head of a long narrow lot that extended inland perpendicular to the shore or road. The advantages of this arrangement for settlers were equal access to the main transport lines, proximity to their land, and variability in land quality with many lots with both swampy shoreline and upland terrace included. As demand for land increased, a second, third, and fourth *rang* were demarcated by the construction of new roads, farther inland but also parallel to the shore or the first road. The best farmland in southern Quebec (and even in a few areas of southern Ontario) had been laid out in this manner before the British takeover. What is popularly known as the French long-lot pattern, therefore, can still be used as a general indication of the better farmlands in southern Quebec (Figure 7-4).

In Quebec, as the rural population grew, migration to the United States or to nearby Cana-

This scene illustrates the typical form of French Canadian agriculture in the St. Lawrence lowland. When this land was distributed by the French crown, it was parceled out in narrow, parallel strips extending inland from the river. Each "long lot," as the fields are called, has access to the river, or to a road built parallel to the river at some distance to the interior. This small, linear village (about 65 miles from the city of Quebec) lies along the first road inland from the river, visible at the bottom of the photograph. The rather irregular topography and the presence of numerous rocks in virtually every field indicate that farming is not easy here, nor is it especially remunerative. (Lawrence R. Lowry)

dian cities drew off some of the increase, but many French Canadians were encouraged to remain true to their rural traditions and expand the settlement frontier onto the Shield or into the equally marginal land of the northern Appalachians. During the nineteenth century, other parts of Canada, more suited to wheat production, began to exert a comparative advantage over even the better land in southern Quebec. The emphasis of production shifted, and dairying, supported by hay and oats, became the most important farm pursuit. Many specialty crops, such as vegetables, sugar beets, fruits, and poultry, were also significant. As jobs became available in resource industries and factories, much of the marginal land has been abandoned and allowed to revert to forest.

The rural landscape in Ontario and in the less well-endowed or less accessible lowlands of southern Quebec was laid out by the British in a square or rectangular pattern. Individual homesteads tend to be more dispersed under such an arrangement. Although French Canadians gradually moved into many of these areas during the nineteenth and early twentieth centuries, the pattern of landholdings was rarely changed from the initial British form.

The pattern of agricultural change is similar in southern Ontario, but the more southerly location and the more extensive lowland generated a

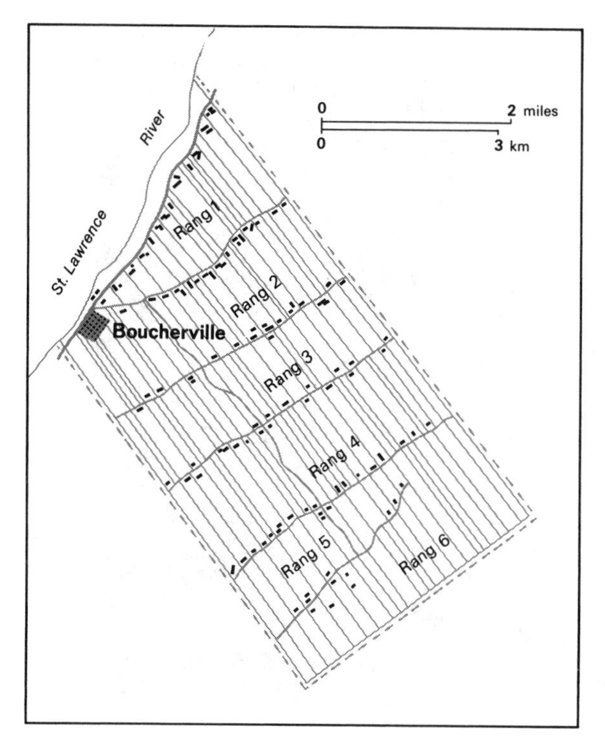

Figure 7-4 Long-lot farms in Quebec. (From *A Regional Geography of North America,* second edition, by G. S. Tomkins, T. L. Hills, and T. R. Weir, copyright 1970, Gage Publishing Limited. Reprinted by permission of Gage Publishing Ltd.)

somewhat different result. Increasing demand for food in the early Canadian core and in the eastern United States led most new farmers in southern Ontario to grow wheat during the early 1800s as their main cash crop. By midcentury, however, wheat yields declined sharply as the alfisols lost their initial fertility under constant wheat production. Farming became diversified, but wheat remained prominent in Ontario agriculture until the Prairie Provinces were settled. Mixed farming, with both feed crops and meat animals produced, became the primary farm type until about World War II. Innovations in storage, transport, and daily operation during the 1920s and 1930s gradually encouraged a substantial proportion of farms in the region to undertake dairying. Dairying is dominant near the urban areas of Windsor and Hamilton and on the Niagara Peninsula and in the nonurban eastern counties, but livestock farming exceeds dairying in value of production near Toronto, London, and Kitchener.

There are also a number of significant specialty crops grown in southern Ontario. Along the middle Lake Erie shore and to the northeast, sandy soils support production of high-value, flue-cured tobacco, the high annual returns for production of this crop offsetting to some extent the anxieties about its long-run demand.

Even less sure of the future are those who depend on the fruit produced within the region. Located between Lakes Ontario and Erie and at a relatively southern latitude (for Canada), the Niagara Peninsula possesses a set of climate and soil conditions approaching the optimum for some fruits. Winters are moderated and spring blossoming is retarded because of the peninsula's position between the lakes. Grapes, peaches, and sweet cherries are the most important of the fruits grown in the region. Future fruit production does not appear secure, however, as urban areas on the peninsula expand outward to accommodate their growing populations. A study of land-use change in the Niagara Fruit Belt concluded that "most of the urbanization has been occurring in the most intensive fruit-growing areas"[1] (Figure 7-5). More serious for the long-run agricultural productivity of the region, a great deal of the urban development has occurred on the best fruit soils. There have been attempts to preserve what remains, but it has not proven easy to control rural-to-urban land-use change in the United States or Canada (see Chapters 4 and 16). The long-term success of such attempts is still unclear.

In recent decades, farmers in the core region of Ontario have experienced the same pressures as those in southern Quebec to increase their applications of capital with a consequent mechaniz-

---

[1] Ralph R. Krueger, "Urbanization of the Niagara Fruit Belt," *The Canadian Geographer*, Vol. 22 (Fall 1978), p. 189.

The Niagara Peninsula, especially the land along the southwest shore of Lake Ontario, contains many fruit orchards and grape vineyards. Although this region continues to be an important source of apples, peaches, and grapes in Canada, urban activities—such as the subdivision development shown in this photo—are encroaching rapidly on the agricultural uses. (Wayne Farras/The Regional Municipality of Niagara, Niagara Planning and Development Department)

ation and increased specialization of production (see Chapter 12 for more on this trend). In both sections of the Canadian core, this will probably lead to further land abandonment along the margins of the region and to further intensification of production within the most favored areas of production in the core itself.

Urban and Industrial Dominance

In addition to being the region where the distinctive cultural dualism of Canada is most strongly represented and the area of earliest productive agriculture, the portion of Canada defined as the national core also has contained the most complex pattern of urban and industrial development in the country. As Canada's population began to spread beyond the Great Lakes to the prairies and the Pacific Coast, southern Ontario and southern Quebec remained the source of much development capital, most of the migrating population, and the primary beneficiary of the industrial growth that accompanied the

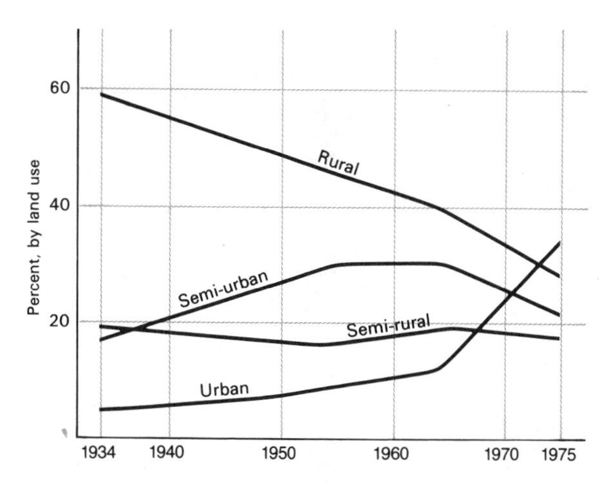

Figure 7-5 Change in land use in the Niagara fruit belt, 1934–1975. (After Krueger.) The rapid conversion of valuable fruit-producing land to urban uses is a great concern in this section of Canada.

economic stimulation accompanying the nation's expansion. The manner in which this industrial and urban growth occurred, however, was variable and reflects differences within the core. Differences in the responses of each section of the core to stimuli arising in the United States have also been important.

Both portions of the core have urban-industrial centers that far exceed the size of all others in their province: Montreal in southern Quebec and Toronto in southern Ontario. Greater Montreal, with a population in 1986 of 2,914,464, benefited from important military and commercial site advantages during the colonial period. The original setting lies at one end of an island in the St. Lawrence River, adjacent to both the confluence of the Ottawa River with the St. Lawrence and the first major rapids on the St. Lawrence, and at the northern end of the Hudson–Lake Champlain lowland. Thus, a strong defensive position was combined with control of two major water routes to the interior. Montreal's even greater advantages of situation, along with a municipal administration that exploited these advantages, became more important stimulators of growth during the nineteenth and twentieth centuries. Its location on the river at the point shipments were required to circumvent the rapids was important to early commercial development. Montreal surpassed the city of Quebec in population between 1825 and 1830. But true dominance in southern Quebec was not assured until the decade of the 1850s when the river was bridged and Montreal was connected by railroad to Portland, Maine, in the east and to Toronto in the southwest. Access to the southeast enhanced Montreal's position at the northern end of the Lake Champlain–Hudson River valley and increased its connections with New York City.

Montreal was a commercial, financial, and transportation center by confederation in 1867, and this put the city's economy well along a self-reinforcing growth spiral of industry attracting labor and a larger market that in turn drew more industry that drew more labor, and so forth. A distinctive result was the development of a large non-French population in the city. Most of the financiers and industrialists in the city were of British origin, while most of the labor force was French Canadian. As economic opportunities attracted rural French Canadians from the countryside, by 1871 the French-speaking population exceeded the English-speaking in number. This history has meant that Montreal retains a much stronger bicultural character than does Quebec. This latter city at the head of the St. Lawrence estuary had fewer of the advantages that stimulated Montreal's industrial growth. And while Quebec's metropolitan area population had grown to 614,736 by 1986, it still retains the strong French Canadian flavor expected of a cultural, religious, and provincial administrative center.

Significant urban growth in southern Quebec outside of the immediate Quebec and Montreal spheres is a relatively recent phenomenon. In every case, growth of the new centers can be traced to resource-based industrial development. Some, like Trois-Rivières (120,095 in 1986), combine access to Shield resources—pulp timber and, more recently, hydroelectric

The city of Quebec was founded at the mouth of the St. Lawrence atop a high river bluff to control the river traffic. During the centuries that followed, the city's commercial, cultural, administrative, and industrial activities gradually overwhelmed its original limited purpose. These activities now support more than one-half million people in the metropolitan area. (N.F.B. Photothèque O.N.F.)

power—with imported materials like the bauxite processed at nearby Shawinigan. Other centers, such as Thetford Mines (34,989) and Sherbrooke (123,672), have grown through utilization of local minerals (asbestos at Thetford) or abundant local labor and diversified industrial investment. In all cases, Montreal dominates through transport and financial connections.

Toronto, with a metropolitan population of 3,377,279 in 1986, is southern Ontario's equivalent to Montreal and its urban competitor within the national core. At first glance, Toronto appears to have few of the natural site or situation advantages that would support the growth of such a metropolis. True, it is located on Lake Ontario, possesses a good harbor, and lies at the head of the Toronto Passage, a lowland route to Georgian Bay in the north. But other sites along the lake shore, perhaps at either end of the lake, would seem to provide equivalent advantages to urban growth. Toronto was not located at a focus of natural access as was Montreal. And Toronto did not possess Montreal's early momentum of urban growth.

Toronto has been able to match, and in some ways exceed, Montreal's economic growth because it has several advantages of function and location not present in Montreal and that can be added to its natural harbor resource and lake accessibility. It was decided shortly after the beginning of British rule in southern Ontario that the region's capital would be located at York (now Toronto) rather than at Kingston, Niagara, or London, so Toronto became its province's political capital while Montreal did not. In order to administer properly, the capital had to be well connected by road, and later by rail, to the rest of the territory. Toronto's political function, therefore, provided the city with a network of land transportation that made it the natural focus of all Ontario prior to confederation. Of equal importance, this land transport network which focused on Toronto drained a larger and more productive agricultural region than was accessible to Montreal. Toronto's superior hinterland provided it with an alternative economic base to Montreal's strong commercial growth potential.

These early stimuli for urban growth were further increased by developments in the United States. As long as the bulk of both the commercial and industrial activity in the United States was concentrated along the northeastern seaboard, Montreal's greater proximity was to that city's advantage in spite of Toronto's good access via the Erie Canal route through the Mohawk and Hudson River valleys. During the late nineteenth century, however, as the centers along the southern Great Lakes began to grow in industrial capacity and the U.S. manufacturing core extended into the country's Midwest, Toronto's location on the Great Lakes and easy accessibility by rail through Windsor-Detroit and Niagara-Buffalo assured it of continued urban and industrial expansion.

The relative strength of Toronto's hinterland and the region's proximity to major U.S. centers of manufacturing are also shown in the variety and intensity of urban-industrial concentrations around the western end of Lake Ontario, in cities adjacent to the southern Michigan industrial concentration, and in cities between these two areas that were well connected to both by railroads. At the extreme western end of Lake Ontario, and with a 1986 metropolitan population of 568,471, Hamilton developed a large iron and steel complex by utilizing iron ore brought from the Lake Superior region through the Welland Canal and, more recently, ore brought from the Shield in Labrador via the St. Lawrence. Coal was also imported inexpensively from the Appalachian region by direct rail connections through Buffalo. The complex of foundries and rolling mills undoubtedly encouraged development of further industry within Hamilton as well as transport and farm machinery in nearby St. Catharines. Hydroelectric power from Niagara Falls, good water access to both the St. Lawrence and the interior Great Lakes, and abundant agricultural production have attracted such industries as chemical, textile, metals refining, and food processing to this region.

Around the western end of Lake Ontario lies a small crescent-shaped region of especially intense economic activity and spatial interaction. This is Toronto's immediate hinterland and a region that has been very important in that city's continuing economic dominance. Between the automobile center of Oshawa, through Toronto and Hamilton, to the uncomfortable blend of high-value agriculture and chemical and metal-fabricating industries on the Niagara Peninsula, the land is occupied intensely and productively. Sometimes called by the Amerindian name Mississauga, the region has been widely recognized by the more memorable appellation of "the Golden Horseshoe." As attractive as this place-name may be, there is some danger of misunderstanding by the unwary. The horseshoe shape is apparent, but the "golden" label refers only to the money-generating potential of the region's activities. As geographer Maurice Yeates put it, "The word 'golden' can hardly refer to the environment, for much of it is cement, steel, and macadam, and although areas

Toronto's skyline continues to demonstrate the city's continued growth as an industrial, commercial, and communications focus. The Skydome baseball stadium, adjacent to the CN tower, is only part of the recreational complex developing along Toronto's harborfront. (Metropolitan Toronto Convention and Vistors Association)

of beauty exist, there has been little regard for preservation."[2] This portion of the Canadian core also has the highest per capita income levels in the country. Thus, the advantages of cheap transportation, good access to raw materials, energy, and markets, proximity to concentrations of labor and supportive industry, and the early growth of Toronto and Hamilton have combined to stimulate urbanization and manufacturing throughout the Golden Horseshoe. In many respects, this region possesses much of the promise and most of the problems of Megalopolis (Chapter 5), although on a far smaller scale.

The southwestern Ontario industrial clusters also take advantage of good accessibility, resources, proximity to other manufacturing activity, and, to some degree, local mineral resources. Petroleum deposits with commercial value were discovered in Oil Springs, near Sarnia, shortly after the middle of the nineteenth century. A refinery at Sarnia by the beginning of the twentieth century and subsequent connection of the city by pipeline with the large petro-

[2] Maurice Yeates, *Main Street* (Ottawa: Macmillan of Canada, 1975, p. 165.

leum producing regions in Alberta made this city of 84,483 (in 1986) a center of petrochemical industries. The complex interindustry linkages of production and supply spilled the southeastern Michigan automobile-producing region across the Detroit River into Windsor (251,136 in 1986). There the industrial mix is supplemented by food and beverage industries. In the cases of both Sarnia and Windsor, location on the continent's major industrial waterway supported growth within the manufacturing sector of the Canadian economy.

Several additional industrial centers have developed between the Toronto-Hamilton area in the east and Windsor and Sarnia in the west. Such cities as Guelph, Kitchener-Waterloo, Brantford, and London are all located on major east-west railroad connections across the southern Ontario peninsula. Limited more to local resource advantages than the other manufacturing clusters, the industries in these cities are involved in food processing, in activities dependent upon large and accessible supplies of wood, or in a manufacturing process, such as electronic machinery production, drawn to the labor supplies of established urban centers.

With the preceding discussion in mind, the urban and industrial differences between the Ontario and Quebec portions of the Canadian core should be clear. Although both portions possess a single dominant urban center and smaller industrial complexes located near good transportation routes and accessible to local resources, the industrialization process is more widely spread across southern Ontario. In a similar manner, Ontario also contains more large cities, after Toronto, than does Quebec, after Montreal. One consequence of these differences has relegated Quebec to a relatively low ranking of all provinces in terms of average family income (see Figure 8-5). The spatial variation of income in southern Quebec is undoubtedly great. Many sections of Montreal and Quebec City as well as other urban centers contain small regions of population with very high annual income. In other parts of these cities and across wide sections of rural Quebec, other people reside and labor each year to earn incomes that can be called no more than marginal. In these rural-urban income contrasts, southern Quebec is no different from other major regions of the continent. However, Quebec possesses only a moderate overall average income even though it occupies a substantial portion of the national core of Canada where incomes might be expected to be higher.

One of the essential differences between the Quebec and the Ontario portions of the national core has been the relative ease of access that the Ontario part has had until recently via the Great Lakes to a large region of intense economic activity: the U.S. portion of the interior manufacturing core. This relative advantage changed somewhat when the St. Lawrence Seaway was completed in 1959. Through construction of locks and canals large enough to handle ocean-going vessels, the Great Lakes were opened to direct oceanic shipping. The Seaway also permitted access between southern Quebec and the industrial interior of the continent, thus reducing the apparent advantage enjoyed by southern Ontario.

After almost 30 years of operation, however, the impact of the Seaway on the Canadian region most directly involved with the St. Lawrence (southern Quebec) has not been extraordinary. In fact, southern Quebec may have been hurt by the Seaway. Considerable industry is located along or near the St. Lawrence, but much of this was present before the Seaway was opened. Most of the benefits from the Seaway have been enjoyed by ports around the Great Lakes, including southern Ontario, for it greatly increased their sources of supply and potential markets. These benefits flowed to both the U.S. and the Canadian portions of the continental manufacturing core, emphasizing its international character. But a major consequence for Montreal has been the loss of many transshipment activities and the potential industry associated with break-in-bulk freight handling. Clearly, the increased amount of traffic that now moves up and down the St. Lawrence can have no real benefit for southern Quebec, because it does not stop.

REGIONAL ORGANIZATION AND CONTINENTAL INTEGRATION

Two additional characteristics of the Canadian core region should be considered. One involves the spatial organization of the region—how and why places in southern Ontario and Quebec are tied to each other through their interaction. The other is related to the general economic integration of the Canadian core with the U.S. manufacturing core. Some of the integration is visible in transportation lines, the flows of people and goods across the political border, and so forth. But what may be even more important are the less obvious financial exchanges that represent corporate investment and control, primarily by U.S. interests over Canadian companies.

Regional Organization

Implicit in much of the earlier discussion of the content and diversity of the Canadian core is the idea that places within the core are important to the Canadian economy at least partly because they are located where they are. Quebec City benefited early from its location at the head of sea navigation but later suffered due to poor rail connections to other places. Montreal was built on the St. Lawrence where water transport was broken by rapids, a site from which New York City was directly accessible. Windsor developed across the Detroit River from the larger center of Detroit. London, located about halfway between Toronto and Windsor, was also in the middle of a good agricultural area and where two major land transport lines crossed. Similar notes could be made for all of the larger urban centers. Some, like Guelph, remained small settlements, the stimuli for their growth appearing in the twentieth cuntury. Others, such as Kingston, have gone through a growth-decline-growth sequence as changing external factors altered the importance of the cities' locations and, thereby, its development support. Clearly, knowledge of the location of economic activity is often critical to an understanding of the reason for a city's existence.

The Canadian core, as discussed, is partly a culture hearth, partly a political focus, and partly the center of the nation's economic structure; it is the diversity of culture in the region and the way these differences have been translated politically that often threatens to divide the national core. But as part of a larger continental core, it is primarily the economic and urban aspects of this region that have tended to hold it together. The components of this region do not exist in isolation, possessing specific characteristics and linked to other places only by their relative location. Instead, each place is tied to other places by the movement of people, goods, or ideas. This pattern of interrelationships is fundamental to the location of economic activities and to the general manner in which regions develop. Because of its explicitly spatial nature, the pattern of economic interrelations is said to be an expression of the *space economy* of the region.

Examination of a region's space economy is useful because it allows us to deal with the full set of interrelated parts with emphasis on the relations rather than on the place characteristics alone. One way in which the notion of a region's space economy could be approached is through the principles suggested by A. K. Philbrick.[3] He called the theme underlying his suggested principles areal functional organization instead of space economy because his principles pertain to noneconomic functions as well, but they can be used to analyze a region's economic interconnections. In this way, they can be considered an approach to an understanding of the space economy. Philbrick's principles deal with the manner

---

[3] Allen K. Philbrick, "The Principles of Areal Functional Organization in Regional Human Geography," *Economic Geography*, Vol. 33 (1957), pp. 296–336.

in which economic, social, and political functions are organized by people within the areas in which they operate. Our concern here is with the application of these principles to the organization of urban-economic functions in the area of southern Ontario–southern Quebec.

Four of Philbrick's original principles are useful to us. First, the spatial pattern described by nearly all human activity has a place of concentration or emphasis (focus) connecting otherwise unrelated places. For an individual, the focus may be a residence or a workplace; for a business, the focus may be company headquarters; a city is the focus for its hinterland; and so forth. This principle states the way in which human activity is organized. Through this organization, the places where this activity occurs can be said to possess *focality*. The second principle states that the sites on which human activity focuses are localized. This simply recognizes that each focus of activity has a unique and discrete position on the earth's surface and that position can be viewed as relative to other places where activity occurs. Third, these nodes of activity are interconnected, and the interconnections define units of organization that are more complex than individual activity. Through this principle, Philbrick was observing that lines of transportation and communication are the means by which an area is organized into a spatial system. The fourth principle deals with the fact that different activities may have their focus in the same location, either because of the complexity or quality of their respective hinterlands or because they share mutually beneficial advantages of proximity to each other. The fourth principle also recognizes that some places are the focus for a greater variety of activities than other places. Large urban centers usually have more functions than small centers. Those that have a wider range of functions are said to be of a higher order than the few-function places. Philbrick's fourth principle, therefore, states that there is a hierarchy of functional organization in which lower-order places are connected with (or focused on) higher-order places as well as nearby places of the same order. In other words, each city tends to have a set of smaller places tied to it by transport and communication links and, in turn, to be one of another group of cities focusing on an even larger metropolis. Note that nothing is said here about *where* these cities are or should be located. Such questions about the reasons for the location of a particular city are more often addressed by considering *site* and *situation* factors (see pp. 96–99). The concern throughout our consideration of a region's space economy is the identification of the way in which areas are organized and tied together by means of interconnected functions.

The functional organization of the Canadian core area illustrates many of these patterns (Figure 7-6). If we refer to a hierarchy of complexity in the organizational levels where human activity occurs, the simplest level is referred to as low-order or first-order. Succeeding levels of organizational complexity are called higher-order sites. In the Canadian core, first-order individual residences and second-order villages are scattered throughout the settled area. Higher-order places emphasize the linear shape of the core region extending between Quebec and Windsor and the urban-economic differences between the two provinces in the core. The core's functional organization (and its space economy) is dominated by the two nodes of Toronto and Montreal. Each fifth-order city is the focus of a pattern of fourth- or lower-order cities (smaller and economically less complex than either Montreal or Toronto). The functional connections emphasize the cultural and provincial differences between Montreal and Toronto. The only exceptions are Cornwall's link with the much closer Montreal and Ottawa's interconnections, as Canada's political capital, with both major cities. In addition, these large fifth-order centers are strongly connected with each other by road, rail, air, and telecommunications (thereby defining the main axis of core organization) and with fifth- , sixth- , and seventh-order foci in the United States and around the world.

The region's pattern of functional organization also illustrates dramatically the relative pau-

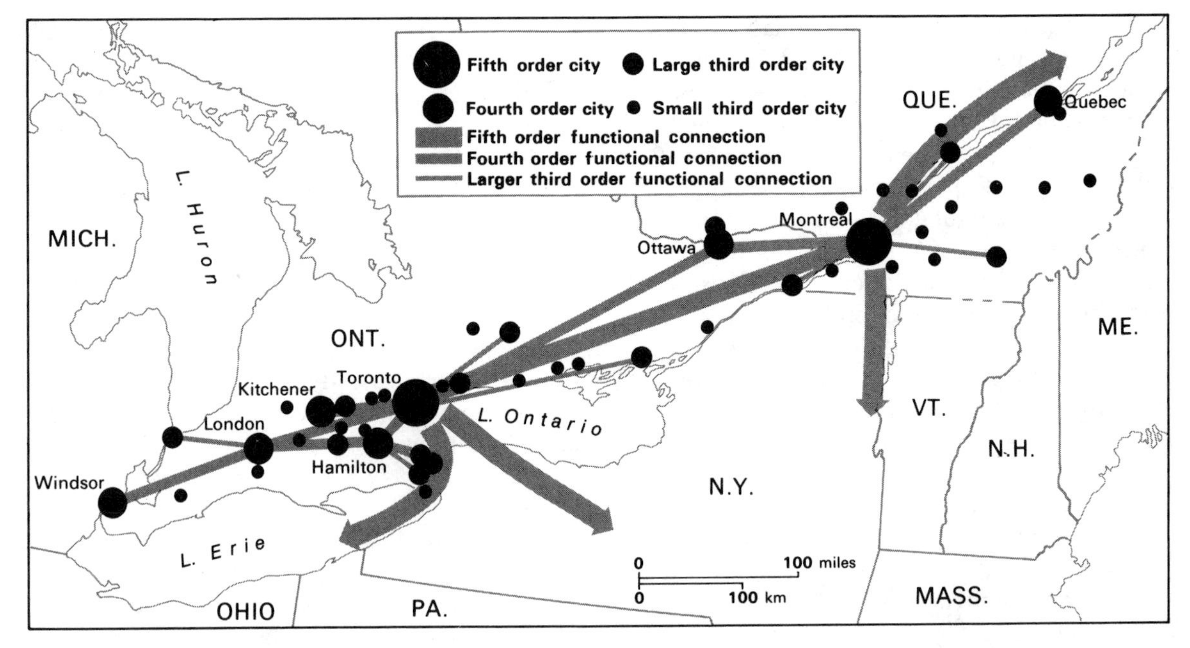

Figure 7-6 The functional organization of the Canadian core. Toronto and Montreal are the focuses for their respective regions, but there is clearly more urban development in Toronto's immediate hinterland of southern Ontario than in Montreal's equivalent hinterland of southern Quebec.

city of urban-economic development in southern Quebec when compared with southern Ontario. Quebec is the only fourth-order city in the province, while four have developed in Ontario between Toronto and southern Michigan. This difference in economic base is shown again with third-order nodes. Aside from Hull, Ottawa's twin city across the Ottawa River, southern Quebec has only two of the larger third-order centers, although Cornwall is close enough to Montreal to be caught in its functional organization. Southern Ontario, on the other hand, can support nine such nodes, focusing on Hamilton, London, Kitchener, or Toronto. It is only when considering the small third-order centers (less than 30,000 population) that the Quebec portion of the core equals Ontario. These towns are agricultural market centers, contain small-scale industrial supply activities or have received strong economic stimulation too recently to have yet grown large.

Regional Integration

A significant facet of the Canadian core's space economy that does not receive appropriate emphasis by this discussion of functional organization is the region's position within the continental core. Most of Canada's national core is also part of North America's great manufacturing core region. The Canadian portion is joined to the larger United States part by commercial traffic along the Great Lakes and the upper St. Lawrence River just as these waters separate the two parts. Canada's core is an integrated section of the continental core through road, rail, and air connections focusing on Toronto and Montreal and on the lower-order centers in the Niagara Peninsula and in extreme southwestern Ontario.

A more subtle form of continental economic integration has also been taking place. Since the turn of the century but more rapidly after World War I, American business and industrial firms have invested heavily in the Canadian economy. Only 14 percent of the foreign capital invested in Canada in 1900 originated in the United States with most of the balance being British. The U.S. proportion had increased to 53 percent by 1926 and surpassed 80 percent (and $25 billion) by 1966. By 1988, the total U.S. investment had increased to more than $66 billion. This invasion of U.S. capital, largely an attempt to circumvent tariffs imposed on U.S. manufacturers by Canada, also reflects an increasingly strong U.S. economy and a gradual weakening in the British economy. By establishing branch plants in Canada, U.S. firms gained access to Canadian markets and British markets without paying the import penalty of tariffs.[4] In some cases, this meant rapid development of Canadian Shield resources or construction of new manufacturing plants to employ Canadian workers and supply the Canadian market.

Most American investment has been in the mining and the manufacturing industries. More than 50 percent of Canada's mining industry, especially mineral fuel mining, is owned or operated by parent U.S. firms; fully 40 percent of Canada's manufacturing industry, especially the automobile industry, is controlled from the United States. And because Canadian manufacturing is concentrated in the southern Ontario-Quebec core region, these two provinces have drawn special attention from U.S. investment capital. As early as 1972, nearly 20 percent of the total taxable income from manufacturing generated by U.S.-controlled companies in Canada occurred in Quebec and more than 60 percent of such income occurred in Ontario (Table 7-1).

This level of U.S. capital investment has been viewed with considerable alarm by some Canadians. Much as Quebecois nationalism reflects French Canada's strong concern and resistance to cultural assimilation by the larger Anglo culture, a broader Canadian nationalism has been expressed reflecting fears of economic assimilation by the larger American interests. Canadian legislation was passed to limit the penetration of foreign (mostly U.S.) capital, but some fear these restrictions are too limited and too late. Also relevant to Canadian concerns is the fact that while U.S. economic involvement is widespread, a considerable share is concentrated within the heart of Canada's national economy—within its national core region.

Canada's desire to remain economically independent of its large neighbor to the south are strong, but the bonds that bring these two countries together should not be overlooked. The political boundary between the two countries is, in fact, a selective and semipermeable barrier. For example, it is very easy for citizens of one country to enter the other for recreation or business. The boundary may be indicated by no more than a strip of paint across the road and a few brief questions from a customs agent. A truly integrated continental economic core region is one result of such a transparent border. But another consequence has been intense cultural and economic influence in Canada from the much more populous United States. More than one quarter of all U.S. exports in 1986 were sent to Canada, with the flow of goods and materials from Canada to the United States comprising more than three quarters of Canada's total exports. Canada's population is politically and culturally distinct from its large southern neighbor, but it is only 10 percent as large as the American population. In spite of its internal divisions, the Canadian core region is strongly influenced by and integrated within a much larger continental culture and continental economy.

---

[4] Although this movement of investment capital may not seem especially surprising given the proximity of these two countries and the relatively open border between them, the rate of investment was unusual for noncolonial powers prior to World War II. It is now a phenomenon that is significant worldwide with great volumes of capital moving into the United States as well as from the United States to other countries.

Table 7-1 Percentage of Taxable Income from Manufacturing, 1972, by Region and Control of Company

Company Control	Atlantic Provinces	Quebec	Ontario	Prairies	British Columbia	Total
United States	2.7	18.9	62.2	9.6	6.7	100
Canada	3.6	28.8	43.2	9.8	14.4	100

Source: W. Clement, *Continental Corporate Power* (Toronto: McClelland & Stewart, 1977), Table 63, p. 298.

ADDITIONAL READINGS

Bradbury, John H. "State Corporations and Resource Based Development in Quebec, Canada: 1960–1980." *Economic Geography*, 58 (1982), pp. 45–61.

Cartwright, Don. "Language Policy and the Political Organization of Territory: A Canadian Dilemma." *The Canadian Geographer*, 25 (1981), pp. 205–224.

Easterbrook, W. T., and M. H. Watkins. *Approaches to Canadian Economic History*. Toronto: McClelland and Stewart, 1967.

Foley, James, ed. *The Search for Identity*. Toronto: Macmillan of Canada, 1976.

Innis, Harold A. *The Fur Trade in Canada*, rev. ed. Toronto: University of Toronto Press, 1956.

Marseau, Jean Claude. *Montreal in Evolution*. Montreal: McGill–Queens University Press, 1981.

Royal Commission on Bilingualism and Biculturalism. Preliminary Report, Preliminary Report and Books I and II. New York: Arno Press, 1978.

Skidmore, Darrel R. *Canadian American Relations*. Toronto: Wiley of Canada, 1979.

Sussman, Gennifer. *The St. Lawrence Seaway: History and Analysis of a Joint Water Highway*. Montreal: C. D. Howe Research Institute, 1978.

Watson, J. Wreford. "Regions of Canada: Central Settled Canada." In *Canada: Its Problems and Prospects*, pp. 130–177. Don Mills, Ont.: Longmans Canada, 1968.

West, Bruce. *Toronto*, rev. ed. Toronto: Doubleday Canada, 1979.

Whebell, C. F. J. "Corridors: A Theory of Urban Systems." *Annals of the Association of American Geographers*, 59 (1969), pp. 1–26.

CHAPTER 8

The Bypassed East

If you look at a map of the eastern seaboard of North America, you might notice a lack of large cities along the coast north of Boston. Few major overland routes extend inland from this coast, and interior cities are generally smaller than are those along the ocean. This area, comprising the Atlantic Provinces of Canada plus northern New England and the Adirondacks of New York, can be referred to as the Bypassed East (Figure 8-1).

The Bypassed East is near, even astride, major routeways, but not on them. Ocean transportation can easily bypass the region. There is little advantage for oceanic trade to pass through the harbors of the Bypassed East. Fine harbors farther south along the coast have easier and more direct access to the U.S. interior. To the north, the coastline of the Atlantic Provinces is hollow, indented by the wide, deep estuary of the St. Lawrence River. Jacques Cartier sailed up the St. Lawrence hoping that he had found the westward sea route to Asia from Europe. At the present site of Montreal, he found rapids blocking his path, and he abandoned his search. What he found was an easy passage deep into the heart of what was to become Canada. It was to be Montreal, not one of the many fine harbors of the Atlantic Provinces, that became the major eastern port for Canada. Like northern New England, the Atlantic Provinces were settled early, but as Canadian settlement pushed westward, these provinces were increasingly isolated from the regions of growth and development.

The Bypassed East, then, is in a transportation shadow (defined as an area of limited transportation development located near an area of much greater facility availability). This has led to slow regional economic growth and even stagnation, for geographic accessibility is a key element in the growth prospects of a place (see Chapters 5 and 6).

As discussed in Chapter 1, we have placed southern and northern New England in two separate regions. The strong image of New England as a unified region is held almost universally outside the area and often accepted within it, but this is an historical carryover. Certainly, it was once a section of the United States with considerable uniformity in economy and institutions. Today, however, southern New England is a part of metropolitan America. Northern New England, for the most part, is not. It is much more like Canada's Atlantic Provinces in their characteristic landscapes, problems and prospects, even though these two parts of the Bypassed East are separated by an international boundary. There is perhaps no better example of the fact that regional boundaries can and do shift over time, although that shift may not be recognized by many observers if the previous region was clearly identified.

Lobster boats bobbing gently in the harbor, their pots stacked neatly along the shore. Weathered yet attractive frame houses of the seaside fishing village disappearing upslope in the cold, encompassing fog. A pervasive sense of solitude and isolation. To many, this harbor scene from Stonington, Maine, captures the essence of the Bypassed East. (William Johnson/Stock, Boston)

THE PHYSICAL ENVIRONMENT—THE JOY AND THE HINDRANCE

> The summer folk call it paradise mountain,
> While we call it poverty hill.

These lines from a well-known folk song are not about any part of the Bypassed East, but they could be. Much of this part of the continent is

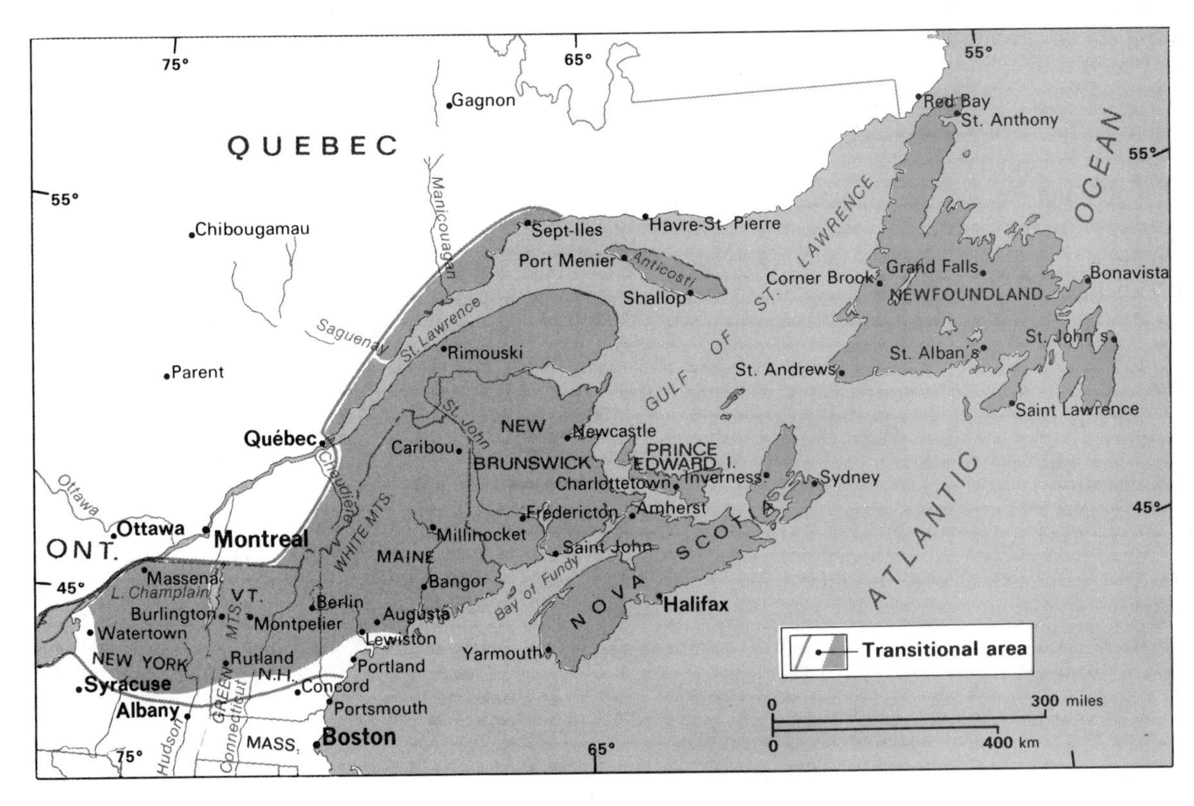

Figure 8-1 The Bypassed East.

beautiful. The Presidential Range in the White Mountains of New Hampshire contains some of the most rugged topography in the eastern United States. The extensive shoreline thrusts out into the Atlantic and meets the ocean's waves with a heavily indented coast that mixes dramatic headlands with many small coves bordered by rocky beaches. Large empty areas, almost totally lacking in settlement, are only hours away from some of the largest cities on the continent.

Major Metropolitan Area Populations in the Bypassed East, 1986	
Halifax, N.S.	295,990
St. John's, N.F.	161,901
Saint John, N.B.	121,265
Burlington, Vt.	108,213[a]
Lewiston–Auburn, Maine	105,259[a]

[a] Indicates 1990 population

Topography

Most of the Bypassed East is a part of the northeastern extension of the Appalachian Uplands. However, the structure of the area bears little surface resemblance to the clearly delineated Blue Ridge–Ridge and Valley–Appalachian Plateau system of the southern Appalachians (see Chapter 9).

The Adirondacks, in northern New York, are a southern extension of the Canadian Shield (see Chapter 18). This broad upland was severely eroded by continental glaciation, so that the surface features are generally more rounded than angular. Although elevations in the Adirondacks are not great, the areal extent of this highland is considerable.

A large upland plateau covers most of New England. This upland is old geologically and has also been heavily eroded by the actions of moving water and ice. One result of this long history of erosion is that elevations throughout the region seldom top 1500 meters (4600 feet). Widespread scouring by continental glaciers during the Ice Age rounded most of the hills and mountains across the plateau. Only where elevations were high enough to remain above the moving ice can one find more rugged mountains.

The two major mountain areas of northern New England are the Green Mountains of Vermont and the White Mountains of New Hampshire. The Green Mountains are lower in elevation, less than 1500 meters (4600 feet) at their highest, and were completely covered by ice during the Pleistocene, with the result that their tops are well rounded. The White Mountains, by comparison, rise to 1900 meters (6200 feet) and their peaks rose like islands above the ice. Thus, their upper slopes are rugged and steep, with an appearance far more like some of the younger mountains in the western part of the United States. Farther south, where the upland plateau has been heavily eroded by flowing water, several isolated peaks stand well apart from the major mountain areas to the north. The largest of them is Mount Monadnock in southern New Hampshire. Its name is given to all such areas of hard rock that have become low, isolated mountains as the surrounding rocks were removed by water erosion. Mt. Katahdin, an equally dramatic *monadnock*, dominates the landscape of its portion of central Maine.

Although northern New England (with New York) is a land that draws character from its mountains, its people find their homes and their livelihoods in the valleys and lowlands. The largest such areas are the Connecticut River Valley between New Hampshire and Vermont, the Lake Champlain Lowland along the northern Vermont–New York border (basically an extension of the Ridge and Valley area of the southern Appalachians), and the Aroostook Valley in northern Maine. This last valley resulted from the erosion of an area of soft rock by the St. John River. A number of smaller lowlands border the seacoast, and innumerable streams have sliced the plateau throughout the area. A detailed map of the distribution of population would look like an inverted map of elevation, with most people in the lowlands (Figure 8-2).

The mountains of the Atlantic Provinces are generally even lower and more rounded than those of northern New England. Elevations do not rise much above 700 meters (2200 feet). The

Figure 8-2 Empty areas of New England. (Adapted from *The Geographical Review*, 44 (1954), Lester E. Klimm.) This map of areas distant from any paved road or dwelling unit is based on data from the early 1950s. A 1990s map would look little different. The forest lands of northern Maine and mountainous areas throughout the area are beyond the margins of most settlement.

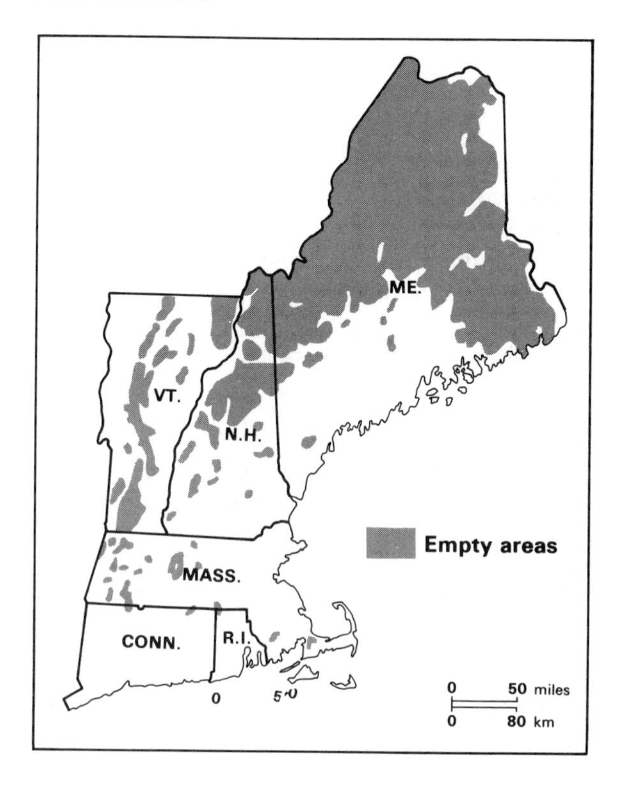

surface patterning of the southern Appalachians is missing almost totally here, although the basic grain of the topography does follow the same northeast-southwest trend. In the southern Atlantic Provinces, uplands in peninsular Nova Scotia and interior New Brunswick are separated by a wide lowland. Although most of it is ocean covered, it includes Prince Edward Island (PEI to Canadians).

As in northern New England, the lowlands of the Atlantic Provinces are the home of most of their people. The central lowland, which includes Prince Edward Island, is underlain with soft rocks, limestones and sandstones. The Annapolis Valley, in southern Nova Scotia, is a part of the same structure that underlies the Connecticut River Valley. These Triassic lowlands reappear frequently along the interior East Coast, even as far south as central North Carolina. The larger cities throughout the provinces, and nearly all of the population of Newfoundland, are clustered around the area's harbors.

Climate

This region is no place for those who desire a dry, warm climate. It is a place where polar, continental, and maritime weather systems meet, and the result is a climate that is seldom hot, often cold, and usually damp. Because of its location on the eastern side of the continent, the wind systems tend to push continental conditions into the area and to limit the maritime impact on the location, especially away from the coast. The substantial climatic difference between the coast and interior is further increased by higher inland elevations.

The Labrador current that flows southward along the Bypassed East is cold. Even in late summer, only the most intrepid swimmers are willing to dip themselves into its waters for more than a short time. Still, the climatic conditions along the coast are moderated substantially by proximity to the water. The growing season near the coast is as much as 70 days longer than the

Along the Cabot Trail on Cape Breton Island, Nova Scotia. Much of the coastline of the Bypassed East is rugged and dramatic, a delight for tourists but a difficult place for agriculture or urban growth. (Nova Scotia Tourism)

interior average of 120 days. Average midwinter temperatures at coastal sites are often 3° to 6° C (5° to 10° F) higher than at nearby interior locations. Midsummer temperatures, in contrast, are slightly higher in the interior than along the coast.

The maritime influence brings frequent cloud cover and fog, particularly along the southern coastline. Southeast winds blowing across the warmer waters of the Gulf Stream and then over the colder waters of the Labrador Current closer to shore create a dense fog by the time they reach land. This presents a substantial climatic problem, for the fog and cloud cover serve to cool temperatures further during the summer. It is consequently difficult to grow many crops that require more summer heat and sunlight than is usually available.

Almost all parts of the region receive substantial precipitation, usually between 100 and 150 centimeters (40 to 60 inches) annually. Precipitation is usually scattered evenly throughout the year. Snowfall is generally substantial, with most places receiving between 25 and 50 percent of their total moisture in the form of snow. Most interior locations average at least 250 centimeters (100 inches) of snow annually. Winter snow cover near the coast is sporadic, with frequent thaws and bare ground, but snow covers the ground for three to five months each winter inland.

POPULATION AND INDUSTRY IN A RUGGED LAND

The Bypassed East is not an easy place in which to live and work. Its harsh climate, hilly terrain, and thin, rocky soils limit agriculture, except in a few particularly blessed locations. Few mineral resource deposits of substantial size have been found until recently. Coupled with a small local market and relative isolation, this has limited the development of manufacturing. The advantages that the area does offer thus become relatively more important.

Early Settlement

This was not always the Bypassed East. Its foreland location, jutting far out into the Atlantic Ocean, meant that its shores were among the first parts of the New World encountered by European explorers and settlers. The first European settlement in Canada was opened by the French in 1605 at Port Royal on the Bay of Fundy. Less than a decade later, the English settled in Newfoundland, where they quickly established a series of fishing villages. By the midseventeenth century, many of the small harbors of central and southern Maine housed British villages. Settlement was kept out of the interior by the local American Indian population until the middle of the eighteenth century. The Indians in the lower St. Lawrence Valley were aided by the French in hopes that the British would be kept as far from French settlements as possible.

These early European settlers utilized two prime local resources. The rich fishing banks off the Atlantic Provinces—possibly visited by European fishermen well before Columbus made his first voyage to the New World—were immediately important. The banks, shallow areas 30 to 60 meters (100 to 200 feet) deep, in the ocean at the outer margins of the continental shelf, underlie waters that are rich in fish (Figure 8-3). Their shallowness allows the sun's rays to penetrate easily through much of the water's depth. This encourages the growth of plankton, a basic food for many fish. Cold-water fish, such as cod and haddock, are abundant. Using this nearby resource, the early European settlers soon began a substantial export of salted cod.

The other prime resource of the region was its trees. England had already lost much of its forest cover to the expansion of agriculture. As England began to develop an important navy and merchant marine fleet, it was forced to turn to the forests of Scandinavia for its ship masts and lumber. This was an unsatisfactory arrangement, for it made England dependent on a foreign power for one of its vital military and economic resources. Consequently, the English

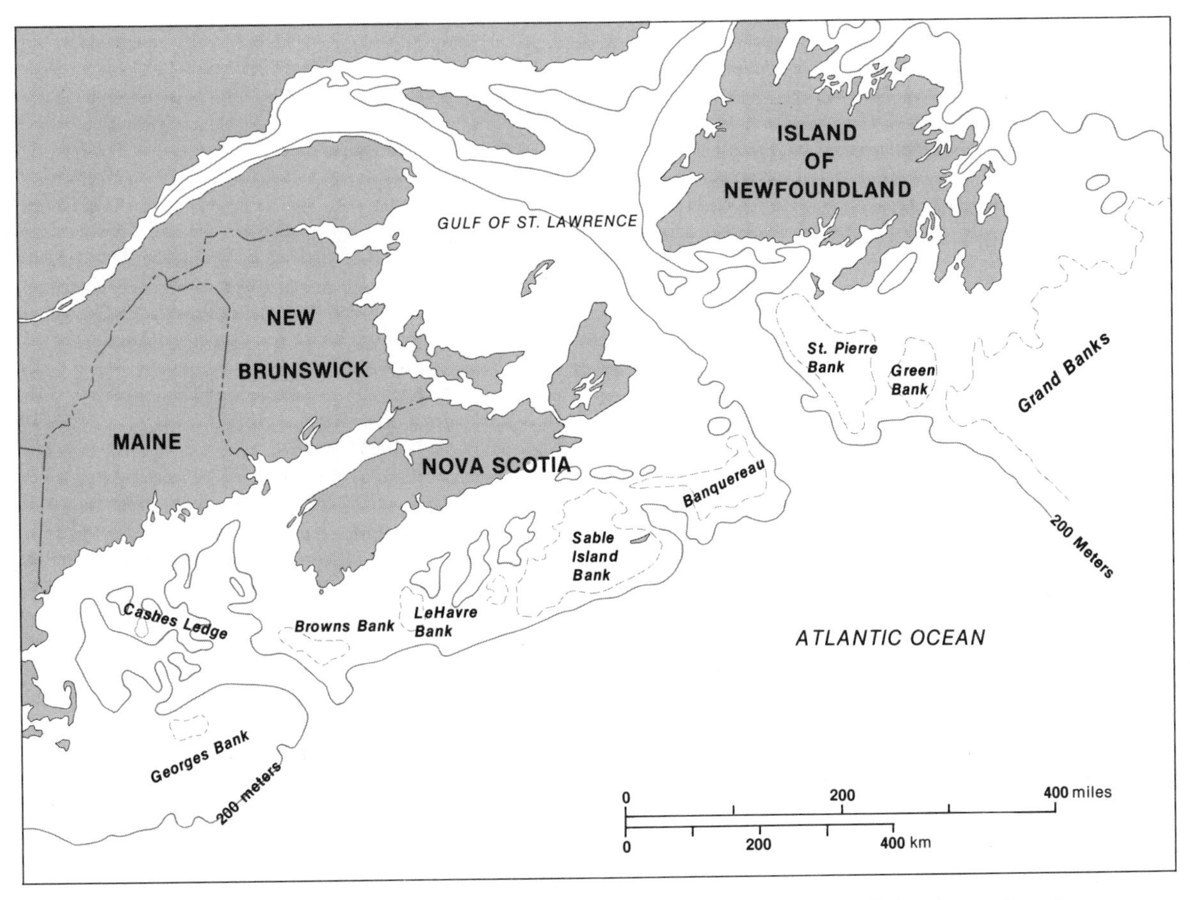

Figure 8-3 Offshore banks. These large areas of shallow, cold ocean off the shore of much of the region have been recognized for 500 years as among the richest ocean fishing waters in the world.

sought an alternative source within their territories.

The white pine was dominant in the forests of New England. A magnificent tree, it reached heights in excess of 60 meters (190 feet) and stood straight. Its wood was clear, light yet strong, and easily cut. Almost all of the virgin forests are gone now, and the second- and third-growth forests that remain are short and insignificant in comparison. A few lone giants remain, in some isolated areas, small groves that remind us of the forests that once covered the region. Forest resources allowed Maine to become a center for ship construction. The best of the white pines were reserved for the ship masts of the Royal Navy. Still, there was more than enough for all.

Agriculture was the third major occupation of the early settlers, but farms tended to be small and production limited. Early farming was primarily a subsistence activity. By the end of the 1770s, however, most of the lowlands south of Newfoundland, and even much of the sloped land, was in farms.

The peak of agricultural development in northern New England probably came just after the start of the nineteenth century. But two developments elsewhere in North America soon began to pull people off their farms, first in a trickle, then in droves. One development was the opening of the West. In the United States, settlement moved beyond the Appalachians onto the rich farmlands south of the Great Lakes early in the century. Then, in the 1820s, the construction of the Erie Canal, and later other canals farther west, made the markets of the East Coast more accessible to western farmers. The poor farms of upper New England rapidly lost what little market they had to crops imported from places like Ohio and Indiana. New Englanders left their farms and joined the migration westward, exchanging bad land for good. Although an accurate count is not possible, it is estimated that about 1 million New Englanders moved off the region's farms and headed west during the nineteenth century. It was said that the only crop those New England fields could produce was rocks and that their only export was people.

The pattern was much the same in the Atlantic Provinces. The settlement of the middle St. Lawrence River Valley in southern Quebec and especially the rapid expansion of settlement in southern Ontario in the decades after the American Revolution provided superior agricultural land (see Chapter 7). Construction of the Welland Canal to bypass Niagara Falls in 1829 increased the accessibility of southern Ontario.

A second blow to the region's agricultural fortunes also occurred during the late 1700s and early 1800s with the development of manufacturing in southern New England, where the Industrial Revolution began in the United States. Industrial growth created a great demand for labor. That demand was first met with New England farmers seeking the higher wages and steady income offered by manufacturing employment. An increase in child and female labor, particularly in the textile mills, further enhanced the value of manufacturing work over farming.

Agriculture Today

Agricultural decline has continued across most of the Bypassed East for the last 150 years. Today less than 10 percent of the land in the three states of northern New England is in farms; 100 years ago, the amount was closer to 50 percent. Until the last decade or two, many northern New England towns had patterns of population decline that lasted for a century or more. Farming retreated off the slopes, allowing them to return gradually to forest. Even in the valleys, soils were often too infertile, the climate too cold, and the farms too small for successful agricultural production. One result is a widespread pattern of agricultural land abandonment and expansion of scrubland and woods.

The pattern of decline is perhaps not quite as great in the Atlantic Provinces, but it is still substantial. From a peak late in the nineteenth century, total land in farms has declined by about two-thirds, and improved land has fallen by about the same proportion. Productivity remains poor compared to the rest of Canada, and regional declines in acreage and in production remain far larger than for other parts of the country.

Where farming in the Bypassed East remains important, it tends to specialize in single-crop production and to be concentrated in a few favorable locations. For example, the acid soils of Washington County, in northeastern Maine, support one of the country's major centers of wild blueberry production. Only Maine, New Hampshire, and Nova Scotia produce wild blueberries. While agriculture is found in a number of other locations, four significant areas of agricultural production in the region deserve special notice (Figure 8-4).

St. John–Aroostook Valley. This area of northeastern Maine and western New Brunswick is the Bypassed East's newest major agricultural area. Commercial farming did not become important until late in the nineteenth century. It was soon one of the leading potato-producing areas in the United States and Canada. The silty loam soils are ideal for potato growth, and the short growing season encourages a superior crop that is used widely elsewhere as seed potatoes. Large-scale, mechanized farming pre-

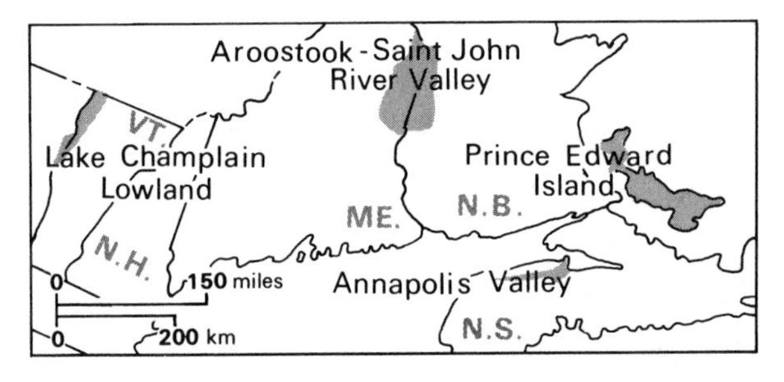

Figure 8-4 Major agricultural areas of the Bypassed East. Most of the region's farming is concentrated in its few lowlands of substantial size.

dominates; French-Canadian farm workers are employed heavily during the harvest season.

The valley's potato growers have gone through a difficult period during the past several decades. Changing American eating habits led for a time to a declining market demand for potatoes. A result has been frequent overproduction with consequent depressed market prices. More recent increases in consumption have emphasized baking potatos from Idaho and frozen and processed potato products. Western potato growers, primarily in Idaho and Oregon, created well-organized merchandising operations, developed many new kinds of manufactured potato products, created effective advertising, and used coordinated marketing. Valley farmers, proudly independent, lost many of their traditional markets to these other areas. Today the valley's agricultural economy is depressed and production is down. Many farmers have gone out of business. Others have tried to diversify their crops, but so far they have met with limited commercial success. Indeed, today poultry and eggs, mostly from large producers in south-central Maine, account for half of the state's agricultural income. That is double the potato's share. Still, conditions are changing. One of the world's largest producers of frozen French fries is in the St. John Valley in New Brunswick.

The Lake Champlain Lowland. Megalopolis represents by far the largest concentration of demand for agricultural produce in the United States. Many of these needs can be supplied by crops grown anywhere. Such products may be relatively low-bulk, high-value crops, like tea, which can be shipped great distances from overseas growing areas with only a small percentage increase in price. Or they may be crops, like wheat or corn, which can be handled roughly without damage or spoilage, so that they can be shipped great distances in bulk containers. Or they may be processed near the area of production, so that they can easily be shipped and stored, such as frozen and canned vegetables.

Fresh milk meets none of these requirements. It is a relatively high-bulk, low-cost product. It spoils easily and cannot be stored for any extended period. Consequently, urban areas usually rely on nearby agricultural areas for fresh milk. Most parts of the United States and Canada are subdivided into milksheds, or nodal regions that supply a city with its milk needs.

The Champlain Lowland's proximity to Megalopolis gives it a substantial market advantage over more distant areas for milk sales. It is in the milkshed of both New York City and Boston. In addition, its summers are mild and moist, a climatic condition that encourages the growth of fodder crops. These cool summers are also well suited to dairy cows. Vermont has long led the United States in the per capita production of dairy products, and most of that comes from the Champlain Lowland. Dairy farming accounts for 90 percent of the state's agriculture, and much of it is found in the Champlain Lowland. Falling prices and a federal policy to reduce milk subsidies by buying up whole herds and destroying them resulted in a 10 percent loss of Vermont dairy farms in the late 1980s.

Prince Edward Island (PEI). An overwhelmingly agricultural economy is found on PEI. The local economy provides little alternative employment, and most of the workers are engaged in agriculture. Over half the land of the province is in farms, with a quarter in crops. The latter surpassed the cropland share of any other province or state in either country. It may also be that no

Prince Edward Island is a land of gently rolling hills and small farms. Its agricultural economy suffers somewhat because of its isolation from the major markets of eastern Canada. (Information Service, Prince Edward Island)

more beautiful agricultural landscape can be found in North America than this rolling, sloping land of small, green farms. Although agriculture is somewhat diversified, the major crop is seed potatoes. No bridge connects the island to the mainland. The high cost of transporting crops to mainland markets means that most other PEI crops are consumed by its 125,000 inhabitants.

One of the delightful aspects of PEI for tourists is the faintly worn appearance of many of the farmsteads. They have been here for a long time. Declining demand for its principal crop, increased competition from other sources, and too many farms that are too small to be profitable in a modern farm economy are all part of farming on PEI. The comfortable, used, lived-in appearance of the landscape reflects in part an agricultural economy that has stagnated and no longer generates sufficient sales for investments in improvements.

The Annapolis Valley. Nested into southwestern Nova Scotia, protected from northwest winds by a low mountain ridge, the Annapolis Valley is another attractive agricultural area. The valley, about 130 kilometers by 15 kilometers (80 miles by 9 miles), has long been one of the main apple-producing areas of Canada. Once again, the principal crop has suffered in recent years by increased competition from sources closer to its traditional markets, in this case Great Britain and the urban centers of Quebec and Ontario. Today, most valley apples are processed rather than marketed fresh. Crop diversification has been more successful in recent years, especially in the production of produce and dairy products for nearby Halifax. With more than a quarter of a million people, Halifax is easily the largest urban center in the Atlantic Provinces.

Forestry

Parts of northern New England offer a good example of the destructive potential of uncontrolled logging. Limited organized reforestation followed the cutting of trees, so that much of the forest that replaced the original trees is of poor quality for both lumber and pulp production. Neither is it as handsome as the forest it replaced. None of these three New England states is an important lumber supplier today, and production is not increasing significantly. This is an area where the great majority of the land is in trees, so the lack of a larger wood products industry may be somewhat surprising. The slow growth rate for trees in the cool climate plus the continued lack of large-scale corporate operations in much of the area are other contributing factors.

An exception to this pattern of limited output is the pulpwood production of northern Maine. Here, on some of the most inaccessible large tracts of land in the eastern United States—where a small number of private owners control most of the land—forest industries remain important. Many of the roads and recreation facilities in these Maine woods are also privately owned. The area contains some of the finest white-water canoeing streams in the country as well as large numbers of lakes and marshes. Disagreements over forest utilization versus forest preservation has been a source of conflict between the pulpwood producers, on the one hand, and canoeists and conservationists, on the other hand.

Forestry plays a much greater role in the economy of the Atlantic Provinces and is the leading supplier of exports from that area. About 90 percent of the area of New Brunswick, 75 percent of Nova Scotia, and 30 percent of the island of Newfoundland is in forest. Most forestland in the first two is in private hands, unlike the pattern of provincial ownership of forestlands common to most of Canada. Only a small percentage of the labor force actually works in the woods, but the manipulation of the cut wood accounts for nearly half of the entire manufacturing employment of the Atlantic Provinces. Pulp and paper are the key products, with the most important areas of logging located in northern New Brunswick and on Newfoundland. The largest paper mills are on the Gulf of St. Lawrence shore of New Brunswick, close to the river mouths. The logs can be floated to the plants for manufacture, and the finished products can be loaded directly on large ships for transport to market.

Fishing

Fishing remains an important, if troubled, part of the economy of the Bypassed East. Canada is among the world's leaders in the export of fish products. Nova Scotia leads all of the provinces in total fish catch each year, followed in order by British Columbia, Newfoundland, New Brunswick, and PEI. Most of Canada's fish exports come from the Atlantic Provinces. Maine lobstermen account for about 80–90 percent of the total U.S. lobster catch, and the state also leads in sardine production.

There are two kinds of ocean fishing in the region. Inshore fishing, the more important, uses small boats and requires a relatively small capital investment, with lobsters and cod the most valuable catch. Cod accounts for about 40 percent of the total catch in the Atlantic Provinces. Maine and Nova Scotia lobstermen each provide about half the total catch annually in their respective countries. Deep-sea fishing on the banks off the coast requires far larger boats and more capital investment. Most fish caught on the banks are bottom feeders such as cod, flounder, and halibut.

The inshore fishing industry faces several substantial challenges. Pollution is a growing problem in some areas, forcing fishermen to move to offshore locations. The inshore fishermen, with their small investments and small incomes—often living in one of the many small fishing villages scattered along the coastline of Nova Scotia and Newfoundland—cannot easily obtain the money needed to buy and outfit large, modern oceangoing vessels. When they move onto the banks with their smaller vessels, they are in direct competition with large fleets of modern ships, including factory ships that process the fish at sea.

The provincial government of Newfoundland, concerned about the uncompetitive nature of its inshore cod-fishing industry, initiated a program of centralization some years ago by encouraging fishermen and their families to move, at government expense, to larger communities. There they could form larger, more effective cooperative fishing operations and, at the same time, help cut the high cost of providing a variety of government services to the many small fishing communities. The plan has had substantial success; out of a total of 1200 of these small communities (called outports in Newfoundland), 500 have been abandoned. A confounding problem, however, is the already high urban unemployment in the province.

For many years Newfoundland was not a part of Canada. Not until 1949, after several years of disastrously low fish prices in Europe had wreaked havoc with Newfoundland's economy, did residents of the province vote by a narrow margin to join Canada. The former name for the eastern provinces, the Maritimes, was changed to the Atlantic Provinces to reflect the incorporation of Newfoundland into Canada.

Offshore fishermen in both countries face heavy competition from European and Japanese ships. Americans and Canadians have lagged far behind their foreign competitors in turning to larger fishing vessels. Still, the offshore industry is expanding, especially in Newfoundland. In 1977 Canada established a 200-mile (320-kilometer) offshore management zone closed to unrestricted fishing by foreign vessels. The United States had done the same a year earlier. The two countries have had to arbitrate conflicting claims along their mutual Atlantic Coast. The number of registered Canadian fishing boats increased sharply in the following several years.

Minerals Production

Offshore fishing has been threatened recently by the high demand for domestic petroleum in the two countries. Fears of pollution from offshore oil drilling in the rich fishing banks in the United States were overruled in 1979 when the Department of the Interior granted exploration leases to several oil companies. Major resources of petroleum and natural gas have been discovered in the Hibernia field near the Grand Banks off Newfoundland, and natural gas has been found near Sable Island, 300 kilometers (180 miles) off Nova Scotia and Newfoundland. It is hoped that the Hibernia field will supply Canada with its first substantial eastern petroleum production in the 1990s. Development is dangerous in this often stormy area. In the early 1980s one large rig sank in a storm with the loss of over 80 lives.

The possible environmental impact is not the only problem with developing these offshore fields. Development costs will be many billions of dollars. Ownership conflicts between the provinces and the central government are difficult to solve. The Canadian and Newfoundland governments in 1985 reached an agreement regulating extraction and revenue sharing.

Mining other than for offshore petroleum and natural gas in the Bypassed East is not currently of great importance. This was not always the case. Iron ore has been mined in the Adirondacks for more than 100 years, and the reserves there are still substantial, but total output from the mines is relatively small. Although New York is usually among the top five states in iron ore production, only the top two, Minnesota and Michigan, can realistically be called major producing states.

Nova Scotia has some of the larger coal reserves in Canada, but production fell steadily to about 2 million tons annually before reviving somewhat in the late 1970s. There is simply not a large demand for the coal in the Atlantic Provinces; American coal undersells it in the more distant markets of Ontario. The coal has been used as coking coal at a small steel plant in Sydney. Unfortunately, the plant is too small and outmoded to be efficient, and it is kept open as a provincial government corporation. Steel from the plant is limited to the local market. Steel produced elsewhere in larger quantity with much better economies of scale can be shipped to the Atlantic Provinces and still undersell the local product. Again, the isolation of the region from the major markets of the two countries has hurt the local economy.

The igneous and metamorphic rocks of northern New England have made the area an important producer of building stone for years. Many granite quarries operate in central Vermont and along the central coast of Maine. Vermont is also the leading marble-producing state in the country. The value of all of these rocks is small compared to the minerals industries found in other parts of the continent, but it is still an important element in the economy of the two states, especially Vermont.

The largest mining operation in the Atlantic Provinces is based on the development of the iron ore deposits in Labrador near the Quebec border, but it has little impact on the region (Chapter 18). Although these deposits are located in Newfoundland, the ore travels by rail southward to a railhead at Sept Isles, Quebec, where it is loaded on lake vessels and sent up the St. Lawrence River. The only benefit of that mining for the Atlantic Provinces is the tax that Newfoundland collects. Otherwise, its impact is significant for southern Quebec, not the coast.

Cities and Urban Activities

By a slight majority, most of the residents of the Bypassed East are urbanites. However, the region contains few substantial urban areas. The two largest cities of northern New England are Burlington, Vermont, and Bangor, Maine. Each has only about 100,000 residents in its urban area. In Canada, Halifax, Nova Scotia, had 295,000 residents in its urban area in 1986 while St. John's, Newfoundland, and Saint John, New Brunswick, had 162,000 and 121,000 metropolitan residents, respectively. Although substantial at the regional level, these are relatively small places.

The small size of the major regional centers is a good indication of what may be the greatest single reason for the relatively low per capita income levels of the region. Most higher-income occupations in the two countries are urban based, and this area lacks urban occupations. Although they are less than half of the total, a high percentage of the work force is engaged in primary occupations, which are traditionally among the lowest paying on the continent. Because of the absence of a large local market and because of poor access to major urban areas, primary industries have not served as the foundation for the development of a more broadly based manufacturing economy, as they have elsewhere in Canada and the United States. For example, in Vermont in the late 1980s a single IBM plant provided a sixth of the state's manufacturing jobs and paid a third of its corporate income tax. As we said earlier, a majority of the manufacturing jobs in the Atlantic Provinces relate directly to the manufacturing of the raw materials produced in the region.

Several cities in the region do have some locational advantages, and they are growing. Halifax, for example, is eastern Canada's second leading ocean port following Montreal. It became the eastern terminus of what is now the Canadian National Railroad in 1876. During winter, use of the St. Lawrence is limited by ice. Although a channel is cleared through the ice, the volume of traffic using the river is greatly reduced. It is during winter that the ice-free harbor at Halifax is at its busiest. The city has developed a diversified economic base as the regional center for government, finance, and commerce and is the most prosperous center in the Atlantic Provinces. Once called by many Canadians the

Halifax, Nova Scotia, is the largest city in the region. Its port function is most important during the winter, when the St. Lawrence freezes over, blocking access to Montreal. Its value has been enhanced by the construction of modern, competitive container piers such as this one. (Nova Scotia Tourism)

"Gray City" for its drabness, new construction and renovation since 1960 have greatly changed the appearance and atmosphere of the city. St. John's, on the Avalon Peninsula at the far eastern end of Newfoundland, is too far east to serve as a major port of entry for the country, but its excellent harbor promises continued growth as the principal port for eastern Canada's deep-sea fishing fleet. Usually ice free, its harbor is an important winter facility for fishing fleets from many countries as they fish the offshore banks. Saint John, the oldest incorporated city in Canada, boomed after 1890 and the completion of the Canadian Pacific Railway across Maine. However, it is more remote than Halifax and its harbor is of poorer quality. Like Halifax, its port is most active in winter.

THE FUTURE—PERPLEXING

The future of the Bypassed East is not easy to predict. We can make some generalizations about the present, and these can help in trying to anticipate the future. First, let us reiterate a few points. This is an area with a strong continuing economic emphasis on the collection and manufacture of raw materials, expecially fish and wood products in eastern Canada. Although a majority of the region's population is urban, it is substantially more rural than Canada and the United States as a whole. The entire region is in a transportation shadow, near some of the busiest routeways in the world, bypassed in favor of alternative paths that provided more effective penetration into the continent.

An important result of all of this has been regional poverty. In 1985 the average family income in the Atlantic Provinces ranged from $29,629 (Canadian) in Newfoundland, the lowest in the country, to Nova Scotia's $34,394. The average for all of Canada was $38,059. In 1988 the average per capita income in northern New England ranged from a low of $14,976 in Maine to $19,016 in New Hampshire (Table 8-1). The national average was $20,625. Average income

Table 8-1 Average per Capita Income for Selected States, 1960–1981

	1960		*1970*		*1987*	
State	Income	Percentage of U.S. Average	Income	Percentage of U.S. Average	Income	Percentage of U.S. Average
Maine	$1825	83%	$4766	83%	$15,106	92%
Vermont	1864	85	4924	89	15,302	93
New Hampshire	2172	99	5417	96	19,434	118
South Carolina	1396	63	4665	80	12,926	78
Mississippi	1196	54	2547	65	11,116	67
United States	2201		5893		16,489	

Source: U.S. Statistical Abstract, 1979, 1990.

levels in New Hampshire were boosted by the higher incomes in the urban areas along the southern edge of the state. Per capita income averages for the states of southern New England ranged from Rhode Island's $16,793 to Connecticut's $22,761, the highest for any state, except Alaska.

Northern New England

There does seem to be reason to anticipate that economic growth will continue in northern New England. The 1980 national census indicated that the long-term demographic stagnation had ended. Maine, New Hampshire, and Vermont were the only states outside the South and West to grow at a rate above the national average. In the 1980s New Hampshire continued to grow at a rate well above the national average, although Vermont and Maine grew at rates somewhat below the average.

There seem to be several reasons for this general shift in regional population fortunes. One is the gradual northward growth in Megalopolis. As the cities of the urban region expand, as new peripheral areas urbanize and become a part of urban America, and as people search farther outward to find residences away from the large cities, the Megalopolitan periphery has been pushed steadily northward in New England. Again, this trend has especially benefited southern New Hampshire.

Northern New England is also attracting a number of new manufacturing facilities, which are locating in many areas and tend to be light industry with a medium-sized work force. In part, they are locating in the region because the employers and their workers find the small-town and rural environments typical of the area to be good places to live. Also, the construction of several interstate highways into the region during the 1960s has provided greater accessibility to the south and west.

Recreation, Second-home Development, and Retirement. The tourist industry has been northern New England's boom industry since the end of World War II. Magnificent scenery, rugged coast, fishing, skiing, canoeing, and just plain driving around looking at the beauty ot the place—all of these are a part of this tourist growth. The region is fortunate enough to have several tourist seasons. There is the summer, the traditional time to vacation for most American families. Next comes the "Fall Foliage Festival," when thousands of automobiles fill the roads so that their occupants can get a glimpse of the oranges, yellows, and reds before they are covered with road dust. In winter comes skiing—"Christmas in Vermont," "White Christmas," and all the rest. Only the spring, the season of mud and flies, has not been fitted into the travel habits of Americans.

Nearly year-round tourism is a cornerstone of the economy of Northern New England. Located near Megalopolis, its many attractions (such as this ski resort near Stowe, Vermont) are within an easy day's drive of perhaps 50 million Americans. (George Bellerose/ Stock, Boston)

The economy of the Adirondacks area is also heavily dependent on tourism. Lake Placid, home of the 1932 and 1980 Winter Olympics, is but one of many ski areas. The state of New York oversees much of the area through its Adirondack State Park, the largest state park in the country. Although the state encourages outdoor recreation on its lands, many local residents argue that the growth should be limited by restricting construction of new tourist facilities.

Second homes, long a part of the life-style of the well-to-do in America, have in the last 30 years become the goal of much of the rest of America as well. Strung along the seashores and around the lakes, and strewn across the mountains is a growing collection of vacation homes. They are occupied by the owners for a few months or a few weeks each year and then rented for as much of the rest of the time as possible to help pay the purchase and upkeep costs. In a number of the counties of northern New England, there are more of these part-time dwellings than there are permanently occupied houses. No one is quite sure what the impact of this is or will be, but at a minimum, it has flourished in the local construction industry. As with tourism, this second-home development arises from the region's proximity to Megalopolis. Nearly a quarter of America's population lives within an easy day's drive of northern New England. Residents of Quebec and Ontario also find it a nearby and important recreation area.

Finally, many of the coastal communities of Maine, the small college towns of Vermont and New Hampshire, and old villages throughout the region have become popular retirement centers. These retirees, although generally not rich, are not poor, and they do make a significant contribution to the regional economy.

Eastern Canada's Prospects

The future prospects of the Atlantic Provinces are more clouded than those of northern New England. The Canadian government has recognized the region as a problem area. It set up a regional development commission to try to identify solutions to the problem of regional poverty and to help in their implementation. The Atlantic Development Board was created not only to advise on regional economic problems but also to oversee federal infrastructure investment. In 1969 the government established the Department of Regional Economic Expansion (DREE) to encourage employment growth in the slow-growth areas of the country. While it existed (the program was eliminated in 1982), roughly half of all DREE funds were given to the Atlantic Provinces, with Quebec getting most of the rest. Included in the overall program were direct payments to private industry encouraging them to locate and helping them survive in less developed areas as well as grants to improve such things as transport facilities and water and sewer systems and funding support for worker development programs.

It is difficult to assess the impact in the Atlantic Provinces of this considerable federal concern. Personal incomes began to converge with the national average after 1950. The process of convergence may have intensified somewhat in the late 1960s, although the recession of the late 1970s and early 1980s was especially hard on the Atlantic Provinces (Figure 8-5). Outmigration has slowed, and it may have turned around in New Brunswick and PEI. Larger urban areas in the Atlantic Provinces are growing fairly rapidly. The area's relative economic condition is improving and will probably continue to do so.

The Atlantic Provinces do not have many of the apparent advantages that explain the recent population growth of northern New England. They are not near any substantial urban region and are definitely not going to benefit from urban overflow. The improved accessibility that the interstate highway system provided to northern New England would be difficult to develop in Atlantic Canada, partly because the distances are greater and partly because the band of Canadian territory that joins them to the rest of Canada is so narrow. High-quality, modern highways have been built, but they do not carry much traffic. Canadians from elsewhere do not find their Atlantic Provinces so attractive for recreation or retirement that they are willing to travel there in large numbers. Although tourism is increasing substantially, especially to Nova Scotia and the beaches of PEI, many of the region's tourists come from the United States.

Still the Bypassed East

It is possible that the Bypassed East is a region in the process of disintegration. Northern New England is in close proximity to Megalopolis. Millions of people sitting almost at its doorstep want to move in—at least for a visit. And with a growing overland connection into southern New England and New York, northern New England is now increasing in population far more rapidly than the country as a whole. Many of the residents of the area may not want that growth, for it will change the reasonably quiet style of their lives and, later, the appearance of their landscape. The long-term pattern of outmigration from the Atlantic Provinces has slowed. There has been a substantial increase in nonprimary jobs, and that pattern too is expected to continue.

As we have seen, however, at the same time that this shift in population fortunes is occurring, northern New England's relative personal

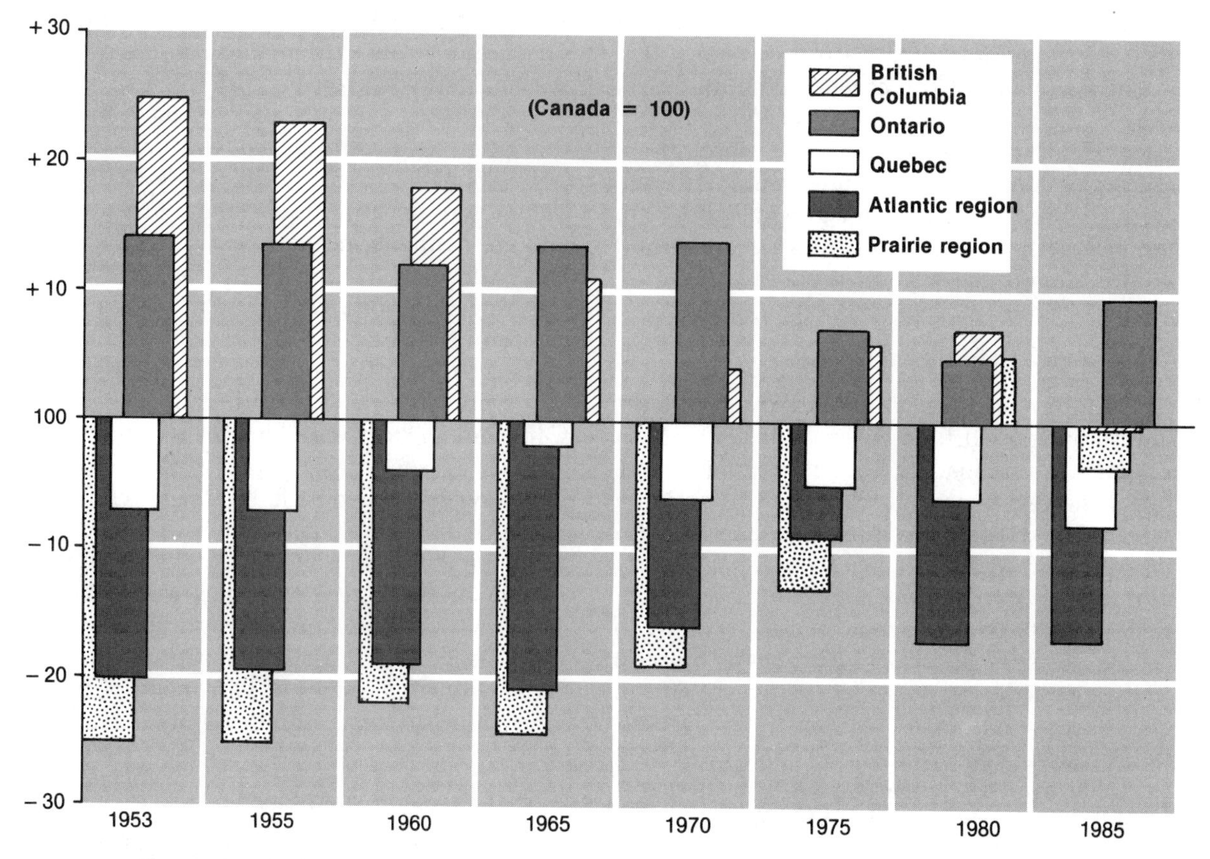

Figure 8-5 Regional income trends by province. New resource developments in the Prairie Provinces and federal policies to encourage growth in low income areas, particularly Quebec and the Atlantic Provinces, contributed to a trend of lessening regional disparities in the late 1960s and much of the 1970s. The recession of the late 1970s and early 1980s was particularly difficult for eastern Canada.

income level remains below the national average and that of the Atlantic Provinces has increased only slowly and with massive federal investment. These provinces remain the poorest of Canada's regions, and they will probably continue to be so for many years, unless petroleum and natural gas development create major revenues.

There is some justification for cautious optimism about the future of the region. Certainly many analysts in eastern Canada, and most in northern New England, anticipate improvement in the region's economy. Real economic growth, however, will be slow, and the region's relative inaccessibility will be one important reason for this slow growth.

ADDITIONAL READINGS

Backus, Richard H. (ed.). *Georges Bank.* Cambridge, Mass.: MIT Press, 1987.

Clark, Andrew H. *Acadia: The Geography of Early Nova Scotia.* Madison: University of Wisconsin Press, 1968.

Day, Douglas (ed.). *Geographical Perspectives on the Maritime Provinces.* Halifax: St. Mary's University, Department of Geography, 1988.

Hale, Judson. *Inside New England.* New York: Harper & Row, 1982.

Hornsby, Stephen J. "Staple Trades, Subsistence Agriculture, and Nineteenth-Century Cape Breton Island." *Annals of the Association of American Geographers,* 79 (1989), pp. 411–434.

Jorgensen, Neil. *A Guide to New England's Landscape.* Barre, Mass.: Barre Publishing, 1971.

Macpherson, Alan G. (ed.). *The Atlantic Provinces.* Toronto: University of Toronto Press, 1972.

Macpherson, Alan G., and J. B. Macpherson (eds.). *The Natural Environment of Newfoundland.* St. John's: Memorial University of Newfoundland, 1981.

Nutting, Wallace. *Maine Beautiful.* New York: Bonanza Books, 1924.

CHAPTER

9

Appalachia and the Ozarks

The Appalachian Uplands, stretching from New York to Alabama, and the area of the Ozark-Ouachita mountains are separated by some 475 kilometers (300 miles) of land that includes the Mississippi Valley. They are actually two parts of a single physiographic province (see Figure 2-1). We place these two territories in the same region because of their topographic similarity and because, in both sections, there is an unusually close association between topography and human settlement.

Early European settlers, when they reached the shores of colonial North America, heard tales of a vast range of high mountains to the west. As they moved westward up and into those mountains, they discovered that the elevation of the mountains had been exaggerated. Westerners of the United States of today often hold a disparaging view of the Appalachians, suggesting that they are not worthy of the term "mountains." Their attitude is in some ways understandable. Crests in both the Rocky Mountains and the Sierra Nevada rise more than a mile higher than the loftiest Appalachian peaks. Only in a few small areas do the Appalachians or Ozarks approach the dramatic mountain vistas so common in the western mountains.

Although the distinction between such features as mountains and hills is far from exact, most who concern themselves with such questions would agree that much of the Appalachian and Ozark topography should be called mountainous. Local relief is greater than 500 meters (1600 feet) in many areas, and it is sometimes greater than 1000 meters (3200 feet). Slopes are often steep. Even today these mountains create major problems for transportation development.

Whatever the features are called, the rugged landscape of this region has played a major role in its settlement history. The human geography of Appalachia remains closely intertwined with its topography. Without those mountains, the area would merely be a part of several adjoining, or coexisting, areas, such as the Deep South. With them, Appalachia and the Ozarks exist as a distinctive and identifiable American region (Figure 9-1).

Appalachia and the Ozarks are dominated by mountains and deep river valleys. Sometimes strikingly beautiful, often heavily scarred by human activity, but always there, this ruggedness does much to shape the character of the region. (Tennessee Valley Authority)

A VARIED TOPOGRAPHY

The ruggedness that characterizes Appalachia is far from uniform. Appalachia is composed of at least three physiographic provinces, each having a distinctive topography (Figure 9-2). There are also substantial differences in the pattern of human occupation in each of these three sections. These subareas are arranged in parallel belts lying roughly northeast-southwest.

The easternmost belt is the Blue Ridge. Composed of ancient Precambrian rocks, this section has been severely eroded. Its highest elevations are currently only a fraction of their former levels. Erosive activity has been greatest along the eastern margins of the Blue Ridge. The Piedmont section of the Atlantic southern lowlands bounds the Blue Ridge along Appalachia's eastern side from New York to Alabama. Although outside Appalachia, the Piedmont is in fact a part of the same underlying rock structure as the Blue Ridge. On the Piedmont, the former mountains have been eroded into a rolling upland plain that, in appearance, is not at all like the mountains to the west. These visual differences

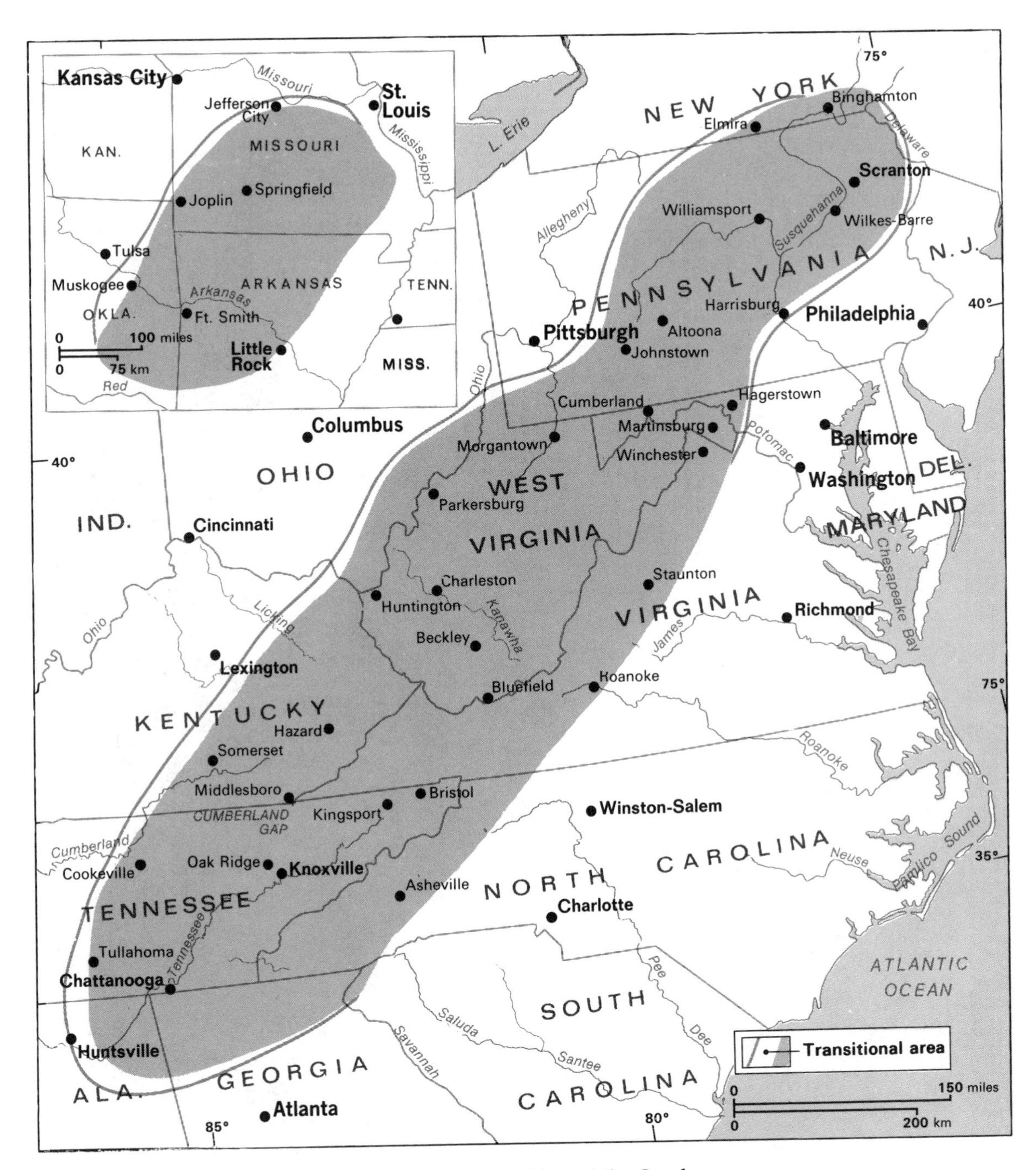

Figure 9-1 Appalachia and the Ozarks.

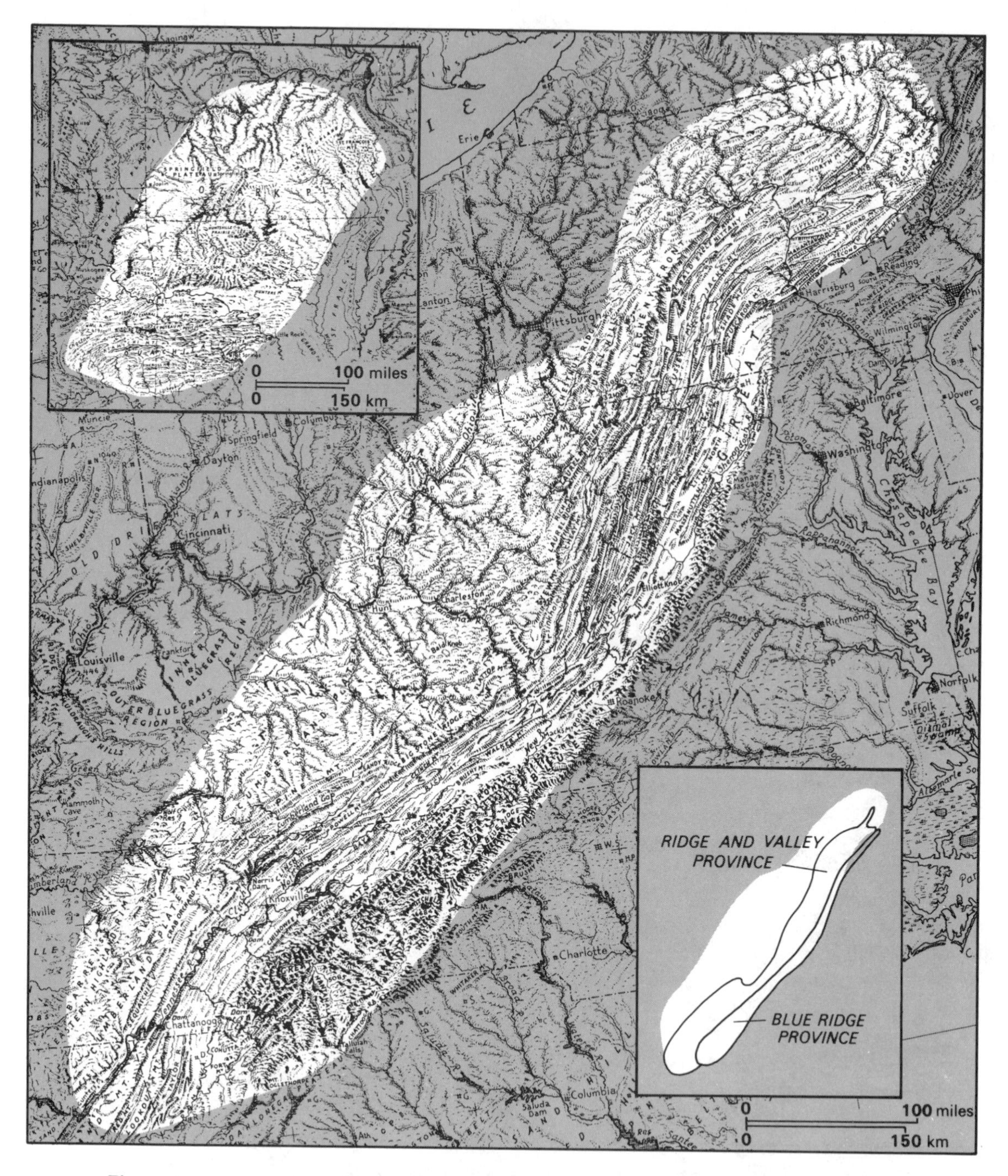

Figure 9-2 Topography of Appalachia and the Ozarks. (Reproduced from Raisz, *Physiographic Diagram of the United States.*) The linearity of the topography of the eastern Appalachians and southern Ozarks is sharply contrasted with the jumbled surface to the west and north. This difference emphasizes the importance of underlying rock structure on topography.

Major Metropolitan Area Populations in Appalachia and the Ozarks, 1990	
Pittsburgh, Pa.	2,056,705
Knoxville, Tenn.	604,816
Lexington-Fayette, Ky.	348,428
Johnson City–Kingsport–Bristol, Tenn.–Va.	436,047
Chattanooga, Tenn.–Ga.	433,210
Huntington–Ashland, W. Va.–Ky.–Ohio	312,929
Charleston, W. Va.	250,454
Johnstown, Pa.	241,247
Springfield, Mo.	240,593
Fort Smith, Ark.–Okla.	177,911

plus the major differences in settlement encourage most observers to treat them as parts of two separate regions.

The Blue Ridge generally increases in elevation and width from north to south. In the south, especially south of Roanoke, Virginia, it is the most mountainous part of Appalachia. The changes in elevation from the Piedmont onto the Blue Ridge are usually abrupt and substantial. In Pennsylvania and Virginia, the Blue Ridge is a thin ridge between the Piedmont and the Great Valley to the west; along the North Carolina–Tennessee border, it broadens to a width of nearly 150 kilometers (95 miles). The Potomac River, the James River at Roanoke, and the New River in northwestern North Carolina provide the best of the few low gaps through the mountains south of Pennsylvania. Movement to land beyond the Blue Ridge was funneled through these openings for many decades. Stonewall Jackson was able to surprise Union forces several times during the Civil War with his unexpected ability to charge large numbers of men across the Blue Ridge quickly.

To the west of the Blue Ridge, one encounters the Ridge and Valley section. This is part of the great expanse of sedimentary rock beds that lie between the Blue Ridge and the Rocky Mountains. The eastern edge of these beds has been severely folded and faulted. A result is the characteristically linear topography of the Ridge and Valley.

The Ridge and Valley section averages about 80 kilometers (50 miles) in width. It is occupied by many ridges following the northeast-southwest trend of the region. They usually rise 100 to 200 meters (300 to 600 feet) above the separating valleys. Gaps in the ridges are relatively infrequent and usually have been created by rivers that cut across the grain of the area. Several miles wide, the valleys provide some of the best farmland in Appalachia. The ridges throughout this section generally are composed of relatively resistant shale and sandstone, and the valleys usually are floored by limestone, which is more easily eroded than the ridge materials in these humid areas. By comparison, limestone is quite resistant to erosion in dry areas (see Chapter 14). Limestone, when broken down by erosive activity, provides needed minerals for a good agricultural soil.

Between the Blue Ridge and the first of the ridges of the Ridge and Valley section is the Great Valley. Running virtually the entire length of the region, the Valley (which is hilly rather than flat in most areas) is historically one of the important routeways in the country. In Pennsylvania it may be called the Lebanon or Cumberland Valley, in Virginia it is the Shenandoah Valley, and in Tennessee it is the Tennessee Valley. Whatever its name, the Valley as a routeway has tied the people of Appalachia together more than any other physical feature save the mountains themselves.

The westernmost part of Appalachia is the Appalachian Plateau, also known as the Cumberland Plateau in Tennessee and the Allegheny Plateau from Kentucky northward. The Plateau is bounded by a steep scarp (slope) on the east called the Allegheny Front. The Allegheny Front was the most significant barrier to western movement in the country east of the Rocky Mountains. The topography of this region has been created largely through stream erosion of the horizontal beds of the interior lowland. Erosion of the plateau created a rugged, jumbled topography, with narrow stream valleys bordered by steep, sharp ridges. The northern portion of the Allegheny Plateau, in New York and

Pennsylvania, was covered and somewhat smoothed by continental glaciation during the Pleistocene. The result is a rounder, gentler appearing landscape. Except for limited areas, such as portions of the Cumberland Plateau, level land is scarce. Most communities are forced to squeeze themselves into small level spaces in the stream valleys (Figure 9-3). The differences between the Ridge and Valley region and the Plateau region are clear to anyone flying over the land. The symmetry—parallel ridges and valleys—of the former suddenly gives way to the rumpled, apparently patternless landscape of the Plateau.

The Ozarks-Ouachita uplands follow a topographic regionalization broadly similar to the Appalachians, with the "grain" now east-west instead of northeast-southwest. The Ouachita Mountains to the south exhibit a series of folded parallel ridges and valleys similar to the Ridge and Valley section of Appalachia. They are separated from the Ozarks section by the structural trough of the Arkansas River Valley, itself similar to the Great Valley. The Ozarks is an irregular, hilly area of eroded plateaus much like the Appalachian Plateau section.

THE APPALACHIAN PEOPLE

Early Settlement

The mountains and the forests catch the eye of casual visitors to Appalachia, but its people have

Figure 9-3 Hazard, Kentucky. The population of most of Appalachia is concentrated in its often narrow valleys. The shaded wooded areas correspond closely with uplands or areas of substantial local relief; cleared land most often marks the valleys of this rugged region.

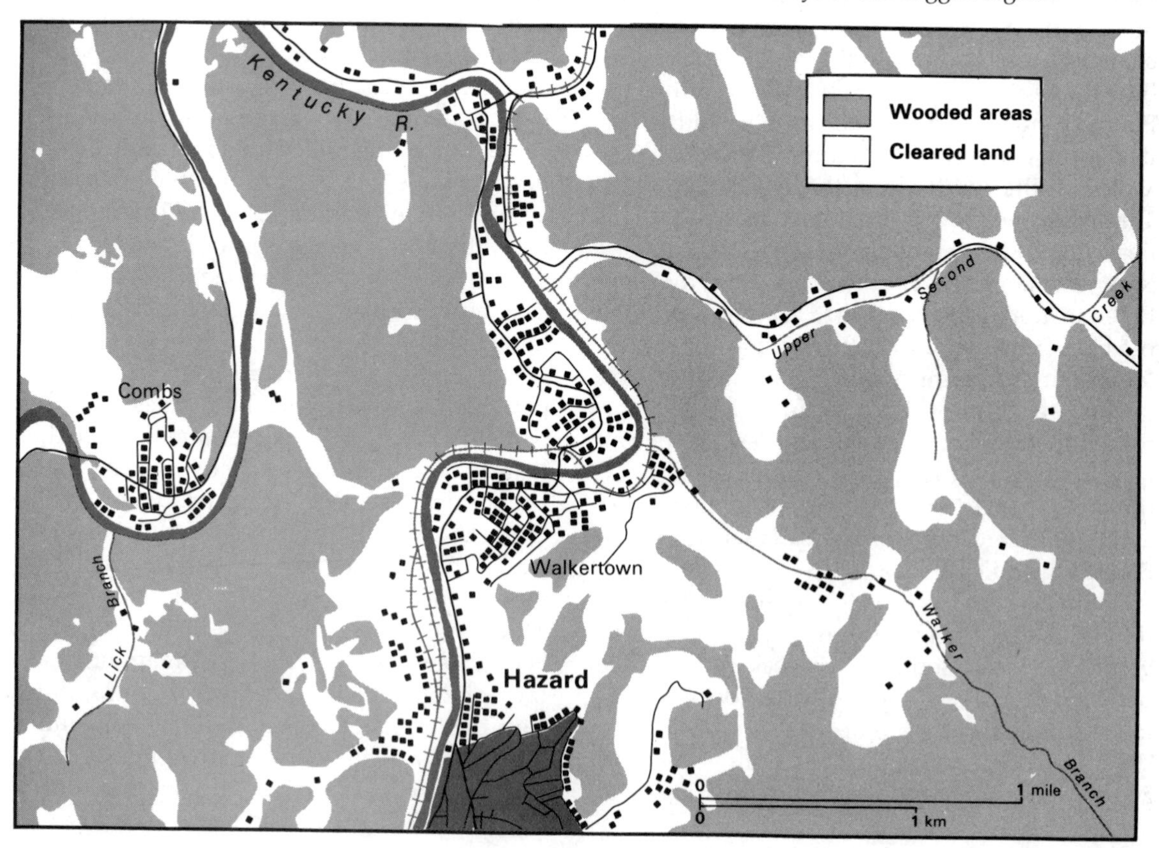

given the area much of its special character. The Appalachian "hillbilly," a person made up far more of myth than of reality in today's United States, is a necessary part of those mountains to countless Americans. Popularly characterized as independent, proud, ignorant, and poor, the hillbilly is as much a part of the popular image of Appalachia as is the rural black in the Deep South. Without the hillbilly, this is just another rather scenic, rugged part of the country. With the hillbilly, it occupies a distinctive spot in the American folk landscape.

Settlers did not push through the Blue Ridge into the Appalachian Highlands until late in the colonial period, 150 years after initial occupation of the East Coast. The easiest and first-used passageway into the Great Valley and the mountains beyond was in southeast Pennsylvania, where the Blue Ridge is little more than a range of hills. Many Pennsylvanians found the mountain lands to the north and west inhospitable. Consequently, they gradually spread their settlement down the Valley into Virginia. Most of these people were either German or Scots-Irish. (The latter are Scots who had earlier settled in northern Ireland as part of the Protestant British crown's program to secure its claim to Ireland.) Both groups were soon joined by others moving inland from the southern lowlands.

Then, late in the eighteenth century, people began settling the valleys and coves of the surrounding highlands. For reasons that are still in some dispute by historians, the Scots-Irish played a major role in this movement. Some argue that these people were a particularly footloose and independent breed, constantly searching for the freedom of the frontier. Others maintain that they merely happened to represent the main immigrant segment of the population at the time, and hence they were the most likely contributors to frontier settlement. Whatever the case, the Scots-Irish, joined by many people of English and German background, dominated early highland settlement.

The land they chose was poor in comparison with areas farther west. Its ruggedness, coupled with the cool upland climate, rendered most of the region unacceptable for the plantation economy. Only in some of the broader lowlands, such as the Shenandoah and Tennessee valleys, did a few sizable plantations develop. One consequence was that many of the counties of the southern Appalachians have always had small black populations.

When American settlers came to this area in the late eighteenth and early nineteenth centuries, the region provided adequate settlement potential for smaller farms. About 10 to 20 hectares (25 to 50 acres) of cleared land was about all a farmer could handle. Such plots were available in the stream valleys. The forests teemed with game, wood was plentiful, and animals could graze in the woods and mountain pastures. By the standards of the time, this was reasonably good land. Thus, although the more rugged areas of the region were ignored by early settlement, a large farming population soon occupied the mountains.

Much of the region gradually grew more isolated and separate from other areas. As flatter, richer agricultural land to the west was opened and grain production was mechanized, the small Appalachian farm became increasingly marginal economically. Even famous pathways through the region, such as the Cumberland Gap at the western tip of Virginia and the Wilderness Road from there to the Bluegrass Basin of Kentucky, were in fact winding and difficult, to be bypassed whenever possible. East-west travel between the northeastern seaboard and the Great Lakes area followed the route of the Mohawk Corridor and the flat lakeshore of Lake Ontario, thus avoiding the northern Appalachian Uplands. There was no easy passage at all across the southern Appalachians. Major railroad lines skirted the area; their owners preferred a longer, but easier, route. Atlanta's early growth, for example, was largely a result of its role as a rail junction at the southern edge of the mountains. Rural residents living in areas of rugged topography and far from large urban centers were particularly affected by limited transportation facilities. A study conducted in Kentucky in the 1960s found that limited accessibility, or geographic isolation, was a major element in the lack of economic opportunity for the state's Appalachian residents (Figure 9-4).

Appalachia, particularly southern Appalachia, was slow to develop any substantial urban pattern. In part, it shared with the rest of the South an emphasis on agriculture that continued well after other regions of the country had begun their rush toward manufacturing and urban living. Also, the products of Appalachia were few, and the demand for the goods and services of cities was limited. In addition to these characteristic southern conditions, however, was added the paucity of transportation. The region seemed to exist largely outside a well-identified pattern of development that was common elsewhere: transportation improvements fostered city growth, and city growth encouraged transportation improvements. Even by the standards of the South, Appalachia had neither.

One major result of this lack of both plantations and urban development in the southern Appalachians was that few new migrants were added to the early Scots-Irish, English, and German settlers. The area remained overwhelmingly northwest European in ethnic background and, increasingly, conservative Protestant in religion. These people tended to stay where they were, and as time passed, their attachment to their family, community, and land grew.

Southern Appalachian Cultures

This regional immobility led to the development of a cultural distinctiveness uncommon in the rest of the nation. In a nation where migration, education, and commerce were adding more and more disparate elements to the total population and yet increasing its interregional similarity, Appalachia became increasingly un-

Figure 9-4 Isolation in Eastern Kentucky. (Adapted from Bowman and Haynes, *Resources and People in Eastern Kentucky*, Johns Hopkins University Press, 1963.) The greater isolation of the Appalachian uplands is contrasted with the accessibility of the Bluegrass Basin and the Ohio River valley. Appalachian highways are usually in the larger valleys.

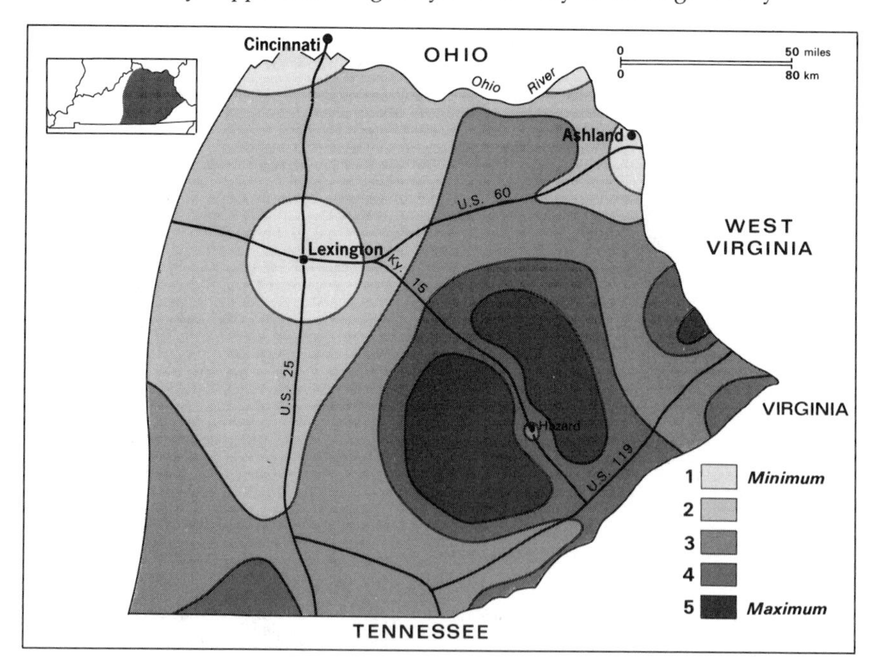

usual by simply remaining the same. Many inhabitants of the region grew uncomfortable in any other place:

> The fierce loyalty of mountain people to home is mostly a loyalty to the only culture in which they feel secure and which operates in ways they know and appreciate. . . . For mountaineers, moving is a kind of death to his way of life. It cuts him off from his sustaining roots.[1]

Henry Caudill, in his sensitive look at one section of the mountains in *Night Comes to the Cumberlands*,[2] noted that in West Virginia during the Great Depression, few native West Virginians left the state whereas more than three quarters of all foreign-born coal miners left to go to larger metropolitan areas.

It is time to take a closer look at these Appalachian people. We already know that they are overwhelmingly white, Anglo-Saxon, and Protestant. They are also relatively poor. Appalachia is easily the largest predominantly white, low-income area in the country. Large swatches of the region—perhaps most notably a band from West Virginia through eastern Kentucky and Tennessee—have high percentages of families living at or below the federally defined poverty level (Figure 9-5). This level, established by the U.S. Department of Agriculture, indicates how much income a family of four should have to meet what is considered a minimally acceptable standard of living. In 1987, the minimum threshold was $11,611 for a family of four. In some areas, especially eastern Kentucky, Appalachia's major coal-producing area, much of the blame for this poverty can be attributed to a great decline in the regional demand for labor. Labor demand decreased as coal mining was mechanized following the outbreak of World War II.

The region's people are also conservative in attitude. Many of the country's most conservative Protestant churches trace their roots to Appalachia. Others are found where mountain people have moved and taken their religion with them. Appalachia is certainly part of the Bible Belt, as any Sunday traveler there will attest after sampling the array of religious programs on the radio. Politically, most elected officials are decidedly conservative, although strands of rural populism are found. One interesting element of the region's politics is the strength of the Republican Party, which has survived and even flourished there. The hill-lands of such states as North Carolina, Georgia, Tennessee, and Alabama long supplied many of the South's small trickle of successful Republican politicians prior to the recent surge in electoral success for the party in the South. Similarly, the Missouri Ozarks have often sent Republicans to Con-

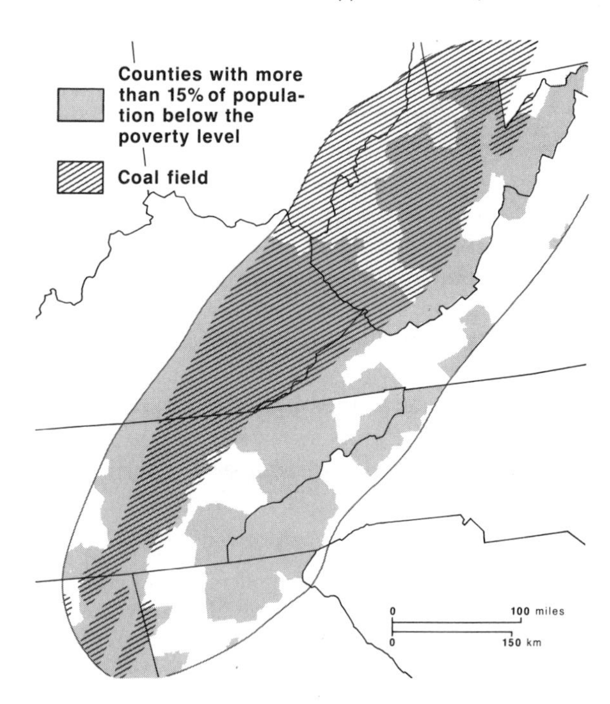

Figure 9-5 Poverty counties. Appalachian counties with more than 15 percent of the population below the poverty level are usually in the south and are coal producers.

---

[1] Jack E. Weller, *Yesterday's People: Life in Contemporary Appalachia* (Lexington: University Press of Kentucky, 1965), p. 86.

[2] Henry M. Caudill, *Night Comes to the Cumberlands* (Boston: Little Brown, 1962), p. 179.

gress. A strong belief in the central importance of the family and the individual have survived massive involvement by the federal government in the region's economic affairs. Some call this regional willingness to accept federal aid, while denying the government's authority, hypocrisy. It is more likely simply a reflection of the population's fundamentally conservative nature.

Appalachia's legendary isolation, where the mountain people living deep in mountain coves and far up steep valleys had almost no contact with the outside world, is certainly gone. The reasons for this change are of key importance to the region and will be discussed shortly. As Thomas Ford noted in *The Southern Appalachian Region: A Survey,*[3]

> As a general rule, at least in our time, the technological aspects (of a culture) are the first to change, followed more slowly by adaptations of social organization to new techniques. Most resistant to change are the fundamental sentiments, beliefs, and values of a people, the ways they feel their world is and should be ordered. So it is not implausible to suppose that the value heritage of the Appalachian people may still be rooted in the frontier, even though the base of their economy has shifted from subsistence agriculture to industry and commerce and the people themselves have increasingly concentrated in towns and cities.

Ford's comments remain apt today. The economy of the area has changed greatly, but the attitudes of its people have not. Many regional leaders hope that these basic attitudes and beliefs will never change.

This comment on the continuation of provincialism in the face of great economic change suggests some further words on the nature of isolation. Many have related this provincialism to residential isolation in a land of steep slopes, twisting valleys, and poor or nonexistent roads. At one time, this may have been a valid argument. Today, however, the provincialism is bred of the strong bonds of family and community. Topography is only an incidental factor. These bonds tie their members together and lessen their association with others.

Most areas of Appalachia are also among the lowest in the nation in terms of the percentage of the total population born in another state. Appalachia's population is basically one that has been relatively little affected by outsiders.

REGIONAL UNITY AND DISUNITY

By now you may have noticed that most of our attention has been devoted to that part of this region that stretches from West Virginia to northern Alabama, with much less discussion of the northern Appalachians from Pennsylvania north and the Ozarks-Ouachita uplands. This has been deliberate. The southern portion is the most clearly Appalachian region, and the one that most Americans recognize as Appalachia. Actually, much of what has been said here about the region's residents fits the Ozarks as well. The stereotyped Ozark hillbilly are as much a part of the legends of the Ozarks as are their brethren in Appalachia, and the legends are equally inappropriate to both areas. Regional attitudes are similar. Even the ethnic background of the settlers is nearly the same.

The northern Appalachians are far less clearly associated with the broader region. Certainly they share the mountainous topography, and some of the early developmental problems created by steep slopes were also common. Beyond that the similarities grow more tenuous. First, although the northern Appalachians are by no means an area of wealth, poverty is far less evident than it is farther south (see Figures 1-3, 9-5). Also, more recent immigrants followed the early northwestern European settlers into the area. This is especially true in Pennsylvania and northern West Virginia where coal mining attracted many East European migrants in the late nineteenth and early twentieth centuries. In

---

[3] Thomas R. Ford, *The Southern Appalachian Region: A Survey* (Lexington: University Press of Kentucky, 1962), p. 29.

Narrow mountain valleys and coves, poorly connected to outside paved highways by often-eroded dirt roads, continue to be home to many of Appalachia's poorest people. (U.S.D.A.)

New York's highlands, settlers came largely from New York and New England, a different source than that for Pennsylvania and the upland South. Many cultural patterns in the northern Appalachians, with religion a notable example, are not at all the same as those of the southern highlands. Fundamentalist churches are less common; in many counties, especially in Pennsylvania, Catholics and members of various Eastern Orthodox churches are in the majority (Figure 3-6).

Transportation within the northern Appalachians soon became far better than that in the southern Appalachians. In part, this was because the mountains were less continuous and lower and, thus, more easily breached. Also, as the upper Midwest boomed, the northern Appalachians became the center of the continent's major belt of commercial and manufacturing growth (see Chapter 6). Transport lines connecting the eastern and western portions of the manufacturing core region soon ribboned through the mountains. The economic consequence of this was far more development within the northern Appalachian area, especially in central and western Pennsylvania and New York, as compared to the southern Appalachians.

This leads naturally to the question, "Why include the northern section in Appalachia at all?" Some argue that the inclusion is recent and only developed when the federal government began pumping money into poverty areas to encourage economic growth, with the northern states asking to be included in order to obtain part of the money. A more fundamental reason is that this northern section is distinctly different from the areas that surround it, regardless of political boundaries; it is linked to southern Appalachia by the pattern of mountains and relative poverty they share. West Virginia is the only state situated entirely within Appalachia. Every other state's capital, largest urban area, and major areas of economic development are beyond the margins of the Appalachian region. Only Tennessee provides a partial exception to this general pattern, having both Knoxville and Chattanooga within the region. Both, however, have about half Nashville's 975,000 people.

The relationship that exists in Appalachia between the Appalachian sections and the rest of each state is a significant aspect of the region. In the South at the time of the Civil War, there seemed little reason for the mountain people to join the rest of their states in secession, because they had little of the plantation economy and way of life. Much of the population of Virginia's mountains chose to withdraw from the state as a consequence of its withdrawal from the Union and they formed West Virginia in 1863. There was strong opposition to secession in the Appalachian sections of much of the rest of the South as well. Still, the ties within each state were strong, and in spite of considerable dissension, the Appalachian sections chose to accompany their states out of the Union.

This is one indication of the impact that the state of residence has on most Americans. The states fund public education; a part of that education is usually the development of a sense of loyalty through exposure to the history and politics of the state. State aid and state programs are important aspects of local economic conditions. Today state universities often provide success in athletic competition, a success felt and shared by people across the state. Most newspapers, and to a lesser extent radio and television stations, restrict their attention to one state primarily and their material contains heavy doses of state news and information. Because the state does provide so much, the activities of its government are important for everyone in the state. In short, the state's boundaries can be viewed as something of a container, serving to unite what may otherwise be disparate areas. For Appalachia, this means that the portions of its states that are outside the margins of the region have a substantial influence within the region. This, too, is a major element in the set of differences between northern and southern Appalachia. The states of the South share many similarities (Chapter 10), just as do those of the Northeast, and the two areas are different from one another in many ways. These distinctions help differentiate the northern and southern Appalachians.

ECONOMIC AND SETTLEMENT PATTERNS

The national image of Appalachia is unquestionably that of a predominantly rural area. In some ways, this is valid. The urban percentage for the region is only about half the national average. A majority of the population is classified by the Bureau of the Census as either rural or rural nonfarm residents (the latter are people who live in rural areas but have urban occupations). Parts of Appalachia, most notably in eastern Kentucky, have the highest rural population densities in the United States. Many counties have rural densities of 60 or more per square kilometer (150 or more per square mile). Clearly, Appalachia's high rural density is not supported by a large-scale, commercial agricultural system. Rather, small farms and minerals dependency (primarily coal) are the keys to this dense rural population.

Agriculture

Appalachia is the primary region for owner-operated farms in the United States, with Kentucky and West Virginia leading the country in that category. One study in an Appalachian county in North Carolina looked at all farms with more than 6 hectares (15 acres) of possible cropland and found that an astounding 99.3 percent were owner occupied. Without any important commercial crop in Appalachia, there was little early growth of farm tenancy, and that pattern has remained. This is a land of independent small farmers, however small the farm income may be.

It has been suggested that the best description of the condition of Appalachian agriculture is that it is the result of too few resources divided among too many people. The average farm in Appalachia is small, containing only about 40 hectares (100 acres). Furthermore, the rugged topography, poor soil, and short growing season in much of Appalachia have resulted in a limited amount of available cropland and a greater relative emphasis on pasturage and livestock. Because the fields are small and scattered in the valleys, the efficient use of large farm machinery is nearly impossible. The net result of all this is that farm incomes are low. A great many of the region's farmers have become part-time operators, turning to other jobs to provide supplementary income that will allow them to remain on the farms.

The growth in farm size common in most other parts of the United States results from the advantages of increased mechanization and the availability of substantial tracts of good land. Land must be reasonably flat for mechanization (Chapter 12). In Appalachia, such large tracts of level land are seldom available. In parts of the region, average farm size is actually declining as farms are split through sale or inheritance. These smaller farms are a good indication of the importance of off-farm income for many Appalachian farmers.

The type of agriculture found in most of the region is called general farming. In fact, this is the major general-farming region in the country. This term is really a catchall that covers a multitude of products and simply means that no discernible product or combination of products dominates the farm economy. Extensive animal husbandry of some form is the most common and probably best agricultural use of the steep slopes. A number of specialty crops, such as tobacco, apples, tomatoes, and cabbage, are locally important in some valley areas, with small plots of tobacco being the most common cash crop in the southern Appalachians. Another important cash crop was corn, which was illegally distilled into the region's famous moonshine and rushed to lowland markets by runners who are now legendary. High production costs, alternative job opportunities, and the expansion of the legal sale of alcoholic beverages have combined to cut moonshine production drastically in recent years. There are indications that marijuana may have replaced moonshine as the region's primary contraband export. Corn remains the region's leading row crop, but it is normally used on the farm for animal fodder.

There are important exceptions to this pattern of semimarginal agriculture within Appalachia. The Shenandoah Valley of Virginia, for example, early was called the breadbasket of Virginia. Competition from wheat grown in the fertile grasslands of the Deep South and Great Plains forced the Valley out of the national wheat market in the late nineteenth century. Although winter wheat is still grown, hay and corn for fodder and apples are now the Valley's major crops with turkey raising also locally important; it remains a productive and beautiful farming region. Dairying and apple production are important in the many valleys of central Pennsylvania. The Tennessee Valley is a substantial agricultural district, with fodder crops and livestock most important. Still, these relatively large productive areas are exceptions.

Coal

Farming, although important, is obviously an inadequate explanation for the large rural population of Appalachia. Over much of the region, farming's chief partner is coal. Almost all of the Allegheny Plateau is underlain with a vast series of *bituminous coal beds* that, taken together, comprise the world's largest such coal district (see Figure 2-7). Coal seams, with thicknesses of as much as 3 meters (10 feet), are interbedded in the flat-lying sedimentary rocks of the plateau. The coal seams have been exposed by the same streams that have, through their erosive activity, created the rugged topography of the plateau. The existence of these beds along a certain elevation on a slope is clearly evident on topographic maps, where symbols for mines often extend for many miles along the same contour line.

The coal of Appalachia became important shortly after the American Civil War. It was the development of new types of coke-burning iron and steel furnaces that created this demand because coke is processed from bituminous coal. The thick coal seams of southwestern Pennsylvania and northern West Virginia provided the fuel for Pittsburgh's rise to its status of Steel City during this period. As the nation turned to electrical power in the present century, coal from Appalachia provided fuel for electric-generating facilities along much of the East Coast and in the interior manufacturing core.

After the better part of a century of growth, the coal industry fell into a period of decline following World War II. Production dropped as petroleum and natural gas replaced coal as major fuel sources. This coupled with the decision by the leadership of the United Mine Workers Union to support mine mechanization to maintain the importance of coal by minimizing its cost led to drastically high unemployment in many coal-mining areas. Outmigration resulted, and many coal counties lost a full quarter of their population between 1950 and 1960. Others felt themselves too old, too poorly educated, or simply too fond of where they lived to move. The resulting economic depression, blending with the poverty common to Appalachia, created areas of particularly severe problems, with eastern Kentucky again the best example.

More recently, production levels increased from around 1960 to record levels in the 1980s (Table 9-1). The great decline in North America's steel industry (after electricity generation, the country's largest user of coal), initiated by the economic recession of the early 1980s, reduced the demand for Appalachian coal. In the mid-1970s, the steel industry used approximately 75 million tons of coal annually; in the late 1980s it used only half as much. Today the United States is producing more coal than ever before, but an increasing share is coal with a lower sulphur content from the west (see Chapter 14). Nearly 100 million tons of Appalachian coal is exported annually, shipped almost entirely through the port of Norfolk, Virginia. The thick seams along with mechanization make this coal cheap on the world market, and it finds ready buyers. Within the United States, growing power demands coupled with the continuing concern over availability and cost of petroleum supplies and the safety of nuclear power has reemphasized the need for coal in electric-power generation. The federal government discouraged the use of coal in power plants during the 1960s because of atmospheric pollution, but it now actively encourages coal use to lessen American dependence on foreign petroleum suppliers (Figure 9-6). New generating plants (with high cooling towers and smoke stacks) use huge quantities of locally mined coal to produce electricity, much of which is transmitted to areas outside the region, and they are an increasingly visible part of the landscape of the central and northern Appalachian Plateau.

The proven coal reserves of the United States are great; they will last at least three centuries and, at current use levels, perhaps much longer. Declining domestic liquid petroleum supplies (estimated reserves have declined steadily for nearly two decades) have emphasized the importance of coal. Pilot projects to extract natural gas from coal have been successful, although at present the process is expensive. It is anticipated that gas produced in this way will become cost competitive as technology improves, production levels grow, and as natural gas from other sources becomes more expensive. There is also hope that petroleum will eventually be extracted from coal at competitive costs.

Appalachian coal is mined in several different ways. Underground or shaft mining was used

Table 9-1 Bituminous and Lignite Coal Production, Major Mining States, 1947–1988 (in thousands of short tons)

State	1947[a]	1960	1970	1988
Eastern States				
Alabama	19,048	13,011	20,560	29,704
Illinois	67,680	45,997	65,119	57,826
Indiana	25,449	15,583	22,263	30,084
Kentucky	84,241	66,847	125,305	156,490
Ohio	37,548	33,957	55,531	32,414
Pennsylvania	147,079	65,425	80,491	74,234
Virginia	20,171	27,838	35,016	47,062
West Virginia	176,157	118,944	144,072	138,794
Western States				
Arizona	—	—	132	11,144
Colorado	6358	3607	6025	14,818
Montana	3178	313	3477	38,306
New Mexico	1443	295	7361	20,704
North Dakota	2760	2525	5693	28,160
Texas	—	—	—	52,868
Wyoming	8051	2024	7222	196,938
Total United States	630,624	415,512	602,932	997,384

Source: U.S. Bureau of Mines, *Minerals Yearbook,* Washington, D.C.: U.S. Governmental Printing Office, 1948, 1960, 1970. Energy Information Administration, *Quarterly Coal Report* (Oct.–Dec. 1988). Washington, D.C.: U.S. Department of Energy, 1989.

[a] Year of maximum production prior to the 1970s.

first and is still quite important, especially in the northern parts of the region. Modern underground mining techniques—huge mobile drills and continuous mining machines rip the coal out of the seams and then deposit it on conveyor belts for the trip to the surface, now mean that tons of coal per minute can be removed from a seam. Surface or strip mining, which is far less expensive as long as the coal seams are near the surface, has increased greatly in importance. In the central region (primarily eastern Kentucky, western Virginia, and southern West Virginia) where the most important producing section is today, large machines remove the rocks along a slope above a coal seam and then simply lift off the uncovered coal. In the past, the waste rock (overburden) was simply dumped down the slope. Extraction along several seams on a slope by this method creates a peculiar, stepped appearance that looks, from a distance, almost like a series of increasingly smaller boxes piled on top of one another. In areas of somewhat flatter terrain, such as southeastern Ohio, massive scoop shovels lift off the overburden and dig out the coal seam.

About half of the coal mined in Kentucky and most mined in Ohio and Alabama is from strip mine lands. Still, most of the coal from Pennsylvania, Virginia, and West Virginia and two-thirds of that from Appalachia as a whole is from shaft mines. By comparison, nearly all of the coal from the western states of Wyoming, Texas, Montana, New Mexico, and North Dakota is from surface mines.

The first important coalfield in Appalachia was not the bituminous fields of the plateau. Their exploitation was preceded by operations in the anthracite field at the northern tip of the Ridge and Valley in Pennsylvania. Anthracite is a much harder, smokeless coal that was impor-

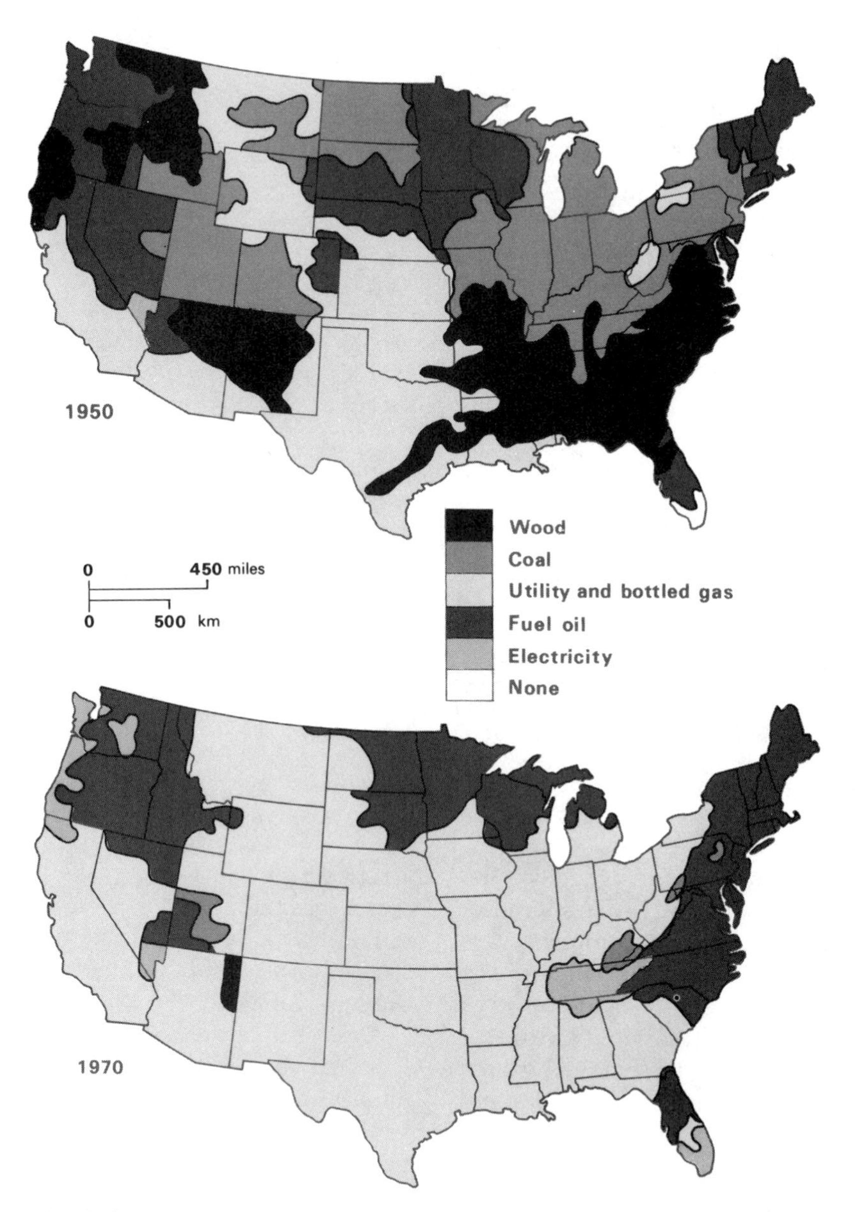

Figure 9-6 Primary home heating fuel, 1950 and 1970. Coal has been largely abandoned as a home heating fuel in the United States, with natural gas (where available) its usual replacement. What would a doubling or tripling of natural gas and petroleum prices do to this pattern?

tant in home heating. Anthracite was also a major fuel for ore smelting until techniques to produce coke from bituminous coal were developed during the Civil War. The decline in use of coal for heating, coupled with the lack of alternative uses for anthracite, led to an economic depression in the anthracite belt that has lasted since the peak period of anthracite production during World War I; although substantial anthracite reserves remain, production today is minimal. More anthracite may possibly be mined as national coal consumption increases, but, because the seams are fractured and folded (this is a part of the Ridge and Valley), mining costs are high.

Coal has been very much a mixed blessing to the people of Appalachia. It has long been the economic mainstay for large parts of the region and has employed, either directly or indirectly, hundreds of thousands of workers. It is difficult to imagine the economic structure of Appalachia without coal. Still, the list of problems brought to Appalachia by coal is long. Tens of thousands have died in mine-related accidents. Black lung, the result of many years of breathing too much coal dust, has affected countless others. Both problems are mostly associated with underground mining. The recent recovery of production in response to increased market demands has been accomplished mainly by greater mechanization and unemployment remains widespread. Mining employment has increased since the early 1970s but remains at less than half the number employed in the late 1940s. Many manufacturing plants have been located in the area in response to the available labor force. However, most pay low wages and many attract a labor force that is largely made up of women. Consequently, the wife is now the employed worker in many families whereas the husband remains at home. Most mineral rights are held by corporations who obtained them early and at low price. While a number of Appalachian states have either initiated or increased surcharges on coal mined in their state, coal taxes remain low, and most coal profits leave the region.

The impact of coal mining on the physical environment of Appalachia has been great. Strip mining is much less hazardous to miner's health than shaft mining. Accident rates per ton of coal mined are lower, and the risk of black lung from breathing coal dust is greatly reduced. Still, strip mining in particular creates great gashes across the landscape. Every coal-mining state has passed laws requiring the reclamation of strip-mined lands. In 1977 the federal government passed the Surface Mining Control and Reclamation Act. The act requires mining companies to prove that they can reclaim the land before a mining permit will be issued and mandates that once the mining operation is finished the land must be restored so that it is useful for the same purposes for which it was used before the mining was started. Land is now returned to something like its original contouring and replanted with vegetation. Pennsylvania has been the leader in regulating strip-mine reclamation. Still, it is extremely difficult, if not impossible, to return rugged topography to its former shape; until recently, few have even tried. Many argue that the cost of land reclamation is so high (estimated at $7500 to $25,000 per hectare ($3000 to $10,000 per acre) in the late 1980s) that restoration costs hurt the mining companies and damage the regional economy. The arguments over strip mining activity have become a classic environmental controversy.

Current reclamation legislation focuses on new mining activity, ignoring what has already happened. Acid mine drainage has killed the fish and plant life in thousands of miles of the region's creeks and streams. In many places, housing has been crowded into narrow valleys that bend and contort settlements like a pretzel, to the detriment of both the people and the landscape. Damaging floods have been a repeated problem in many of these locations. Even the rich greenness of Appalachia's summer vegetation often cannot hide the gray of vast slag heaps and coal dust. A beautiful part of America has been damaged in a severe way. Full recovery will be a matter for the centuries.

Coal mining can be very disruptive. This Pennsylvania town is interlaced and nearly surrounded by the scars of strip mining and piles of waste material brought to the surface from shaft mines. (Grant Heilman/Grant Heilman Photography)

Other Mining Activities

Three other Appalachian mining activities deserve mention because of their local importance. The tri state district in the Ozarks, where the borders of Oklahoma, Kansas, and Missouri meet, has long been a major area of lead mining. Southeastern Missouri, outside Appalachia, has produced lead for over 250 years, and surface mines there remain the country's most important. Missouri has supplied most of the lead ever mined in the United States and currently produces more than three quarters of the total national production.

The first oil well in the United States was drilled in northern Pennsylvania in 1859 and that state led the country in production through most of the nineteenth century. Today the area supplies only a small part of the nation's crude oil needs, but it remains an important producer of high-quality oils and lubricants.

Finally, southeastern Tennessee is the most important remaining area of zinc production in the United States. In addition, several mines around Ducktown, near the North Carolina and Georgia borders, are the only major copper producers east of the Mississippi River.

REGIONAL DEVELOPMENT PROGRAMS

The Tennessee Valley Authority

Like coal, Appalachia's rivers have also been a mixed blessing to the region. Some of the streams have been important transportation routeways, and water power was used by the earliest gristmills and sawmills. These rivers also had a darker side, for they frequently flooded their narrow valleys during periods of heavy rain. The southern highlands are the moistest area of the continent east of the Pacific Coast. Out of a desire to control one of these rivers, the Tennessee, the largest and perhaps most successful regional development plan in American history was implemented.

The Tennessee River, with its headwaters in the mountains from Virginia to Georgia, was once a river that flooded as often as any in the South. It also flowed through some of the poorest parts of the inland South. President Franklin D. Roosevelt, early in his first term, conceived a plan to harness the river and to use it to improve the economic conditions of the entire Tennessee Valley. Many contended that the federal government had no constitutional authority to create such a regional development scheme. As a result, the Tennessee Valley Authority (TVA) was first given the charge to develop the Tennessee River for navigation, a right clearly granted to the federal government as a part of its control over the interstate commerce. Today a 3-meter (9-foot) barge channel exists as far upstream as Knoxville.

In a sense, most of the other activities of the TVA can be viewed as logical extensions of the initial commitment. Navigation development included the construction or purchase of a series of dams to guarantee stream flow and reduce flooding. As long as the dams were there, it was natural to include water-power facilities with them. Today most of the more than 30 dams controlled by the TVA on the Tennessee and Kentucky rivers have power-generating facilities (Figure 9-7). A number of these dams were developed privately and TVA control was acquired subsequently. The TVA became the electricity supplier to a broad region centered on the river. As the power needs of the region grew, demand exceeded the capacity of the water-powered facilities. The TVA turned to the construction of other power sources. Today, about 80 percent of the electricity it produces comes from thermal plants, including 10 that burn coal, and several nuclear-powered facilities. The TVA uses nearly 50 million tons of coal annually and is now Appalachia's largest coal user.

TVA power was once cheaper than most alternatives in the country, although that is no longer the case. The inexpensive electricity attracted to the Valley a few industries that are heavy users of power. Most notable, perhaps, was the large aluminum processing facility at Alcoa, south of Knoxville. The country's first atomic research facility was placed at Oak Ridge, west of Knoxville, partly because of the availability of large amounts of power there. Knoxville; Chattanooga; and the tricities of Bristol, Johnson City, and Kingsport in Tennessee are all substantial manufacturing centers. The TVA also became a principal developer and producer of artificial fertilizers, another heavy power-consuming industry.

Proponents of the TVA argue that much of the significant manufacturing growth of the area over the last four decades came as a direct result of the work of the authority. This ignores the fact that other Southern areas, such as the Piedmont from North Carolina to Alabama and the lower Mississippi River Valley, also experienced major increases in manufacturing during the same period. Such TVA contributions as flood control, cheap water transport, and power production probably did have an impact on regional manufacturing growth. Still, it seems reasonable to assume that a substantial increase in manufacturing would have occurred here even without the TVA. One consequence of the comparatively

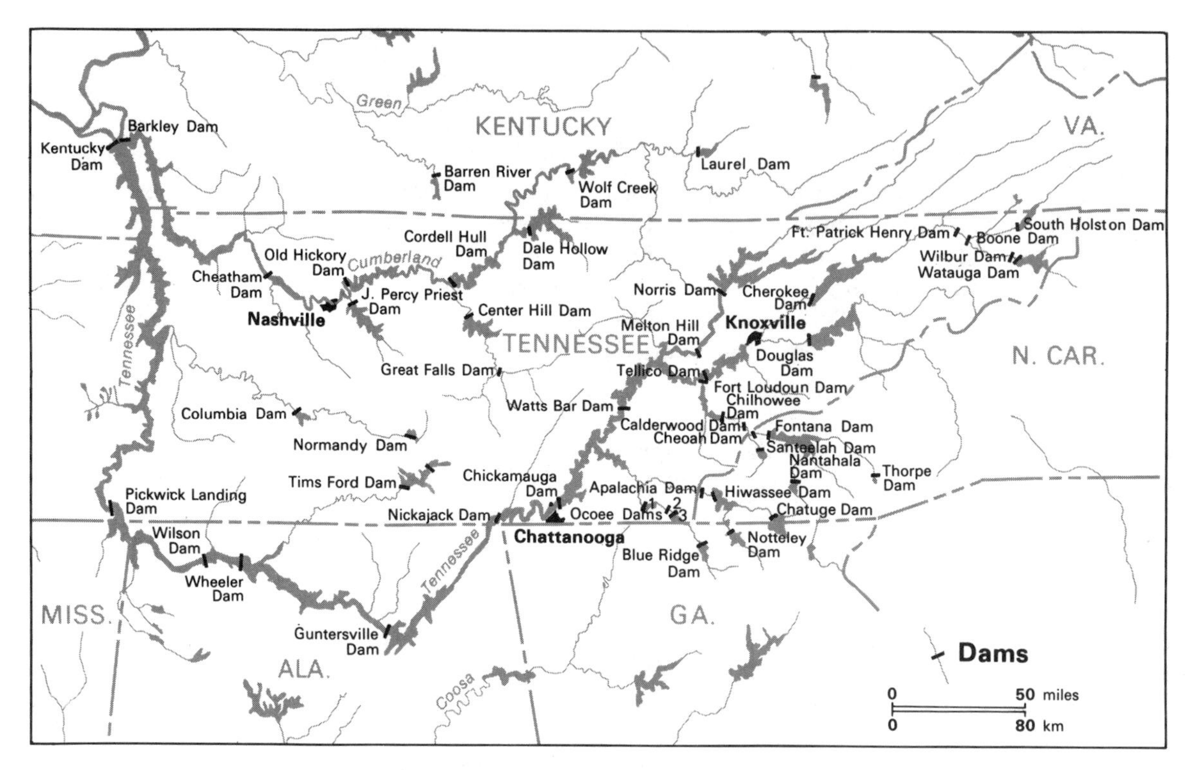

Figure 9-7 The Tennessee Valley Authority. The Authority has long been a major influence on life in the Tennessee Valley, with the more than 40 dams on the area's streams perhaps the most obvious aspect of that influence. Tennessee is a national leader in the number of power boats per capita.

early development of heavy manufacturing in the Valley is that the region experienced a decline in industrial activity in the early and mid 1980s similar to that found in the manufacturing core (see Chapter 6).

The series of dams had an obvious flood control value. Above the dams, the TVA initiated a major program to help Valley farmers control erosion at the farm. A careful inventory of the agricultural potential of the region was taken, thousands of farm ponds were constructed, and advice was offered on other types of erosion control. The goal was to hold part of the floodwaters at the farm and to slow the rate at which the lakes were filling with silt.

In addition to the water itself, the authority also owned 520,000 hectares (1.3 million acres) of land along parts of the rivers. Major public recreation areas were developed on some of this land, and recreational boating was encouraged on the water. This is now a substantial recreation facility, with one of the finest inland water-recreation potentials in the country.

The TVA must be viewed as having easily reached its basic goal to change the economy of the area. Frequently criticized by outsiders, it is generally well regarded within the Tennessee Valley itself. With nearly 35,000 workers, the TVA is also one of the region's more important employers. Tennessee, once one of the South's poorest and least industrial states, remains a regional leader in manufacturing.

Pineville, Kentucky, in the spring of 1977. Flooding is an ever-present problem on the Appalachian plateau, where the towns and streams crowd together in narrow, steep-sided valleys. (Billy Davis/The Courier Journal)

Appalachian Regional Commission

In 1965, Congress passed the Appalachian Redevelopment Act, an extension of the Area Redevelopment Act of 1961. It created the Appalachian Regional Commission (ARC). Responsible for an area that extends from New York to Alabama, the commission has spent several billion dollars in a program to improve the region's economy. Its explicit goals are quite different from those of the TVA, its primary thrust being to improve highways in Appalachia in the hope that this will decrease isolation and encourage manufacturers to locate in the region.

The principle of the government investing in transport development to encourage regional economic growth is well established in the United States. The Erie Canal, opened in 1826, was funded by the state of New York, and canals were built by the federal government in Ohio, Illinois, and Indiana during the 1830s and 1840s. More than 40 million hectares (100 million acres) of federal land were given to western railroads to help finance their construction. Virtually the entire road and highway pattern of twentieth-century America was built by local, state, and federal agencies. The costs involved in transport development are great, and economic returns resulting from construction are usually long-term. Thus, private investment in transport development is unusual. It has been more usual in other forms of communication--most notably, telephones--but only because of governmental restrictions that guarantee a monopoly over the area served.

The ARC is different from the TVA in several other important ways. First, it involves federal-state cooperation, with each level of government paying a portion of the total costs. This, incidentally, helps explain the large area contained within the project. The states were given the task of defining the target area for the commission. No one wanted to be excluded; eventually the entire physiographic region was included. The ARC also worked to improve public and vocational education and regional economic planning and to maintain the quality of the physical environment.

The Arkansas River Navigation System

One other regional development effort has been important for the region. The Arkansas River Navigation System constructed during the 1960s and 1970s and dedicated in 1971 established a 3-meter (9-foot) navigation canal up the Arkansas River from its confluence with the Mississippi River to Catoosa, Oklahoma, just downstream from Tulsa. The result has been an increase in barge traffic and the production of hydroelectric power from the many dams constructed to stabilize the river's flow. However, the project was extremely expensive, and many argue that its economic returns do not justify the cost of its development.

APPALACHIA'S FUTURE

The problems faced by the region are well known. The term Appalachia itself widely connotes an environment of isolation and poverty. Certainly Appalachia and the Ozarks are not likely to become part of the manufacturing core of the country, and few in the region really want that. Poverty will remain a problem especially in the mining areas.

Still, there is a very real sense of change in portions of the region. Parts of the southern highlands in Georgia, the Carolinas, and Tennessee have witnessed a boom in recreational and second-home construction. A rapid increase in the area's land values attest to this development. Some communities, such as Highlands and Cashiers in North Carolina, are summer (and often winter) havens for the well-to-do. Their manicured, well-cared-for golf courses and communities stand in striking contrast to the common image of a slightly ramshackle Appalachia. The Blue Ridge and Shenandoah Valley of Virginia have experienced a similar boom, mostly fueled by residents of southern Megalopolis. The Ozarks and Ouachita Mountains are dotted with new second-home and retirement communities. The long-term pattern of outmigration from the region, while not ended entirely, has surely been reduced. The gap between per capita income levels in the region and the United States, still very real, has nevertheless narrowed. There seems to be reason to hope that, economically, the worst is over for much of Appalachia and the Ozarks.

ADDITIONAL READINGS

Appalachian Regional Commission. *Appalachia.* Washington, D.C.: ARC, 1967–.

Arnow, Harriett L. *Hunter's Horn,* New York: Macmillan, 1949.

Batteau, Allan (ed.). *Applachia and America: Autonomy and Regional Dependence.* Lexington: University Press of Kentucky, 1983.

Caudill, Harry M. *Night Comes to the Cumberlands.* Boston: Little, Brown, 1963.

Eller, Ronald D. *Miners, Millhands, and Mountaineers: Industrialization of the Appalachian South, 1880–1930.* Knoxville: University of Tennessee Press, 1982.

Grant, Nancy L. *TVA and Black America.* Philadelphia: Temple University Press, 1990.

Jones, Loyal (ed.). *Reshaping the Image of Appalachia.* Berea, Ky.: Berea College Appalachian Center, 1986.

Marsh, Ben. "Continuity and Decline in the Anthracite Towns of Pennsylvania." *Annals of the Association of American Geographers.* 77 (1987), pp. 337–352.

Ulack, Richard, and Karl Raitz. "Appalachia: A Comparison of the Cognitive and Appalachian Regional Commission Regions." *Southeastern Geographer,* 21 (1981), pp. 40–53.

Watts, Ann DeWitt, "Does the Appalachia Regional Commission Represent a Region?" *Southeastern Geographer,* 18 (1978), pp. 19–36.

Whisnant, David E. *All This Is Native and Fine: The Politics of Culture in an American Region.* Chapel Hill: University of North Carolina Press, 1983.

Wigginton, Eliot (ed.). *Foxfire,* books 1–9. Garden City, N.Y.: Doubleday (Anchor Books), 1975–1988.

CHAPTER 10

The Deep South: A Culture Region

New England is provincial and doesn't know it, the Middle West is provincial, and knows it, and is ashamed of it, but, God help us, the South is provincial, knows it, and doesn't care.

Thomas Wolfe

The South contains clear contrasts between its past and its future. Within this region, which has modern, thriving cities such as Charlotte, North Carolina, live people who also self-consciously maintain attitudes and traditions not too different from those of their great-grandparents. (Charlotte Convention and Visitor's Bureau)

The South is a region singular in the depth and persistence of its regionality. It has maintained a deeper sense of itself as a culture region than perhaps any other part of the United States. Other regions are identified as distinct cultural territories, of course. The "Down East" of upper New England, the Plains and Prairies, the Southwest, the Pacific Northwest, and other such "regions" (see Figure 1-5) possess cultural characteristics and views of themselves and each other that contribute to their senses of territorial distinctness.

If we were to map for the entire United States the many expressions of habit, attitude, and behavior relevant to a definition of culture, "the South" would often appear as a characteristic territory. As Thomas Wolfe observed, Southerners are acutely aware of their region; some resent it, but most are proud of the identification and gain sustenance from it. The South's regionality is reinforced by the nearly uniform opinion of people living elsewhere in the country that this region is distinctive.

The view from outside the region—until recently largely negative, critical—is partly a function of the region's distinctness; that, in turn, is largely a function of past socioeconomic characteristics typical of the region and the institutions that developed to maintain those characteristics.

In spite of the consistency with which the South is seen as a uniform culture region, this view is far from accurate. Strong differences also exist within the South. The Gulf Coast, the Southern highlands, the Georgia-Carolinas Piedmont, and many portions of the northern interior South possess their own versions of Southern culture, each at variance with the others. True Virginians know how they differ from Georgians, and both recognize their differences with Southerners from Tennessee or Mississippi. But all are also clear about and find comfort in the special "Southernness" they share. Newcomers to the region (and outsiders) do not often recognize the basis for this comfort the culture provides, but those reared within the region know. Peninsular Florida and coastal Texas, locationally the most southern territories of all, are probably the least imbued with the region's cultural flavor and might well be excluded from the region. The source of this flavor is not easy to identify geographically, although its association with a regional core is reflected in the label, Deep South.

The list of cultural characteristics that give the South its character is long, as it would be for any region. The character cannot be appreciated by such a list, however, or by looking for confirmation of what might be expected, but only by passing through the landscape and living among its people with your senses opened. Throughout much of the Deep South, one can still stroll down a dusty red dirt road in the summer with an open field on one side and a mixed, but mostly pine woods on the other. A green wave of kudzu vine cascades across the edge of the woods and extends tentative ribbons of vegetation out into the road. As recently as 30 years ago, the single-story, broad-porched frame house with the painted metal roof might have contained a family preparing a customary evening meal of ham hocks and snap beans, sweet potatoes, and early turnip greens, followed by

Major Metropolitan Area Populations in the Deep South, 1990			
Atlanta, Ga.	2,833,511	Birmingham, Ala.	907,810
New Orleans, La.	1,238,816	Jacksonville, Fla.	906,727
Norfolk–Virginia Beach–Newport News, Va.	1,402,107	Richmond–Petersburg, Va.	865,640
Charlotte–Gastonia–Rock Hill, N.C.–S.C.	1,162,093	Greenville–Spartanburg, S.C.	640,861
Nashville, Tenn.	985,026	Raleigh–Durham, N.C.	735,480
Memphis, Tenn–Ark.–Miss.	981,747	Baton Rouge, La.	528,264
Greensboro–Winston-Salem–High Point, N.C.	942,091	Little Rock–North Little Rock, Ark.	513,117
		Charleston, S.C.	506,875
		Mobile, Ala.	476,923

cornbread and molasses. The house appears pretty much the same today, but higher incomes and the spread of grocery store chains have greatly reduced the dietary distinctiveness of this region.

References to the Deep South are references to whatever is felt to be the source area for Southern culture, with depth in this case implying cultural intensity. Non-Southerners often use the term geographically, with "deep" implying a more southerly location, but this is incorrect conceptually. The states of Georgia, Alabama, Mississippi, Louisiana, and South Carolina are not the only area within which Southern culture is found. In addition, many elements of this culture originated in or continue to be strong elsewhere in the larger region. One style of Southern country music, for example, is largely focused on Nashville, Tennessee, another style carries the name of the Bluegrass Basin in Kentucky, and a third, Old-Timey, is associated with the southern Appalachians. The blues are a legacy of Memphis, Tennessee. Gospel music is prevalent throughout most of the South, although there

Regionally distinctive house-types may suggest some aspects of a region's cultural underpinnings. The older house on the left was enlarged by its owners as changing needs were given form in additions to the original simple boxlike structure. Note how the newer house on the right, different as it may be in detail, has many features similar to the appearance of the "developed" older house. (Stephen S. Birdsall)

are several styles. Like many sports, stock-car racing has a clear regional concentration. It is a highly popular, typically "good-old-boy" Southern tradition that flowered within the South and continues to be more intensely related to that region than any other in the country. Even though it has been exported widely to other regions, this part of Southern culture is almost entirely a North and South Carolina production.[1]

The region of Southern culture (the Deep South) can be viewed as a geographic composite of beliefs, attitudes, patterns, habits, and institutions associated with the entire group of Southeastern states (Figure 10-1). Some of this composite is absent, or at least diminished, in parts of the territory, but much of what is seen as Southern is an outgrowth of early patterns that were present throughout. Arguably, we exclude the southern Appalachians from the Deep South, primarily because the former shares themes of resource dependence and government intervention with northern Appalachia (see Chapter 9). In addition, however, the mountain portion of the South experienced less directly much of what made the Deep South distinctive.

Many of the early patterns in the region and the changes in them that are currently underway are explicitly geographic; many others have definite geographic consequences. In comprehending the South as a culture region, therefore, we first examine the source of the early patterns and the ways in which they changed and became clearly defined. We then discuss the external pressures on the old cultural patterns and the implications of these new alterations for the region. In other words, when considering the South as a culture region, we need to understand the origins of the culture and the modifications to it that have occurred if we are to make sense of what exists now and to guess what the region is likely to be in the near future.

[1] Richard Pillsbury, "Carolina Thunder: A Geography of Stock Car Racing," *Journal of Geography*, Vol. 73, no. 1 (January 1974), pp. 39–47.

THE HERITAGE

Southern culture has its roots in the region's location and the economy generated in that location. The earliest European colonization in North America was commercial and exploitive. The first century of European settlement was almost entirely meant to provide the means of and support for that exploitation.

The coastal plain of North America south of Delaware Bay, and especially that south of Chesapeake Bay, contained many areas that appeared ideal for agricultural exploitation. The long, hot summers, regular rainfall, and mild winters permitted settlers a selection of crops complementary to those grown in Great Britain and northern Europe. The large number of rivers that crossed the plain, navigable to small boats at least, allowed settlement to expand freely between the James River in Virginia and the Altamaha River in Georgia. Population densities remained low throughout most of the region, with urban concentrations larger than the village size limited to port cities (Norfolk, Wilmington, Charleston, Savannah) or the heads of navigation on the main rivers (Richmond, and later Columbia and Augusta). The strong rural and agrarian elements of Southern culture, therefore, developed from the beginning of white settlement in the region, establishing a pattern that did not weaken until late in the nineteenth century and remained significant until after World War II.

There is more to early Southern agrarian culture, however, than relatively low average population densities, an absence of large cities, and the presence of a supportive physical environment. European exploration during the fifteenth century and later was motivated by economic purpose more than by human curiosity and a search for knowledge. The greatest return for the effort expended in settling Caribbean islands

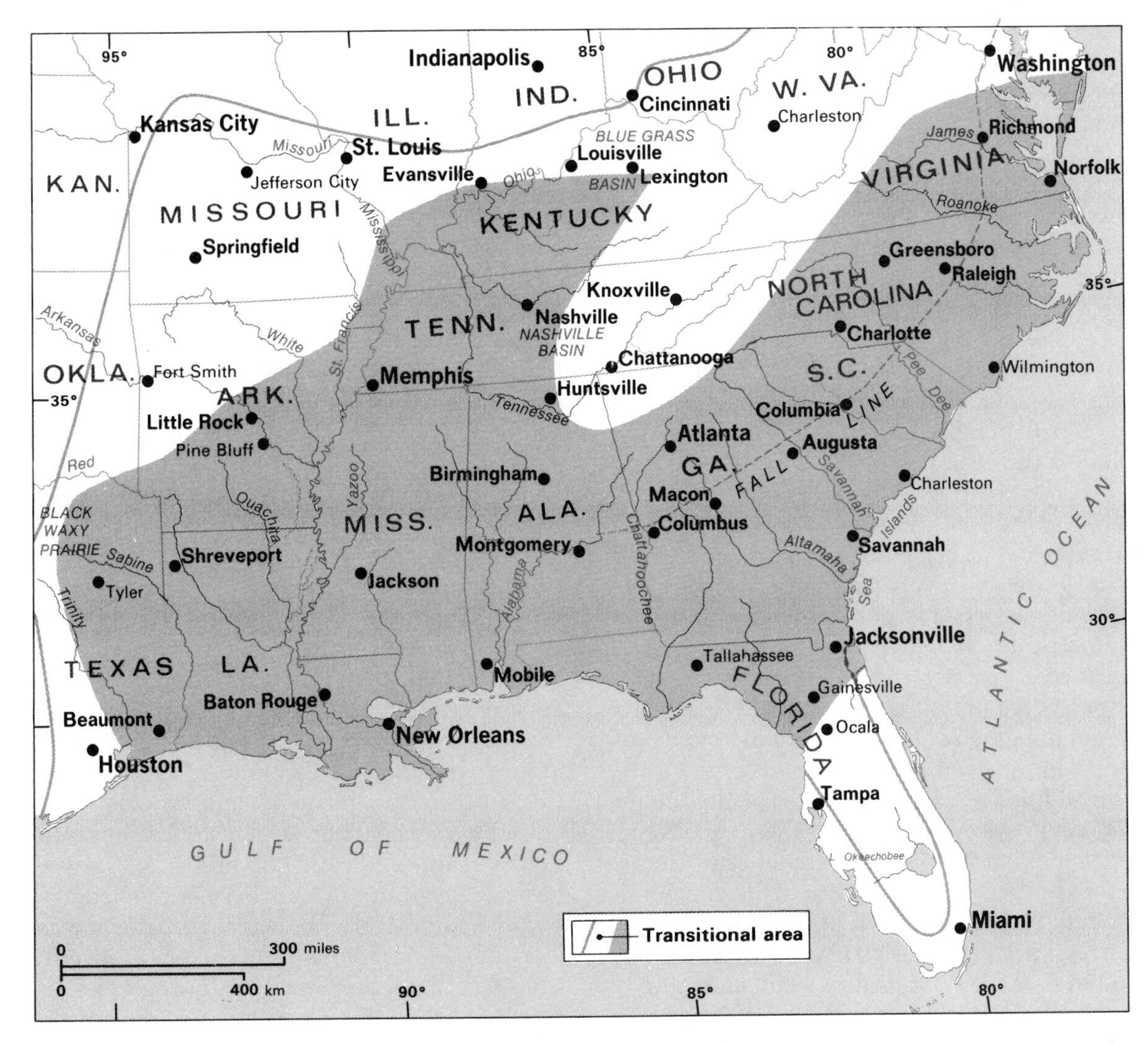

Figure 10-1 The Deep South.

and the Atlantic southern lowlands was through highly structured cash crop agriculture. The plantation organization, already well established throughout the Caribbean, came gradually to dominate the early Southern colonial economy. Production of tobacco along the James River and to the south in northeastern North Carolina, and production of rice and indigo in and around the many coastal swamps in the Carolinas and Georgia were important from 1695 onward. Cotton production grew slowly in importance until about 1800 and spread rapidly inland thereafter, from the initial concentrations on the Sea Islands between Charleston and Spanish-held Florida. Although privately held small farms were numerous, the plantation form of organization was successful enough that it was carried westward with cotton production and reached its most prevalent form in Georgia, Alabama, Mississippi, and Louisiana during the first half of the nineteenth century. Tobacco was similarly carried westward into Kentucky and Tennessee by settlers migrating from Virginia and North Carolina.

In contrast to the individually owned and family-operated farm, a plantation was a business enterprise. Its purpose was production for cash return and export. As a result, the South's spatial organization was weakly developed with small market centers serving as collection and transshipment points; the largest cities, containing a wide variety of economic activities, were very few in number. The transportation network accompanying this pattern was one that simply allowed the inland products to be moved most directly to the coastal export centers; interconnections between the smaller marketplaces remained few (see Figure 6-2). A major consequence was relatively high rural isolation for most of the region's population and a continuation or development of distinctively local allegiances. This isolation is common to rural populations, but it was especially pervasive throughout the South and contributed to the region's unashamed provincialism.

In addition to a strong rural provincialism, the plantation system generated other, equally significant contributions to Southern culture through the mix of cultural backgrounds brought by the immigrant populations entering the region. Large-scale plantation agriculture required a sizable annual investment, and much of that investment was in the form of labor. The combination of low population densities, cheap land, and early opportunities led to the heavy use of slave labor in the South. Once established, this practice continued to restrict population immigration because potential settlers and urban workers found freer opportunities in the North. Since early in the nineteenth century, therefore, the South has had a lower proportion of its population that was foreign born than any region of the country. Because significant immigration to the United States and Canada from countries outside the British Isles did not occur until the 1840s, the overwhelming majority of Southern whites are of English and Scots-Irish descent. The accompanying stimulation of new skills and knowledge from non-British cultural sources that altered the mix of habits and attitudes elsewhere in the country throughout the nineteenth century was also lacking from Southern culture and contributed to an entrenchment of cultural expressions distinctive to this region.

Several important ethnic variations on the dominant black and Anglo white theme can be found within the South's population. The two long-term resident populations that are neither Anglo nor African in ancestry are the Cajuns of southern Louisiana and several Indian (Native American) groups. The name Cajun is undoubtedly derived from Acadian, for this Catholic, French-speaking population is descended from French exiles from Acadia (now Nova Scotia and New Brunswick). The rural Cajun population stayed in southern Louisiana and remained culturally distinct from the rest of the South in spite of France's sale of the Louisiana Territory in 1803 and the gradual integration of the remainder of the state into Deep South culture. Most Indian groups were removed from the South in much the same ruthless manner and at the same time as in the Midwest, but several significant exceptions remain. There are also numerous smaller, identifiable groups of Native American heritage, but the largest are the Lumbee in southeast North Carolina, remnants of the once powerful Cherokee along the Blue Ridge in southwestern North Carolina, the Choctaw in central Mississippi, and the Seminoles in southern Florida, all of whom managed to elude removal until federal efforts to this end decreased.

While not directly tied to the plantation system, another strong element of Deep South culture took root in the agrarian, isolated communities and homesteads that gradually spread across the region. The South's population has long been characterized by its persistent adherence to various evangelical Protestant religions. Small, unpretentious church buildings still dot the countryside, consistently drawing their congregations every Sunday from the scattered rural and small-town populations. Although Methodist, Episcopal, and other Protestant congregations are numerous across the region, it is the Baptists who have numerical dominion in the South (Figure 3-6).

Freewheeling, energetic Baptist evangelists entered the Southern Piedmont and hill country hard on the heels of the mideighteenth-century settlers and found them eager listeners. The religion was adopted through an intensely emotional, but personal, conversion experience that was compatible with frontier hardship, isolation, and individualism. As a group, Baptists were also resistant to the rigidities of formal institutional organization, and this too permitted easy expansion of the religion across the sparsely settled South. Any individual who was spiritually "born again" and felt the "call" to preach could do so without either the formal approval or denominational constraint of a church hierarchy. And because most nineteenth-century immigrants to the United States settled elsewhere in the country, the religious pattern that was established in the South by the early 1800s was not affected by the religious affiliations of later arrivals.

The heavy use of African slaves in the Southern colonies lies at the crux of both additional

Religion is important in the Deep South. There is more diversity in this small town in northwestern Georgia than in much of the region, for Baptists predominate, but to have this many active congregations in a town of only about 600 people suggests a high degree of support. (Bruce Roberts/Photo Researchers)

components of Southern culture. One impact, heavily discounted until recent years, was the transfer of many elements of the African cultures to the region and the amalgamation of these elements with those of the white population. The first Africans arrived in Virginia in 1619, only ten years after the initial James River settlement had been established and a year ahead of the *Mayflower's* arrival at Plymouth. Although slaves were not imported in large numbers until the early eighteenth century, blacks were present in the region and were part of its organization and social environment from the beginning. By the time of the first U.S. census in 1790, blacks comprised 39.1 percent of the population of Virginia, 43 percent in South Carolina, and 25.5 and 35.4 percent in North Carolina and Georgia, respectively. Even Maryland's population was 33 percent black in 1790. As socially and psychologically isolated and disoriented as imported Africans may have been, many of their original cultural mannerisms were retained for a while and transmitted to Southern whites. Their impact on patterns of speech, diet, and music in the South is undisputed. Undoubtedly, other more subtle contributions to the dominant white culture also occurred. And Southern culture is richer for the amalgamation.

The other indisputable cultural consequence of the black slave presence in the region is less positive. For a slaveholding situation to exist effectively and over a substantial period of time in an overtly Christian society, there must be no empathetic identification of the slave owner with the slave. If the owner were to "put himself" in the slave's position, the impact and implications of the slave's subservience in the relationship would have to be recognized. By extension, then, to justify the enslavement of other human beings, it is necessary to consider the entire enslaved group as decidedly and perhaps permanently inferior. The acceptance of this view of blacks by Southern whites was no different from the dominant European view until late in the eighteenth century. By the turn of the nineteenth century, however, opposition to slavery had gained strength in those regions where it was of less importance. Slavery was a pervasive institution throughout the Southern states by this time. Changes in white attitudes toward the supposed inferiority of blacks within the South, therefore, were small; instead, justification of slavery became more intense and self-righteous within the region as pressure to eliminate it arose from outside.

Much of the white South's antiblack attitude seems to have been tied to proximity and numbers as well as to economic aspects of the master-slave relationship. More than 90 percent of all blacks in the United States lived in the South until the beginning of the twentieth century and were present throughout the region in large numbers except in the Appalachians and peninsular Florida. Events outside the South during the middle of the twentieth century revived the argument that whites with strong antiblack attitudes existed across the country; the opportunities for expressing these attitudes were simply greater in the South.

As a point of contrast, however, it should be noted that the long and intimate relationship between whites and blacks in the South seemed to produce a relationship between the two groups that was not apparent outside the region. While basic animosities remained strong, as long as both sides knew what to expect from the other under most circumstances the day-to-day interactions between individuals could often be conducted with little apparent tension. That tensions existed and frequently flared into violence cannot be denied and underlines the basic inequality of the relationship. Nevertheless, this greater intimacy between the races, not easily comprehended by non-Southerners, was as much a part of the distinctive Southern culture as the region's long history of violence between blacks and whites.

By the outbreak of the War Between the States, the geographic pattern of population settlement and economic organization had changed dramatically from the region's colonial beginnings. As new areas were taken from the Indians in Georgia, Alabama, and Mississippi, settlers moved westward from the Carolinas and eastern Georgia. Complete plantation operations, including entire slave populations, would often be transferred to the rich new land. Families who passed through the Appalachians from Maryland, Virginia, and North Carolina settled into the more fertile farming areas of Tennessee (such as the Nashville Basin), Kentucky (the Bluegrass Basin), and along the major rivers in these states. Arkansas, Missouri, and southern Illinois also received settlers from these areas. Many of the attitudes and behavior patterns that identify Southern culture were transplanted with the migrations.

Although slave densities were generally lower in the so-called "border states," blacks were present in higher proportions in the zones of better commercial agricultural land. Thus, they were concentrated in the Louisville and Bluegrass Basin sections of Kentucky, in the Nashville Basin and on the loess plains of western Tennessee, and throughout the tobacco and hemp plantation country along the Missouri River westward from St. Louis. In a similar manner, plantations existed in the panhandle of northern Florida and on the interior highlands almost as far south as present day Ocala. Louisiana and the plains of eastern Texas became western extensions of the Deep South's plantation economy during the 1850s.

As the Civil War erupted, the South was composed of a strongly rural population. Except for a few major collecting points for the region's agricultural produce, urban development was limited numerous villages and small towns. The larger cities were almost all located on the coast or at major transfer points along interior waterways. Transportation and communication networks were sparse, with the best connections between the relatively few scattered large urban centers.

The biracial population pattern was uneven. Although blacks were likely to be found in almost every county outside the Southern highlands, they comprised a large proportion of the total population only in the original plantation areas (south-central and tidewater Virginia, northeastern North Carolina, and tidewater Georgia and South Carolina) and in the new lands most suited to large-scale cotton production (Piedmont South Carolina, the inner Coastal Plain of Georgia, Alabama's Black Belt, the Mississippi Valley, the loess plains south of Memphis, and the black-soil prairies of eastern Texas).

Supported by slave labor, production of plantation cotton in particular had become so successful that the region's economy was dominated by this one crop. Other crops were grown but primarily as a local food supply or a decidedly secondary cash alternative. Tobacco, rice, sugar cane, and hemp, for example, continued to be grown in areas suited to their production. In 1860, cotton dominated not only the South's economy but also, at least in terms of export income, the entire country's; slightly over 60 percent of the total value of goods exported from the United States during that year was from cotton. Currently produced in significant quantity outside the South, cotton still ranked ninth in value of U.S. agricultural exports in 1986.

With the loss of the Civil War, the South's economic underpinnings were very badly damaged. Railroads were torn up and equipment confiscated, shipping terminals disrupted, and most of the scattered industrial base destroyed. Confederate currency and bonds were suddenly worthless. Cotton stocks awaiting postwar sale in warehouses and ports were confiscated by Union forces. Farms and fields were in disrepair, and implements and livestock were often stolen or lost. The slave labor supply was formally eliminated and large landholdings broken up or heavily taxed. Investment funds had either been used up during the war or were drawn off during the decade after the war. While the large black population was no longer slave, most

The bustle of commercial activity, fundamental to New Orleans, is caught vividly in this 1851 lithograph of the city. Riverboats and ocean schooners vie with smaller vessels and rafts on the Mississippi while many dozens of each huddle tightly along the riverbank to load or unload their goods. ("Bird's Eye View of New Orleans," 1851 by John Bachmann/I. H. Phelps Collection of American Historical Prints, Print Division, New York Public Library)

blacks gained very little economically during the early decades following their emancipation. To an overwhelming degree, they continued to live very near their prewar locations.

As a region, the South had lost much politically, militarily, and economically. However, the new geographic patterns that emerged and continued to evolve without significant disruption during the next three quarters of a century were tied in large part to underlying cultural patterns that neither the war nor the subsequent economic turmoil had weakened. Having lost almost everything except the land itself, certain aspects of white Southern culture became more intense. These aspects were reflected in many of the geographic consequences that followed the regionwide disruption of the Civil War, consequences that reverberated throughout the nation and that laid the foundations for the current geography of the South.

THE CONSEQUENCES

The first half-century following the Civil War was a period of readjustment for the South, during which the regional culture and economy had to become reconciled to the changes imposed by the war. The white population proceeded through several alternative reactions to the emancipated status of the large black population before finally settling into institutionalized segregation. Blacks, for their part, experienced changes in opportunity that were largely out of their control until more than a half-century after the Civil War. This was also a period during which Southern attitudes and feelings of isolation from the remainder of the country became even more inflexible than they had been in the antebellum period. Each of these consequences interacted with the others and were reflected in geographic patterns.

Persistent Poverty

The disintegration of the antebellum economic organization led to very difficult times for most of the South's population during the 12-year Reconstruction (1865–1877). Quite aside from the destruction of transportation and manufacturing capacity, the plantation economy had become refined to the point of rigidity and overdependence on slave labor. While the North had undergone the beginnings of industrialization and was coming to expect change as normal, this adaptability was not valued in the South before the war. After the war, a continuation of intense exploitation was necessary to meet heavy taxation and other costs of rebuilding. The resource most available for exploitation continued to be the land; thus, cotton production remained dominant in the region's economy.

The other factors necessary for production, however, were much less available. In addition to land, both labor and capital are required. Local capital was scarce, with much of it consumed by the war effort or drawn off after the war by the North through taxation. With credit in short supply, interest rates increased sharply and farmers, both large and small, found themselves continually in debt. This tended to perpetuate the general Southern dependence on agriculture and maintained cotton and tobacco as the most important cash crops.

The availability of labor was complicated by emancipation. No longer slave, the black population was not forced to work for white farmers for social reasons. The dream of Yankee-supplied economic independence gradually faded, however. With few jobs available in the small towns of the region, most rural blacks were forced to make whatever arrangements they could with the remaining white landowners. *Sharecropping* became the means of subsistence and the way of life for these blacks just as it was for many poor whites who had lost their land. Once this pattern was established for the black population, it was enforced with "Black Codes" that restricted black movements outside the agricultural areas and with a continuation of low educational opportunities. Sharecroppers were frequently in debt to the landowner (just as he or she was in debt to the bankers and merchants) and were not permitted to end the sharecropping arrangement until their debts were paid.

For most rural blacks, slavery was exchanged for an agrarian existence not far above the subsistence level or far from indebtedness to white landowners or local merchants. Some did own the land they worked, but the proportion of blacks operating farms who were also owners was small (25 percent in 1900), and this number increased only as nonowners left agriculture altogether.[2]

Even when they owned their land, black farmers were hampered by poor access to credit, farm sizes too small to be highly productive, and the antiblack aspects of the regional culture. These conditions created and maintained very low annual income levels among the overwhelming majority of rural blacks in the South. When considering the distribution of poverty in the region, therefore, the distribution of the rural black population was a strong explanatory factor. Even though the South's economy and population distribution experienced major changes during the decades after World War II, this association between the proportion of a county's population that was black (Figure 10-2) and the median family per capita income (see Figure 1-3) remained clearly visible in 1980.

Poverty has not been exclusively an experience of rural blacks in the South. Even during the antebellum period, a majority of the white population were small farmers; relatively few had the capital to create a large landholding and the expertise to operate it successfully. Cotton plantations were established on the best agricultural land during the first few decades it was

---

[2] For a clear and thorough discussion of this, see James S. Fisher, "Negro Farm Ownership in the South," *Annals*, the Association of American Geographers, Vol. 63, no. 4 (December 1973), pp. 478–489.

Although this house's clean exterior and yard suggest great care and effort by its occupants, complete maintenance and a full rehabilitation are just as clearly beyond the means of this low-income, rural Southern family. Compare the form of this house with that of those shown on page 225. (U.S.D.A.)

available for settlement. The pattern of black population reflects the location of this land. By the same token, poor whites were unable to compete for land in the main cash crop areas or for employment with the slave population. Most poor whites, therefore, settled in the lower-quality agricultural areas north and south of the Black Belt; in the peripheral Southern Appalachians; or among the large plantations, but on locally poor land or in nonagricultural employment. After the Civil War, hard times pushed many of the rural poor whites into sharecropping arrangements little different from those affecting blacks.

About 1880, the patterns began to change. The distribution of poverty was changed little, but the environment for economic opportunities in the South entered a new phase. During this decade, manufacturing experienced very rapid development, perhaps more rapid than in any prior 10-year period, led by growth of the cotton textile industry. Largely a New England indus-

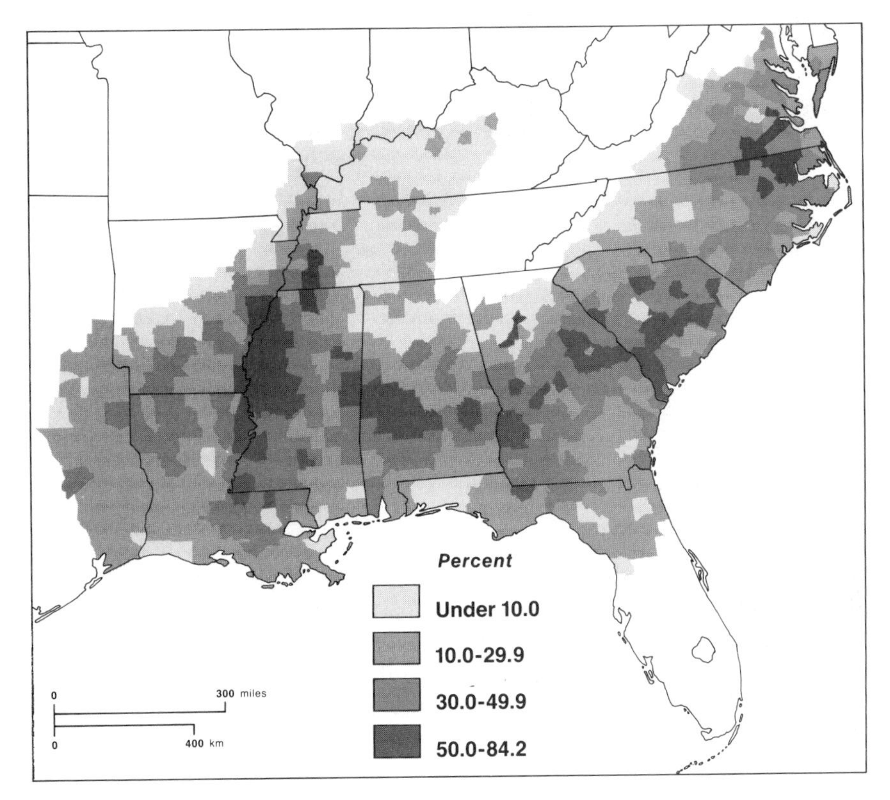

Figure 10-2 Percentage of county population black, 1980. In spite of considerable out-migration, the distribution of racial balance in the South in 1980 was very similar to that in 1860. Changes in the pattern to 1990 are not likely to be significant.

try prior to the turn of the twentieth century, cotton textile manufacturing began to shift southward several decades earlier. It grew much more rapidly in the South after 1880 than anywhere else in the country. By 1929, 57 percent of the nation's cotton textile spindles were in the South, over two and a half times the share existing in 1890.

Within the region, textiles and the related apparel industries grew most strongly throughout the piedmont of North and South Carolina and Georgia. The shift of these industries into the region can be attributed to changes in labor availability, resource proximity, technology, and the inertia costs of location. The low-skill, labor-intensive industries originally situated along the northeastern margins of the continent's manufacturing core region began to experience a gradual loss in labor availability as workers shifted toward higher-skill, higher-paying employment. The high levels of *underemployment* in the piedmont South offered an alternative: a large pool of labor available for the textile and apparel industries. Natural and synthetic fiber industries began to appear in the region to produce the raw material for cotton and synthetic textile manufacturers just as the textile industries provide the raw material for apparel manufacturing. Taking advantage of proximity, the growth in textile and apparel manufacturing across the Carolina Piedmont and in northern Georgia was followed by an increase in the number and output of fiber industries.

In addition to all this, as technological improvements occurred in these activities, New England plants became increasingly obsolescent. Internal costs of production and locational inertia costs grew for the Northern manufacturers as taxes, land, and power costs were driven up by higher-value industries. This had not yet occurred in the Southeast.

There was much more to the relocation of textiles, therefore, than a desire to be close to the raw material source. In fact, much of the new industry located outside the primary cotton-producing regions (Figure 10-3). There was a degree of resistance to any type of manufacturing industry in the best agricultural areas. This resistance, when combined with a large surplus labor force and active campaigns by local leaders, led the cotton textile industry to be drawn to northern Alabama, northern Georgia, and the Piedmont sections of North and South Carolina. Counties in these areas had predominantly white populations and showed their determination to keep it that way. The two chief results of this were that (1) only some parts of the South experienced the economic and cultural stimulus of early industrialization, and (2) the black portion of the region's population were not well located to benefit from these changes. While it would be a mistake to overestimate the impact of this development on general income levels (textiles are a low-wage industry, thus the appeal for cheap labor), that there was *any* manufacturing growth in the South during this period is frequently overlooked.

Figure 10-3 Cotton production, 1909. Cotton acreage remained high in most Southern states well into the twentieth century, a pattern reinforced by heavy regional debt, cultural inertia, and the apparent absence of alternatives.

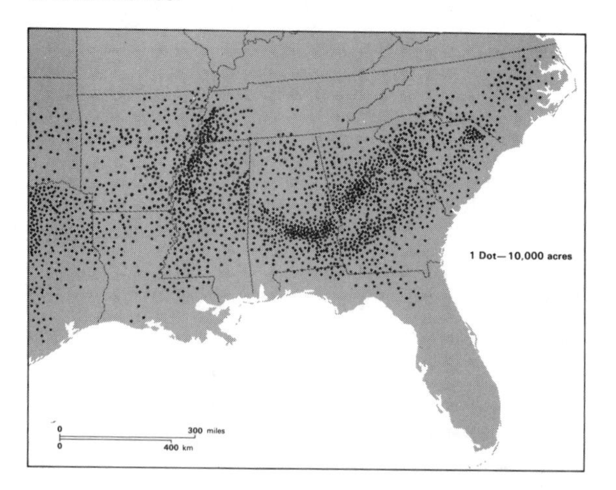

Cotton textile manufacturing was not the only new source of industrial opportunities during the 1880s and 1890s. Reconstruction of the region's railroads and other public improvements stimulated the flow of money and the development of railroad towns. Cigarette manufacturing began to be focused in the tobacco regions of North Carolina and Virginia following technological advances. With the establishment of a new federal land policy and a strengthened railroad network, the South's very large timber resources began to be exploited. Much of the timber was taken out as a raw material, but furniture manufacturing in North Carolina and Virginia and (after 1936) pulp and paper manufacturing throughout the South also were an outgrowth of the exploitation. These industries continue to be important in the South (see Figure 10-4).

It was also during the last quarter of the nineteenth century that the South's most important iron and steel industry received a series of stimuli. An iron industry had existed in the region for many decades in small and dispersed smelters. By the mid-1870s, however, technological improvements in iron making had led to the rise of Chattanooga, Tennessee, as an important center of iron production. At the same time, a large deposit of high-quality coking coal was discovered near Birmingham, Alabama, and exploitation of the seam was begun before the end of the decade. Numerous iron-making companies and iron- and steel-using industries accumulated in and around Birmingham and Chattanooga. These two cities combined with the transport focus and subsidiary industries in Atlanta to form an important industrial triangle by the end of the century.

The abundant pine forests of the South produce about two-thirds of U.S. pulpwood. Pulpmills such as this one at New Bern, North Carolina, transform wood waste to pulp through a process that requires wastewater aeration, as can be seen in the foreground. Note the heavily wooded, flat landscape. (Courtesy Weyerhauser)

This development was significant in the economic geography of the South, although the critical importance of this triangle's appearance was weakened by the economic and political context of the period. First, the development was significant because of the way in which iron and steel production tends to draw other manufacturers dependent on steel. Second, these industries are also not as low-skill and low-wage as textile and tobacco product manufacturing. And, finally, this centrally located region of nonagricultural economic development could have been an industrial focus for the South as a whole, stimulating increases in labor skills, income levels, and general economic welfare through each city's connections with other major urban centers.

This did occur to some degree, but discriminatory shipping rates imposed on Birmingham-manufactured products dampened the beneficial effects considerably. Because of low labor costs and the fortunate proximity of raw materials in the Birmingham area, steel could be produced there at much lower cost than in Pittsburgh, the North's primary center of steel

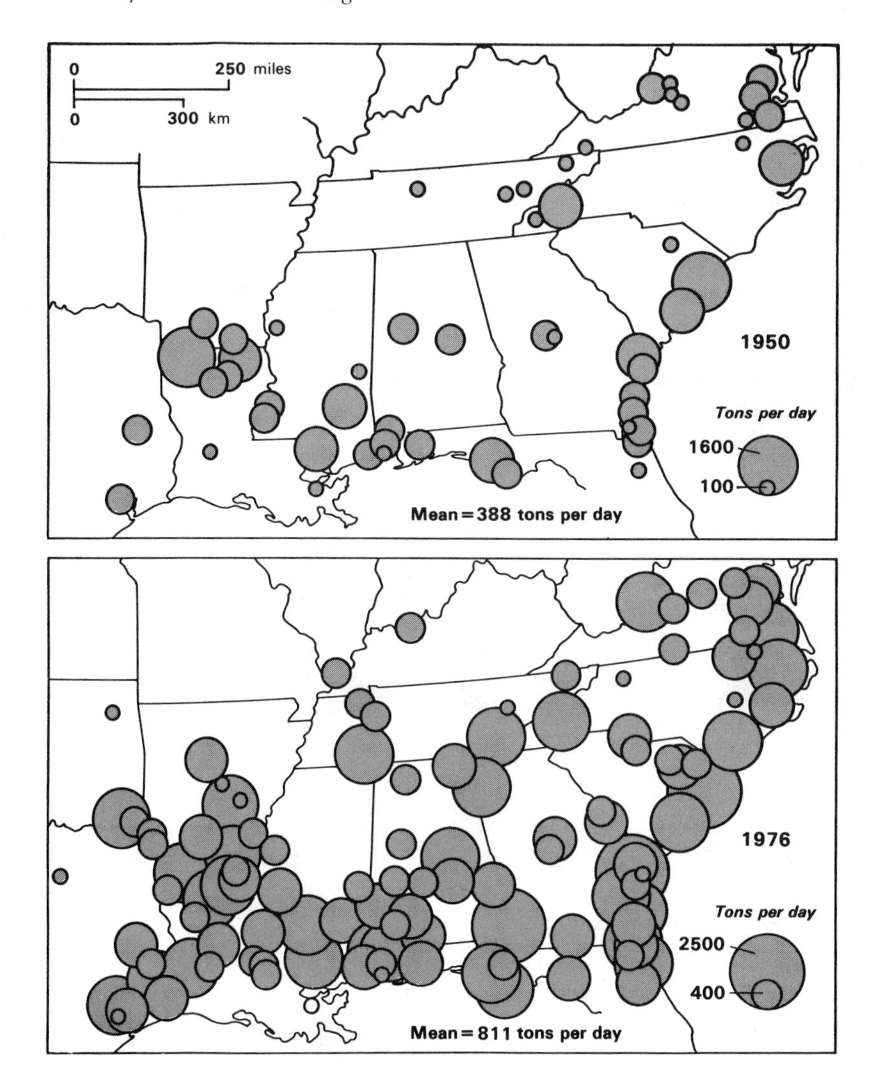

Figure 10-4 Pulp mill production, 1950 and 1976. (Courtesy of J. F. Hart.) The tremendous growth in the exploitation of Southern forests during the third quarter of this century is indicated by changes in pulp production. Most increase in production has occurred on the Atlantic and Gulf coastal plains.

production. Without interference from the federal government, however, the United States Steel Company imposed and enforced what came to be called the Birmingham differential, or a pricing policy of "Pittsburgh plus," on all users of Alabama steel. Under this policy, consumers were required to pay the price of steel at Pittsburgh plus $3.00 per ton ($2.72 per metric ton) plus the regular freight cost from Birmingham. Even though this discriminatory pricing practice was eventually ruled illegal and stopped, the policy severely restricted the competitive cost advantage of Alabama steel during the rapid economic expansion decades of the early twentieth century and contributed to the slow growth of Southern industry. One way to view this would be to consider Birmingham and the South in general as having been much farther away from the national economic core region than in fact they were. Atlanta firms, for example, found it to their advantage to order steel from Pittsburgh rather than Birmingham, even though a Pittsburgh-Atlanta shipment had to travel more than four times the distance of a shipment from Alabama.

In summary, changes in Southern economic development were irregular and often slow for more than half a century after the Civil War. The region was far from resource poor, and yet its population continued to have an average annual income far below the national level. Gradual industrial growth helped raise income levels somewhat, but agricultural incomes remained low and most industrialization was of the low-wage variety. Geographically, the greatest growth in manufacturing employment occurred in the Piedmont section of Virginia and the Carolinas and in the Atlanta-Birmingham-Chattanooga triangle. The Gulf Coast also began to experience rapid growth as local resources were developed (Chapter 11). The number of white families who remained in poverty was large, but an even larger proportion of the black population lived at these levels. Most blacks not only resided outside the areas experiencing industrial growth and found themselves locked into restrictive agricultural arrangements, but they also experienced increasing racial repression at the same time that economic opportunities were increasing for whites.

The Institutionalization of Racial Segregation

A good deal of variation in interracial attitudes existed in the South prior to the Civil War. Even under the assumptions of inherent inequality required by the institution of slavery, it is clear that most whites were not comparable to Simon Legree, just as most blacks would not fit the Uncle Tom image. Following the war, however, antiblack feeling gradually spread and intensified among whites until it pervaded almost every aspect of Southern life.

During the early years after the war, antiblack activities were localized. Some former slaves were able to gain title to good agricultural land; to enter professions as educators, scientists, lawyers, and doctors; and to become elected to political office. The bulk of the population struggled under the economic constraints of the period, but the rapid achievements of many belied the label of inferiority placed on blacks as a group.

By the late 1880s, segregation laws began to be passed requiring the racial separation in more and more aspects of Southern life. This followed by about a decade the departure of federal troops. Restrictive laws, called Jim Crow laws, were passed in each Southern state in rapid succession during the 1890s, until virtually total legal segregation was established.

Formal segregation had many geographic expressions. It was by definition the separation of the total population into two or more societies. It was often necessary for members of these two groups to meet, communicate, work near one another, and in other ways have their lives connect, but segregation's purpose was to maintain a physical distance between members of the different groups wherever possible. Two sets of schools were operated. Two sets of restaurants, recreation facilities, park benches, drinking fountains, rest rooms, and other points of potential contact had to be constructed and maintained. Housing was separated into white areas and black areas. Entry into certain occupations was restricted, with the second-class group finding only the less desirable employment available to them. In the South, both overt and covert restrictions were placed on black efforts to vote. Blacks were almost totally disenfranchised well before World War I.

The geography of the Deep South, therefore, became much more complex because there were two very different human regions in the South at the same time, one black and one white. In some cases, such as residential patterns in the western South, the two areas of human activity hardly overlapped at all in the urban places. But in other cases, such as employment, the regions overlapped greatly in a geographic sense, even if they did not come into contact in social and economic ways. When individual resistance to the discrimination arose, violent repression was an effective method of removing the resister. A propensity for lynching became an ugly component of white Southern culture during the early decades of this century.

In spite of a nineteenth-century Supreme Court ruling permitting segregation as long as equality was ensured, *de jure segregation* did not lead to facilities and opportunities for whites and blacks that were "separate but equal." Basically, there was no desire or commitment to make the facilities equal; whites presumed themselves superior to blacks. While this presumption probably differed little from that of many whites outside the region, Southerners used the force of law to institutionalize the belief. The South was not a wealthy region, and, in allocating scarce financial resources, black facilities received much less than half of the budget for many years. By 1930, the difference between white schools and black schools, for example, was striking (Table 10-1). Every Southern state except Tennessee and the border states of Kentucky and Maryland spent more than twice as much per pupil on white schools as black schools, with white schools in South Carolina and Mississippi receiving well over *five* times as much per pupil as black schools. These figures suggest some of the pressures that led to the eventual reversal of the "separate but equal" doctrine by the U.S. Supreme Court in the 1954 decision *Brown* v. *Board of Education of Topeka.*

Outmigration

A third geographic consequence of the war and the antebellum Southern heritage, intimately connected with the patterns of persistent poverty and racial repression, was one of the country's major streams of population redistribution. For almost 50 years following the end of

Table 10-1 Selected Characteristics of Racial Inequality in Southern States, 1930

State	Percentage Black	*Annual School Expenditures per Pupil ($)*		*Number of One-Teacher Schools per 1000 Pupils*		*Percentage Illiterate*	
		White	Black	White	Black	White	Black
Alabama	35.7	27.43	8.31	2.03	5.04	4.8	26.2
Arkansas	25.8	22.75	10.42	4.89	4.95	3.5	16.1
Florida	29.4	34.59	10.56	n.a.	n.a.	1.9	18.8
Georgia	36.8	24.64	6.53	1.48	6.49	3.3	19.9
Kentucky	8.6	22.22[a]	21.90[a]	7.17	8.99	5.7	15.4
Louisiana	36.9	36.41	8.78	0.79	4.02	7.3	23.3
Maryland	16.9	49.41	33.89	1.74	4.52	1.3	11.4
Mississippi	50.2	30.60	4.79	1.51	6.72	2.7	23.2
N. Carolina	29.0	28.76	12.70	1.22	3.13	5.6	20.6
S. Carolina	45.6	37.53	6.65	1.07	4.27	5.1	26.9
Tennessee	18.3	28.97	15.02	3.42	5.02	5.4	14.9
Texas	14.7	38.76[b]	16.02	1.75	4.19	1.4	13.4
Virginia	26.8	29.44	13.06	2.89	4.59	4.8	19.2

Source: Charles S. Johnson, *Statistical Atlas of Southern Counties* (Chapel Hill: University of North Carolina Press, 1941).

[a] Does not include Louisville.

[b] Includes salaries of superintendents.

the Civil War, the slow trickle of black migrants who left the South increased very little. Lack of information about alternatives outside the region, initial postwar improvements in local opportunities, the impoverished and low-skill characteristics of most Southern blacks of that period, and perhaps simple inertia combined to keep black outmigration below the level of natural increase. Thus, 91.5 percent of all U.S. blacks resided in the South in 1870 and 89 percent in 1910, while the Southern black population increased by almost 4 million (Figure 10-5). During the next decade, the number of black emigrants increased sharply, and the subsequent half century revealed a continuation of this trend.

The numerous factors that accumulated about the turn of the century to "push" blacks out of

THE MIRAGE
—Temple in the New Orleans *Times-Picayune.*

This 1923 editorial cartoon from the New Orleans *Times-Picayune*, demonstrates the Southern viewpoint regarding Northern labor recruiters. The sharp-eyed "outsider" attempts to sway "contented" black labor with false stories. In fact, black labor was not contented, and wages *were* higher and hours shorters in the North. (Ettie Temple/Historical Pictures Service.)

Figure 10-5 Changes in regional distribution of the U.S. black population, 1870–1980. Following World War I, the South's black population began to leave the region in very large numbers, a level of out-migration that continued until the 1970s when the departures declined sharply.

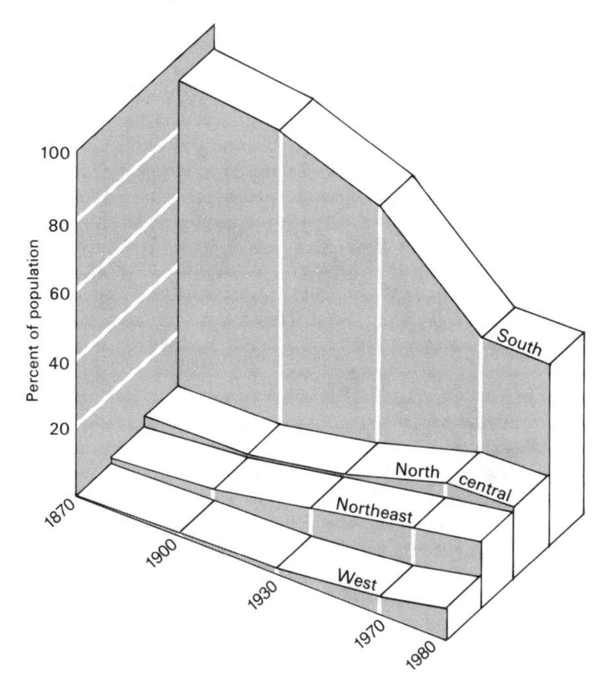

the region have already been discussed; Jim Crow laws, violence, and near-subsistence economic conditions are the most prominent. A further reduction in economic support caused by a severe boll weevil infestation coincided with the onset of World War I. The war led to a strenuous effort by the North to "pull" blacks (and poor whites) from the South.

Prior to 1914, national industrial expansion had depended on millions of European immigrants to meet the large demand for labor. More than one-third of the U.S. population in 1910 was foreign born or had at least one parent born outside the country (Chapter 3). When the war shut off this supply, an alternative was found in the large unemployed and underemployed

Southern labor pool. Press campaigns and labor recruiters were sent into the South to glorify the virtues and opportunities of Northern manufacturing employment and Northern life in general. And once the outflow was begun in earnest, relatives, friends, and former neighbors provided the information and often the financial support required to encourage others to undertake the long move to the North.

The migration of blacks from the South was not uniform. Almost all who left the region moved to urban places, but not all destinations were equally served by a given origin area. Those from the Atlantic states traveled for the most part to Northern cities east of the Appalachians: Washington, D.C., Baltimore, Philadelphia, New York, Hartford, Providence, and Boston, for example (Figure 10-6). Blacks from the South-Central states, Alabama, Mississippi, Kentucky, Tennessee, Arkansas, and Louisiana, moved north along the major transportation lines to St. Louis, Indianapolis, Gary, and Chicago, or to eastern interior manufacturing cities such as Cincinnati, Detroit, Cleveland, Pittsburgh, and Buffalo. The largest and most southerly cities tended to receive large numbers of blacks first, while smaller industrial centers often did not possess large black populations until after World War II. The more distant Western cities, such as Denver, Portland, Seattle, and the California cities, also experienced a later growth in their black populations than did those

Figure 10-6 Major black migration streams, 1965–1970. Migration from the South formed distinctive "streams" of population movement flowing between states in the region and destinations in the North and West.

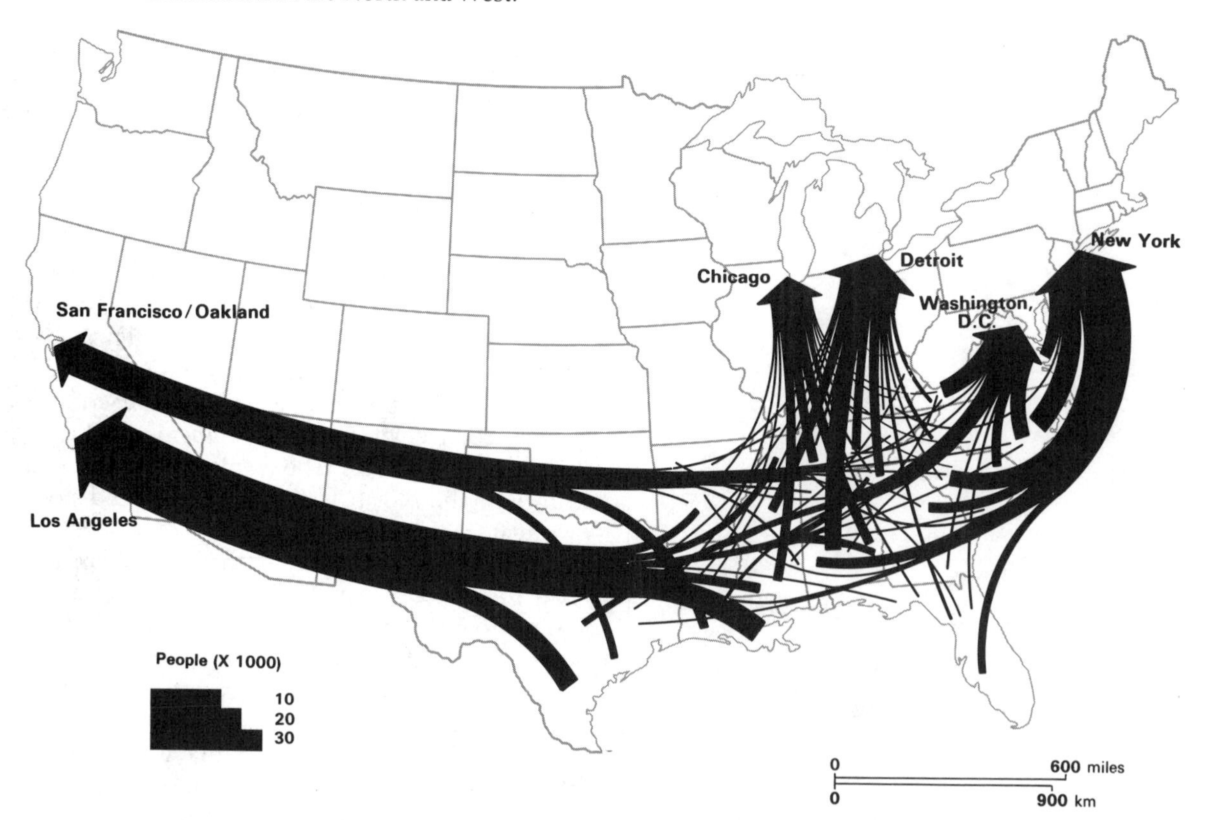

along the East Coast, with most western growth also occurring after World War II; blacks migrating westward directly from the South moved primarily from Louisiana, Texas, Arkansas, and Mississippi. In each case, the factors of distance and the predominant pattern of transportation can be seen as the basic explanation for these geographically distinctive migration streams.

The Southern economy might not have suffered from this exodus if the population involved had not also been selective. After all, the average Southern black did receive very low wages if employed, and after World War II sharecropping became less necessary and much less efficient as Southern agriculture gradually mechanized. Most blacks who left, however, were between the ages of 18 and 35. Raised in the South, this group's most economically productive years were then spent outside the region. Many of those who remained behind were in their later productive years, retired, or not yet in the labor force (Figure 10-7). Racial limitations on opportunities in professional occupations also resulted in a loss of many of the most highly trained young people from the region.

Sectionalism

A fourth major consequence of the Civil War and the South's antebellum cultural heritage was an intensification of the sectionalism already felt in the region. Partly because of the slavery issue, the South felt itself a region apart even before the war. The four years of turmoil, fought almost entirely within the South, the loss of the war, and the repressive aspects of Reconstruction all reinforced the sense of regional difference. The South is the only part of the nation to have ever suffered occupation by a conquering army, and it has taken a century and a great deal of economic growth to temper the bitterness that followed.

The war and Reconstruction were also instru-

Figure 10-7 Comparative age/sex distributions: United States and Mississippi, 1980. These graphs, called age-sex pyramids, suggest that Mississippi blacks left the state in significant numbers after 1920, and that there had been some departure of whites from World War II onward, as well.

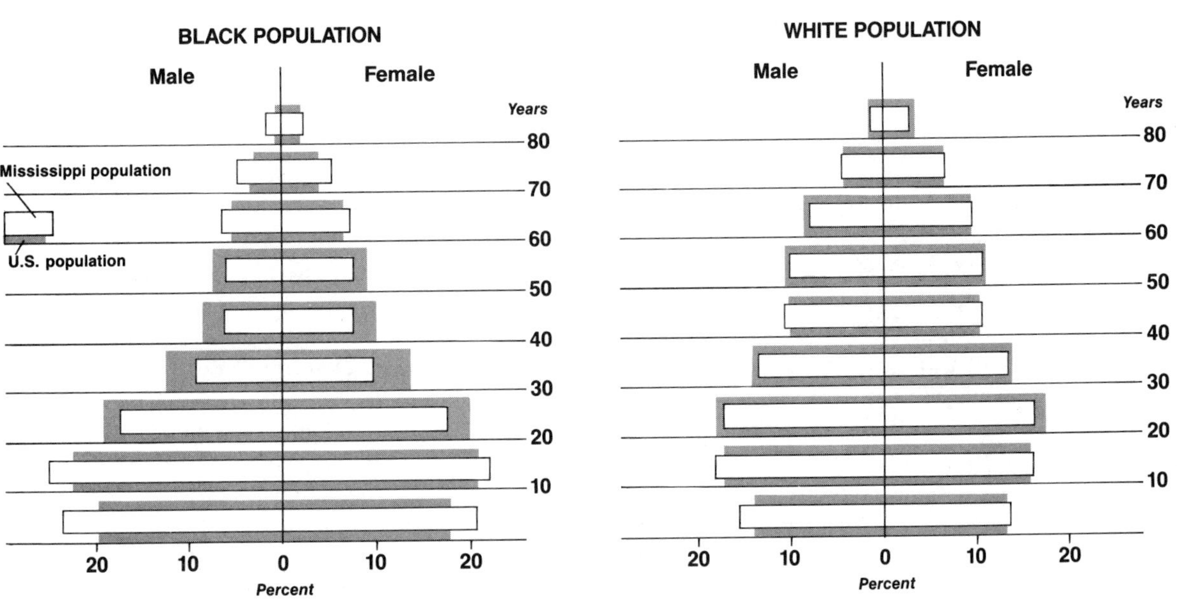

mental in unifying Southern whites. The direct impact of blacks on politics during Reconstruction was not large, but politics became imbued with the race issue by the 1880s. This provided politicians and voters with an oversimplification and distraction from significant economic issues that remained widespread until the 1960s and surfaces occasionally to this day.

The sectionalism of the South is easily visible in its political behavior. The "Solid South" was a term that indicated that the entire region voted as a bloc and often in direct contradiction to otherwise national trends. The Civil War and Reconstruction were associated with the North and the Republican Party, so Southern whites became stubborn opposition Democrats. Even after blacks were disenfranchised and the Northern tenor of both parties changed, the South continued to vote Democratic, at least at the local level. When the national party became so different that Southern whites could no longer tolerate the ideological connection, the explicit sectional label "Southern Democrats" became common. The South was clearly considered a distinctive region.

Although national political changes and Southern cultural changes have made the South no longer solidly Democratic, even at the state level, the 1968 formation of the American Independent Party was a continuation of Southern sectionalism. While receiving votes from every state, the viewpoints presented by presidential candidate George Wallace that year were basically old-fashioned Southern viewpoints, and the only states he carried in the election were Southern states. Even with a successful Southern candidate for the presidency in 1976, most political observers agreed that the white South viewed President Carter with a mixture of sectional pride for his Southern origins and Southern culture and sectional disapproval for his obvious departure from these origins and culture in appealing to the national electorate. The non-South's near obsession with President Carter's regional background during the first half of his presidency did nothing to allay this sectional feeling within the South.

The issue of sectionalism can be extremely subtle but is nevertheless a manifestation of the distinctive Southern culture. Interwoven with the other major strands that made up the region during the century that followed the Civil War—widespread poverty, explicit racism, and a continued agrarian and small-town orientation—the Southerners' sense of their region as having always been different from the rest of the country created an isolation and regional consciousness that reinforced whatever differences there were. As Thomas Wolfe pointed out in the quotation at the beginning of this chapter, the Southern sense of region was strong and conscious. Perhaps because of the strength of this regional identification and the somewhat disdainful attention it received from non-Southerners, the slow erosion of its potency was little noticed for many years. Even after it was clear that change was occurring, observers continued to refer in sectional terms to the rise of the "New South." Just what was new about the "New South's" geography?

THE ONSET OF CHANGE

The New South is a modification of the South that preceded it. Present spatial and regional characteristics have been built on patterns that evolved over decades and, in some ways, over centuries. The key to recent changes lies in the gradual loss of regional isolation. Prior to World War II, most of the South's population and certainly its leadership appeared to believe and react to events as though the region had successfully seceded from the United States, almost as though the South was a separate country reluctantly required to continue dealing with a Northern neighbor. The result accentuated certain cultural and geographic patterns that had existed very early in the region. Since the later 1930s, however, and especially since the end of World War II, trends and pressures external to the South began to infiltrate the region and break down its isolation. Some of the changes induced in the South's geography were caused by factors

Located near the center of the Southeast, Atlanta's role as a major transportation focus for the region has stimulated rapid growth since World War II. (Atlanta Convention and Visitor's Bureau)

that affected the entire nation's economy and have no clear relation to regional location. Some changes have followed from specific federal intervention in regional affairs. And some characteristics of the New South have come about through modifications to the distinctive Southern culture. All have eroded the isolation that set the South apart from the rest of the country.

Economic Reorganization

Changes in the South's economic structure since the mid-1930s have been dramatic and of national significance. The mid-Depression economy of the South was little different from that of 1870: dominantly agrarian, producing raw agricultural products primarily for export, capital deficient, supported by heavy use of animal power (usually mules) and hand labor, operated through sharecropping and tenant-farming arrangements and a regionally distinctive crop-lien system. Such industry as existed was largely low-wage or oriented toward narrow local markets. The region's urban structure continued to reflect this orientation with small market centers, railroad towns, textile mill towns, and county seats representing the pervasive urban form in the South.

By the 1980s, in approximately half a century, tremendous changes had occurred. The South's economy is no longer primarily agrarian. By the early 1950s, over half of the region's labor force was engaged in urban-based, nonagricultural employment, and the proportion in agriculture has continued to decline. This has paralleled a sharp increase in manufacturing employment and employment in service activities. Further, the industrial mix in the South has shown a strong trend toward diversification; no longer is Southern manufacturing limited to the early stages of raw materials processing.

Within agriculture, diversification also occurred, although that is not agriculture's most dramatic transformation. Among the traditional cash crops produced in the region, cotton remains the most important, with others including tobacco, sugar cane, peanuts, and rice.

Throughout the tobacco regions of North Carolina and Virginia where the crop is flue-cured, the distinctive curing barns are fast becoming relic features on the landscape as they are replaced by more efficient sheds that look little different than abandoned truck trailers. Thus, landscape changes are instituted by the adoption of new technology. (Left, courtesy: W. K. Collins, N.C.S.U. Crop Service Extension Specialist; right, Ray Wilkinson III, Raleigh, N.C.)

However, all of these crops, except rice, have been under strict federal acreage controls since the 1930s and 1940s, with rice under controls since about 1951. Total output remained at high levels as growers reserved their best land for cash crop production, but the area producing cotton, for example, is only a shadow of its former size (see map appendix; compare with Figure 10-3). This areal shrinkage was supported by the decay of old cotton-ginning institutions in sections of the former production area. The decline is most pronounced in the Carolinas–Georgia Piedmont and the Black Waxy Prairie of eastern Texas.[3]

Cotton is no longer exclusively a Southern product. Irrigated cotton is grown in the Texas Panhandle, along sections of the Rio Grande, and in Arizona and California. Within the South, soybeans replaced cotton as the most valuable cash crop produced. By 1981, the soybean crop earned Southern farmers nearly $3.5 billion in that year, while cotton lagged far behind at $1.1 billion.

While cotton dominance declined, livestock industries and other crops, such as soybeans, increased sharply. Beef production improved greatly after World War II as farmers improved pastures with better grasses and fodder crops and with higher fertilizer applications. At the same time, new cattle strains were developed to survive and thrive in the hot, humid Southern summer (see Chapter 11). Within the last 30 years, national broiler and chicken production has become industrialized and concentrated in the South. Clearly a regional industry, approximately 75 percent of commercial broilers pro-

[3] For an excellent discussion of this, see Merle C. Prunty and Charles Aiken, "The Demise of the Piedmont Cotton Region," *Annals, the Association of American Geographers*, Vol. 62, no. 2 (June 1972), pp. 283–306.

Chicken broiler and egg production require heavy investment and large-scale operations today. This laying house in Mississippi is almost identical to those found in other production regions across the South. (U.S.D.A.)

duced in the United States were Southern in origin (see Map Appendix). Chicken prices have remained relatively low through the application of modern production and management techniques. Highly automated production in large volume with opportunities to share new technology and a well-developed, spatially concentrated marketing network combined to continue the concentration of this agricultural industry in the South.

Even more dramatic than the diversification of agricultural production have been the transformations in the means of farm production. Agriculture is no longer an animal-power, hand-labor operation. Wherever possible machinery has been applied to the production process just as in manufacturing and in agriculture elsewhere in the country, and regional agriculture is now much more efficient than before. It is traditionally assumed that as Southern agriculture became automated, the underemployed and marginal farmer were *pushed* from the land as unneeded excess labor. One study, however, argued the opposite.[4] Urbanization and nonagricultural industrialization prior to 1960 were said to have led to largely voluntary migrations in search of better economic opportunities; because

[4] Merle C. Prunty, "Some Contemporary Myths and Challenges in Southern Rural Land Utilization," *Southeastern Geographer*, Vol. 10, no. 2 (November 1970), pp. 1–12.

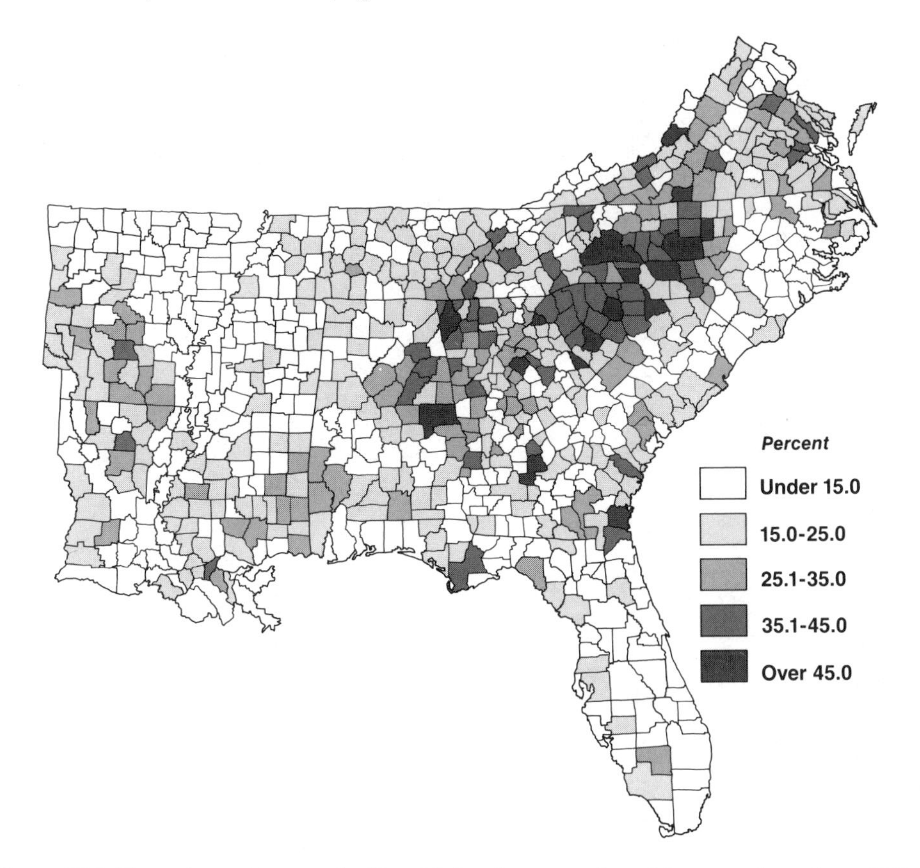

Figure 10-8 Percentage of nonagricultural labor force employed in manufacturing, 1950. Except for portions of the piedmont and other scattered pockets in the South, manufacturing employment remained low across most of the region in 1950.

of this labor outflow, farm mechanization was required in order to maintain production levels. The distinction between these alternative perspectives is important because it can affect our view of social organization and policy development. The belief that farm populations were pulled from the land by greater opportunity creates a very different viewpoint from the belief that these farmers were pushed out in spite of whatever wishes they may have had.

Associated with the mechanization of Southern farming, the traditional sharecrop system has almost disappeared since the mid-1930s. Accompanying this change was a sharp increase in the average farm size in the South. Again, traditional wisdom saw this increase as a reflection of sharply dropping farm *ownership*, but Professor Prunty argues convincingly for another explanation.[5] Since 1870, the U.S. Census has counted tenant and sharecropped subunits as separate farms. With the end of the sharecrop system, therefore, and largely because of increased urban and industrial opportunities, the average farm size increases and what appeared to be a reduction in farm ownership were exaggerated by elimination of the small sharecropped farm units. Furthermore, much of the rural population that could leave the countryside for urban employment has now already

[5] Ibid.

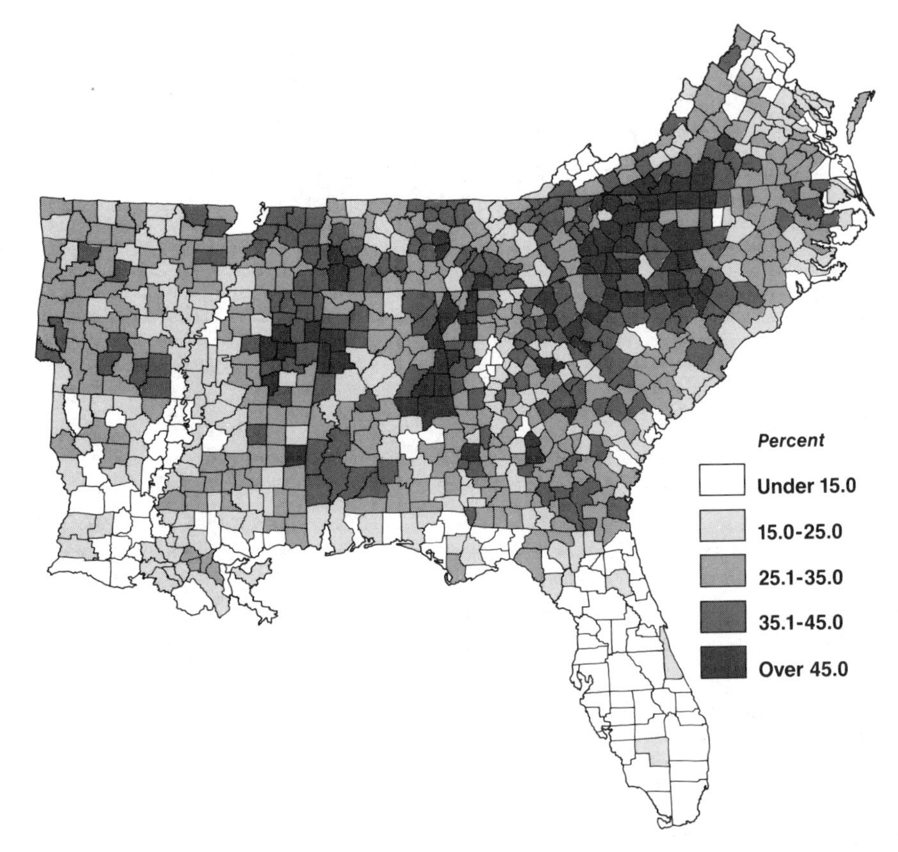

Figure 10-9 Percentage of nonagricultural labor force employed in manufacturing, 1980. Within one generation of the pattern in Figure 10-8, the South had experienced a tremendous growth in manufacturing employment. By 1980, few parts of the region were without a significant manufacturing work force.

done so. Those who remain are either successful full-time farmers, part-time agriculturalists, or individuals with few formal skills who manage to eke out a living at very low levels of income.

Of those who left the rural South, virtually all migrated to urban centers; not all of these centers, however, have been located in the North and West. Rural-to-urban migration *within* the South increased rapidly during and after World War II as the region's economy participated in the post-Depression expansion. This city-bound movement included members of both races, suggesting again that the pull of increased opportunities dominated social push factors or was at least important enough to partially forestall the exodus of blacks from the region. In 1940, there were only 35 cities with populations greater than 50,000 in the South. By 1950, the number had increased to 42 cities, a jump of 20 percent, and by 1980 it had reached 75, an increase of 114 percent over 1940. Many other small Southern places, more accurately "villages" than "towns" at the outset of this period, have developed a certain vitality from the larger growth centers. While still strongly rural in character, the trend in the South is toward a level of urbanization typical of the United States as a whole.

The pull to the cities was stimulated by industrial growth and a diversification that promises to match that of Southern agriculture and to produce an industrial mix typical of the United States as a whole, as well. The proportion of the nonagricultural labor force in manufacturing jobs increased greatly and in virtually every part of the region (Figure 10-8 and 10-9). The traditional industries, such as steel (Alabama), tobacco products (North Carolina and Virginia), and textiles (northern Georgia, the Carolinas, and southern Virginia), remained regionally important for a period but less dominant as other kinds of manufacturing activity appeared. In recent years, these traditional industries have experienced serious declines, but even in this they have mirrored the national economic structure. Synthetic textiles and apparel industries, the former in the Carolinas and the latter primarily in northern Georgia, widened activities even within this broad industrial category. Chemical industries expanded rapidly along the Gulf Coast. Furniture production in the central Carolina Piedmont increased, and other wood-processing plants (paper-container plants and plywood manufacturing) became more prominent throughout the eastern and Gulf coastal plains. Shipbuilding was continued at Norfolk and begun at several sites on the Gulf Coast; aircraft production at Marietta, Georgia, drew skilled labor and higher wages to the Atlanta area.

Most significantly, as the average Southern consumer earned higher wages, the regional market increased enough to draw many consumer-goods manufacturers into the South. This increased the demand for nonagricultural labor, spreading the income further and strengthening the local market, and so on into the growth cycle common to developing regions. As this trend continues, some of the low-wage industries that remain concentrated in the region (Table 10-2) are likely to be driven out by the demands of high-wage activities for the limited labor supply. As a more skilled labor force is demanded, improvements in education will be essential or the transition to a higher-wage economy will be slowed. Overall, economic activity in the South is gradually shifting into something that more closely reflects the patterns found in other regions. Even though the South continues to contain many areas with very low per capita income, the persistent poverty of the region's population has begun to diminish.

Government Intervention

Perhaps the most significant changes in the South were caused, or at least triggered, by direct governmental involvement, from the federal level until the 1970s when state initiative increased. This intervention occurred in economic as well as in social matters. Furthermore, the economic intervention has most often been as a specific stimulus to development and not merely a kind of regional welfare payment to cover income deficiencies. The regional development characteristics of the Tennessee Valley Authority (TVA) are an excellent example for a portion of the South (see Chapter 9). Federal expenditures in the South have been uneven, with North Carolina, Arkansas, and Louisiana receiving less than average outlays from Washington, D.C., while Virginia, Texas, and Mississippi received more than the average state in per capita terms.

The South's rapid industrial growth since the mid-1930s is a consequence of a growing regional market, gradually demanding and able to pay for more goods and services. But the question remains: Why did the market expand? Per capita incomes in the early 1930s were about one-third the national average, and the economic structure had changed very little since Reconstruction, a full 60 years earlier. Professor Prunty proposed that the federal Agricultural Adjustment Acts (1935 and later) provided the main stimulus to the market growth.[6] Although

---

[6] Merle C. Prunty, "The Agrarian Contribution to Recent Southern Industrialization," in *The South: A Vade Mecum*, James R. Heyl, ed. (Washington, D.C.: Association of American Geographers, 1973), pp. 108–118.

Textile and apparel industries have a high machinery-to-labor ratio, require a large amount of floor space, and demand relatively low skills from their labor force. The women tending these nylon spinning machines also represent the largely female work force in such low average wage industries. (Ray Ellis/Photo Researchers)

Table 10-2 Index of Average Hourly Wage in Manufacturing Industries, 1952–1989 (U.S. average = 100)

State	1952	1957	1962	1967	1972	1977	1989
Alabama	79	86	86	85	85	87	90
Arkansas	69	71	69	71	73	76	82
Florida	76	79	83	84	85	82	86
Georgia	73	75	74	78	80	79	89
Kentucky	90	96	95	96	97	101	103
Louisiana	85	94	96	96	97	102	109
Mississippi	66	68	69	72	73	74	79
North Carolina	73	70	69	72	73	73	84
South Carolina	73	70	70	73	74	76	85
Tennessee	81	81	81	81	81	83	92
Virginia	81	79	80	81	81	83	96

Sources: Calculated from U.S. Bureau of Labor Statistics, *Employment and Earnings, 1939–1974,* and *1991* and *U.S. Statistical Abstracts,* 1978.

aimed at farmers throughout the country, the consequences for the largely rural South were especially strong.

Before the Agricultural Adjustment Acts took effect, the prices that farm products could demand were set to a great extent by supply and demand in the international marketing place. To the South, this meant that prices for Southern cotton, for example, fluctuated partly according to the production success or failure in other cotton-growing areas of the world. More important, farm labor in the cotton South (basically tenant and sharecrop farmers) was in competition with hard-pressed cotton producers along the Nile or Indus rivers, or elsewhere in what was still largely a colonialized world economy. When agricultural wages and prices were adjusted upward under the Agricultural Adjustment Acts to reflect more accurately national industrial wage differentials, the sharply improved market in the South for manufactured goods initiated the upward development spiral still affecting the region.

In an act of federal intervention much more widely recognized as significant to the South's social structure, the U.S. Supreme Court struck down in 1954 the racial "separate but equal" doctrine permitted almost 70 years before. Changes in the South's social geography were initiated by this decision, changes that reverberated in every other part of the country where race affected opportunity, and the repercussions are far from settled almost 40 years after the decision itself.

The geographic and institutional consequences of racial segregation were not quickly affected by simply declaring the practice illegal. The antiblack component of white Southern culture had been consciously nurtured for too many years for change to occur rapidly, but even in this the South is considerably different now than it was prior to 1954. In some parts of the South, new schools, still to be segregated, were built for blacks during the 1950s in an attempt to show that their facilities could be equal. But public facilities were gradually required to prove compliance with the Supreme Court's judgment either through direct testing by blacks or by serving the public according to a specific black-to-white ratio.

Throughout this period and accompanying the South's economic revitalization, many blacks migrated from rural areas but stayed within the region. By 1980, more than two-thirds of the South's black population were living in the region's major metropolitan areas. Although the black metropolitan population has been increasing, migration by whites into the South from other regions, also mostly into the metropolitan areas, has kept the proportion of metropolitan population that was black from growing even more rapidly.

Outmigration continued at a high rate, as discussed earlier, but by the early 1970s there was a slight decline in the overall level of black outflow. The region's increase in economic opportunities also began to be somewhat more available to blacks than previously in the South's urban areas. The 1980 Census of Population confirmed that more blacks were entering the South than were leaving it, and proof that this trend continued is expected when detailed results from the 1990 census are available. This suggests that regionally based racial "push" factors have become insignificant in national population redistribution. The South, in other words, may no longer be seen as any more threatening racially than other regions. It also confirms the growing economic maturity of the South, for migration in the United States has been associated traditionally with the attraction of economic opportunity. And, finally, it suggests that Southern culture, already containing a strong elements introduced by its black population, may continue to benefit from greater than average black cultural contributions.

Cultural Integration

With the possible exception of this last speculation, a thread common to many of the South's changes since the mid-1930s is the gradual decline of its regional distinctiveness. Very simply, much of what has been occurring in the South is making it more and more like the rest of the eastern half of the country. Economic diversity is replacing simple dependency on agriculture. There are indications that the region's supply of low-wage labor is almost exhausted; new industry and service activities will have to compete more actively and may continue to force wages upward slowly. A significant infusion of Northern migrants, especially to regional metropolitan growth centers, has made some of these cities less distinctively Southern in culture and more clearly just urban. Although the issue may be argued vigorously and at length by many native Southerners, the leading city of the region, Atlanta, has lost much of what once made it regionally distinctive. Atlanta is now much less a Southern city than a national city, an American city. If this transformation has happened to Atlanta, can Charlotte, Richmond, and Nashville be far behind?

Racial discrimination was never peculiar to the South, although its expression was often more blatant and institutionalized than elsewhere. With the end of most overt, and all legalized, forms of racism, the South has at least reached the level of interracial attitudes found through the rest of the country, although some of the old institutions occasionally resurface.

As the South experienced these changes, the region's sense of isolation and sectionalism also declined. Not only have some regional characteristics become more "national," but the North and West have also demonstrated an affinity for what Southerners recognize as their own. Improved communications have broadened the national exposure of many Southern writers, poets, artists, and others who draw deeply on the region's distinctive cultural heritage, and often the regional origin is not recognized by non-Southerners. Black cultural contributions are also more widely dispersed. Politically, Southern elections and administrations are no more tumultuous than elsewhere and hardly more race oriented. The full range of the political spectrum is represented among Southern elected officials, although the majority tend to continue some of the traditional orientations. Since the Civil Rights Act of 1964, blacks have had less difficulty in registering to vote and in exercising that right. All of this has reduced the defensiveness of Southerners somewhat and therefore reduced the sectionalism of the region's population. Southerners generally remain aware of their cultural distinctiveness, but the grounds for this self-awareness are disappearing. Before too many additional decades pass, it may well be that the region's cultural provincialism will be no stronger than that found in other parts of the United States and Canada.

AGRICULTURAL CASE STUDY THE YAZOO-MISSISSIPPI DELTA

The Yazoo-Mississippi Delta is a microcosm of several aspects of the Deep South even though the smaller region has a history modified by its physical geography. The Delta, as it is known in Mississippi, occupies an eastern portion of the alluvial flood plain of the central Mississippi River (Figure 10-10). There are small local variations in relief, but the overwhelming visual impression is that of an almost perfectly flat landscape. Filled in since the last Ice Age by glacial outwash and flood-deposited alluvium, the Delta drops only about one-third of a meter (1 foot) of elevation for every 3.2 kilometers (2 miles) of the 320 kilometers (200 miles) between Memphis and Vicksburg.

Poor internal drainage, frequent floods by the Mississippi, and the presence of Choctaw

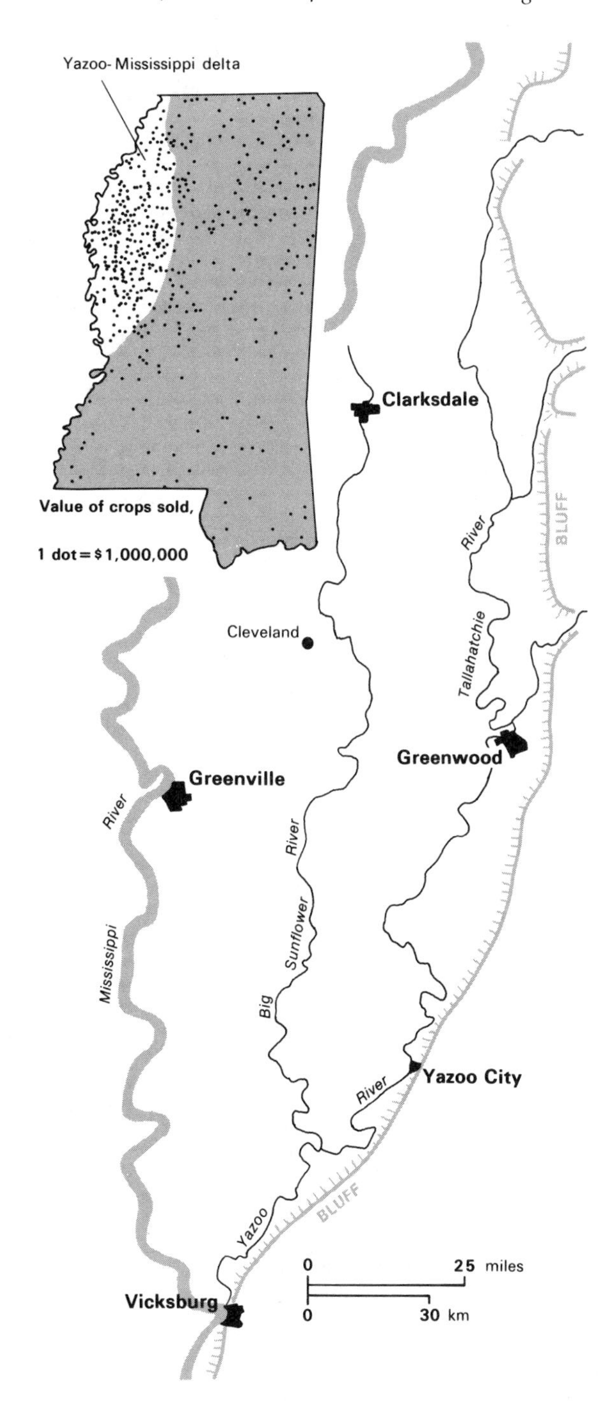

Figure 10-10 Yazoo-Mississippi delta. With rich soil and flat land, subject to flooding and used intensively for agriculture, the delta continues to have very few urban places.

and Chickasaw Indians kept European settlement in the Delta tentative for many years. However, after the 1830s, the lure of very fertile soil, proximity to river trade, and the profits to be made from cotton gradually drew plantation agriculture into this rather hazardous region. As elsewhere in the South, with cotton production came large numbers of slaves. This was so exclusively plantation country that by 1860 blacks outnumbered whites in the entire Delta by more than 3 to 1 with some areas exceeding a 10:1 ratio. Development of the land was slow, however; ravages of the Civil War and frequent floods kept the region largely wilderness. The Delta was inundated by floods, for example, in 1858, 1862, 1865, 1867, 1868, 1871, and 1874. The tremendous productive pressure on the South's land during the decades after the war and political pressure by railroads and land speculators during the last quarter of the nineteenth century and the first quarter of the twentieth finally attracted federal efforts to control the flooding. While these efforts were only gradually successful, the Delta attracted population again (both white and black) as more land was cleared, swamps were drained, and new areas were planted in cotton. Cotton acreage increased almost 500 percent between 1880 and 1920. Towns within the Delta remained small, with Greenville the largest at 11,560 in 1920 (the only town larger than 8000 population). While no longer slaves, the Delta's majority black population continued to labor in the fields as sharecroppers.

An especially severe flood in 1927 followed shortly by the national distress of the Depression led to more decisive governmental intervention in the Delta economy. Large landowners recovered financially with the Agricultural Adjustment Acts and expanded acreages began to be developed, but the amount of land that could be planted in cotton was controlled from Washington. It was soon found that rice and soybeans could be very profitable additions to the traditional cotton

Land in the delta is very flat and most is intensively used. In the fall, mechanical harvesters scuttle across the large fields gathering in the year's crop. This may be soybeans, as shown, or with different equipment, it may be cotton or rice. (U.S.D.A.)

economy, especially since both crops did well on land too wet for cotton.

Agricultural expansion during the 1930s and 1940s, however, occurred just as Delta blacks increased their migration out of the region. During the half-century after 1920, the black population in the Delta declined 33 percent. Landowners responded with increased mechanization as elsewhere in the country. Cotton had always been an especially labor-intensive crop, so the transformation of agricultural operations was more extreme. And because mechanized cultivation, harvesting, and the application of fertilizer and pesticides require considerable capital and large-scale operations, the most successful landowners often expanded at the expense of the smaller farmers.

In addition to this array of features—dependence on agriculture, a large, poor, black sharecropper population, an absence of major urban centers, a decline in the relative importance of cotton, and changes stimulated by federal intervention and by mechanization—the Delta also highlights Deep South patterns in its grudging industrial growth. Later to appear and less intense—even by 1980—than elsewhere in the larger region, a few new manufacturing activities began to surface in the Delta after World War II. The success of agriculture (plus the valid fear of flood) probably combined to retard industrial entry for many years. Successful farm mechanization eventually generated a surplus labor force even with heavy outmigration, and, in turn, stimulated modest industrial development, most of it related to processing agricultural products; manufacturing employment grew modestly from 6491 in 1955 to 15,079 in 1970, but then grew to more than 25,000 by 1980.

Very rich agriculturally, increasingly diverse economically, possessing a great deal of residual poverty (especially among its majority black population) in spite of the presence of both old and new wealth, the Yazoo-Mississippi Delta is a distinctive region, but just as clearly, it is representative of the larger Deep South.

ADDITIONAL READINGS

Aiken, Charles S. "A New Type of Black Ghetto in the Plantation South." *Annals of the Association of American Geographers.* 80 (1990), pp. 223–246.

Carney, George O. "Country Music and the South: A Cultural Geography Perspective." *Journal of Cultural Geography,* 1, 1980, pp. 16–33.

Comeaux, Malcolm L. *Atchafalaya Swamp Life: Settlement and Folk Occupations,* Vol. 2. Baton Rouge: Louisiana State University, School of Geoscience, 1972.

Cromartie, John, and Carol B. Stack. "Reinterpretation of Black Return Migration and Nonreturn Migration to the South, 1975–1980." *Geographical Review,* 79 (1989), pp. 297–310.

Gottmann, Jean. *Virginia in Our Century.* Charlottesville: University of Virginia Press, 1969.

Hartshorn, Truman A., et al. "Metropolis in Georgia: Atlanta's Rise as a Major Transaction Center." In *Contemporary Metropolitan America,* Vol. 4. Cambridge, Mass.: Ballinger, 1976.

Johnson, Merrill L. "Postwar Industrial Development in the Southeast and the Pioneer Role of Labor-Intensive Industry." *Economic Geography,* 61 (1985), pp. 46–65.

Kovacik, Charles F., and John J. Winberry. *South Carolina.* Boulder, Colo.: Westview Press, 1984.

Reed, John. *Southerners: An Essay in the Social Psychology of Sectionalism.* Chapel Hill: University of North Carolina Press, 1982.

Rubin, Louis D. *William Elliot Shoots a Bear: Essays on the Southern Literary Imagination.* Baton Rouge: Louisiana State University Press, 1975.

Shuptrine, Hubert, and James Dickey. *Jericho: The South Beheld.* Birmingham: Oxmoor House, 1974.

Wheeler, James O., and Ronald L. Mitchelson, "Atlanta's Role as an Information Center." *Professional Geographer,* 41 (1989), pp 162–172.

Winberry, John J., and David M. Jones. "Rise and Decline of the Miracle Vine: Kudzu in the Southern Landscape." *Southeastern Geographer,* 13, 1973, pp. 61–70.

CHAPTER

11

The Southern Coastlands: on the Subtropical Margin

The southern margins of the United States can be divided into two approximately equal sections. One half, the Southwest Border Area (Chapter 15), shares a long land boundary with Mexico and includes an extensive inland area that has experienced many influences from that neighboring country. The other half traces the coastline eastward from the mouth of the Rio Grande to North Carolina and includes Florida's peninsula (Figure 11-1). Both stretches are southerly in latitude, and they share a small area of overlap in southern Texas, but the Southern Coastlands are as distinct from the Southwest Border Area as are any other two adjacent regions on the continent.

The Southern Coastlands comprise a distinctive region for two primary reasons (although a third factor has also been an important in developing the region's character). First, the region has a humid subtropical environment. The southern tip of Florida lies only 160 kilometers (100 miles) from the Tropic of Cancer, with the chain of coral islands called the Florida Keys extending another 65 kilometers (40 miles) south. The warm waters of the Gulf of Mexico also contribute a strong maritime influence to the coastlands' climate. The region has a clear appeal to visitors and potential residents, and its agriculture is distinctive specifically because of this environment.

Second, the region derives much of its character from its location at the continental margin. The environment, of course, is related to this aspect of location, but the region's role in generating U.S. trade patterns with the rest of the world and the distinctive industrial pattern along this coastal region also help to define the region. As in other regions, there is a good deal of potential conflict between activities that depend directly on the unblemished character of the natural environment and those that treat the environment merely as a setting within which the activity must be carried out. Thus, the two distinctive components of the Southern Coastlands' regional geography—subtropical environment and location on the continent's southern margin—and the seeds of conflict they contain provide the theme for this region.

The third factor relevant to the character of the region is its position between the United States' Deep South and Latin America. This position on the margins of the South has influenced some of the cultural attributes of the Southern Coastlands and continues to create difficulty in attempts to sharply define the inland margins of the latter region. By restricting our view of the Southern Coastlands to its coastal influences and leaving discussion of the larger region's culture to Chapter 10 where it is treated more fully, this southeastern margin of the continent simply does not "fit" well, economically and to some extent environmentally, with the rest of the South.

The cultural influences on the region from Latin America were buffered for a long time by the water body separating most of the coastlands' population from their non-Anglo neighbors. In contrast to the Southwest Border Area, Latin influence was more muted in the Southern Coastlands until well after World War II. The growth of the population of Cuban heritage in southern Florida and the intensification of trade

---

Imagine a wide, clean beach with a scattering of sunbathers and swimmers enjoying the sand, sun, and surf while relaxing on holiday. This image is as much a part of the southern coastlands as the one shown. The tension between these two landscape images will continue to define the region for decades. (B. J. Nixon/Tenneco, Inc.)

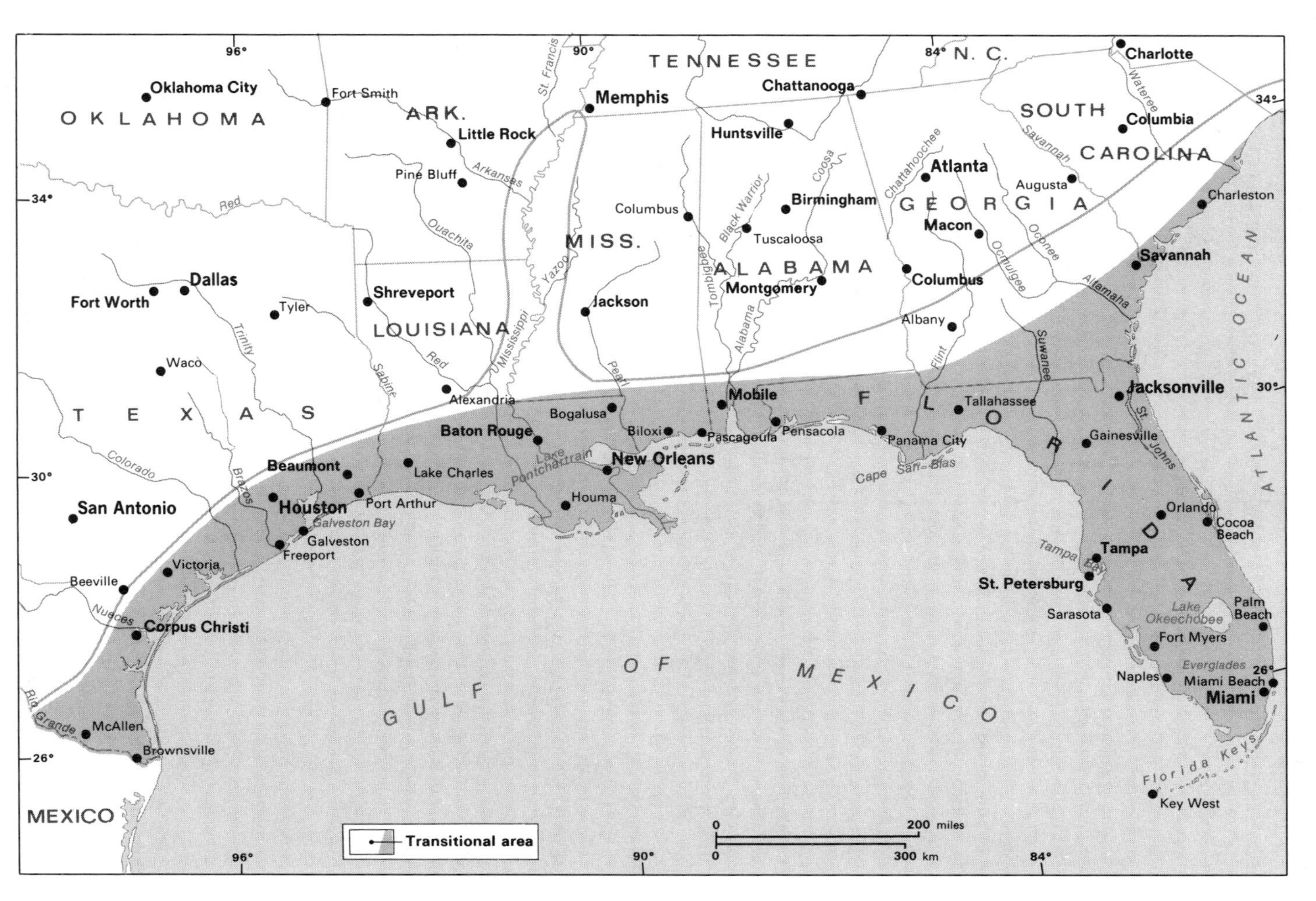

Figure 11-1 The Southern Coastlands.

Major Metropolitan Area Populations in the Southern Coastlands, 1990	
Houston, Tex.	3,301,937
Tampa–St. Petersburg–Clearwater, Fla.	2,067,959
Miami–Hialeah, Fla.	1,937,094
Ft. Lauderdale–Hollywood–Pompano Beach, Fla.	1,255,488
New Orleans, La.	1,238,816
Orlando, Fla.	1,072,748
Jacksonville, Fla.	906,727
West Palm Beach–Boca Raton–Delray Beach, Fla.	863,518
Baton Rouge, La.	528,264
Charleston, S.C.	506,875
Mobile, Ala.	476,923
Lakeland–Winter Haven, Fla.	405,382
Beaumont–Port Arthur, Tex.	361,226
Corpus Christi, Tex.	349,894

of all sorts between Latin America and the United States, much of it flowing through southern Florida, has emphasized the distinctive character of this region within North America.

THE SUBTROPICAL ENVIRONMENT

Of the several components that make up a physical environment, climate has the greatest impact on the human geography of the Southern Coastlands. A humid subtropical climate, the long growing season, mild winter temperatures, and warm, humid summers all contribute to patterns of human activity associated with the region.

Only in southern California and southwest Arizona, north of Mexico, and Hawaii is the average length of the growing season equivalent to that in the Southern Coastlands. Measured as the number of days between the last killing frost in the spring and the first frost in the fall, almost the entire region experiences 9 months or longer of potential growth for its agricultural crops (Figure 11-2). In much of the region, in fact, the average growing season is in excess of 10 months, and the southern half of the Florida peninsula is not likely to experience frosts every year. In addition to the long warm season, almost the entire region receives precipitation abundant enough for most agricultural activities. An average rainfall in excess of 125 centimeters (50 inches) is received by all of the Southern Coastlands east of Houston, Texas, with most of the rain falling in the summer between April and October. Thus, abundant rainfall can also be expected in the period during which sunlight is plentiful and warm temperatures support plant growth.

Agriculture

There are two primary consequences of the presence of this set of climatic conditions in North America. First, as long as other agricultural conditions are met, such as fertile soil, appropriate gound drainage, and control of insect pests, farmers can grow their crops without fear of an early frost until late in the fall. In a few locations, with careful selection of complementary crops and judicious planting, it is possible to harvest two crops in one growing season, and some vegetable farmers are achieving even more. Second, however, and even more important, is the opportunity for production of specialty crops that can be grown in few other locations in the United States and Canada. Citrus fruits—produced in significant quantities outside the region only in southern California and Arizona—rice, and sugarcane are the most important specialty crops produced in the Southern Coastlands. All three contribute greatly to the local economies of the region. Citrus is especially important in Florida, with significant production in southern Texas as well. Only in the production of lemons and navel oranges does California exceed Florida's citrus output (Table 11-1).

Citrus production has been an important contributor to Florida's economy since it was first introduced to the region by Spaniards during the sixteenth century. Although the areas of pri-

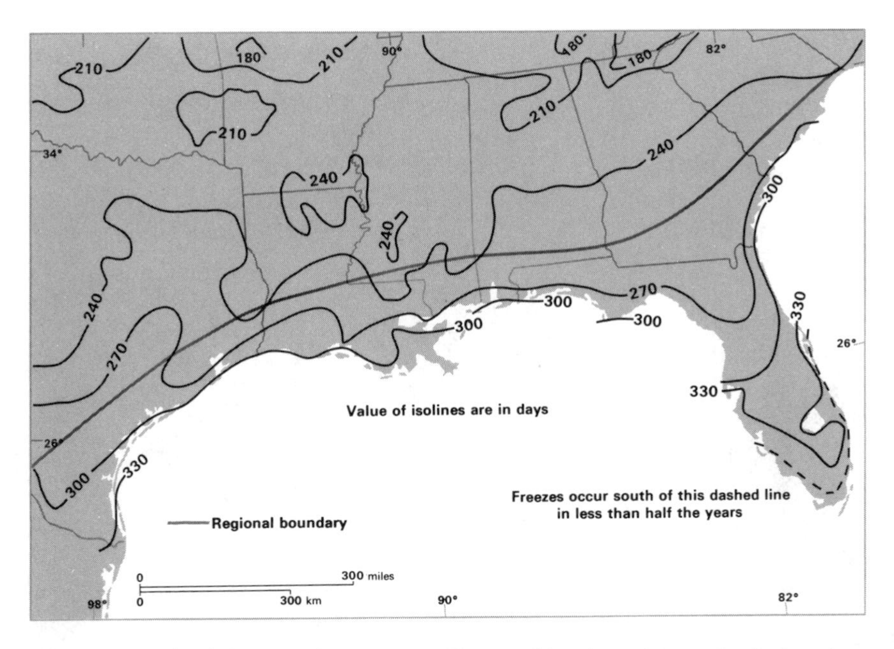

Figure 11-2 Length of the growing season. The combination of a southerly location and maritime proximity leads to a long growing season and, across most of the region, sufficient precipitation for reliable specialty crop farming.

mary production have gradually shifted southward along the peninsula's interior, the bulk of citrus is produced south of 29° north latitude and north of Lake Okeechobee. The southward movement is the response of growers to the yearly threat of frost, but swampy soils and standing water in the Everglades have prevented most citrus production from moving farther south into the effectively frost-free margin of Florida. About half of Florida's citrus production remains in the region between Tampa and Orlando (see Map Appendix).

Oranges and grapefruit are the most important of the seven major citrus fruits grown in the state. Orange production increased slowly but steadily in recent decades to 1978 when 7.4 million tons were harvested; this dipped to 5.23 million tons of oranges harvested in 1987. Since 1945, an increasing share of the total orange crop has been processed rather than sold as fresh oranges. The proportion processed had reached about 80 percent of the total output as early as 1970. By processing the crop (mostly into frozen concentrate), a sizable industry has developed in Florida. This spreads the benefits of this specialty crop among a larger proportion of the state's population than if only fresh fruit were shipped northward. In addition, processing allows year-round sales instead of limiting returns to the immediate harvest period. Grapefruit are produced in an area almost coincident with orange production, but total demand is lower and output is only about one-fourth that of the more important crop. Significant orange and grapefruit farming in the Southern Coastlands is also found under irrigation in the extreme southern portion of Texas between Brownsville and McAllen.

Since citrus are tree crops, a major share of the production costs are associated with the harvest. The fruit must be hand-picked, often from the top of a long ladder. Large amounts of short-term labor are required during the harvest season, a feature of citrus farming that annually

Most first-time visitors to central Florida are struck by the immense orange groves encountered for mile after mile across the gently rolling landscape. (Florida News Bureau, Department of Commerce)

draws thousands of migrant laborers to the dense, parklike groves for periods of intense physical effort. Housing facilities for temporary workers are often cheaply constructed, and while a day's labor may pay a respectable amount, the workday itself may be long. The conflicting goals of producer and worker have not been resolved, and, so far, attempts to develop mechanical pickers for citrus crops have been unsuccessful. The results are a substantial labor force whose members cannot earn enough at one location to satisfy their annual living requirements and an industry that is required to devote more than one-third of its total production expense to hired labor.

Although citrus crops are grown in the Southwest as well as in Florida and Texas, sugarcane production is exclusive to the Southern Coastlands in the mainland United States. Cane is the tropical source of sugar and exceeds beets as the primary source of the world's sugar supply. Within the United States and Canada, however, the temperate climate beet crop has been a more important source of domestic sugar production. Sugarcane is, in fact, a perennial crop requiring more than one year for full maturity, and it is not at all tolerant of frost. In addition, water requirements for cane are high, needing a minimum rainfall of about 125 centimeters (50 inches) per year. Both temperature and water requirments would seem to preclude growth in the continental United States except under irrigation, but moderately large quantities of cane are grown in Louisiana and Florida. The necessary water is available naturally in Louisiana and supplemented by irrigation in Florida, and immature cane is harvested early, especially in the Louisiana locations. Even though

Table 11-1 Production of Citrus Fruits, in Florida and California, 1987 (thousands of tons harvested)

Citrus Fruits	Florida	California
Oranges (all)	5233.5	1986.5
Tangerines and mandarins	72.5	23.9
Tangelos	92.6	36.0
Grapefruit	1816.8	290.4
Kumquats	0.1	0.1
Lemons	22.6	628.4
Limes	37.4	5.9

Source: U.S. Department of Commerce, *1987 Census of Agriculture* (Washington, D.C.: U.S. Government Printing Office, 1990), Vol. 1, Part 51, Table 28 (county).

the sugar content in cane is significantly lower before it matures fully, federal import controls and price supports ensure that returns are usually high enough to justify the agricultural effort. In Florida alone, the 1987 value of sugarcane production was approximately $383 million. Even though federal controls on cane production were removed in 1974 (and accompanied by a rise in world sugar prices), higher levels of production are not anticipated in the Southern Coastlands because cane farming there is not competitive on the world sugar market.

Rice is less demanding than sugarcane in its climatic requirements. Given sufficient water, rice will mature within one growing season at a rate roughly in proportion to the amount of heating it receives during the summer. Within the Southern Coastlands region as we have defined it, irrigated rice is grown in Louisiana (0.9 mil-

Florida's citrus harvest remains labor intensive. Whatever the fruit, it is picked by hand, one at a time, from the small trees, as are the oranges shown. (Courtesy Florida Department of Citrus)

Growing conditions for sugar cane are far below optimum along the Gulf Coast, but it can be grown. The use of machinery in large-scale operations permits profitable production where smaller scale and hand-labor would not. (U.S.D.A.)

lion tons in 1987) and in Texas near Houston (0.8 million tons). The transitional nature of our regional boundary is emphasized by a rice-producing region far to the north of the Gulf Coast in the Mississippi Alluvial Valley sections of Arkansas and Mississippi. These portions of Arkansas and Mississippi are dissimilar to the Southern Coastlands in most respects, but large volumes of this warm climate crop are grown in those states (2.73 million tons in Arkansas and 523,000 tons in Mississippi in 1987; see also Map Appendix).

In addition to the specialty crops, portions of the Southern Coastlands are also among the continent's primary regions of vegetable production and have developed, since World War II, a strong beef industry. The cold winter continental climate experienced by the large population living in North America's urban manufacturing core region has complementarity with the Southern Coastlands' subtropical environment. This complementarity has led to production of a wide variety of vegetable crops in the coastlands. Most fresh vegetables reaching the urban markets during the winter are grown in Florida and the southern margins of the other Gulf coastal states, although some fresh vegetables are shipped from the extreme southwestern United States. Proximity to the northeastern markets undoubtedly provides the Southern Coastland producers with a distinct locational advantage over California vegetable growers.

In a somewhat different manner, the environmental possibilities for beef production have been realized only after a good deal of effort and innovation. The long growing season and abundant rainfall do suggest the potential for high-quality pasture once the land has been cleared. The combination of high temperatures and high humidity, however, produces poor natural grasses. More important until rather recently was the impact of cattle tick fever. Even with improved pasture, the better European cattle breeds succumbed to tick fever, and the native fever-resistant cattle did not yield good beef even on the better grasses. Success of the industry was finally achieved after a long series of efforts between World War I and 1950 to eradicate the cattle-fever ticks (it is hoped), improve pasturage, and crossbreed hardy Brahman bulls imported to Florida from Texas with improved domestic cattle (Table 11-2). The Florida beef industry has become an important contributor to that state's economy. And because these changes are still relatively recent, it is likely that the beef industry in Florida is still capable of a considerable increase.[1] While some of these improvements have been adopted in other, noncoastal sections of the Southeast, it is in the Southern Coastlands, and especially central and northern Florida, that the benefits are most strongly felt.

Table 11-2 Number of Cattle and Calves Sold, Florida, 1944–1987

Year	Number Sold
1944	216,236
1954	658,617
1964	798,615
1974	966,002
1987	1,025,178

Source: U.S. Department of Commerce, *Census of Agriculture* (Washington, D.C.: U.S. Government Printing Office, various years).

Although climatic conditions in the region are, in the balance, favorable for agricultural activity, soil conditions and quality are much more variable and less clearly beneficial. The soils range from the fertile but very poorly drained muck of the Louisiana coast and Mississippi Delta to extremely sandy soils in north and central Florida. To complicate the pattern, sections of Gulf coastal Florida and the extensive Everglades region are largely swampy muck soils or poorly drained sands, while the coasts of Texas and of Georgia and South Carolina may have either marsh soils or sandy soils, depending on local conditions. Areas of muck in Louisiana have proven to be extremely productive, especially for sugarcane and rice, once they have been drained. But attempts to drain Florida's swamps have met with more failure than success and have, in any case, been strongly opposed by conservationist groups. The economic (and technical) failures in Florida are primarily a result of extremely low relief variations. Conservationist opposition is based on the widespread disruption of the complete ecological system of the vast marshlands that would result from the draining. A lowered *water table* permits extensive encroachment of salt water, exposes the grasslands to fire, and upsets the delicate life-support balance for wildlife, shellfish, and vegetation.

In sharp contrast to a need for drainage in many parts of the region, much of the remainder of the Southern Coastlands benefits from heavy irrigation. The central highlands of Florida, for example [at about 30 to 46 meters (100 to 150 feet) above sea level], are underlain by sandy soils with moderately poor to very poor water retention capacity. The productive citrus growing region and vegetable areas have yielded annual output as much as 10 times more valuable when the crops are irrigated than when precipitation is depended on as the sole source of moisture. With this degree of improvement possible and the technological capacity available for its achievement, the distinctive subtropical environment of the coastlands has been developed agriculturally beyond the levels found in much of the interior Southeast.

Amenities

In addition to the economic benefits derived through agricultural specialization, the Southern Coastlands' subtropical climate also benefits

[1] For a study of the early years of beef industry development, see W. Theodore Mealor, Jr., and Merle C. Prunty, "Open-Range Ranching in Southern Florida," *Annals of the Association of American Geographers*, Vol. 66, no. 3 (September 1976), pp 360–376.

the region by its contrast with the more northern and continental climate of the North American interior. As personal disposable income and old age financial security increased in the United States and Canada during the decades following World War II, recreation and retirement became major industries in those areas that could attract large numbers of people. In an article calling attention to this trend and studying its early significance, Edward Ullman[2] argued that as the economic reasons for migration become relatively less important, people will choose to live where they find it "pleasant"; he suggested that regions with an outdoor climate most similar to a comfortable indoor climate would be most preferred. The two regions in the United States and Canada best approximating this climate are coastal southern California and what is defined in this chapter as the Southern Coastlands, at least in winter months. The widespread adoption of air conditioning provides summers with nearly equivalent appeal.

Even by the early 1950s, when Professor Ullman wrote, the importance of amenity factors in stimulating the regional growth of Florida and the Gulf Coast was clear; since then their effects have increased. Statistics show that Florida is viewed as one of the prime states for retirement by many approaching that stage in their lives. In 1950, 12.4 percent of Florida's population was over 60 years of age. By 1985, the proportion over 65 was estimated to be more than 17.7 percent. In comparison, Ohio and Connecticut, located in the continental manufacturing core, contained populations in 1950 of 13.7 and 13.3 percent, respectively, above the age of 60, and proportions of their populations 65 or older in 1980 of 12.1 percent for Ohio and 13.1 percent for Connecticut.

Although the amenity of a mild winter climate draws large numbers of retired persons to the Southern Coastlands, a larger direct economic advantage is generated by tourist activities in the region. New Orleans has been a strong attraction for vacationers because it mixes the opportunities of a metropolitan urban area with a culturally distinctive African-American flavor, all set within the mild subtropical climate of the Southern Coastlands. Between New Orleans and Mobile, Alabama, the coastal section of Mississippi, locally referred to as the "Mississippi Riviera," has recently experienced a tourist boom with construction of numerous hotels, motels, restaurants, and artificial beaches. Biloxi is an especially notable example.

Greater general affluence and reduced transport costs permit more of us to enjoy the environmentally pleasant places. However, this greater use has almost always meant deterioration of the environment. Thus far, the prime alternatives to this dilemma have been to place restrictions on individual "freedom of choice" or to do nothing and let the desired places become less pleasant than they had been. Many areas in the southern half (and western region) of the United States are currently facing this dilemma. (Berry's World © 1977 by NEA, Inc.)

[2] Edward L. Ullman, "Amenities as a Factor in Regional Growth," *Geographical Review*, Vol. 55 (1954), pp. 119–132.

The most dramatic tourist magnet in the region, however, has been Florida. With long beaches facing both the Atlantic and the Gulf of Mexico, this state has drawn winter vacationers from the urban centers of Megalopolis and the industrial interior for decades. Luxury hotels lining the oceanfront in Miami Beach and luxury estates along the coast at Palm Beach proved to be only the cores of an expanded tourism-based urban development extending over 160 kilometers (100 miles) along the southeastern coast of the peninsula. Strong competition and an aging physical plant has diminished the appeal of many of the older hotels of the region, but the impact of their earlier magnetism had already functioned to transform the face of the land in coastal southern Florida. Spatially less continuous pockets of recreational development have sprouted from Palm Beach along Florida's Atlantic coast as far north as Jacksonville. And although the Gulf side is less well endowed with attractive beaches, considerable growth has occurred from Tampa Bay southward to Naples.

The demand for subtropical amenities has become so strong that the development of recreation resources has spread well north along the Atlantic margin into coastal Georgia and coastal South and North Carolina. Sections of the Georgia and South Carolina coasts, such as Georgia's sea islands and Hilton Head, Charleston, and Myrtle Beach in South Carolina, are experiencing equivalent development booms.

Not all tourist attractions depend on resources provided by nature. The decision by Walt Disney to construct Disneyworld, an eastern equivalent to his highly successful Disneyland in southern California, has drawn millions of out-of-state visitors to south-central Florida. Many other new attractions were drawn to this part of the state, especially around Orlando, by the promise of tourist traffic and the possibility of tourist expenditures, a promise that became self-fulfilling. The magnet provided by this central Florida recreation complex has stimulated considerable spillover growth in the region. It is expected to be the inland link between Florida's east coastal and west coastal urban clusters. By the end of the century, a small megalopolis—or coalescence of urbanized areas—is expected to extend from Miami to Tampa–St. Petersburg via Orlando and contain approximately 13 million people. Although not the only reason for this expectation, the more than $18.6 billion spent by tourists in Florida during 1985 is a clear indication of economic power.

Hazards

Although the subtropical environment of the Southern Coastlands provides a variety of advantages through its complementarity with the rest of the continent, this environment is not entirely beneficent. Agriculturally, successful vegetable production has encouraged growers to attempt to produce crops year-round. Thus, when an occasional midwinter frost does reach into southern Florida, considerable crop damage can occur. Similarly, Florida citrus is harvested between October and late spring, and a winter freeze can hurt the ripening fruit. Lake County, Florida, citrus growers, for example, usually expect the first frost about December 15 each year but may be faced with a hard freeze as much as three weeks earlier. Smudge pots, oil heaters, and other temporary (and partially effective) devices are used in attempts to protect the growers' investment. Less well publicized is the potential damage of these untimely wintery blasts for sugarcane growers in Louisiana. Ironically, the shortness of the frost period and general mildness of the freezes also create problems for agriculture generally, because, although controlled by expensive chemical means, insect and parasitic pests thrive during the long growing season and create a constant threat to farm production.

More erratic, more sporadic, more dramatic, and locally more destructive along the Southern Coastlands are the region's hurricanes. These large cyclonic storms are generated by intense solar heating over large bodies of warm water. When the proper set of conditions has been created, a large low-pressure atmospheric "pocket"

The coastal resources of peninsular Florida are well suited to recreation, as illustrated in this section of Fort Lauderdale where the landscape has been modified to permit maximum access to the water. (Florida Department of Commerce, Division of Tourism)

results. Warm, moist air flows into the low-pressure cell and is lifted vigorously upward until the air-pressure "sink" is very deep and creates a powerful pull on more warm, moist air. The earth's rotation and spheroidal shape mean that air flowing toward the hurricane's center will establish in the storm a broad counterclockwise rotational pattern. Depending on areal extent and the atmospheric conditions involved, the result of all this might be a large but local thunderstorm, a gale, or a full-fledged hurricane.

Hurricanes can affect settlement far north of the Southern Coastlands, but their chief concentration is in that region. When a storm does move onshore in some section of the heavily populated region of Megalopolis, severe destruction can occur, as in 1955 when hurricane "Diane" struck New England. But because of their exposure toward the tropical waters of the Caribbean Sea and the southern Atlantic, Florida and the entire Gulf coastal lowlands to Brownsville, Texas, are extremely vulnerable to hurricane damage. The potential for damage was demonstrated dramatically in 1989 when hurricane "Hugo" came ashore just north of Charleston, South Carolina.

The same features that provide the region's amenities and make it attractive to tourists and retirees—subtropical climate and sea coast ori-

The Gulf Coast is especially vulnerable to the tremendous destructive power of hurricanes. Although the effects of these large storms may be felt across very large areas, as is shown by this 1972 hurricane, the most severe damage is done to the coastal region as the storm moves ashore. (NOAA)

entation—are those that support the creation of hurricanes and permit such great damage when they strike land. Because these storms are an accepted feature of the region and weather satellites and other forecasting tools are available, preparations to withstand the greatest force of the wind and rain can be made early. And because the heaviest damage is usually limited to a relatively narrow swath as the storm moves onshore, many portions of the region have not been affected for years. On the other hand, because hurricanes are so variable in occurrence and strength, settlements have spread in spite of warnings into coastal areas that are very much exposed to the dangers of a large storm. And the long period between direct coastal strikes at any specific site by hurricanes tends to make newcomers complacent about the tremendous force of major storms.

The region is also gradually succumbing to what might be called a slow-acting hazard. If ignored or not dealt with adequately, this problem will pose a considerable hazard to many in the region. Portions of the Southern Coastlands provide an example of the long-term environmental impact of rapid and widespread development.

As surprising as it may sound, major sections of the region will soon be faced with a situation in which water demand exceeds water supply. Even though most of the region receives abundant rainfall and no portion of the region is more than a short distance from a seacoast, there is clear evidence that water is already being withdrawn from some local sources faster than it can be recharged.

The potential problem has begun to appear most obviously where there are major urban developments at some distance from rivers which might be used as a water supply. Consider for a moment the options available to those who must provide water to growing municipalities. If rivers are too distant and the land is too flat to make construction of large reservoirs feasible, the remaining possibility is to withdraw water from the water table by drilling wells and pumping it as needed.

In central Florida, this is the primary method used. The region's rains are usually sufficient to recharge the *water table.* However, even the normally abundant rainfall of the region is not always enough. The tremendous growth in population and business between Cape Canaveral and Tampa with the greatest concentration in Orange County (containing Orlando) has meant that roads, parking lots, homes, motels, amusement parks, and all other facets of tourist development have spread across the land at an amazing pace. Water that falls in these built-up areas is carried away through storm drains and does not percolate down to the water table. When combined with the increased withdrawal, the land surface loses some of its support from below (as the water table drops), and it can settle, sometimes very suddenly. The abrupt appearance of a *sinkhole* that swallowed an automobile dealership in Winter Park, Florida, brought national attention to this potential hazard, if not its underlying cause. Other, generally smaller, cavities have appeared in urban areas of central Florida since the Winter Park experience. While we do not suggest that the state will dissolve in some catastrophic manner, this does demonstrate that the repercussions of development for an environment are not always immediate or obvious. The short-term solution to the problems caused by excessive withdrawal of water are not obvious, either. And in the meantime, urban forms continue to reach ever further across the landscape.

ON THE MARGIN OF THE CONTINENT

The subtropical character of the Southern Coastlands is a function of its location along the southern margins of North America. But there is more to the significance of location than the physical environment found at that place. Just as urban places can be studied in terms of their absolute and relative locational characteristics, or site and situation as discussed in Chapter 5, analysis of other kinds of places can be enhanced by examining both types of locational characteristics.

The Southern Coastlands' absolute location is probably best reflected in the subtropical nature of the physical environment. The region's relative location, however, its position with respect to, and spatial relations with, other places may be more effectively understood in terms of two characteristics: first, as an exchange area between the continental interior and overseas territories, especially those of the Caribbean and Middle and South America beyond, and second, as a geological region of considerable economic importance. Both characteristics are direct consequences of the region's location on the continental margin. The first is a function of economic and cultural exchange, while the second is more related to factors of resource location and exploitation. Both have generated consequences that help define the variability that exists within the Southern Coastlands.

Trade

The coastline of the Gulf of Mexico has only a few good-quality harbors suitable for large-volume trade activities. A shallow, emergent coastline containing many *high-action beaches,* much of the coast itself is either backed by extensive swamps or partially shielded behind offshore bars. If passage can be made through gaps in the sand bars, protection from rough seas is gained. Coastwise shipping uses this protection in the *intracoastal waterway system.* However, most bays behind the bars are too shallow to provide good anchorage for the large ships engaged in transoceanic trade. Most of the larger ports have developed on the edge of large river estuaries along the coastline, or a short distance inland from the mouths of rivers emptying into the Gulf or the Atlantic (see Chapter 3). Pensacola (Florida), Mobile (Alabama), and Galveston and Corpus Christi (Texas) have all developed in the shelter of large bays, each of which also bears the name of the port city. Jacksonville (Florida) on the St. Johns River, Brownsville (Texas) on the Rio Grande, and New Orleans (Louisiana) on the Mississippi River share riverine port characteristics. The largest single city in the Southern Coastlands, Houston, was not originally a port city but has become one through construction (beginning in 1873) and repeated improvement of the Houston Ship Channel across shallow Galveston Bay.

Perhaps more important than physical facilities, a coastal location must be accessible to inland sources of exchange items before trade can be generated. Each of the bays providing good harbor facilities is the outlet for a river draining a part of the interior, but these rivers vary greatly in navigability. All assisted early settlement expansion and some are still navigated by small barges. All have had their access to the continental interior strengthened by rail connections with major inland markets, or the river flowing into the coastal harbor has been improved for better navigation. Jacksonville, for example, was an early terminus for railroads entering Florida from Georgia. It also was the main focus for a hinterland that extended westward into the state's "panhandle" and south into the agriculturally rich central highlands. Other ports were too distant or too low-quality to be attractive alternatives. A result was that Jacksonville was well established even before highway connections reinforced its locational bases for growth. Mobile's access to the interior has even been extended to the Ohio River via the Tennessee-Tombigbee waterway.

When discussing the accessibility of Southern Coastland cities to their hinterlands, New Orleans is in a class by itself. The city was founded very early in the European colonial period by the Spanish, soon afterward succeeded by the French, as a southern counterpart to Quebec and Montreal on the St. Lawrence. New Orleans was both a control point and a shipping focus for the entire Mississippi River system. The Mississippi was navigable (with proper caution) by shallow-draft paddle-wheel steamers far to the north, well into the agricultural heartland of the continent. The river's main tributaries were also navigable and extended the single waterway system both westward into the Great Plains (see Chapter 13) and eastward into the continent's manufacturing core region (Chapter 6). New Orleans' site, within a large river meander and on the low-lying river delta only a few feet above mean sea level, meant that flooding was an annual threat often realized. But the city's situation was of such tremendous advantage to all associated with trade that its population grew early in the nineteenth century and has remained large. New Orleans, in fact, was easily the largest urban place south of the Ohio River and east of the Rocky Mountains until after World War II.

New Orleans lies midway between the recreation region of the eastern Gulf and the industrial centers of the western Gulf. The French colonial heritage is consciously maintained in the city's French Quarter, or the *Vieux Carré.* A strong non-Anglo cultural imprint is apparent outside the Quarter as well, but the distinctive mix of local Creole, Cajun, and European cuisines, a wealth of jazz and Dixieland music performances, legitimate eighteenth-century architecture, and an anything-goes reputation has drawn millions of tourists to the city. A casual visitor to New Orleans, there for its recreational opportunities, may be surprised by the heavy barge and ship traffic on the river (it is the busiest port in North America) and by the heavy industry supported by this traffic.

The other major city in the western portion of the Southern Coastlands, the largest in the en-

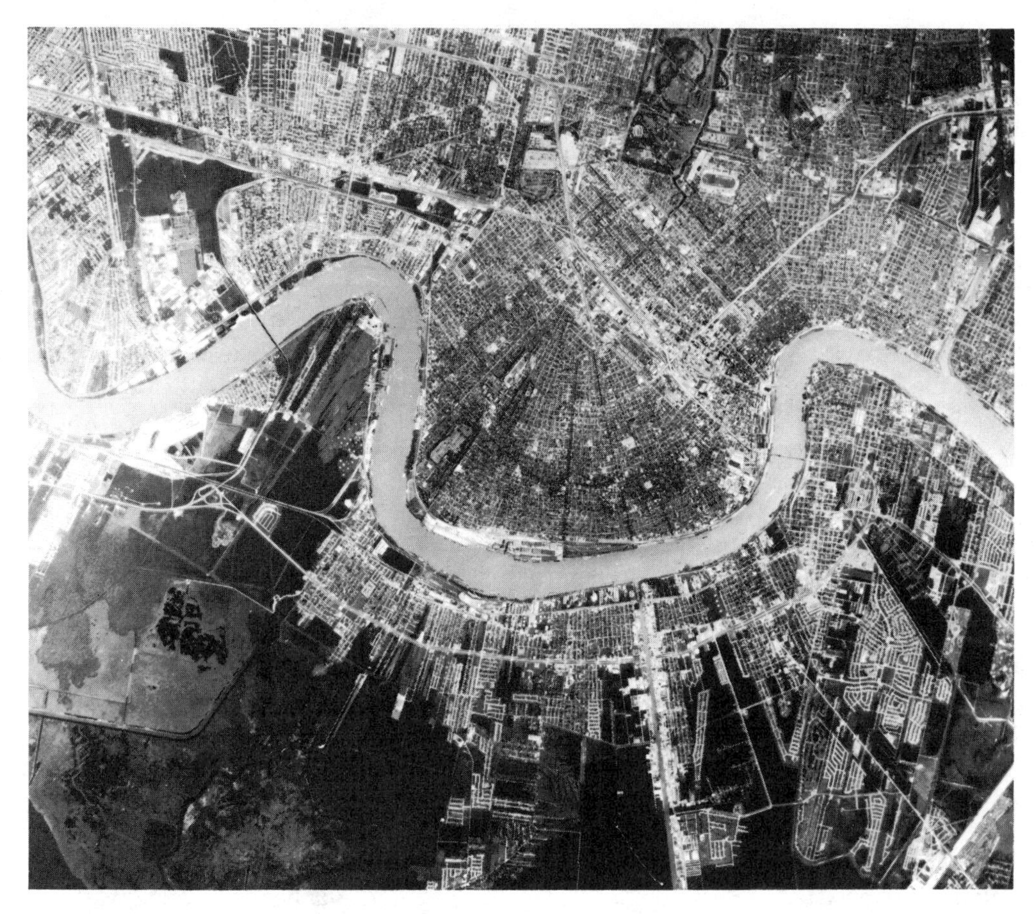

The high-altitude vertical photo of New Orleans dramatically demonstrates the city's intense urban development within a large meander and the trade and transfer point advantages along the lower Mississippi River. (NASA)

tire region by 1970, provides a contrast with New Orleans. Houston is a new city in many ways, while New Orleans has a long history of cultural and economic importance. Not actually located on a river, or even on Galveston Bay, Houston's port connections became more important to growth after World War II with the rise of the local petrochemical industry. Houston possesses a high degree of focality for most of lowland Texas (see Figure 3-4), but these overland connections, too, came after the city's economic growth, stimulating it further; they did not precede it. Regional mineral resources and Gulf Coast access have been the primary geographic factors in Houston's rise to urban prominence.

Resources

The true physical margin of North America as a continent does not coincide with the seacoast. A shelf of land below sea level extends beyond the coastline. In some cases the shelf extends only a few miles, but along much of the Atlantic Coast and in the Gulf of Mexico, the edge of this shelf may lie more than 80 kilometers (50 miles) from shore. Mineral exploration along the coast

The fog-enshrouded spires of St. Louis Cathedral overlook Jackson Square in the very heart of New Orleans' French Quarter. The French cultural inheritance has been preserved here and is one of the attractions that draws many thousands of visitors to the city eacy year. (Courtesy of the Greater New Orleans Tourist and Convention Commission)

from the Rio Grande to the Mississippi River mouth led to the discovery of an extensive series of petroleum and natural gas deposits, both onshore and off (Figure 11-3).

When the Gulf Coast oil field was brought into production during the early 1900s, Houston was still a moderate-size city of less than 75,000. By 1930, the city's population had multiplied four times to almost 300,000. Through the Great Depression years and World War II, Houston continued to grow, doubling its population again by 1950. And by the time the 1980 census was taken, the city itself had grown to 1,595,138 to rank it fourth in the United States behind New York, Chicago, and Los Angeles. Located about midway along the long coastal arc between the Mississippi and the Mexican border, Houston also lies at the coastal apex of the Texas Triangle joining Dallas–Fort Worth and San Antonio. With Dallas ranked 7th in population in 1980 and San Antonio ranked 10th, connections with these major inland growth centers and the large cotton exports that originate in eastern Texas have also contributed to the locational strength of Houston.

The search for additional petroleum deposits along the Gulf Coast was extended seaward before midcentury. The petroleum companies' success created new problems, of a political nature, even while economic problems were temporarily alleviated by the discoveries. In overcoming the technological difficulties of drilling for and extracting petroleum using platforms far beyond sight of land, conflicting claims between state and federal governments over jurisdiction of continental shelf resources flared for the first time. Texas and Louisiana had immediate stakes in the argument, and it was clear that other coastal states, especially those bordering the Gulf, would also be interested before too long. The federal form of political organization in the United States permitted a series of litigations between the individual states and the national government. One result of this complex series of court cases has been variable jurisdiction over continental shelf resources in the Gulf of Mexico by the Southern Coastland states; Florida and Texas are sanctioned to claim up to 9 miles seaward, while Louisiana, Alabama, and Mississippi are limited to 3 miles. The claims by Texas and Louisiana were initially the most significant because of the resources already known to exist off their coasts.

The continuous rapid increase in domestic consumption of petroleum, natural gas, and petroleum products led the national government, during the early 1970s, to open commercial bidding on offshore tracts between Biloxi, Mississippi, and Tampa Bay, Florida, and beginning in the early 1980s off the Atlantic Coast. Opposition to exploratory drilling in the eastern section of this region has been very strong from environ-

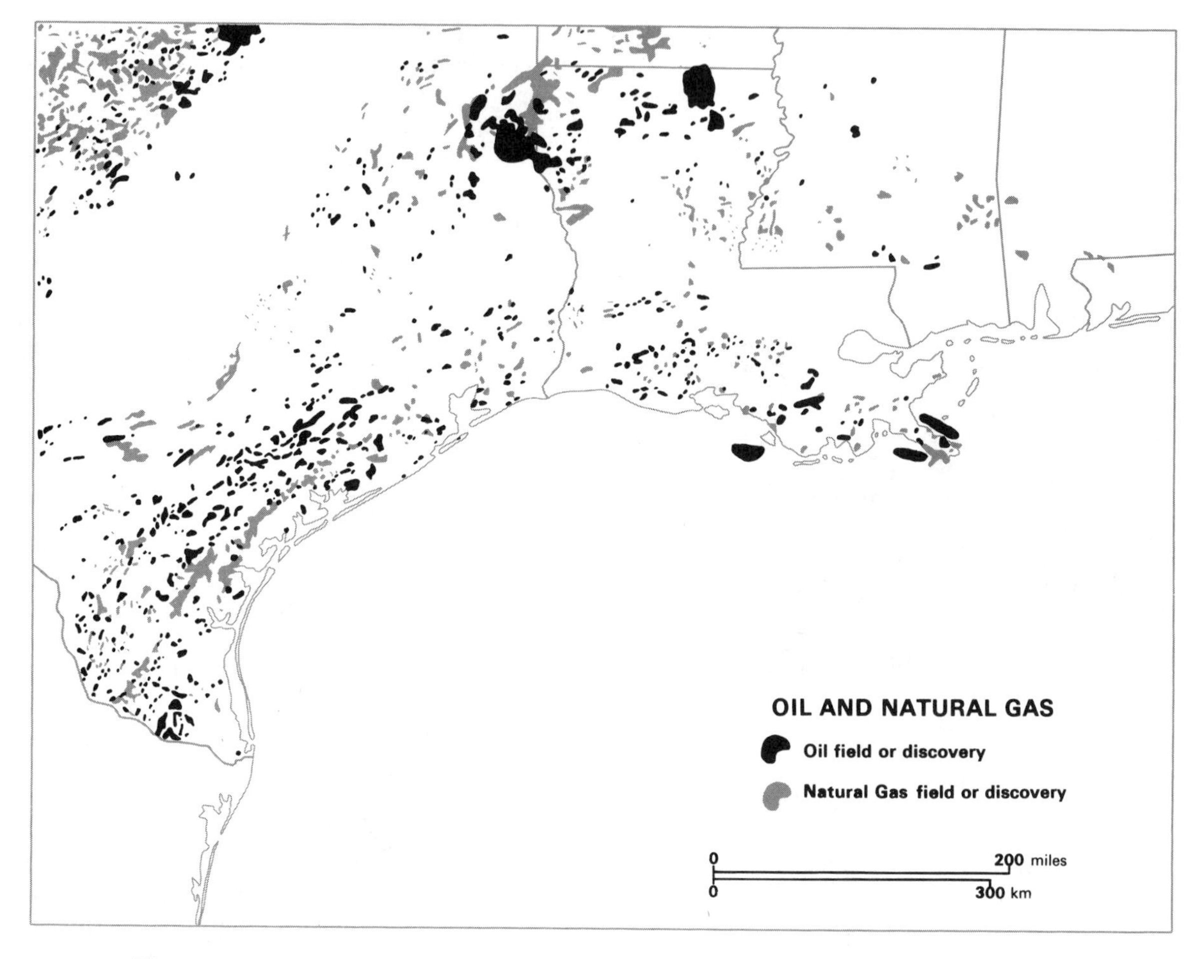

Figure 11-3 Petroleum and natural gas deposits: Western Gulf Coastlands. Given the coastal location of these resources, the long arc of petroleum and natural gas deposits paralleling the Louisiana-Texas Gulf Coast has generated a large refining capacity at Texas and Mississippi River ports.

mentalist and fishing groups and from other economic interests already present but based on the coast's amenity characteristics. Florida, naturally, has led the opposition, a reflection of its very long Gulf coastline and its extremely strong orientation toward recreational resources. While oil spills or leakage could have disastrous short-term effects on certain tourist activities, many conservationists argue that even more serious consequences are possible. Extensive mangrove swamps, coastal marshlands, and commercial fishing beds are in a delicate ecological balance, and they could be badly damaged for long periods if oil drilling mishaps or shipping accidents polluted them. The classic disagreement between those who would develop consumable resources as they are needed and those who wish tc preserve the balance of the natural environment is not likely to be resolved easily along the Gulf Coast. The dispute also illustrates the regionalization within the Southern Coastlands, for the western and eastern portions of the region each represent opposing interests.

So extensive are the petroleum deposits along

More than 40 miles inland from the Gulf of Mexico on the vast coastal plain of Texas, industry near the head of the Houston Ship Channel is supplied by ocean-going vessles. (Bob Daemmrich/The Image Works)

the coast between northern Mexico and the Mississippi River that the resource importance of the Southern Coastlands would be a national priority even if these deposits were the only mineral resource in the region. Texas and Louisiana are currently two of the three leading petroleum-producing states (with Alaska), far exceeding the two next most important, California and Oklahoma. And while Texas and Louisiana possess large producing fields well inland from the Gulf, the coastal fields are major contributors to both states' totals.

Natural gas, often found in conjunction with petroleum deposits, is another resource in the Gulf. For many years, gas was burned off as an unwanted by-product of petroleum production. But the energy potential of this resource was eventually realized and generated a search for and utilization of natural gas independent of petroleum exploration. The numerous scattered Gulf Coast deposits of natural gas are spatially intermingled with the region's long arc of petroleum deposits. Pipelines to carry the gas radiate from the main coastal production centers to the primary consumption points across the continent and within the manufacturing core.

Although a less common mineral association, the geologic formations of the Texas and Louisiana coastlands that contain petroleum and natural gas contain two additional minerals of economic value—sulfur and rock salt. The subsurface rock contortions found where petroleum and natural gas have been trapped in economically recoverable deposits were formed in this region by the gradual upthrust of large *salt domes*. Far less valuable than either of the mineral fuels, rock salt is nonetheless mined in large quantities in southwestern Louisiana. More valuable than salt, although not so much as petroleum or natural gas, is the sulfur found in the *caprock* covering many of the salt domes. The large sulfur deposits at Beaumont, Texas, and across the state boundary near Lake Charles, Louisiana, supply all U.S. needs. Additional deposits both inland and beneath the continental shelf indicate an abundance of this mineral for many years. Also of national significance is phosphate production from substantial deposits in Florida.

Industrial Development

Petroleum and natural gas extraction do not usually generate, by themselves, strong local urban and industrial growth. The process of exploration and drilling requires specialized and expensive equipment but not the large variety of support materials or labor demanded by many mining operations. Once the crude oil or gas is being pumped, only routine attention is required. However, as was dramatically illustrated for North Americans who had forgotten their own earlier experience—first in the 1970s when oil prices rose sharply, and again in 1990 following Iraq's invasion of Kuwait—large-volume petroleum production can generate tremendous quantities of capital in a short time. Although both mineral fuels of the region can be transported cheaply and reliably through pipelines and by oceangoing tankers, locally accumulated wealth has attracted a wide variety of industries able to use the minerals near their production sites. Petroleum refineries were built outside all major ports from Corpus Christi, Texas, to Pascagoula, Mississippi, with the most intensive concentration around Houston, Beaumont, and Port Arthur.

Of broader development impact than the refining process itself are the industries that depend on the refinery output for their own existence. The petrochemical industry is one of the most important and rapidly growing enterprises in highly industrialized countries such as the United States and Canada. Natural gas and petroleum products are used as chemical components for a great many products in everyday use. Items ranging from plastics, paints, and antifreeze to fertilizer, insecticide, and prescription drugs have their origin in chemical plants located along the western Gulf Coast (Figure 11-4). In addition, other chemical industries not tied to petroleum and natural gas production, such as those producing sulfuric acid, superphosphate fertilizer, and synthetic rubber, are major consumers of sulfur and salt. The coincidence of the basic minerals within a region possessing large capital investment capabilities has supported rapid economic and population growth, especially in urban centers such as Houston. Just as quickly, when declining oil prices reduce the investment capital available, the economy of the region lags.

The consequences of the drop in oil prices through the 1980s for many urban centers in Texas, Louisiana, and Oklahoma are not atypical of economies that are dependent on the income derived from a single product. The local economy stalls. Businesses caught with heavy debt loads (undertaken to take advantage of what seemed rich entrepreneurial opportunities) struggle to cover interest payments. Banks that loaned the investment capital are unable to get it back and unable to pay the depositors who are now themselves in need. Significant numbers in the labor force become unemployed, and those who can move elsewhere for jobs. Although Houston was much larger in 1988 than a decade earlier, the loss of more than 115,000 jobs in manufacturing and construction indicate the root of that city's economic struggle. The consequences for individuals are hinted at in the 10 percent decline in median sales price for existing single family homes between 1984 and 1986. There is a great deal more to Houston's economic geography than oil and natural gas, but the capital generated by these resources remains a key.

There is also more to the location of industry, however, than the availability of capital and proximity to raw materials, even petrochemical industries. The two large oil fields extending from southeastern New Mexico to central and eastern Kansas, for example, have produced vast quantities of petroleum, natural gas, and capital but with much less industrial development than along the Gulf Coast. The difference

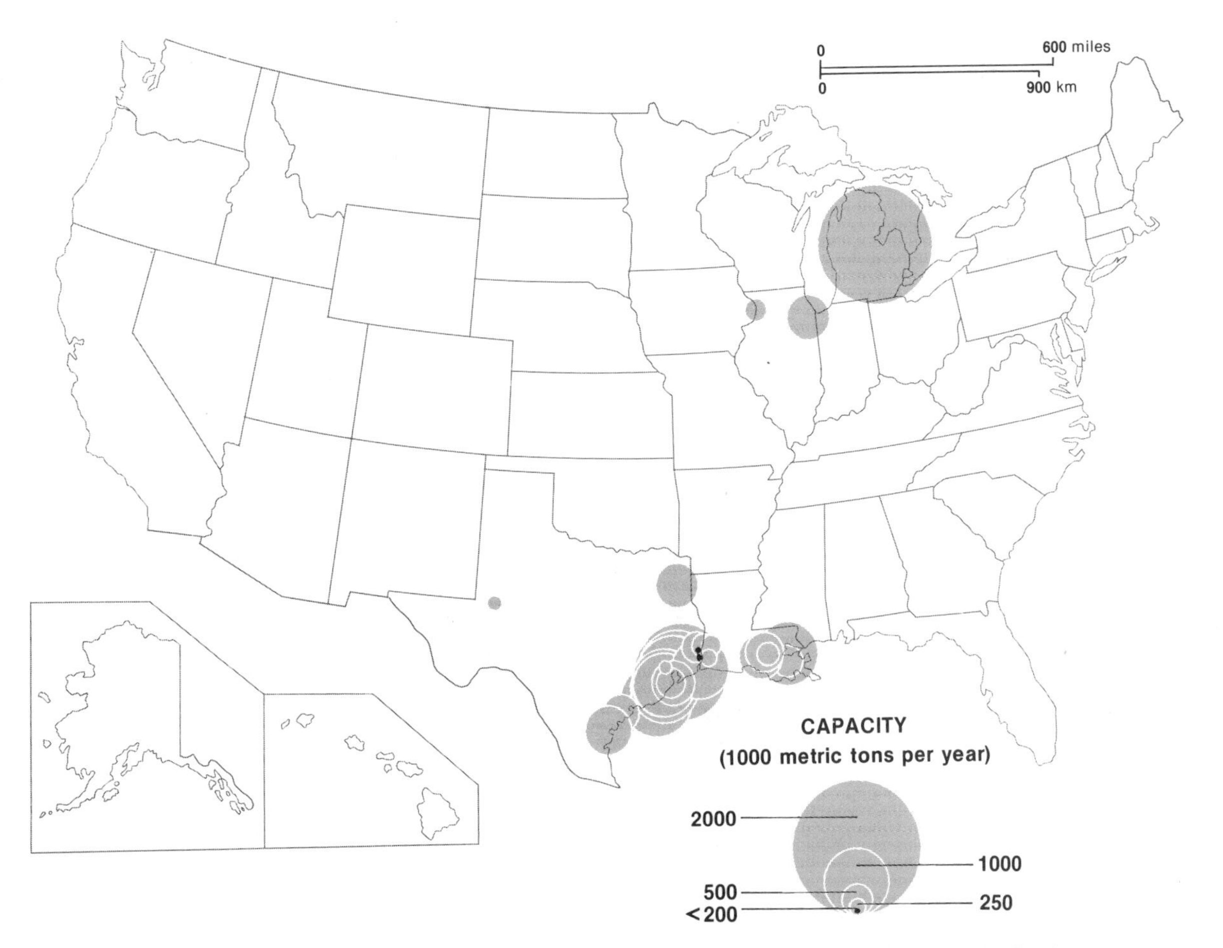

Figure 11-4 Installed ethylene capacity, 1989. Except for a large Dow Chemical facility in Midland, Michigan, production of this basic petrochemical is heavily concentrated along the western half of the Southern Coastlands, where large stocks of petroleum and abundant investment in refining capacity are located.

is largely a matter of accessibility. The Southern Coastlands, again, are on the continental margin and as such constitute the line of exchange between water transport and land transport. Further, since water transportation is cheaper per ton-mile than land transport, even though slower, shipment of the finished products from the Gulf Coast can be accomplished efficiently by ocean carrier to the ports of Megalopolis and by barge via the intracoastal waterway and the Mississippi River system to the manufacturing core.

Just as coastal location can facilitate shipment out, products and raw materials can be moved more efficiently to Gulf Coast industries. The most significant raw material import along this coast is bauxite. This ore, from which aluminum is eventually refined, is originally of low aluminum content. Ideally, the ore is refined near the site of its extraction because it is not efficient to move the large proportion of the ore that will be discarded, but the refining process requires large amounts of electricity. Since most bauxite imported by the United States originates on the energy-poor Caribbean island of Jamaica or the South American counties of Surinam and Guyana, the ore can be shipped at relatively low cost by freighter to Gulf Coast ports. Once there, abundant local fuel permits refinement of the ore to levels of aluminum content that can be transported economically to various markets across the continent. Only at Arvida, Quebec, and Kitimat, British Columbia, is imported bauxite ore refined in significant quantities in North America outside of the Gulf Coast.

The distinctive character of the Southern Coastlands, then, is strongly divided. While the entire region possesses a subtropical environment, the eastern half of the coastlands has been developed largely with emphasis on the advantages of such an environment. The rapid decline in precipitation west of the Sabine River has limited the agricultural possibilities of the Texas Gulf region. Recreation and retirement are possible along the coast west of the Mississippi, but these possibilities have not been exploited as strongly. Mineral exploitation and the consequent industrialization have provided more obvious economic returns in the western portion of the coastlands, and this form of development has taken the lead in this western part of the region.

The three major urban centers also illustrate the region's internal contrasts. Houston and the Miami area have both grown very rapidly in recent decades, but one has been based on manufacturing and trade, and the other on recreation and travel. More recently, Miami's Cuban-American population has stimulated strong financial and trade connections with Latin America, broadening that city's economic base as well as its cultural flavor. Locationally between the two, New Orleans can claim both industrial and recreation importance, though this is an uneasy mix, as we have seen, so New Orleans has not grown as rapidly as either Houston or Miami. By 1975, the industrial and mineral exploitive character of the western Gulf began to encroach on the eastern Gulf. While petroleum costs remained high, pressure on the recreational character of coastal Florida increased. The decrease in oil prices during the 1980s relieved the pressure somewhat, but potential remains strong.

CULTURAL—ECONOMIC CASE STUDY

LATIN AMERICA IN FLORIDA[3]

Florida is a finger of land pointed at the Caribbean. Comprised of a peninsula that reaches southward approximately 350 miles (560 kilometers) from the main body of North America, the state is extended along a chain of islands known as the Florida Keys to a latitude almost 50 miles (80 kilometers) farther south and within 95 miles (152 kilometers) of Cuba's north coast. Along the southeast coast of the peninsula, between the Everglades and the Atlantic Ocean, a sprawling megalopolitan cluster has grown around and north from Miami. The main themes that characterize the Southern Coastlands can be found uniquely

[3] Much of the data in this section have been taken from Thomas D. Boswell and James R. Curtis, *The Cuban-American Experience* (Totwa, N.J.: Rowman & Allenheld, 1984), and Antonio Jorge and Raul Moncarz, *The Political Economy of Cubans in South Florida* (Coral Gables, FL: North-South Center, Uniterisity of Miami, 1987).

There has been a significant Cuban presence in southern Florida for a long time, but the influx of immigrants from that island country since the late 1950s has transformed large sections of the major urban areas in the region. This street corner in Miami provides vivid proof of the Latinization of the metropolitan area. (Michael Heron/Woodfin Camp Associates)

represented in the unparalleled cultural tranformation of the greater Miami region since 1959.

Miami was viewed as a valued recreational destination from its early decades of growth following construction of Henry Flagler's railroad south from Jacksonville in 1896. The region's subtropical location was especially attractive to residents of the major urban centers of the northeastern seaboard, and cities along Florida's southeast coast grew rapidly in order to provide the setting and the services desired.

Through the 1950s, at least, Miami's attractions were treated as passive characteristics of place. It was enough to go and to enjoy what was there. The city's weather was warm throughout the winter; its coastal location provided easy access to long beaches and warm tropical waters; the harbor required only modest improvement to ease the travel of cruise passengers from North America to the attractive and culturally distinctive Caribbean islands nearby. These characteristics of the Miami region's subtropical location were recognized and exploited, but it was not until the 1960s that the region's location at the southern extremity of the continental margin led to the full integration of Miami into a hemispheric system of finance and commerce.

Key to the change in Miami's use of its location is the Cuban immigration to southern Florida in 1959 following the successful revolution in Cuba led by Fidel Castro. There had been Cubans in Miami for generations; it is the natural point of immigration, if only because of proximity. Many who came from Cuba during the earlier decades, however, moved on to the more dynamic labor markets in and around New York City. At the time of the 1950 census, more than 45 percent of all Cuban-born residents of the United States were living in New York, with only slightly more than half of that proportion residing in Florida.

Beginning in 1959, Cuban refugees arrived in Miami in unprecedented numbers and with tremendous impact on the region's economic and cultural landscapes. Due to fluctuating emigration policies in Cuba, there have so far been three major waves of immigration from that country. Between the revolution and the 1962 missile crisis, well over 200,000 Cubans migrated into and through Miami. Then, following a 3-year hiatus during which another 56,000 arrived, mostly through third countries, more than 300,000 additional Cuban refugees arrived during the nearly 7½-year airlift between Havana and Miami. Once again, following another 7-year period of restricted emigration, the Mariel boatlift brought another 125,000 Cubans to Miami during a single 5-month period in 1980. There was also some migration back to Miami from other parts of the United States by Cubans who had entered the country in the 1960s and settled initially outside of southern Florida. Overall, during this 22-year period between 1959 and 1981, the Latin American population of the greater Miami area (about 85 percent of whom are Cuban) grew from some 25,000 to almost 700,000, or perhaps 40 percent of the region's total population by the end of the period.

The numbers themselves represent a staggering increase for so short a time, but the broader economic impact on the region is grounded in the characteristics of the immigrants. Cuban refugees who arrived during the initial wave "were overrepresented in the legal and white-collar professions and underrepresented in the primary occupations (such as agriculture and fishing) and in blue-collar jobs."[4] It is easy to overstate this occupational tilt; in fact, the immigrants possessed a very diverse labor structure, but it did favor more professional occupations. Subsequent waves also tended to favor professional occupations, but less so as time went on. The contributions that Cubans were prepared to make to the local economy, therefore, were very broad but weighted toward the professional, managerial, and skilled end of the spectrum.

The effect on the economic geography of Miami has been considerable. The Cuban population was absorbed relatively quickly into financial, commercial, and retail activities. The rapidity with which large numbers settled in Miami created its own "instant" market. Businesses located in other parts of the United States but wishing to expand their activities into Latin American markets began to move at least part of their domestic operations to Miami to employ the Spanish-speaking Cubans. The trans-Caribbean business contacts brought with the Cuban refugees were also attractive to Anglo business interests. Multiplied many times in many ways and over several decades, this pattern has opened up Miami's natural geographic orientation toward the south. Miami has taken more active advantage of the region's extreme southerly location on the continental margin. Southern Florida's major domestic Hispanic competitor for Latin American business, Los Angeles, is largely oriented toward Mexico and is almost

[4] Boswell and Curtis, *The Cuban-American Experience,* p. 45.

2000 miles (3200 kilometers) farther from South America than is Miami. Even Houston is almost 750 miles (1200 kilometers) more distant from South American contacts than is southern Florida.

Beyond the obvious rapid cultural transformation of southern Florida, the addition of hundreds of thousands of Cuban-Americans has also transformed the economic geography of the region. Still subtropical in climate and still attractive to many for its recreational potential, southern Florida is now also an active commercial and financial "hinge" on the continent's margin. In a smaller and culturally distinctive fashion, this region's urban complex is beginning to function in much the same manner with Latin America as Megalopolis has with Europe.

ADDITIONAL READINGS

Balseiro, Jose Agustin, ed. *The Hispanic Presence in Florida*. Miami: E. A. Seeman, 1976.

Blake, Nelson Manfred. *Land into Water—Water into Land: A History of Water Management in Florida*. Tallahassee: University Press of Florida, 1980.

Boswell, Thomas D., and Manuel Rivero. "Cubans in America." *Focus,* 35 (April 1985), pp. 2–9.

Fernald, Edward A., ed. *Atlas of Florida*. Tallahassee: Florida State University Foundation, 1981.

Hilliard, Sam B., ed. *Man and Environment in the Lower Mississippi Valley,* Vol. 19. Baton Rouge: Louisiana State University, School of Geoscience, 1978.

Lewis, Peirce F. *New Orleans: The Making of an Urban Landscape*. Cambridge, Mass.: Ballinger, 1976.

Palmer, Martha E., and Marjorie N. Rush. "Houston." In *Contemporary Metropolitan America*. Vol. 4. Cambridge, Mass.: Ballinger, 1976.

Symanski, Richard. "Across the River Lies Another Country." *Focus,* 37 (Winter 1987), pp. 20–25.

CHAPTER 12

The Agricultural Core: The Deep North

The interior of North America is vast and nearly flat. Between the Appalachians to the east and the Rocky Mountains to the west, there are relatively few pockets of landscape that vary from the gently rolling swells of this dominant countryside. So large is the area encompassed, however, that other characteristics become the means of subdividing the territory into numerous regions. Low average temperatures, a short growing season, and related environmental factors separate the Northlands (Chapter 18) from the remainder of the interior plains. The distinctive cultures of the Southeast (Chapter 10) and the Southwest (Chapter 15) clearly set these regions apart from each other and from adjacent portions of the interior plains.

The balance of the interior is characterized by gradual *longitudinal* differences in the amount of annual precipitation and the ways in which the land is used in light of this rainfall transition. The western portion of this territory is a drier region dominated by grassland vegetation. It has a grand openness that has led these plains to be called the Great Plains and Prairies (see Chapter 13). The eastern half of the interior is no less a plain, but the more humid growing season, different vegetation, the higher *carrying* capacity of the environment, and more numerous natural transportation routes have supported the development here of the continent's agricultural core region (Figure 12-1).

In most directions, the thematic distinctiveness of the agricultural core is not troublesome. The agricultural core does not overlap strongly with Appalachia, with the Northlands, or with the cultural influences of the Deep South. The core's western boundary is more clearly a transition zone, however. Many would argue that the far-flung wheat belts are too significant to both the United States and Canada to be excluded from a region designated the agricultural core of North America. We agree that wheat is tremendously important, but we also consider the two portions of the large agricultural interior too different to treat them usefully as a single region.

There remains a possibility of confusion over differences *within* much of the territory occupied by the agricultural core. As we discussed earlier, the agricultural core overlaps territorially with several other thematic regions (Figure 1-5). Most prominent is the eastern portion of the agricultural core, which appears to occupy the same area as the western half of the continent's manufacturing core region. It bears repeating that this is no geographic accident; much of the urban development and manufacturing capacity of the region was stimulated initially by the demands of agricultural producers or the tremendous volume of agricultural output. From the eastern cities of the region, such as Buffalo, Detroit, and Cincinnati, to the western regional centers of Minneapolis–St. Paul, Omaha, and Kansas City, marketing and shipment centers gradually adapted their activities to include processing and supplying the needed technological hardware demanded by the region's farmers.

The in-between location of this territory is reflected in the variety of names that have been applied to the region. Most common, perhaps, are "the Middle West" or "the Midwest." Widely used and probably a permanent fixture

---

Almost all of the land in the Deep North is used; even the small woodlots are part of the transformed landscape. Note as well the scattered farmsteads and the gently rolling terrain. Crops include corn, soybeans, wheat, and hay as well as pasture. (Erwin W. Cole/U.S. Soil Conservation Service)

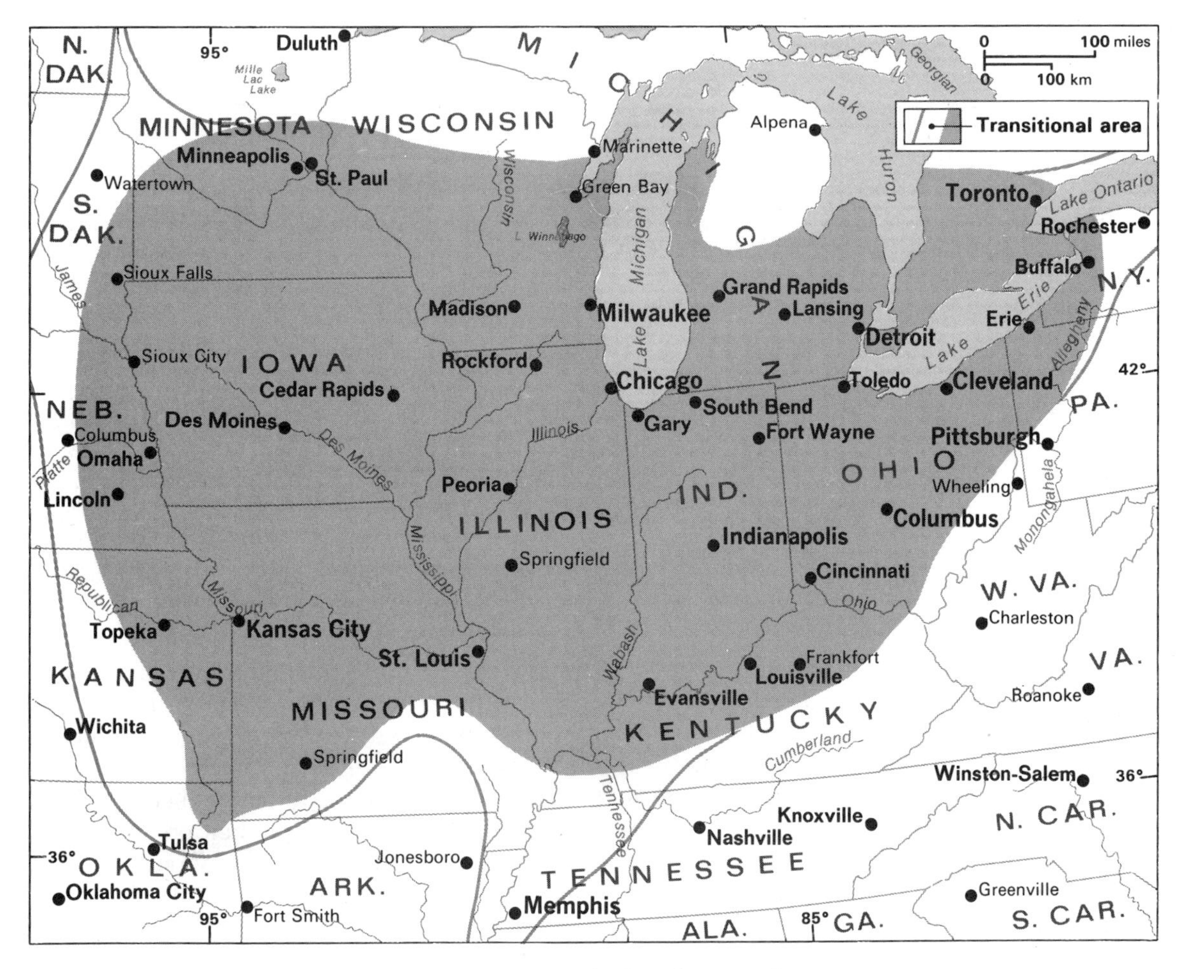

Figure 12-1 The agricultural core region.

among America's geographic place names,[1] this term is basically a carryover from the days of eastern seaboard dominance. It implies that "the West" begins at the Appalachians with the closer Middle West gradually merging into the Far West somewhere in the general area of the Rocky Mountains. The Great Plains are clearly as much in the continent's east-west "middle," and many students, when asked to identify the Midwest, point to Kansas, Nebraska, and the Dakotas. Only habit and the sequence of European settlement prevent Oregon residents, for example, from referring to the region as the Middle East. In spite of this, the Middle West will probably continue to be the most widely used and understood name for the U.S. portion of the continent's agricultural core region.

Even though both themes are not expressed across identical territories, the Middle West is clearly a region of farms *and* factories; of a white, Protestant, dispersed, agricultural and small-town population *and* a multiracial, multiethnic,

[1] James R. Shortridge, *The Middle West: Its Meaning in American Culture* (Lawrence: University Press of Kansas, 1989).

Major Metropolitan Area Populations in the Deep North, 1990[a]	
Minneapolis–St. Paul, Minn.–Wis.	2,464,124
St. Louis, Mo.–Ill.	2,444,099
Kansas City, Mo.–Kans.	1,566,280
Louisville, Ky.–Ind.	952,662
Omaha, Neb.–Iowa	618,262
Des Moines, Iowa	392,928
Madison, Wis.	367,085
Fort Wayne, Ind.	363,811
Davenport–Rock Island–Moline, Iowa–Ill.	350,861
Lexington–Fayette, Ky.	348,428
Peoria, Ill.	339,172
Appleton–Oshkosh, Wis.	315,121
Evansville, Ind.–Ky.	278,990

[a] For other large metropolitan areas, see table for North American Manufacturing Core, p. 130.

clustered, urbanized, and manufacturing population. These two aspects of the Middle West are essential to an understanding of the entire region. The urban-manufacturing contribution has been discussed in Chapter 6; the agricultural component deserves equivalent consideration.

It may be helpful to think of the regionalization of this section of the United States and Canada by picturing each thematic region as a separate plane, or flat surface, lying over the appropriate portion of the earth's surface (Figure 12-2). The manufacturing plane extends eastward toward Megalopolis. It also contains islands of industrial clusters west of the main region where regional centers support more local, but still significant, manufacturing activities. The agricultural plane can float at a different level over much the same portion of the earth's surface, although more western and not extending as far eastward beyond Ohio and around Lake Erie into southern Ontario. This agricultural plane may be seen to have holes in it where urban and manufacturing functions crowd out almost all direct agricultural production. These holes are largely a matter of scale and definition, for even the most urban counties contain some agricultural production. Cook County, Illinois, for example, is almost entirely filled by Chicago and its suburbs and was the second most populous county in the United States, yet it contained 46,907 acres of farm land yielding more than $22.5 million in farm products in 1987. By placing our emphasis on the agricultural functions, however, the region is essentially rural and small town. As such it possesses a distinctive economic and cultural orientation, an orientation that lends strong support to a distinctive regional label—the "Deep North."

The Deep North, as we have defined it, is a culture region. It is based, as strongly as the Deep South, on an accumulated mix of habits, attitudes, and reactions to the traditional opportunities for livelihood and contact with other groups within the region. The term "Deep North" refers to an area of cultural intensity and pervasiveness, but it does not refer to an area that is farther north than other places. Basically, the Deep North is small-town and rural America specially flavored with the agricultural patterns of this region. The Deep North should not be confused with the urban regional culture of the Middle West, for in spite of their proximity, they are very different regions. The population of the Deep North is politically and socially cautious, yet independent, with an underlying "what's in it for me" suspicion, secure in what has proven successful, and not strongly exposed to the pressures for change found in major urban centers or in the transition zones between regions. "Middle America" is a popular term applied to the region's U.S. population. "Middle" means in between, and as such this population shows neither extreme of America's social, economic, or political spectra. Geographically, "middle" also may imply a degree of isolation, or at least buffering, from the forces of change. By being rural and small town, on the one hand, and located in the continent's interior, on the other, the Deep North possesses many of these features and helps to define the character of Middle America.

Consider the contrast between the Deep South (Chapter 10) and the Deep North in terms of their population composition. The South's

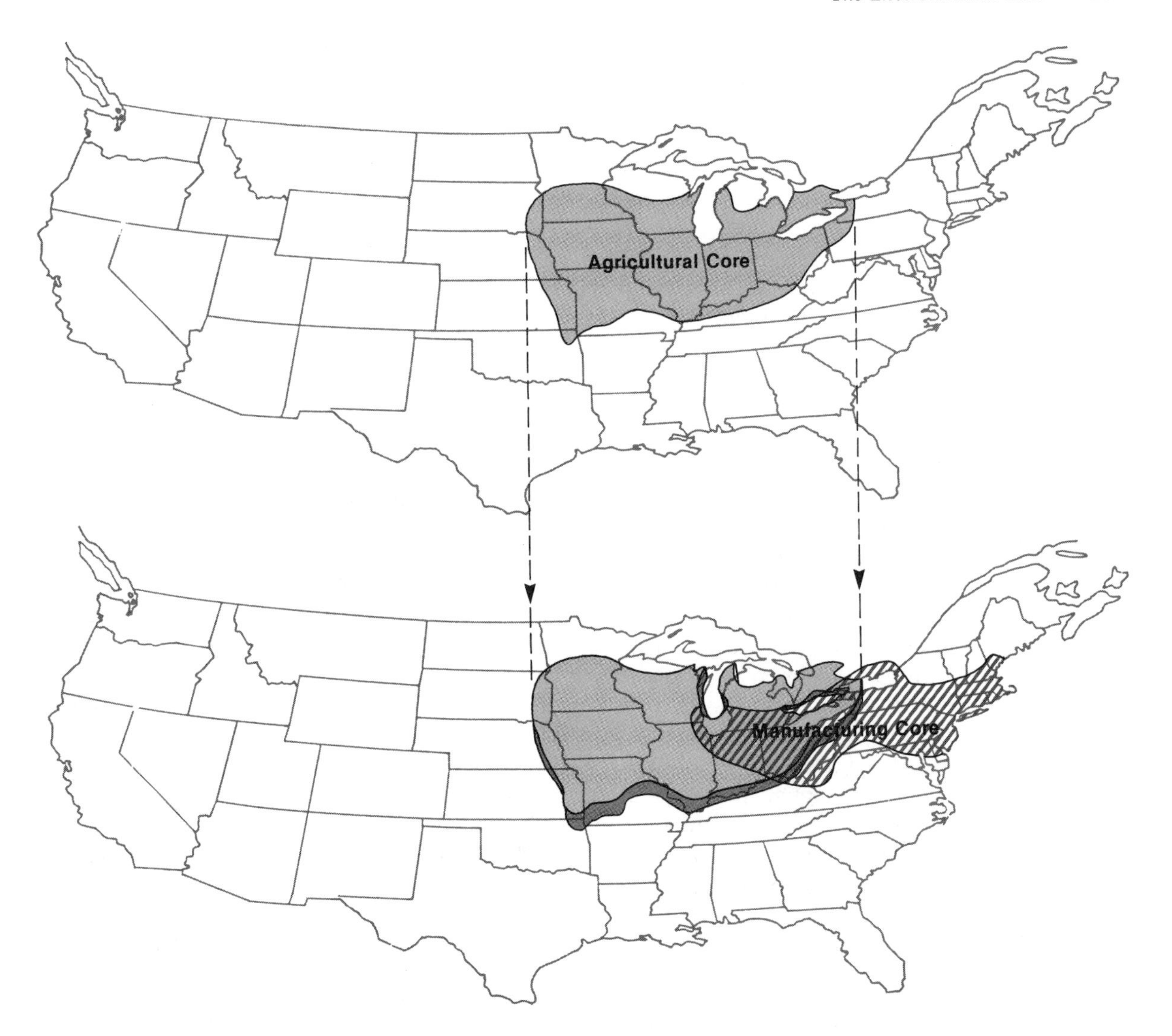

Figure 12-2 Overlapping regions. Occupying much the same territory on the earth's surface, the Agricultural Core is a small town and farm landscape extending toward the Great Plains and Prairies, while the Manufacturing Core is an urban and industrial landscape between the Great Lakes and Megalopolis.

population had very few immigrants from Europe or elsewhere for many decades after the nation's independence; the white population was almost entirely of English or Scots-Irish descent (with a sprinkle of Germans and the notable exception of the French influence in southern Louisiana). The large black population possessed considerable potency in the formation of Deep South culture. The Deep North population (that living in the rural and small-town portions of the Middle West) is predominantly white. It received contributions of foreign-born migrants until late in the nineteenth century. Most of these immigrants, however, originated in Northwestern Europe—Germany, the Netherlands, the several Scandinavian countries, and the British Isles. Later immigrants, from eastern Europe and some of the Mediterranean countries, found the better agricultural land already occupied, so they settled in the nearby metropolitan areas of the manufacturing core. The agricultural core region encloses many large cities, but since their urban function is by definition nonrural, the contributions to Deep North culture by the cities' populations have been indirect or through political confrontation at the state level with agrarian interests and attitudes.

The proportion of local population that is black is small in the Deep North. Perhaps more important, local blacks are nearly "invisible" to many residents of the region. This is not primarily a function of race tolerance by Deep North whites, although that may exist. Instead, it is a result of the combined effects of very low black numbers in small-town and rural areas, large and increasing clusters of blacks in the major urban centers, and a job opportunity structure that relegated the rural region's few blacks to occupations that do not expose them to the rest of the population. During the half-century period of northward migration by Southern blacks, labor demand was higher in metropolitan areas and new jobs were limited in the farming areas and small towns of the Deep North; very few blacks were attracted to settle here. In fact, much of the agricultural core has declined in population since the 1930s. Most new jobs that did become available were filled by local people.

The pattern of Deep North culture, then, is partly reflected in its stability, a resistance to some kinds of change, and a degree of isolation from many of the forces that have created change. In this, there are strong similarities to the Deep South. Also, as in the South, however, recent changes have disrupted this stability, and modifications to the Deep North way of life have become as insistent as elsewhere in North America. The underlying culture in this region is strongly and explicitly tied to the environmental setting. The geographic structure of the Deep North and current alterations in that structure clearly demonstrate this interdependence.

THE ENVIRONMENTAL BASE

The physical environment is naturally very important in determining the character of the agricultural core of the United States and Canada. To some extent, technology has been used in many human activities as a modifier of the natural environment. Urban and industrial activities use technology intensely and are therefore less subject to influences of the local environment. Unless their purpose is specifically to tap some aspect of the environment (e.g., recreation centers or port cities), many of the activities associated with urban location and manufacturing are carried on wherever the nonenvironmental conditions for these activities are met.

In spite of heavy application of technology, agriculture remains intimately associated with the physical environment. Industrial fertilizer, irrigation, and development of new plant varieties have enlarged the areas in which crops can be grown and have increased yields in old areas. New tools and farm machinery have changed the productivity of individual farmers. Controlled environments, such as greenhouses, and research in *hydroponics* indicate that if we are willing to pay the price, agriculture can be almost as independent of nature as urban activities. The costs, however, are immense, and in North America it is not yet necessary to bear them outside the agricultural experiment stations.

The natural mix of environmental characteristics in the Deep North is excellent for certain kinds of agricultural pursuits. After a brief period of wheat dominance, settlers of this region selected a variety of corn-based farming systems that changed little for more than a century. The success of these farming patterns contributed greatly to the stability of this region's economic and cultural forms.

The suitability of the environment for corn production does not mean that it is the most desirable climate on the continent. As one long-term observer of this region has stated,

> Geographers have long extolled the virtues of the Middle West for growing corn, as well they

Farming in the Deep North, as here in Illinois, requires careful application of new methods with traditional crops. The alternative strips of corn and alfalfa reduce soil erosion while maintaining high productivity. Note also the new silos and machinery sheds. (Tim McCabe/U.S.D.A.)

> should have, but for anyone whose aesthetic requirements transcend those of a cornstalk the climate is pretty darned miserable, winter or summer. . . . It's a great climate for making a living, but it's no place to have to live, unless you consider making a living the principal purpose of living.[2]

Although Hart's comment refers to the region's climate, much the same could be said about the relief and, less strongly, about the soils of the Deep North and their strong natural support for an agrarian economy. Even in terms of its natural transportation resources, the agricultural core is outstandingly blessed. The need to share these waterways with urban and manufacturing activities, however, has greatly altered the rivers' original aesthetic benefits.

Climate

The early American and European settlers who moved across the Appalachians and onto the eastern interior plains were less concerned with a pleasant environment than with survival and pursuit of a livelihood. Except for the resistance of the local native population (the Indians) and the usual vagaries of nature, the environment was supportive. Most of Ohio, Indiana, and lower Michigan were covered by a mixed

[2] John Fraser Hart, "The Middle West," *Annals of the Association of American Geographers*. Vol. 62 (June 1972), p. 267, footnote 14.

hardwood forest. The trees encountered were familiar and indicated to the experienced easterner where the best soils were located, and they also provided considerable local fuel and building material. Near the western margin of Indiana and farther into Illinois and southern Wisconsin, the small openings and glades in the forest were larger and more frequent. Except along the rivers and in hillier country, Illinois, Iowa, and parts of southern Minnesota and northern Missouri appeared to be as much open grassland as forest. By the time settlers reached north-central and western Iowa, the dense woodlands had been left miles behind. Only along river courses and in valleys were substantial clusters of trees to be found. The truly barren prairie, essentially treeless, stretched westward from the margin of the agricultural core region to the Rocky Mountains.

This gradual east-west transition from heavily forested land to grassland was encountered by waves of American settlers, and it was an indication of the average annual precipitation they could expect. The relationship is not perfect, for soil and groundwater conditions also govern tree growth even in areas of comparable temperature patterns. Small areas of prairie, for example, were encountered in Ohio and Kentucky, and the Prairie Wedge, a zone of grassland where trees might be expected on purely climatic grounds, reached well into central Illinois. But, in general, the presence of trees indicates adequate moisture for crop growth. Except for the northwest corner of the region and a few sections of Michigan, Ontario, and eastern Wisconsin, the entire agricultural core receives an average of more than 75 centimeters (30 inches) of precipitation each year. The southern margin of the core can expect in excess of 100 centimeters (40 inches) each year.

More important, most of this annual precipitation occurs between the end of April and the beginning of November. Thus, the moisture is received primarily as rainfall during the growing season. Also important to plant growth, the variability of this rainfall over a 10-year period is low. Farmers in the more humid sections of the region are less likely to suffer from devastating droughts than are agriculturalists working in drier sections of the continent. Summer rains do often come in the form of intense thundershowers, occasionally accompanied by damaging hail and high winds, but even in this the region's farmers are less likely to be economically crippled than are those located on the open plains. Rainfall amounts, frequency, and timing, therefore, are all beneficial to the Deep North economic structure.

Crops harvested each year also need sufficient time to grow to maturity before the first killing frost. As the in-between location of the agricultural core would indicate, the region has neither the long growing season of the South nor the short period of crop growth of the Northlands, the latter too brief for reliable production by nineteenth-century farm technology. The average date of the last killing frost in the spring ranges from mid-April across the southern portions of the agricultural core to mid-May along its northern margins. The first fall frost is not expected until late September anywhere in the region and may not arrive for several more weeks in southern Illinois, Indiana, Ohio, and northern Kentucky. The growing season regularly extends over a full four to five months throughout the region, a length sufficient to permit flexibility if late spring rains postpone the customary start of cultivation and planting by a week or two.

In concert with an adequately long frost-free period, growing season temperatures must also be high enough to stimulate rapid crop growth. The agricultural core's interior location in North America is the basic reason why a region at this latitude possesses these temperatures. Like other interior regions, the Deep North has a climate referred to as continental. A *continental climate* is characterized by a wide range in its temperatures through the year. The coldest winter temperatures at a given latitude are often as low as those occurring much farther north at sites closer to a moderating maritime influence. Similarly, summer temperatures can be expected to climb as high as those found at more southerly latitudes. At Peoria, Illinois, for example, located near the center of the Deep North, the average temperature in January is −4°C (25°F), while the average in July is 24°C (76°F). The average annual temperature range, calculated throughout the region as the difference between the mean cold month and the mean warm month temperatures, clearly demonstrates the continental character of this region's climate (Figure 12-3). For the agriculturalist, the high summer temperatures encourage rapid crop growth. As Hart pointed out, the average nonfarm resident of the region is less likely to appreciate the miserable combination of hot days, warm nights, and high humidity. While southerners experience long summers of much the same character, southern winters are milder and refreshingly short when compared to the Deep North's long, often gray, and uncomfortably cold winters.

Relief

Just as the agricultural core's climatic mix is highly appropriate for farming—neither too hot nor too cold, neither too wet nor too dry, and with an adequate frost-free period—the region's topographic relief is properly moderate. The landscape is gently rolling with few areas of either very flat or very hilly terrain. The low relief means that a very large proportion of the total area can be used for cultivation, and fields can be as large as practical for good management without a high risk of erosion. As farm machinery was developed, it could be used throughout the region. The scattered hillocks and stream courses that broke up the unending land swells were obvious locations to maintain woodlot or pasture. The persistently rolling landscape also permitted good soil drainage and, in most cases, restricted swamps to small areas. Where land was especially flat, as in the Great Black Swamp

Figure 12-3 Average annual temperature range. The effects of interior continental location and the modest moderating influence of the Great Lakes are evident on this map of the difference between the average temperatures during the coldest month and the warmest month.

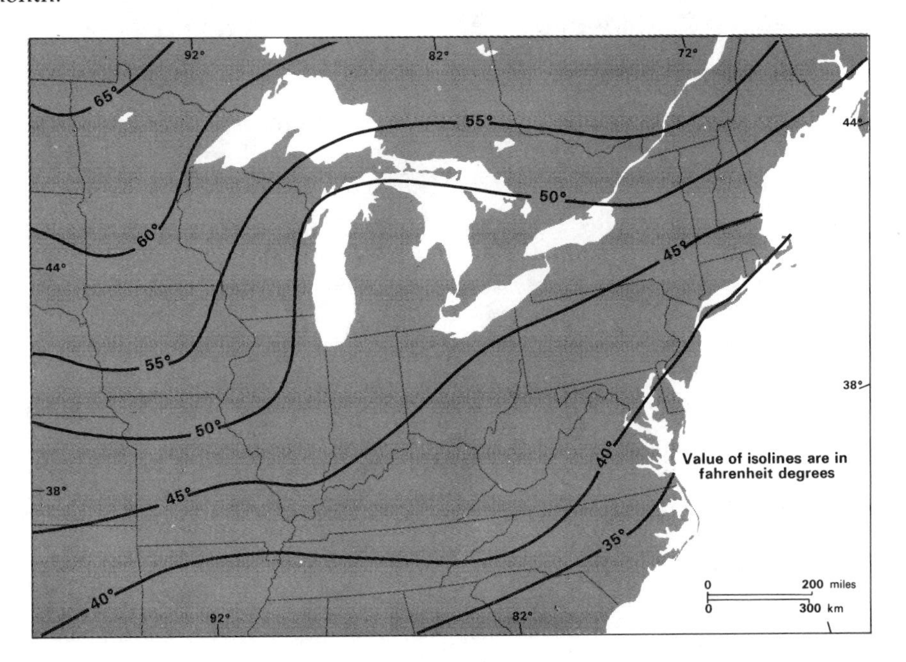

region of northwestern Ohio, or especially hilly, as in the plateau footlands of the Appalachians and Ozarks, settlement was postponed until the abundant "easier" land was taken. Like the climate of the agricultural core, then, the landscape relief was intermediate and highly suitable for good farming.

The highly practical, though unimpressive, landscape that dominates the Deep North is largely a consequence of the same glaciation that created the harbors of Megalopolis (Chapter 5), and the mechanisms of landscape formation were the same. As the heavy ice mass spread outward from its Canadian Shield center, soft sedimentary hilltops were ground down by the weight and movement of the ice. The debris, or *glacial till,* removed in this way became incorporated in the ice sheet, gradually settling out and partially filling the valleys between the decapitated hills. As the glacier fronts later retreated, long, low hills of this debris, the glaciers' terminal moraines, remained to offer a few lines of slightly greater relief for the human populations that followed. The tremendous quantity of meltwater released by the glacial retreat eroded several major river outlets, such as the Illinois River west and south from Lake Michigan and the Mohawk-Hudson river valleys east and south from Lake Ontario. The higher surface altitude of the Great Lakes during this period submerged large areas of what is now dry land south of Chicago, south of Saginaw Bay in Michigan, and the Black Swamp lake plain extending from Toledo, Ohio, to Fort Wayne, Indiana. Much of the land in these areas, especially the last, is a *lacustrine plain* (former lake bed). It is extremely flat and not far above the local water table. While the soils there are deep and rich, settlement was forestalled until effective ground drainage could be developed.

Immediately south of the Ohio River in north-central Kentucky is a large basin many would argue does not belong within the continent's agricultural core. Beyond the limit of glaciation, the Bluegrass Basin, or Bluegrass Plains, nevertheless does extend the region of low relief and highly productive agriculture into the margins of the Appalachian Plateau. The low, rolling relief of this region is primarily a residual *karst* terrain, which is one that has developed over thick limestone bedrock. The limestone is gradually soluble in moving water and permits many major surface features to be worn away. The limestone is also dissolved underground and forms stalactite- and stalagmite-columned caves that can extend for miles. The Mammoth Cave complex southwest of the basin is probably the best known in this region. Small solution holes (*sinkholes*) are often visible at the surface, but the countryside has the same prevalent rolling appearance found throughout the glaciated areas.

The lifestyle emulated by horse farm owners in the Bluegrass Basin of Kentucky offers one image of the old Southern aristocracy in modern trappings. (Kentucky Department of Travel Development)

Soils

The soils of the agricultural core can be described in much the same qualitative terms as the region's climate and relief: they are good, often much better than average, but usually not excellent. As discussed in Chapter 2, soil research has distinguished hundreds of soil types in North America, each varying in terms of their sand-clay-silt content, their water-holding capacity, their acidity, and numerous other variables. By taking a broad view, soils of the agricultural core have been grouped into two basic types. With the major exception of central Illinois and south-central Wisconsin, soils east of central Iowa are alfisols—almost all of which are in the udalf subcategory (see Figure 2-5). Western soils within the region and those through much of Illinois are mollisols.

Alfisols are formed under conditions of moderate moisture and usually in association with coniferous or mixed forests. Although the thin surface soil *horizon* (the A-horizon) is deficient in humus, the B-horizon has not been heavily leached of agriculturally important minerals. In general, the udalfs found throughout the eastern Deep North only require careful plowing (to avoid erosion), some form of crop rotation, and application of agricultural lime (to neutralize the mild soil acidity). With these relatively modest practices, the land remains productive.

Mollisols are formed under different conditions and are the most fertile major soil group in the region. These soils were formed under grasses rather than under forest cover. They range in color from dark brown to almost black, indicating a very high organic content. They also tend to be rather deep, with the surface horizon between 50 and 150 centimeters (about 1.5 to 5 feet) thick. The high organic content is a result of the grasslands' dense, tangled root network. Contributions of dead roots and grass leaves to the matted soil surface keeps organic content high. Mollisols, among the most fertile of all soils, are naturally suited to grain production.

The major soil exceptions to these two broad categories are the alluvial soils, found within the main river valleys and former lake beds, and the swamp soils. Both types of soil are capable of high fertility but often require special treatment. As mentioned earlier, settlement of the Black Swamp in northwestern Ohio occurred late, about two decades after land around its periphery. Today, the area is intensively cultivated and highly productive, but deep drainage ditches around the fields' rims continue to indicate the original soil moisture conditions. Throughout the larger region, glaciation strongly altered prior natural drainage patterns and led to a jumbled landscape of low relief and poor drainage. Many small swamps, lakes, ponds, and bogs are scattered in an uneven pattern and are especially characteristic of Wisconsin and Minnesota. In most cases, these wet areas have not been drained because of the expense involved and the more than adequate quality of the remainder of the region's soils.

Natural Accessibility Network

The waterway pattern in the agricultural core region is excellent. In this factor, the natural environmental conditions provided settlers with highly beneficial transportation opportunities. The natural network of accessibility was supplemented by land connections, some of them linking the region directly with the eastern seaboard population centers. Even prior to the railroads and further development of the road network, however, the river and lake connections within the region permitted easy and inexpensive shipment of goods to these same markets as well as to the main international trade ports. Much of this has been discussed earlier in terms of the continent's manufacturing core region (Chapter 6), but it bears further emphasis because river and lake shipping were very important to the region's agricultural development.

Movement of settlers into the region was earliest along the larger waterways (see Figure 3-1 and 3-2). The southern Great Lakes, the Ohio River, the Illinois, Wabash, and Wisconsin rivers to the east of the Mississippi and the Missouri River westward to Kansas City all provided major routes of entry for settlers and major routes for marketing their produce. The eastern Great Lakes offered more direct shipment through the Mohawk-Hudson routeway to New York City. The entire interior river network funneled into the Mississippi system and, because of the low relief and regular rainfall, was navigable by small boats and barges with very few interruptions throughout.

Detroit grew as a military control point and focus for farm products. This city, whose name literally means "the narrows" in French, is located at the best crossing point between Ontario and Michigan and is also near the entry of the northern lakes into Lake Erie. The southern Michigan hinterland was not as rich agriculturally as that of northern Ohio, however, and Cleveland remained more populous until after 1910 when mass-produced automobile industries transformed the Motor City's economic structure. Located at the Great Bend of the Ohio River, Cincinnati became the main collecting and shipping center for agricultural products from the southeastern portions of the agricultural core as early as 1820 and never relinquished its dominance over that section of the river's trade. Kansas City, at the junction of the Kansas and Missouri Rivers, also experienced early urban growth by handling large quantities of agricultural products in river transit. As discussed earlier, Chicago's location was also beneficial to rapid growth. Near the southernmost end of Lake Michigan, it also lies only a short land distance from the upper Illinois River. The transshipment opportunities provided by this site were supplemented by extensive canalization projects and by land connections built west and south across the rich agricultural core and, later, eastward directly to the growing cities of Megalopolis. The opportunities for industrial growth in these centers are clear. Their initial growth, however, and the growth of many other urban places in the region was a direct consequence of the high-quality agricultural landscape and the natural network of accessibility provided by the region's waterway system.

THE AGRICULTURAL RESPONSE

As the settlement frontier gradually moved westward across the region during the early nineteenth century, it was accompanied by a wave of wheat production for eastern markets. Wheat was a high-value crop, one that was in demand in the population centers, and one that was familiar to both producer and consumer. The bulk of raw wheat was not a great problem for shippers while water transport was continuous, but flour milling soon became established at the points of embarkation (such as Cincinnati on the Ohio River) or sites at which the grain was to change from one transport mode to another (such as Buffalo at the lake terminus of the Erie Canal). Continuous wheat farming was hard on the region's soils, however, and primary zones of production moved westward with the expanding line of settlement.

For farmers who remained behind as wheat production moved on, the next best agricultural product was meat from domestic livestock. Both cattle and hogs were raised, with the importance of one over the other varying from year to year. Hogs grow to market size very rapidly if supplied with sufficient feed; even grain-fed cattle reach a "finished" stage more readily than animals raised on pasturage alone. The intensive hog-raising activities of southern Ohio farmers earned Cincinnati the nickname of Porkopolis by the 1830s. So economically reliable was feed grain and livestock farming that it quickly supplanted wheat production as the dominant farming system throughout the agricultural core. The system was also well suited to long-term operation within the limits of the natural environment, and it could take advantage of the natural waterway network. Alternative farming systems were tried occasionally, but none was as persistently successful.

Corn was the grain that best met the combination of environmental requirements and high economic return. Well adapted to a humid summer climate, corn thrived during the region's long hot days and warm nights. Since even corn could not be grown year after year on the same soil, a three-year crop rotation system became common as early as 1820 in southwestern Ohio; a small grain (usually wheat or oats) and a hay crop (such as clover or alfalfa) were alternated with corn in each field over the rotation period. If a fourth year were needed in the system, the land could be allowed to rest in *fallow* under one of the recuperative grasses and perhaps pastured lightly. Under this system, the land is always contributing in some direct way to livestock production or cash-crop production.

Given that corn is ecologically suited to the region's climate and soils, it has additional economic benefits over other feed grains. Grain yields are high since plants can be grown close together, and each plant produces two or more ears of grain. Furthermore, the large quantity of vegetative matter produced by each plant can also be used as feed with appropriate supplements and cutting. Stalks and leaves were chopped finely and stored in large *silos* to permit fermentation and provide winter feed for the farm's animals. No other plant yielded such quantities of usable feed, and its dominance spread to the limits of environmental support and beyond. It may sound strange for a crop to be grown beyond its environmental limits, but along the northern margins of the Corn Belt, as the region came to be called, where summer temperatures do not permit full ripening of the grain, the plant is cut green and chopped for *silage*.

Throughout the continent's agricultural core, then, a mixed farming operation of crop-livestock production has provided farmers with economic security beyond that found in any other major agricultural region. Localized specialty crop production elsewhere in the United States and Canada can yield higher economic returns, but only in the agricultural core are relatively high returns consistently achieved by so many farmers. Farming by this system is very demanding, for three to four crops plus livestock must be handled correctly throughout the year. It has been suggested that the Deep North respect for, and faith in, hard work can be traced to the rewards of Corn Belt farming that were pretty well assured as long as the system's rigorous organizational and labor requirements were followed.

Although corn has until recently been the most important crop grown in the agricultural core, whether measured by value or volume, its distribution of production is not even (see Map Appendix). Corn is grown in substantial quantities from central Ohio to central Nebraska. Its southern margin of production follows the irregular relief of the Appalachian foothills in Kentucky and southern Indiana and the Ozark highlands in southern Missouri. Although it could be grown farther south on the rich alluvial soils of the Mississippi River valley, it is replaced there by the higher returns of cotton, rice, and more recently soybeans. The northern margins of the Corn Belt extend from the southern Ontario Peninsula across south-central Michigan and Wisconsin to the subhumid plains of southwestern Minnesota. The areas of most intensive corn production remain the fertile mollisol regions of Illinois and Iowa. When comparing the distributions of cash-grain and livestock farming in the region, those areas most suited to crop production appear to have fewer farms strictly engaged in raising livestock. Even in this, however, both specializations occur near one another; the livestock farms are the prime consumers for the specialized corn farmers.

A distinctive characteristic of the central Deep North landscape is a semiregular rectangular field pattern. Although not as rigid as that found across the Great Plains states, the rural geometry of the region is stubborn in its consistency. The original 13 states had developed their internal administrative and ownership boundaries in an unsystematic manner, indicating that settlement preceded governmental organization. The *metes and bounds* system of lot designation relied on visible landscape features, compass directions, and linear measurement to define land parcels. The irregularly shaped results were often subject to confused interpretation and litigation. Through the Ordinance of 1785, the land north of the Ohio River and west of Pennsylvania, known as the Northwest Territory, was delimited according to the regular rectangular *township and range* survey before it was opened to

The square township and range land survey system is portrayed dramatically in this vertical aerial photo. As roads were constructed along the edges of many of the 1-square-mile sections, the regular survey pattern was reinforced and remains little affected by transfers of land ownership. (Aerial Photography Field Office/U.S.D.A.)

settlement by the newly independent United States farmer. A series of east-west baselines and north-south meridian lines were drawn, and township squares 6 miles (9.6 kilometers) on each side were laid out with respect to these lines. Each township was further divided into 36 sections of 1 square mile (2.59 square kilometers) each (Figure 12-4). The 640 acres (259 hectares) in each square mile was subdivided into quarters of 160 acres (64.75 hectares). These quarter-sections were originally designated the minimum area that could be purchased for settlement (at $1.24 per hectare—50 cents per acre), but the minimum was later reduced to 80 acres (32 hectares) and, still later, to 40 acres (16 hectares). The irresistible logic of this system remains visible in the predominantly rectangular road network of most of the United States between the Appalachians and the Rockies. It is also brought to general awareness by Holly-

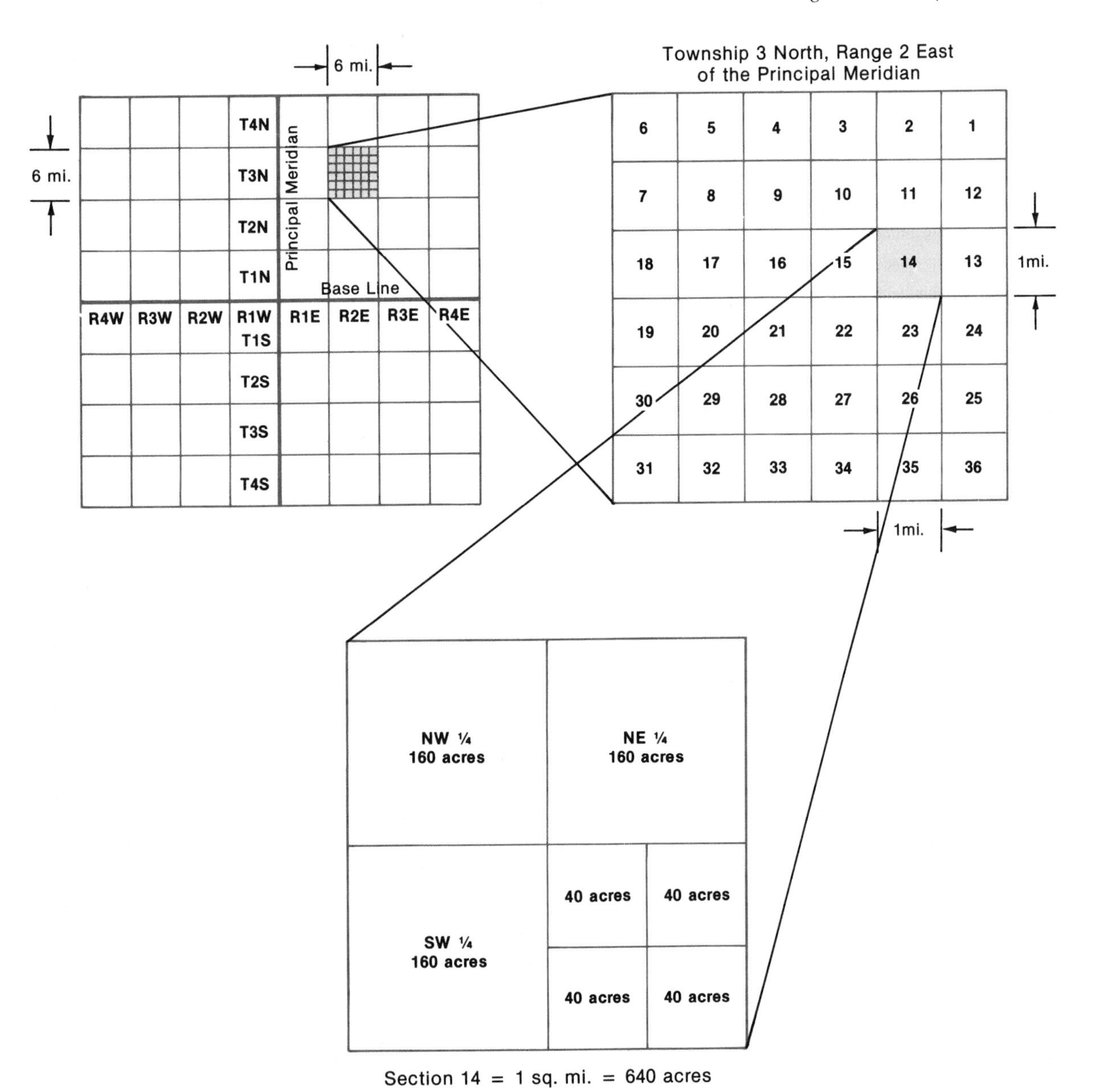

Figure 12-4 Rectangular land survey system. By dividing the land into square-mile sections of 640 acres each, the federal government hoped to standardize the identification of parcels so that land ownership and land transfer would also be more "rational" and result less often in legal disagreements.

wood's dramatic references to "trouble on the south forty," although the statement is usually uttered by a struggling pioneer farmer on the western Great Plains rather than in the Deep North.

While the land survey system and the ecological and economic realities of the Deep North produced a kind of inevitable homogeneity to the landscape, there are portions of the agricultural core that lie beyond the Corn Belt. In Wisconsin and central Minnesota, north of the centers of grain production where the climate prevented feed grain maturation, farmers chose dairy farming as an economic substitute. Corn in silage, other grains such as oats and barley, and abundant hay crops provided excellent support for large dairy herds. When the supply of fresh milk exceeded even the large demand of the nearby cities, it was converted into butter and cheese for more leisurely shipment to more distant markets. This pattern of farm production was not at all unfamiliar to the German and Scandinavian immigrants who settled the area. A similar belt of dairy farming is located on the Ontario Peninsula north of the corn- and mixed-farming region. Within the United States, Wisconsin continues to produce a large proportion of the nation's surplus milk and approximately half of the country's entire cheese output (see Map Appendix).

Another distinctive extension of the agricultural core's boundaries beyond the Corn Belt occurs around the western Great Lakes. Just as in the fruit-growing region of the Niagara Peninsula (Chapter 7), fruit production is possible far to the north of its inland latitude in a narrow band along the Lake Michigan shores of Wisconsin and especially Michigan. The moderating influence of the lake retards fruit tree blossoming in the spring, usually until after the last frost, and also retards the arrival of the first killing frost in the fall. Sour cherry, apple, and to a lesser extent, grape production are all important. A similar effect is found along the southern shore of Lake Erie, especially the few coastal counties in Pennsylvania and western New York, where grape production has been significant for more than a century.

CHANGES IN THE PATTERNS

The entire region was pretty well settled by 1890, and the mixed-farming system based on corn and livestock that had been found to work well in southern Ohio was carried west to the edge of the Great Plains with only local adjustments. Early technological improvements, such as the reaper (1831), the steel plow (1837), and other devices suited to the region's chief economic activity, tended to ensure the system's success. More recent changes, however, have stimulated modifications to the traditional geographic patterns.

Soybean Substitution

One of the more subtle changes in Corn Belt patterns lies in the rise of one crop's importance since World War II. During this period, soybeans have been grown in increasing acreages throughout the United States. As late as 1925, less than 200,000 hectares (one-half million acres) of soybeans were harvested in the country. By 1949, soybean acreage had increased to about 4.5 million hectares (11 million acres) with almost 3.7 million (or 9 million acres) of those in the agricultural core and most of the balance immediately south in the lower Mississippi River valley (Figure 12-5). During the next 20 years, soybean acreage exploded to 16.1 million hectares (41 million acres): plantings in the agricultural core exceeded 10 million hectares (25 million acres) and in the lower Mississippi River valley increased to nearly 4 million hectares (10 million acres). Nationwide acreage had reached a peak of 25.4 million hectares (64.8 million acres) by 1982. Market demand for soybeans remained greater than supply throughout the period of acreage growth, and it began to be described, perhaps too enthusiastically, as the miracle food of the future. Competition in the

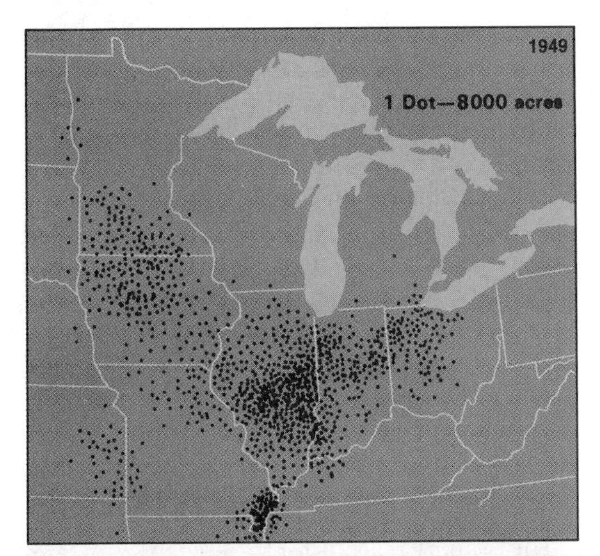

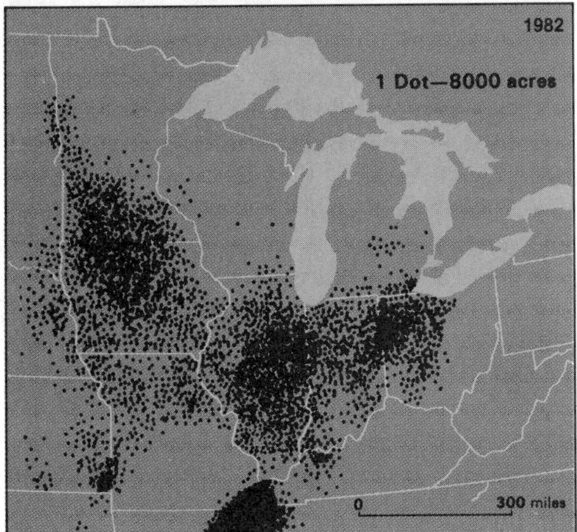

Figure 12-5 Soybean acreage, 1949 and 1982. The decades during which soybean production expanded are clearly evident within what had traditionally been a farming region dominated by corn.

world market from other countries during the 1980s, most notably Brazil, led U.S. farmers to cut back on acreage somewhat and reduced the total area of the 1987 soybean harvest to 22.2 million hectares (56.4 million acres).

The reasons for the tremendous increase in soybean production are several. First, as a *legume,* soybeans act as a soil reconditioner by increasing the nitrogen content of the soil in which they are grown. Second, the environmental requirements for soybean production are rather broad. While vulnerable to some local conditions or climatic variations, soybeans generally may be grown throughout most of the eastern United States, parts of Ontario, and even in areas receiving less than 20 inches of rainfall if irrigation is feasible. Third, soybeans can be used in a variety of ways. In this they are similar to corn. The bean itself can be eaten directly or milled to produce an edible vegetable oil and a meal low in fat but very high in protein. The meal has been used primarily as a livestock feed supplement, but an increasing amount has entered human consumption patterns in place of some grains, as milk substitutes and as meat helpers. And, fourth, the world food and feed situation maintained export demand for soybeans at high levels as well. This has kept prices relatively stable, an encouraging condition to farmers. Only increased production in other parts of the world, such as recent expansion in Brazil, has tempered soybean's tremendous growth in the United States.

This combination of advantages has concentrated a great amount of soybean production in the agricultural core. The traditional three- and four-year rotations gradually gave way to a two-year corn-soybean rotation. In some cases within southern portions of the core, early maturing varieties of soybeans can be planted in the late spring after a harvest of winter wheat, giving the farmer three crops (corn, wheat, soybeans) every two years without significant loss of productivity in any year. The large, almost level fields of the agricultural core that permitted early mechanization of farming operations are well suited to soybean cultivation. Much of the same equipment can be used and the high returns per work-hour contributed to the early acceptance of the crop even before the market developed as intensely as it has. By 1971, harvested acreage of soybeans exceeded that of corn in southern Illinois and northern Missouri and was greater than two-thirds of the corn acreage in western Ohio and the Grand Prairie region of east-central Illinois. In 1979, for the first time ever in the United States, more acreage was planted in soybeans than in corn within the traditional corn-producing sections of the agricultural core. Although the slight decline in soybean production since 1982 may return hay to Deep North crop rotations, soybeans have proven their worth and will remain significant in the region. Once this is recognized more broadly outside the agricultural community, the region may be referred to more accurately as the Corn-Soybean Belt.

Mechanization and Farm Size

A more complex set of changes in agricultural core geography is built on new levels of mechanization and alterations in average farm size. As discussed earlier, the original land survey in the region set the minimum farm size that could be purchased at 160 acres, and then at half or one quarter that amount at various times. After the initial purchase, of course, parcels of land could be broken up and sold in even smaller lots or added to previously established farms. Hart determined that the amount of land in various farm-size categories remained very stable during the first third of this century and argued that this stability was a continuation of farm-size patterns set during the original purchase period.[3]

Whatever the case, by 1900, farm size in the Deep North states showed marked variation: about one-third of the farms were of 73 to 202 hectares (180 to 499 acres), another third were of 40 to 72 hectares (100 to 179 acres), and most of the remainder were smaller than 40 hectares (100 acres) (Table 12-1). The amount of land in farms smaller than 73 hectares (180 acres) began to decline after 1935; the greatest increases occurred in farms larger than 105 hectares (260 acres). By 1964, more than 50 percent of the farmland in these states was in farms larger than 105 hectares (260 acres); fully 1 hectare in 5 was located on a farm larger than 202 hectares (500 acres) in area, a trend that has continued. At the same time that large farms were increasingly important in the region, small farms were disappearing rapidly. Less than 10 percent of the states' farmland remained in farms smaller than 40 hectares (100 acres) by 1964, and only 5.8 percent was in farms smaller than this by 1987. The largest farms were located in a wide band across central Illinois and northern Missouri and generally along the western margins of the region. Smaller farms were more eastern in location—primarily in Ohio, southern Michigan, and eastern Indiana and Wisconsin.

The primary reasons for these changes in farm size are economic and related to mechanization of operations. Deep North farmers traditionally have taken advantage of mechanical innovations to increase their output per work-hour. The large fields and gentle terrain in this region permitted early and continuing use of farm machinery that would have been impossible on smaller farms and erosion-prone hill farms. Individual farmers were able to operate over areas much larger than would have been possible if more hand labor had been required.

After about 1935, however, several changes occurred in the degree of mechanization. These changes had significant consequences for the form of farm operation in the region and eventually for general farming thoughout the United States and Canada. The worst of the depression years were followed by a variety of government-backed rural development schemes; rural credit was loosened, price support and cropping controls were legislated, and public services were extended into the countryside. Tractor-powered machinery became more common on all but the smallest farms, many of which were already marginal.

This application of technology to the agricultural production process is no different than that which occurs in manufacturing, and the results have been the same. Economies of scale favored

---

[3] Ibid., p. 273.

Table 12-1 Proportion of Farmland in Farms by Size Category in Eight Agricultural Core States, 1900–1987

Size Category (Acres)	1900 (Percent)	1935 (Percent)	1964 (Percent)	1974 (Percent)	1987 (Percent)
0–49	6	4	2	2	2
50–99	18	15	7	5	4
100–179	34	35	19	14	10
180–259	36	21	19	14	10
260–499		19	33	31	25
500 or more	6	6	20	34	50

Source: Calculated from John Fraser Hart. "The Middle West," *Annals of the Association of American Geographers,* Vol. 62 (June 1972), Figure 16, p. 273; U.S. Department of Commerce. *1974 Census of Agriculture* (Washington, D.C.: U.S. Government Printing Office, 1978): and U.S. Department of Commerce. *1987 Census of Agriculture* (Washington, D.C.: U.S. Government Printing Office, 1989).

medium and large farms during the early years of farm-size growth. A labor shortage during the early 1940s accelerated the mechanization process, and innovations became oriented increasingly toward large-scale operations. Two- and four-row equipment gave way to six- and eight-row equipment. Storage and shipment operations also became mechanized and more and more attuned to the requirements of large-volume producers. At each stage of improvement, the main benefits could be achieved by the largest producers with fewer benefits associated with medium-size operations. The squeeze on small-scale farmers was completed from the other end as well since larger operators could afford the equipment that would give them the greatest cost efficiency. One consequence was the gradual accumulation of farmland into acreages that could use the new large machinery effectively. Even here another size advantage appears. As farm operations have become larger and more expensive, the demand for land to permit these operations also increased. Land prices increased, and the larger farmers were those best able to obtain credit or afford additions whenever land became available for purchase.

Accompanying these changes in farm size, the amount of land farmed in the region declined gradually. Not all land formerly owned by small-

Large, nearly flat fields in the agricultural core can be managed by small numbers of farmers as long as they have enough equipment. The first rows of corn harvested in this field lie straight outward from the sheltered farmstead located near the western margin of the region. (USDA)

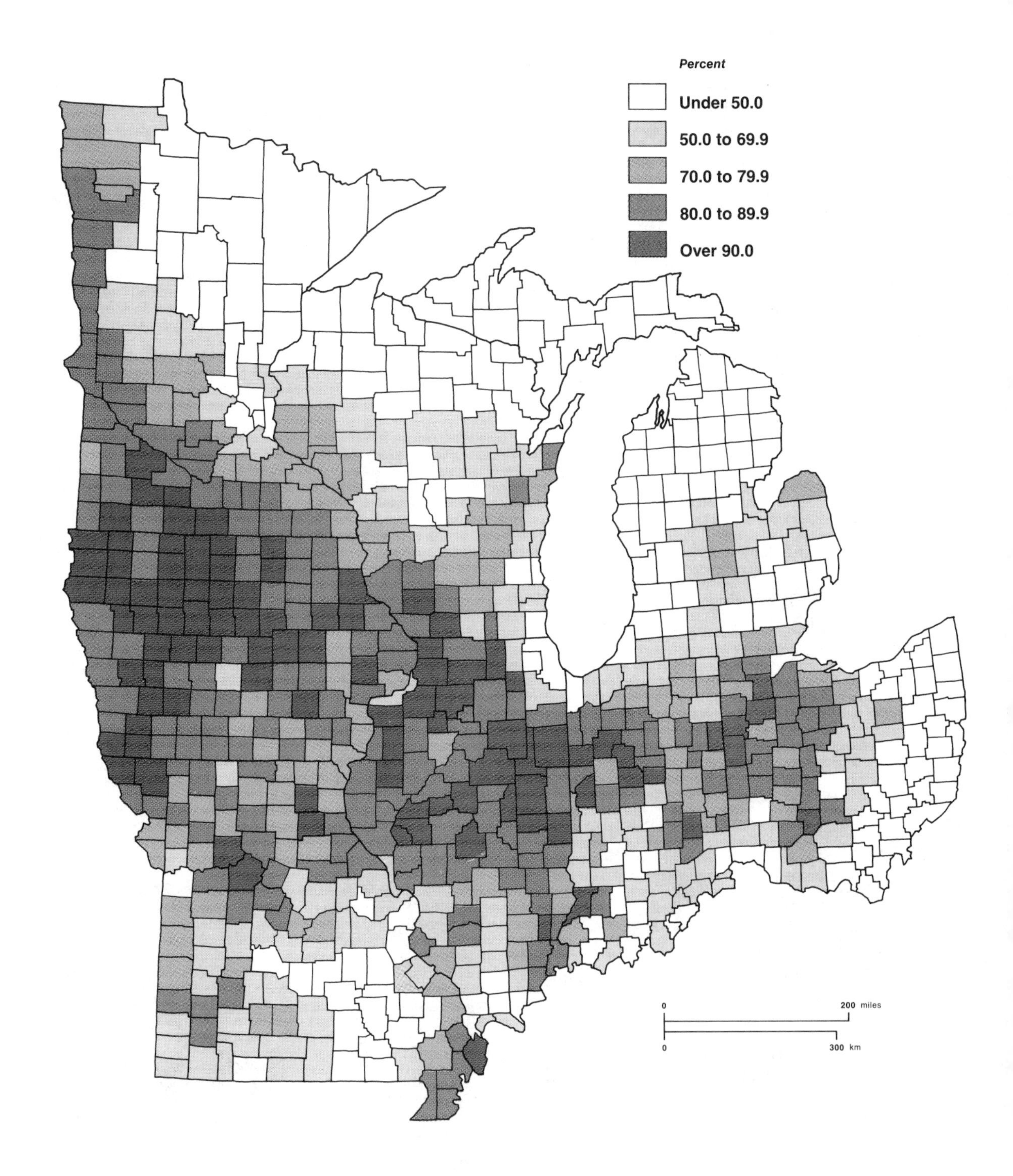
Percent
Under 50.0
50.0 to 69.9
70.0 to 79.9
80.0 to 89.9
Over 90.0
0
200 miles
0
300 km

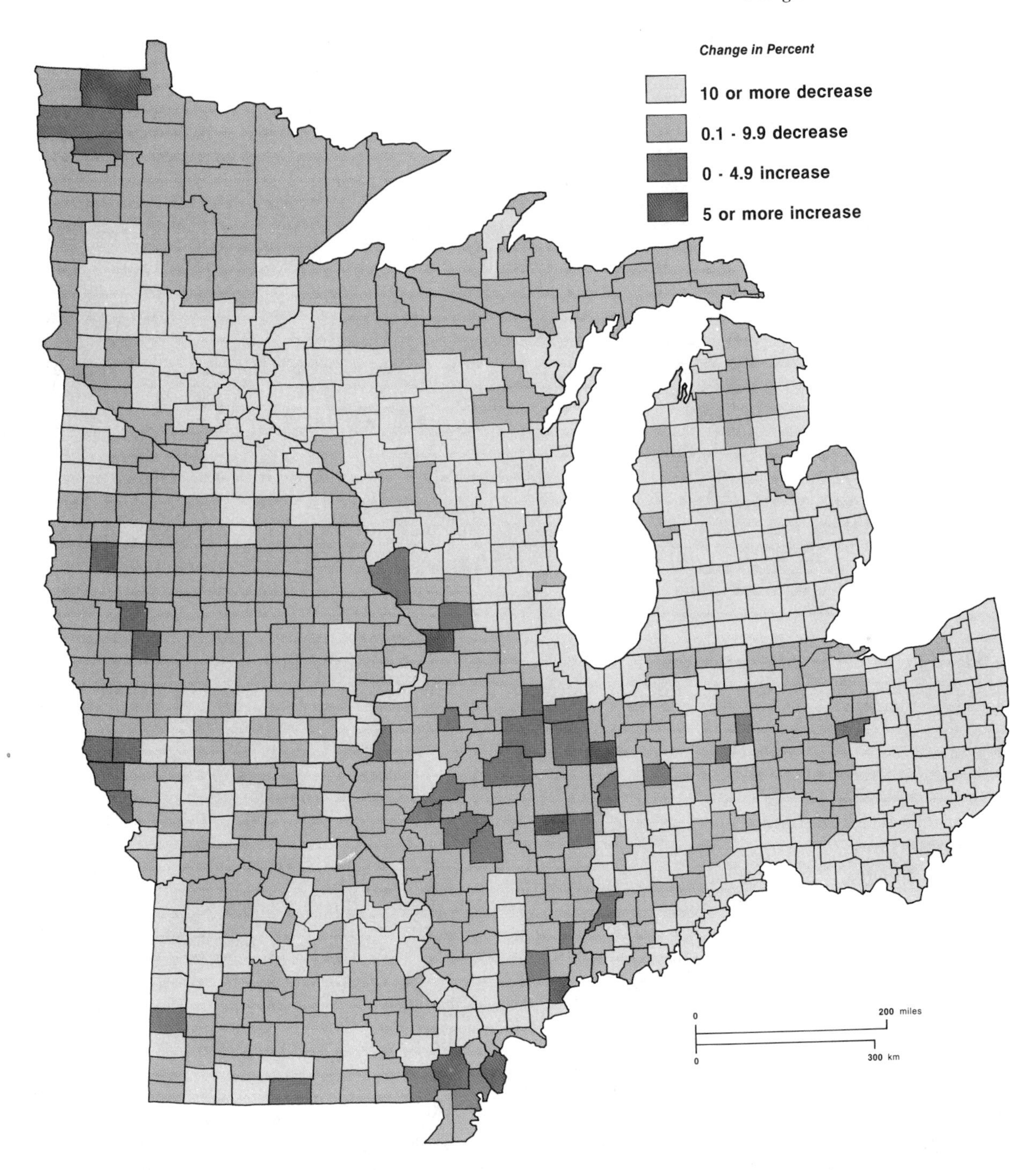

Figure 12-7 Change in percentage of land in farms, 1959–1987. The areas of greatest decline in farmland are those around the edges of the most intensively farmed portion of the region. This suggests that marginal land, satisfactory for farming at the beginning of the period, was taken out of farming as investments were concentrated on better land.

scale farmers was aggregated into larger farm operations. The proportion of county land in farms (Figure 12-6) was above 80 percent across much of the region in 1987, with most of Iowa and Illinois still showing rates above 90 percent. Even so, the overwhelming number of counties in the Deep North had experienced a reduction in land in farms across the previous two decades (Figure 12-7). Only in central Illinois and portions of Iowa within the region and in the Red River Valley in northern Minnesota were there significant clusters of counties showing growth in farm area since 1959. Granted, it is unlikely that counties already having more than 90 percent of their land in farms in 1959 could increase that proportion, but the general pattern was one of a moderate decline in farmland. The greatest reduction in proportion of land in farms occurred around the margins of the region we defined as the Deep North. This supports the argument that a major factor in the overall decline is a consolidation of successful efforts onto the better land.

During the early 1980s and continuing, at a slower pace, throughout the decade, many farm families in the region were forced to leave what had been a way of life for generations. Not every such experience was simply the result of a marginal operator's inability to increase the scale of his or her farm operation, as discussed. Some successful farmers borrowed heavily in the late 1970s, at the very high interest rates of the time, in order to finance the technological change or to lease or buy additional land necessary for larger operations. When the economy slowed and both interest rates and prices for farm products fell, a number of these families were unable to meet their schedule of loan payments. Their equipment and their farmland were lost. Farming has never been an easy or certain business, but those in the agricultural core had developed a distinctive farm culture to take advantage of the many environmental advantages in the region. This experience in the early 1980s, however, showed that even farmers in the heart of the agricultural core could not ride out wide economic fluctuations while attempting to adapt to the necessary long-term structural changes underway in North American farming.

The Family Farm and Ownership Changes

Deep in American folklore, and to some extent in Canadian, is the almost sacred notion of the family farm. Individually owned and family operated, the farm was passed down from generation to generation with only a succession of hired hands included in what remained fundamentally a family operation. A gradual increase in rural nonfarm population did nothing to disturb this image, for was it not the family farm that embodied much of the independence and proximity to nature that was missed by the large numbers of urbanites? Although they knew this image to be exaggerated in some ways and downright untrue in others, farmers of the Deep North fit the pattern until World War II as well as did those in any major region. During the more than 45 years since the war, it has become clear that the fully owned, individually operated family farm is in rapid decline. While still dominant in the United States and Canada, it is disappearing most rapidly from the agricultural core states. The relatively self-sufficient dairy farms in Wisconsin—frequently with an orchard or two, a vegetable patch, a few hogs and chickens to complement the cows, and even a rollicking mixture of children and collie dogs—probably approach the nostalgic image better than do farms in most areas. If the current trend continues, however, only a small proportion of the significantly productive farms in the Deep North will fit the cultural image by the end of the century.

The decline of the family farm is directly associated with the demands of increased farm efficiency. When it was defined by hard work, alertness to detail, and rigid adherence to traditional methods that had proven themselves reliable, efficiency was controlled directly by the farm family involved. As the cost advantages of

Dairy farms in the Deep North, such as this one in Ohio, provide a bucolic landscape that contains all the elements of the region's culture. (Erwin W. Cole/U.S. Soil Conservation Service)

large-scale farming became much more important, national and even international economic swings began to intrude on traditional approaches to Deep North farming. The ability of the farmer to obtain credit on favorable terms, to survive fluctuations in bank interest and crop prices when loans were taken, to use large equipment and to use it on larger farms, shifted the bases on which farm efficiency and agricultural success was established. Personal effort and individual integrity still contribute to a farm's success, but the factor of scale is increasingly critical. Larger operations do better on the average, and midsize family farms are challenged not to follow the smaller farms into extinction. It is difficult to anticipate the long-term impact of this change on Deep North culture.

There are several ways in which Deep North farmers, and those in other sections of the United States and Canada, have already adapted to these changes. As the need for more land per operation increased, some farmers found it feasible to rent or lease additional land rather than purchase it outright. The rented land can be part of a nearby farm or several sections of farmland in scattered locations and still be accessible by the well-developed network of paved rural roads (Figure 12-8). The rented land may continue to appear in the census as a separate small farm operated by what is called a part owner, thus masking the intensity of small operation decline. Other farm operators may be full tenants, choosing to work for the landowner through one of several arrangements. As a U.S.

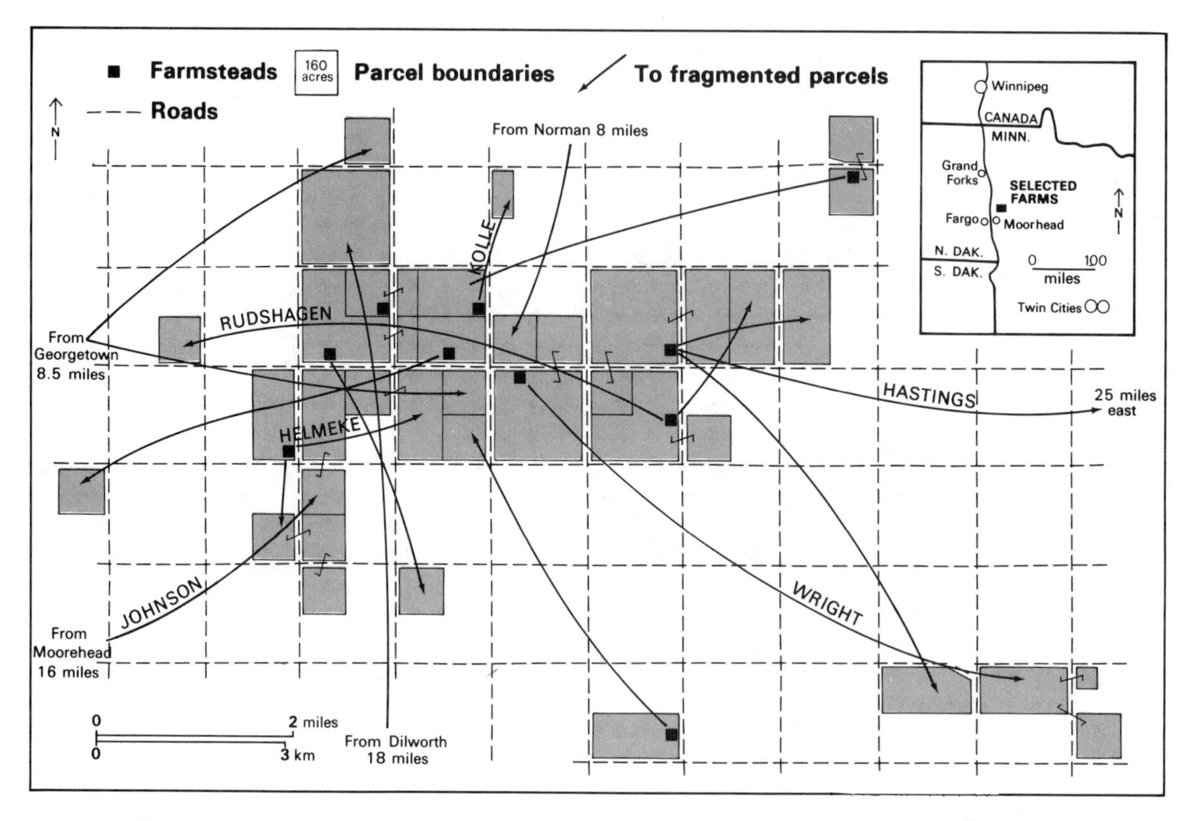

Figure 12-8 Associated parcels on fragmented farms. (Reproduced by permission from the Annals of the Association of American Geographers, Volume 65, 1975, E. G. Smith, Jr.) This example clearly demonstrates some of the scattered, noncontiguous land parcels owned and worked by individual farmers. The pattern has increased since this study was done.

Department of Agriculture study pointed out,

> Farmland investment by nonoperator landlords is similar to stockholder investment in corporations. However, the nonoperator landlord is not synonymous with the "Wall Street" type of investor. Evidence suggests the majority of these landlords are either retired farmers, members of farm families, or closely associated with agriculture through the small rural communities in which they live.[4]

In addition, about one-third of those renting farmland were leasing it from a relative, often as a means of transferring the land from one generation to another.

Within the agricultural core, tenure characteristics reflect the new economic reality. Of the eight Deep North states, the three that lie at the region's center had the lowest proportion of their farmland operated by full owners (Table 12-2). More than 7 acres in every 10 in Illinois, Indiana, and Iowa were operated under part-time or tenant-farming arrangements, a proportion greater than that found in any other state. Average farm-size differences also illus-

[4] Bruce B. Johnson, *Farmland Tenure Patterns in the United States* (Washington, D.C.: U.S. Department of Agriculture, Economic Research Service, 1974), p.4.

Table 12-2 Tenure Characteristics of Farms in Agricultural Core States, 1987

State	*Land in Farms*			*Average Farm Size*			*Market Value per Farm*		
	Full Owner	Part Owner (percent)	Tenant	Full Owner	Part Owner (acres)	Tenant	Full Owner	Part Owner ($1000)	Tenant
Ohio	33.5	54.9	11.5	107	353	198	22.6	82.3	51.1
Indiana	27.7	60.2	12.1	110	441	255	30.7	106.5	60.5
Illinois	19.3	60.4	20.3	141	530	337	33.0	116.7	75.2
Iowa	25.2	54.5	20.3	165	490	295	49.2	138.0	78.0
Missouri	43.8	46.4	9.8	181	519	304	20.4	68.8	43.1
Michigan	33.5	61.1	5.4	111	373	181	24.8	96.1	48.2
Wisconsin	41.4	51.2	7.4	157	338	194	42.4	105.3	64.9
Minnesota	32.0	56.3	11.7	188	511	296	42.3	104.8	65.9

Source: U.S. Department of Commerce, *1987 Census of Agriculture* (Washington, D.C.: U.S. Government Printing Office, 1989).

trate the tenure patterns; farm-size averages are much lower for full-owner operatives in every state. This is repeated again in the distribution of farm product market values per farm. The large accumulation of invested capital through part-ownership agreements has raised the average market value of owner-renter farm combinations to more than twice that of single-owner farms, even in Wisconsin. The ability of part owners to use fully the largest equipment over large average total farm acreages justifies the additional tens of thousands of dollars required by such large-scale farming.

The changes in farm ownership patterns from full ownership or tenant arrangements to part ownership have been striking throughout the region (Table 12-3). Between 1969 and 1987, the practice of renting land to farm (tenancy) has declined the most in the three central states in the core and in Ohio. Full ownership declined in every state, with 1 full owner in 11 no longer in this category in Missouri and almost 1 full owner in 3 in Wisconsin leaving full ownership since 1969. It is part ownership arrangements that have grown during this 18-year period. Except for the slower growth in Missouri, part-time farming in the agricultural core states increased considerably.

SETTLEMENT PATTERNS

The region's visible landscape contains more than large rectangular farm fields, livestock pens, and a network of straight roads. Houses dot the countryside in a distinctive pattern. Small market centers and service communities are located farther apart but still at regular distances from each other. The internal form of many of these small towns reflects the immensely practical survey system used across the rural landscape. The farm population living in

Table 12-3 Changes in Percentage of Land in Farms, by Tenure Characteristics, 1969–1987

State	Full Owner	Part Owner	Tenant
Ohio	−4.6	+9.8	−5.3
Indiana	−6.2	+11.6	−5.4
Illinois	−5.4	+14.4	−9.4
Iowa	−9.2	+15.2	−6.0
Missouri	−4.3	+5.7	−1.4
Michigan	−13.3	+13.7	−0.4
Wisconsin	−18.8	+19.1	−0.3
Minnesota	−9.5	+10.0	−0.5

Source: Calculated from U.S. Census of Agriculture, 1969 and 1987.

the region has been declining, as elsewhere in the United States and Canada (2.0 percent of the total U.S. population in 1987 was classified as in farming, down from 8.7 percent in 1960). A growing rural nonfarm population in many areas, composed of people who have left farming but continue to live in the countryside and others who have come out from the cities and towns to live in a rural setting, exceeds the loss of farm population. The settlement pattern of the Deep North is undergoing a process of change that promises to affect its traditional geography as strongly as increased mechanization has been altering the organization of farming in the region.

Spatial Reorganization and Transport Technology

The gradually declining farm population and the increasingly sophisticated technology at the disposal of the average rural resident have had a consistent and significant impact on village and small-town patterns throughout the United States and Canada. The consequences have been especially persistent in the agricultural core region. There has been an increase recently in the population of rural and small-town America, but the pattern for the Deep North is much less clear. Both trends, the national rural population decline (into the 1950s) and the reversal to rural growth everywhere except in the agricultural core itself since that time, are related to changes in transport technology.

During the 1930s and 1940s, rural roads were paved and individual farmers soon could afford a reliable truck to carry livestock or farm produce to market. Transport costs declined as transport efficiency increased. Similarly, the improved road network permitted more rapid movement; farmers could carry their own produce farther than they had been able to several decades earlier without a significant increase in time or money. As a consequence, farm produce began to be trucked past the local village to one of the larger market centers or livestock yards. Deprived of at least a portion of its consumer population, the smallest communities did not grow in population and in many cases began to lose what small population they had.

An examination of population change in 15 rural counties in northern Ohio and Indiana, between 1910 and 1960,[5] found that there was a direct association between village size and population increase: the larger the town, the greater had been the absolute population increase. For the very smallest communities, those with 1910 populations below about 1000, actual declines in population had often occurred.

In terms of the settlement pattern in the Deep North, therefore (1) the rural landscape was becoming less densely populated and (2) the population was increasingly clustered. Restating this in terms of time and distance, individuals of the 1960s were traveling farther but at no increase in travel time. The utility of the very small rural hamlet had declined because it no longer served much direct economic purpose.

The relationship among travel time, distance, and the process of spatial reorganization was studied by D. G. Janelle in a way that illustrates these changes.[6] He observed that in the past, a demand for accessibility frequently led to an improvement in transportation, usually through some technological development. Further, the transport innovations often improved access between points by reducing the direct cost of travel or the time required to complete the trip. The innovations may have been in either the vehicle of movement (railroad, automobile, truck) or in the routeway. By determining the average travel time between a number of towns and cities in southern Michigan, Janelle found that a trip that required almost 24 hours to complete by land transport in 1840 had been reduced to about 2½ hours by 1900 and about 80 minutes by 1965. In

---

[5] S. Birdsall, unpublished manuscript, 1962.

[6] D. G. Janelle, "Spatial Reorganization: A Model and Concept," *Annals of the Association of American Geographers*, Vol. 59 (June 1969), pp. 348–364.

effect, the *time-distance* between these places had shrunk greatly; they had become much closer together if we measure distance as travel time.

It was also determined, however, that the time-distance between places did not converge at the same rate. The larger the place, the greater the demand for improved access to other locations; small towns and villages do not generate as much interaction with other places as major metropolitan concentrations do. Also the populations of large cities interact most strongly with other large cities. Therefore, improvements in transportation are greatest between the more populated centers. In terms of time-distance convergence, large cities have grown closer together more rapidly than small towns and villages. As discussed earlier (Chapter 6), there are economic benefits to being located close together. Thus, the benefits of time-space shrinkage are enjoyed primarily by those places that are already the most dynamic economically: the larger places grew, while the very smallest communities lost their economic base and declined slowly in population.

A mild contradictory trend has appeared in much of the eastern United States and Canada but may not include the Deep North. There have been indications that since 1970 many rural areas and nonmetropolitan places were no longer losing population and some were actually growing. A mid-decade study of nonmetropolitan cities and towns concluded that there was a good deal of evidence for population growth in all but the very smallest villages, even though growth was more rapid in metropolitan areas. Furthermore, sharp differences appeared between villages in North Central states and Southern states. On the one hand, "most resurgence of growth in very small places appears to have taken place in the South," while on the other, "it is essentially in the North Central states, where there are hundreds of very small incorporated places, that the notion of 'dying' small towns comes closest to reality."[7]

Thus, although the Deep South and the Deep North are both strongly agrarian, they are very different regions. Hardships experienced by Southern farmers for a century or more may have been a prelude to regional population growth now that many of the former difficulties are slowly dissolving (Chapter 10). In contrast, the efficiency that has made Deep North farming so successful may also be the basis for continued population decline in the region.

---

[7] Glenn V. Fuguitt and Calvin L. Beale, *Population Change in Nonmetropolitan Cities and Towns,* Agricultural Economic Report No. 323. (Washington, D.C.: U.S. Department of Agriculture, Economic Research Service, 1976), p. 4 and p. 6.

AN URBAN CASE STUDY

CLARKSVILLE, IOWA

There are a large number of small towns and villages in the Deep North that could be used to illustrate the characteristics of the region. The village of Clarksville, Iowa, is representative of many agricultural core communities (Figure 12-9).

Located in the north-central part of the state in Butler County, Clarksville takes its name from the Clark family who arrived with others, by wagon, from north central Indiana in July 1852. As Iowa became more fully settled in the decades that followed, Clarksville grew slowly but steadily. Its population reached 895 in 1910, 1328 in 1960, and 1424 in 1980. The

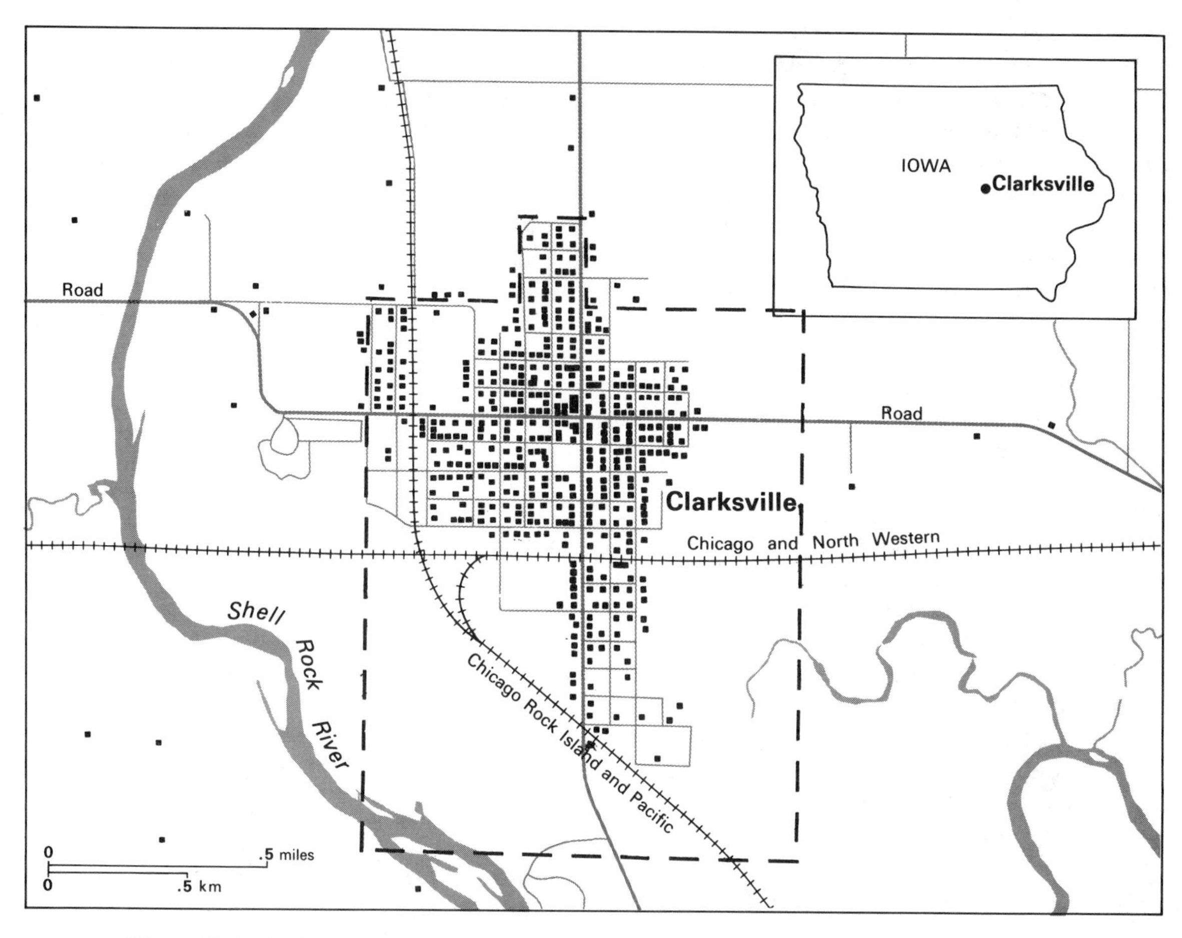

Figure 12-9 Clarksville, Iowa. In addition to the representative features of the agricultural core landscape shown on this map, you are encouraged to use the aerial photograph of this same community on page 311 to enhance the information that might be identified by either source alone.

village is incorporated and has its own post office, bank, and a relatively new high school (completed in 1970), but it has remained basically a service and market center for nearby farmers. Even so, changes in the marketing of farm products led in the early 1980s to the closure of one of Clarksville's rail-side grain elevators. Other advances in technology have meant that the Clarksville *Star*, the weekly newspaper, is no longer printed in town; instead, it makes better economic sense to use the more efficient press in Waverly, a somewhat larger community about 10 miles away.

Clarksville is not served by any of the region's main highways. This has kept the village's growth slow, even for the Deep North, and it contributes to the citizens' open curiosity about the presence of any strangers. Individuals arriving in town by road are recognized immediately and are expected to be there for a reason; few would drive through town when traveling between major destinations. A recent visit to Clarksville

Small towns of the Deep North, such as Clarksville, Iowa, obtain much of their character because they have grown slowly, if at all. The "downtown" remains a few blocks long, the small square residential blocks are occupied by only six or eight homes, and one could easily walk into the surrounding countryside from anywhere in town in less than 10 minutes. (U.S.G.S. High Altitude Aerial Mapping Photography, N.C.I.C. Headquarters)

elicited several friendly inquiries of "Who in town are you here to visit?" and clear surprise at the response "No one; just passing by."

While Clarksville has continued to grow, however slowly, there are indications of a potential for decline in the ratio of female-to-male population and the age structure of that population. There were 757 females in Clarksville in 1980 compared to only 667 males. The median age of Clarksville's population (35.5 years in 1980) was more than 5 years older than the median age for the entire state (30.0 years). And almost one quarter of the town's population was 65 years old or older in 1980. (The state proportion was only 13.3 percent.) It appears that a significant number of the population migrate out following high school—probably in search of greater employment opportunities—and leave behind a somewhat elderly population.

Clarksville's urban form is also typical of Deep North farm communities. The square and rectangular residential blocks have been laid out around the two through streets, Main Street and Superior Street. Topography is ignored in the regularity of this pattern. The town's small business district is clustered around the intersection of Main and Superior, but neither road is so busy that a traffic light is required at the intersection. Homes on large lots retreat outward from this center at low housing densities. With a small six-block exception to the north, the village boundaries form a neat 2- by $2\frac{1}{2}$-mile rectangle, although the political area is far from filled by the village's homes. The square land survey system can also be observed well beyond the village's limits in the rural road pattern. Only the railroads and short sections of the two main through roads deviate from the straight north-south, east-west section boundaries.

Clarksville, Iowa, is representative of thousands of small communities in the Deep North. Stable or slowly growing in population, it has begun to be affected in small ways by changes in technology and the region's economy in spite of its insulation from these changes. A peaceful, attractive, friendly place in which to live, Clarksville offers few of the locational advantages of communities with greater access to local centers of economic activity. It is likely to hold its own to the end of the century, and there is little that suggests significant changes ahead.

ADDITIONAL READINGS

Andrews, Clarence A., ed. *Growing Up in Iowa: Reminiscences of 14 Iowa Authors.* Ames: Iowa State University Press, 1978.

Carter, William. *Middle West Country.* Boston: Houghton Mifflin, 1975.

Dillon, Lowell I., and Edward E. Lyon, eds. *Indiana: Crossroads of America.* Dubuque, Iowa: Kendall/Hunt, 1978.

Fuguitt, Glenn V., et al. *Rural and Small Town America.* New York: Russell Sage Foundation, 1989.

Hart, John Fraser. "Corn: A Photo-Essay." *Focus,* 39 (Summer 1989), pp. 4–9.

Hart, John Fraser. "Small Towns and Manufacturing." *Geographical Review,* 78 (1988), pp. 375–397.

Hudson, John C. "North American Origins of Middlewestern Frontier Populations." *Annals of the Association of American Geographers,* 78 (1988), pp. 395–413.

McAvoy, Thomas T., ed. *The Midwest: Myth or Reality?* Notre Dame, Ind.: University of Notre Dame Press, 1961.

Nelson, Ronald E., ed. *Illinois: Land and Life in the Prairie State.* Dubuque, Iowa: Kendall/Hunt, 1978.

Platt, Rutherford H. "Farmland Conversion: National Lessons for Iowa." *Professional Geographer,* 33, 1981, pp. 13–121.

Raban, Jonathan. *Old Glory: An American Voyage.* New York: Simon & Schuster, 1981.

Raitz, Karl B. "Kentucky Bluegrass." *Focus.* 37 (Fall 1987), pp. 6–11.

Winsor, Roger A. "Environmental Imagery of the Wet Prairie of East Central Illinois, 1820–1920." *Journal of Historical Geography,* 13 (1987), pp. 375–397.

CHAPTER 13

The Great Plains and Prairies

In a desert you know what to expect of the climate and plan accordingly. The same is true of humid regions. Men have been fooled by the semiarid regions because they are sometimes humid, sometimes desert, and sometimes a cross between the two. Yet it is possible to make allowances for this, too, once it is understood.

C. Warren Thornthwaite

Kansas was chosen because it was dull and brown, a striking contrast to the rainbow and the land beyond.

H. Frank Baum (from a discussion about *The Wizard of Oz*)

The historian Walter Prescott Webb suggested in his book *The Great Plains* that the northwest Europeans who settled much of the United States faced three great "environmental encounters"—areas where climatic conditions were so unlike those of their home region that the agricultural crops and settlement patterns developed in Europe were inappropriate. Before those areas could be settled, new crops, new land-use patterns, and sometimes new technological developments were needed. The first of these "encounters" was with the high summer temperatures and humidity levels of the Southeast. The second problem environment was the arid Southwest and interior West. The third was the great continuous grasslands located astride the center of the country (Figure 13-1).

European settlers encountered several environmental problems on the grasslands of the Great Plains and the prairies of Canada. Average annual precipitation was much less than in the East. Within the grasslands, the western margins were much drier than the eastern area. There was also considerable variation in the annual precipitation, a variation that seemed to come in broad cyclic patterns. Violent storms accompanied by high winds, hail, and tornadoes were common. Blizzards with wintery blasts intensifying the cold drove the snow into immense drifts. The hot, dry winds of summer parched the soil and sometimes carried it away in great billowing clouds of dust.

The region's sparse natural water supply would not support tree vegetation except along the stream courses. Many of these streams were small and flowed only intermittently. Eastern farmers, accustomed to a plentiful supply of water for crops and animals as well as ample wood for building, fencing, and heating, had to adapt to quite different conditions in their attempts to settle the Great Plains.

This is a land of large landscapes, where a gentle topography and limited tree cover together open distant vistas to the observer. Farmers in northern sections (this scene is from North Dakota) often plant "wind breaks" of trees and bushes to protect against wind and blowing snow. (U.S.D.A.)

PERCEPTIONS OF THE PLAINS

When Coronado, the Spanish explorer, wandered into south-central Kansas in search of the mythical Cibola, the cities of gold, he was sadly disappointed at the rude Indian villages he found instead. In spite of this setback, he was struck by the richness of the land: "the country itself is the best that I have ever seen for producing all of the crops of Spain," he said, adding that the land was "very flat and black" and "very well watered by the rivulets and springs and rivers." The lack of gold and silver plus the great distances separating the area from the core of Spanish colonial development in central Mexico, however, overcame the attractiveness of the environment, and the Spaniards never occupied any substantial part of the Great Plains.

After the travels of Lewis and Clark from St. Louis to the mouth of the Columbia River and back in 1804 and 1805, additional American exploration was followed by gradual settlement of the Plains. The Americans found a land of seemingly endless, almost flat grasslands. (What schoolchild has not read *Around the World in Eighty Days*, which includes the tale of a covered wagon railcar propelled eastward across the Plains by a sail?) This land, so unlike the wooded lands of the East, was often branded a desert. "Almost wholly unfit for cultivation, and of course, uninhabitable for a people depending

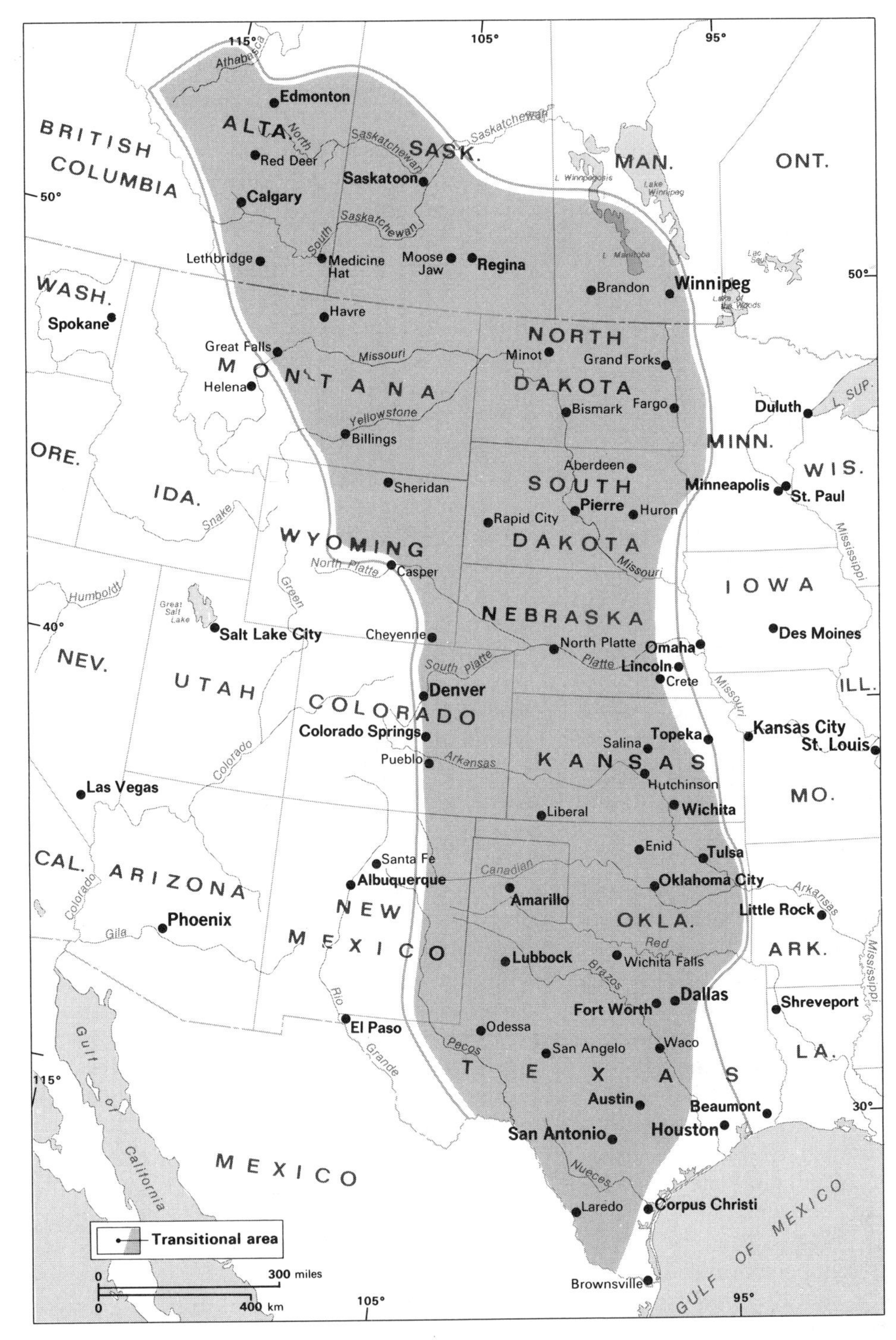

Figure 13-1 The Great Plains and Prairies.

Major Metropolitan Area Populations on the Great Plains and Prairies, 1990	
Dallas, Tex.	2,553,362
Denver, Colo.	1,662,980
Ft. Worth–Arlington, Tex.	1,332,053
San Antonio, Tex.	1,302,099
Oklahoma City, Okla.	958,839
Edmonton, Atla.	789,376[a]
Austin, Tex.	781,572
Tulsa, Okla.	708,994
Calgary, Alta.	671,355[a]
Winnipeg, Man.	623,821[a]
Wichita, Kans.	454,242
Colorado Springs, Colo.	397,014

[a]Indicates 1986 population.

upon agriculture for their subsistence" was the reaction of Major Stephen Long in his 1821 expedition across the region. The Oregon Trail, stretching from Kansas City to the Willamette Valley in Oregon was for many a way of bypassing the unfriendly lands of the Plains for the more familiar demands of the wooded territory in the Pacific Northwest. Some American geographies published during the middle decades of the nineteenth century branded the area "The Great American Desert." For many, that was an acceptable definition.

Coronado and Long were both describing roughly the same part of the continent, the area that we know now as the southern Great Plains. These men differed, however, in their environmental backgrounds, which strongly influenced their perceptions of the area. Coronado was from the dry, basically treeless plateau region of central Spain; he had spent additional time in the equally treeless, dry lands of central and northern Mexico. To him, the Great Plains vegetation, although containing few trees, seemed lush. Compared to much of his experience, the streams were abundant and flowing. Long, a product of the humid, forested eastern United States, had quite a different perspective. To him, the lack of trees was shocking; the streams seemed dry and inadequate. Only gradually did widespread attitudes similar to Long's disappear in the United States; even today, many Americans would still agree with his evaluation.

Such great variation in perception of the quality of the Plains environment is not surprising. Numerous studies of human perception have emphasized the importance of individual experience on perception of an environment. We each evaluate and filter selectively all new information, and previous experience is a major part of that filter. New situations are evaluated within the framework of those past experiences. If the new situation is basically different from those encountered previously, we may not know how to react and that environment will be avoided. Correspondingly, familiar situations require little adjustment and may be readily accepted.

The importance of past experiences on perception can perhaps be best illustrated with a few examples. Students at many universities across the United States have been asked to rank the states of the country in terms of their desirability as residential locations. Some states, such as Colorado, Vermont, Washington, and California are usually ranked high. Southern and Great Plains states are normally found at the bottom of the list. Ironically, one-third of Colorado is in the Great Plains. Those who rank it highly think of Colorado as a mountain state. Beyond these usual extremes, marked variations occur. Students at the University of Minnesota see little difference between Mississippi and Alabama, ranking them both quite low, whereas students at the University of Alabama perceive a great difference between the two, ranking Alabama as the more preferred state, placing Mississippi well below the average. Conversely, Alabama students identify little difference between North Dakota and Minnesota, whereas those at Minnesota see a great gap separating these two states. When asked to rank the states with respect to the quality of their climates, students at the University of North Carolina ranked states across the South high, with rankings basically declining toward the North. Their peers at the University of Vermont reversed that evaluation, placing their own cooler North at the top of the list.

Along a somewhat different line, a series of studies on the perception of natural hazards have found a direct relationship between the frequency of occurrence of a particular hazard and the accuracy of the perception of that frequency. In a study of floodplain development, for example, it was found that inhabitants of areas that seldom experienced flooding paid little heed to its likelihood. Those in areas of frequent flooding gave that possibility considerable emphasis in development planning. In a study of drought perception in a series of counties extending from central Kansas into eastern Colorado, researchers found that farmers to the east, where drought is relatively less frequent, consistently underestimated its frequency. To the west that discrepancy declined, until in eastern Colorado perception and reality coincided.

Previous experience, then, is a major contributor to perception. It is not surprising that Spaniards found the Plains amenable; it was similar to the physical environments with which these explorers were familiar. The differences—perhaps a little more water, a little more grass—were small and positive. For the typical American, the difference was more radical, and for many, such as Major Long, the reaction was strongly negative.

THE PLAINS ENVIRONMENT

The topography and vegetation of the grasslands is among the least varied—some would say most boring—to be found anywhere in Canada or the United States. Early settlers following the Oregon Trail could reach the Pacific Coast in one season of travel, in part, because the grasslands were so easy to cross. The region lies entirely within the interior lowlands physiographic province. The underlying sedimentary beds dip gently. Elevation increases gradually, almost imperceptibly, from east to west. Along the eastern margin the elevation is only 500 meters (1600 feet), whereas in the west, Denver, at an altitude of more than 1500 meters, is the "Mile-High City." Physiographically, the largest portion of the Great Plains is the High Plains stretching along the western margin of the region from the Edwards Plateau of south Texas northward to southern Nebraska. Covered by a thick mantle of sediments that are quite sandy and extremely porous, this section is generally flat. Some areas, such as the Llano Estacado (Staked Plains) in west Texas and eastern New Mexico, are extremely flat. Only along streams, such as at Scottsbluff on the Platte River in western Nebraska or at Palo Duro canyon on the Red River in panhandle Texas, has erosion resulted in substantial local relief. The Lake Agassiz Basin, formerly occupied by the largest of the Pleistocene lakes (see Chapter 2), is another exceptionally flat area and includes much of southern Manitoba, eastern Saskatchewan, and the valley of the Red River of the North in North Dakota and Minnesota.

Not all portions of the region are so unvarying topographically. The most obvious exception is the Black Hills section of South Dakota and Wyoming. A large, dome-shaped area of eroded igneous rock, the Black Hills are associated both geologically and topographically with the Rocky Mountains to the west. Other, smaller outliers, most notably the Cypress Hills in southern Alberta and Saskatchewan, dot the western margins of the northern plains. In southern Texas, the Edwards Plateau, really the southern edge of the High Plains, is heavily eroded into a canyonlands landscape along its southeastern margin where it is adjacent the the coastal plain. In central and northwestern Nebraska, the Sand Hills offer a dense, intricate pattern of grass-covered sand dunes, many of which are well over 30 meters (100 feet) high. The dunes were created by sand blowing along the southern margins of continental glaciers during the Pleistocene. Badlands topography—extremely irregular features resulting from wind and water erosion of sedimentary rock—is widespread on the unglaciated Missouri Plateau from northern Nebraska northward to the Missouri River. The best examples are found in the Badlands Na-

Few regional boundaries are as sharply obvious as that between the Plains and Prairies and the Empty Interior along the Front Range of the Rocky Mountains in central Colorado. This view looks west across Denver to the mountains. (Stephen S. Birdsall)

tional Park of western South Dakota. North of the Missouri River and west of the Lake Agassiz Basin, the glaciated Missouri Plateau, although sometimes flat, is covered with ponds, moraines, and other glacial features. Many visitors to north-central North Dakota are surprised at the many lakes, so important to migratory wildfowl on the midwestern flyway.

To state that the characteristic vegetation of the grasslands is grass would be a deceptive oversimplification, ignoring both the variety and distribution of grasses. Although agriculture has destroyed much of the original grasslands vegetation, the moister eastern portions [areas with more than 60 centimeters (24 inches) of annual precipitation in the north or more than 90 centimeters (36 inches) in the south] were originally a continuous tall-grass prairie, where grasses grew between 30 centimeters (1 foot) and 1 meter (3 feet) in height. Big bluestem was perhaps the most characteristic vegetation type. Grasses were shorter in the drier climates toward the west. Along the western margins of the Plains, continuous prairie grasses gave way to bunch grasses, with grama grass a good example of the dry steppe grasslands. These shorter, more separated grasses could succeed in the semiarid conditions of the western Plains.

The prairie grasses have developed deep, intricate root systems that commonly extend much deeper into the soil than the grass blades reach above, allowing them to utilize available water. The tangled root system made the prairies exceptionally difficult to plow. The first settlers often had to employ "Bonanza Teams," heavy plows pulled by as many as 20 animals, to break the sod. The prairie sod could also be "cut" into large bricks used in the construction of the sod houses commonly constructed during the early period of Plains settlement by Europeans.

The Problems of Climate

Precipitation. Most precipitation on the Great Plains results from the interaction of air masses generated over the northern interior of the continent and over the Gulf of Mexico. Air masses from the Southwest also often interact with the continental air masses, but they tend to contain less moisture and create much less precipitation. The warm, moist tropical maritime air flowing in from the Gulf of Mexico, the prime contributor of

The Plains and Prairies landscape is more varied than commonly imagined. Badlands, such as these in South Dakota, result from the erosion of sedimentary rock by the action of water and wind. (Paul Horsted/South Dakota Tourism)

moisture to the Plains, commonly curves up the Mississippi Vallley and then moves northeast, missing much of the western Great Plains entirely. One result of this pattern is the marked westward decline in average precipitation amounts across the U.S. portion of the region. This is represented on maps by lines of equal annual precipitation, called *isohyets,* which bend more toward the east on the northern Plains. In Kansas, for example, average annual precipitation varies from a moist 105 centimeters (42 inches) in the southeast to a semiarid 40 centimeters (16 inches) in the southwest.

The pattern of precipitation on the Canadian grasslands is markedly different. The east-to-west zonation is replaced by one that runs northeast-southwest (Figure 13-2). The driest portion of the prairies, called Palliser's Triangle after a nineteenth-century surveyor who pronounced the area unfit for cultivation, is found along the Alberta and Saskatchewan border with Montanta. This dry zone stretches perhaps 300 kilometers (185 miles) north of the national border in the west. To the north and east the precipitation is adequate to support a tall-grass prairie. This one, called the Park Zone, is the major section of cultivation on the prairies.

Periods of higher than normal precipitation on the Great Plains result when tropical air masses move northwestward from the Gulf of Mexico, which brings these air masses over portions of the Plains. This provident current is far from dependable, however. Annual precipitation amounts can vary widely, although they usually fall between 80 and 120 percent of the average (Figure 13-3). During severe drought years, such as those during the mid-1930s, precipitation amounts may fall to less than 75 percent of the long-term average. Fortunately for the Plains farmer, the bulk of the precipitation falls during the period of more rapid crop growth, from April to August. About three quarters of the region's total annual precipitation can be expected during this 5-month period.

Major droughts have tended to occur in 20-year cycles, with the last five centering on the

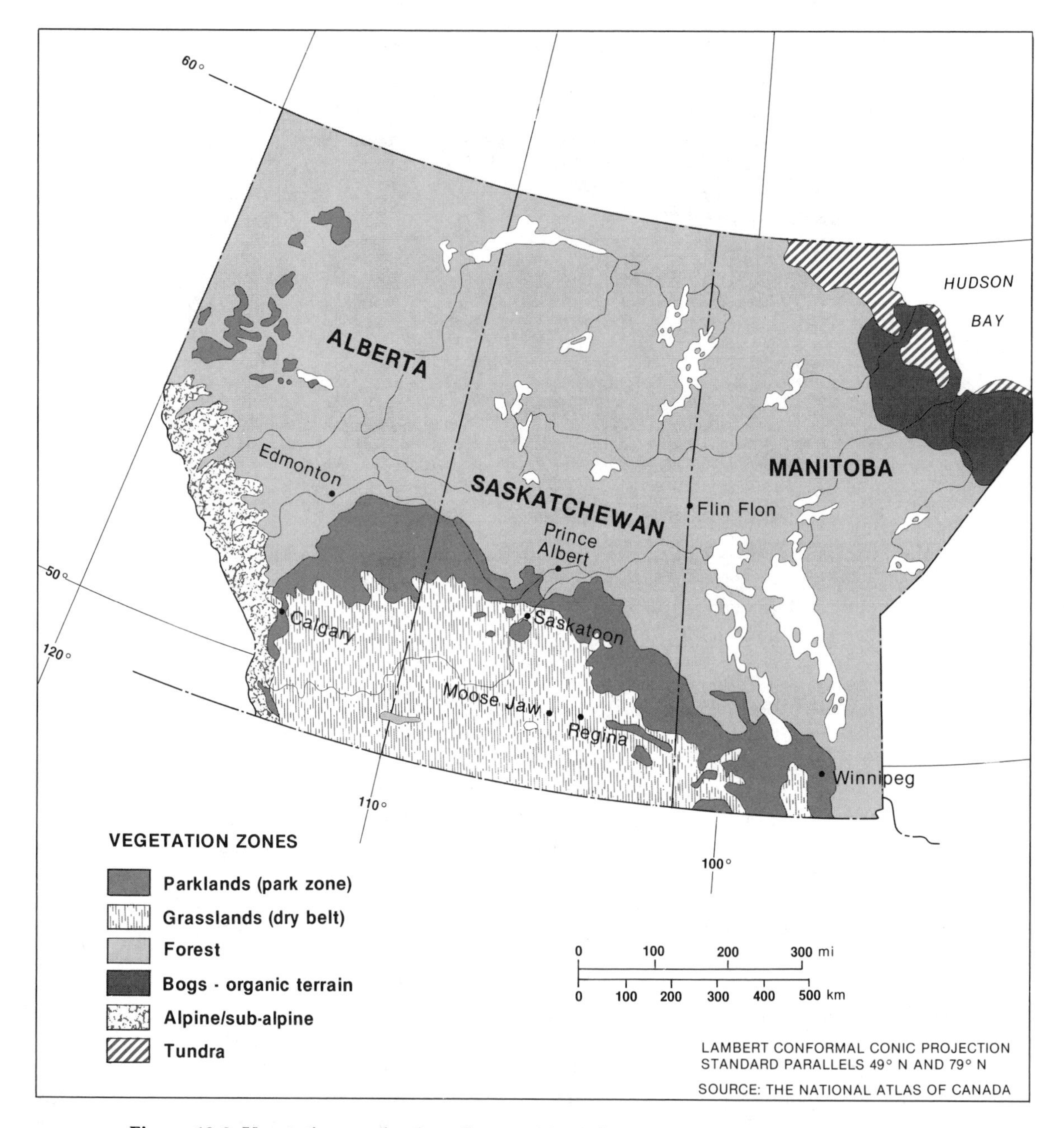

Figure 13-2 Vegetation on the Canadian prairies. The southwestern corner of the Canadian prairies is a dry grassland, and ranching is the primary focus of the rural economy. The better-watered region to the northeast is a major focus of the prairie agricultural economy.

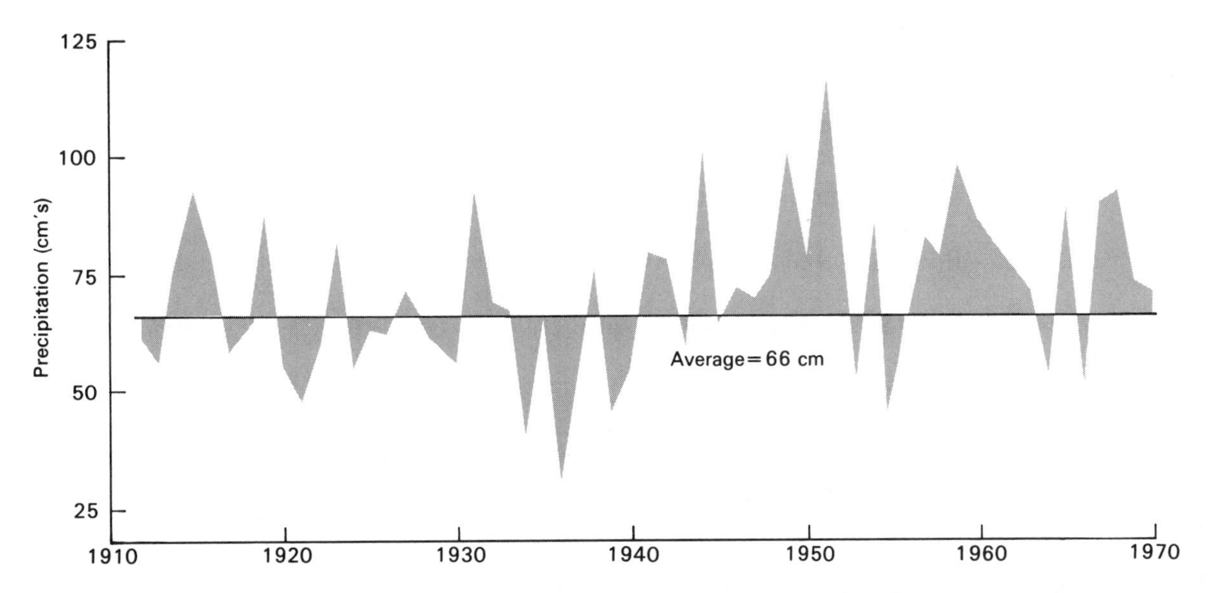

Figure 13-3 Rainfall variability in Crete, Nebraska. Crete is located on the eastern margin of the Great Plains and is thus less affected by annual precipitation variability than more westerly sections of the region. Nevertheless, the 1930s drought, followed by several decades of greater than normal precipitation, is clearly evident.

1890s, 1910s, 1930s, 1950s, and mid-1970s. For farmers operating close to the Plains low-moisture margins, severe long-term periodic variations in precipitation have resulted in times of widespread optimism and expansion followed by years of climatic and economic setbacks. During the period around World War I, when precipitation abundance coincided with high prices for wheat, the premier crop of the Plains, acreage expansion was particularly noticeable. Farmers moved into the drier southwestern margins of the wheat-growing area in Kansas and Oklahoma. The collapse in wheat prices in the early 1920s brought economic recession to much of the Wheat Belt by 1925.

The Great Depression of the 1930s coupled with several years of severe drought on the Plains meant economic ruin to areas of recent expansion as crop production, marginal even in good years, shrunk to nearly zero. One result was the famous Dust Bowl. With the soil-binding quality of the land destroyed by years of repeated cropping, dunes of 3 meters (10 feet) or more in height were built up by the dry winds, and the sky was often laden with dust. A second consequence was the migration of large numbers of destitute farmers from the farms of the southern Plains to the new promised land of California. Great as the soil damage was, the Soil Conservation Service found that the drought years of the 1950s—following as they did years of ample precipitation and high demand for wheat throughout much of the 1940s—in reality resulted in the damage to far more acres by wind erosion than had the 1930s drought.

Some of the region's spring and summer precipitation comes in the form of violent thunderstorms. Hail is occasionally a product of these storms. These frozen pellets, sometimes measuring more than 5 centimeters (2 inches) in diameter, which can look like golfballs or even baseballs, have the power to devastate a crop of mature, top-heavy wheat. Much of the southern and west-central Plains experiences frequent hail storms, with parts of western Nebraska and southeastern Wyoming leading the continent in average annual hail frequency (Figure 13-4).

Wind. Tornadoes, which can have funnel wind speeds in excess of 350 kilometers (200 miles per hour), are another violent result of these storm systems of the Great Plains. Although the area affected by one funnel is small, the frequent occurrence on the central Plains makes them a significant regional hazard (Figure 13-5).

Wind has been a mixed blessing on the Great Plains. Late spring and summer wind velocities on the central and southern Plains are among the highest in interior North America. In the past, this served to maximize the efficiency of windmills in the region. However, the presence of high winds also means that the amount of moisture evaporated and transpired by plants (when sufficient moisture is available) is high across much of the Great Plains. Maximum rates vary from about 165 centimeters (60 inches) annually in Texas to around 75 centimeters (30 inches) at the northern edges of the Plains in Canada; levels depend on average temperature and, therefore, on latitudinal location. Because these rates, called potential evapotranspiration, are high, they limit the effectiveness of precipitation.

Figure 13-4 Average annual number of days with hail. Each year, hailstorms destroy crops valued at many hundreds of millions of dollars. Damages are especially heavy on the High plains.

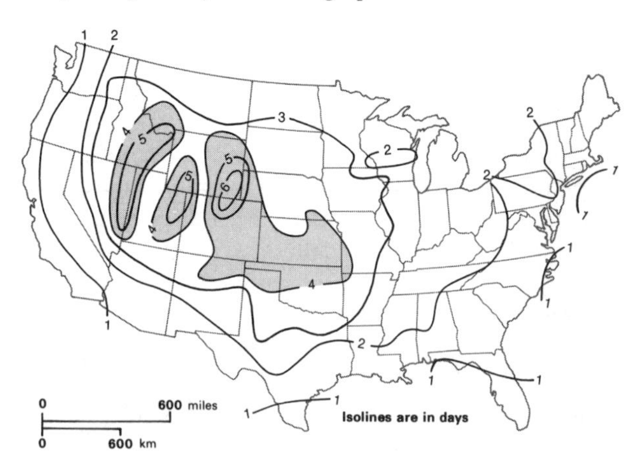

The chinook, a winter wind, occurs when dry, relatively warm air from the Pacific Coast pushes over the Rocky Mountains. As it descends onto the Great Plains, it warms still further and is much warmer than the cold, continental air mass commonly found over the region in winter. The Pacific air temporarily pushes the cold air from the western Plains, and a rapid, dramatic temperature rise results. Partly because of this interesting phenomenon, winter temperatures along the higher western area are slightly warmer than along the eastern edge of the Plains, and cities, such as Denver and Calgary, are able to bask occasionally in balmy midwinter days over 10°C (50°F) not found farther east on the grasslands.

Temperature. As is common with moist continental locations, seasonal temperature variations on the Plains are considerable, and the annual temperature range increases as one moves northward. North Dakota and Texas have both recorded highs of 49°C(121°F), the only places outside the desert southwest in the United States to ever have temperatures in excess of 120°F. North Dakota has also recorded a low of −52°C (−60°F), the only place in the United States east of the Rocky Mountains to ever record so low a temperature. Amarillo, Texas, has an average January temperature of 2°C (35°F) and an average July temperature of 28°C (82°F); the corresponding temperature range for North Platte, Nebraska, is from −5°C (23°F) to 26°C (78°F); and for Winnipeg, it is from −17°C (1°F) to 20°C (68°F).

In addition, the length of the frost-free season, which averages 240 days in central Texas and less than 90 days on parts of Canada's prairies, varies widely around the average from year to year. As with annual temperature range, the variation increases as one moves northward. Along the U.S.-Canadian boundary, frost-free periods in some years are as much as 30 percent higher or lower than the area's 100- to 120-day average.

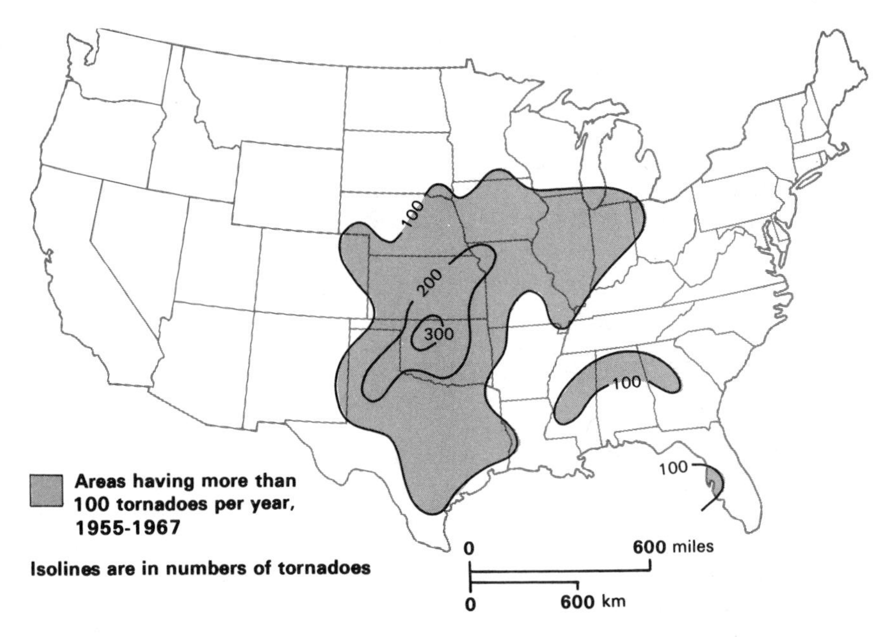

Figure 13-5 Average annual tornado frequency. Tornadoes are far more common on the Great Plains than in any other area of the world. Tornado warning sirens are a normal part of spring and early summer in communities in "Tornado Alley" from northwest Texas to northeast Kansas.

Blizzards. Snow, wind, and cold are all part of one of the most devastating weather elements on the Plains: the blizzard. A blizzard occurs in winter when a very cold polar air mass pushes southward along the Rocky Mountains and onto the Plains, breaking the usual west to east storm pattern. High winds, intense cold, and considerable amounts of snow are associated with these storms. Ordinarily, snowfall amounts are not high in the region, averaging perhaps 100 centimeters (40 inches) annually on the northern Plains, 40 to 80 centimeters (15–30 inches) in central Plains states, and less that 25 centimeters (10 inches) in most of Texas. However, a blizzard can last for several days and bring half of the average winter snowfall. Plains ranchers usually leave their stock outdoors during the winter, and a severe blizzard can block the animal's access to food and result in high animal mortality. It is estimated that as much as 80 percent of the cattle of the northern Plains died during a series of unusually difficult blizzards from 1887 to 1888. The impact of these storms was made even more severe because of a widespread decline in the quality of the steppe grasslands, the result of a decade of overgrazing. Today a blizzard can block highway travel for days and require frequent air drops of hay to stranded cattle.

EARLY SETTLEMENT OF THE PLAINS

Plains Indians

The pre-European occupation of the Plains by the American Indians was limited. Hunting, particularly for buffalo, was the primary economic activity. Most tribes lived along the streams in semipermanent settlements. With no means of rapid long-distance overland movement (the dog was the only domesticated animal in pre-European North America), the Indians could not leave the dependable water supplies of the streams for any long period. This was a substantial problem, for the migration of the great buffalo herds often took this food source far away from the settlements for many weeks.

When the Spanish departed from the southern Plains following their initial explorations, they left some of their horses behind, a "gift" that dramatically altered the life-style of the Plains Indians during the following two centuries. By the time Americans reached the Plains in the early nineteenth century, they found what many have called the finest light cavalry in world history. The horse had diffused throughout the grasslands, and the Plains Indians, no longer restricted to the waterways, freely followed the buffalo migration. This new resource enabled the Indians to thrive as never before. A few tribes, like the Dakota (Sioux) in the north and the Apache and Comanche, had used their well-developed warrior abilities to dominate large areas.

These Native Americans were gradually pushed from most parts of the Plains during the nineteenth century (see Chapter 15). To some extent, this was caused by European hunters who hunted the Plains buffalo to the brink of extinction. Mostly, however, the Indians were simply swept aside by the flood tide of American settlement.

American Settlement

The early American perception of the region as an unpromising and difficult place to settle was not totally wrong. The widespread lack of trees meant that farmers had none of the traditional material used for the construction of houses and barns, for fencing, or for fuel. Water sources were scarce, often rivers and streams had only a seasonal flow. Those who arrived early settled along these waterways. Later arrivals had no ready access to flowing water because they were bound by the existing system of water law that gave all riparian rights to the landowner. This application of English common law in North America had been satisfactory in the more humid areas of the East, but it was totally inadequate on the Plains. The crops that settlers brought with them to the plains often failed, and crop success varied greatly from year to year as precipitation amounts fluctuated widely—this was an unfamiliar climatic pattern in the East. Agricultural production rates were also generally lower, and the 160-acre (64.75-hectare) farm size that seemed so adequate farther east proved to be too small on the Great Plains.

The settlement frontier hesitated along the eastern boundary of the Plains partly as a result of these problems. Settlers tended to bypass the Plains for the Pacific Coast, until technological and land ownership changes made plains settlement more inviting.

Grasslands Ranching. During this hesitation in settlement expansion, an alternative economic system swept across the region with the speed of a prairie fire. An extensive ranching economy had been introduced into south Texas by Spaniards and into east Texas by frontier settlers from the U.S. South. This economy spread from Texas northward as far as the grasslands of Canada's Prairie Provinces during the period from 1867 to 1885.

A series of events and changing conditions permitted this rapid expansion of ranching. The American Civil War years resulted in a large herd of unbranded cattle in south Texas as most of the ranchers went off to war. With the end of the war in 1865, the South's economy was in shambles, and the ranchers were broke—but millions of cattle were wandering around Texas unclaimed. In the North, the war had spurred urbanization and increased the commercial demand for food. Railroads were given the right to construct routes from the settled Mississippi Valley to the West Coast, but they found little potential for profitable freight originating in the territory in between. The U.S. federal government aided the railroad companies with land grants of tens of millions of acres along the railroad rights-of-way, and the railroads actively encouraged agricultural settlement along their routes to generate transportation demand. Even so, until farmers arrived and until rail lines were completed to the Pacific Coast, business was limited.

Great herds of cattle were driven northward from south Texas to railheads in Kansas both for shipment east and to stock the huge, relatively unsettled Plains region. The possibilities of open-range cattle ranching were soon obvious. Within a decade the system spread northward across the Plains and into Canada. By 1880, perhaps 5 million head of cattle had been moved from south Texas to the Kansas railheads or to the ranges of the central and northern Plains.

The open-ranching economy collapsed rapidly in the late 1880s. Widespread overgrazing, competition from the superior beef of expanding cattle-raising operations in the Midwest, a slipping national economy, the disastrous winter of 1887–88, and a rapid influx of farmers onto the Plains combined to end this short period of American history. (At least one television western series, "Gunsmoke," had a longer tenure than the period it attempted to portray.) The open-range, unimproved ranches were pushed to the drier western side of the Plains or were forced into a more restrained fenced operation.

The Agricultural Frontier. A series of innovations during the two decades following the Civil War also gradually eliminated, or lessened in severity, the problems encountered by new settlers. Barbed wire, developed commercially in the 1870s, provided an effective alternative fencing material to take the place of the missing wood supply. For a time, dwellings constructed of sod provided adequate housing. The prairie sod was difficult to plow because of its intricate system of grass roots, but when lifted off the soil like peat and used as a building material, it would resist erosion for years. Nevertheless, these "soddies" were relatively short-lived, and most settlers replaced them as soon as possible with frame homes. Lumber was brought in by the railroads, which were under construction all across the Plains by the 1870s. The development of a simple windmill and mechanical well-drilling devices meant that sufficient water could be obtained locally for humans and animals, and for irrigation. The technology for deep-well drilling had long been known, but neither in northwestern Europe nor in the eastern parts of North America had windmills ever been used much. It was the widespread adoption of this technology on the grasslands that led to its subsequent acceptance across most of rural America. Grain farming also became increasingly mechanized during this period, and an individual could cultivate far more land than had been possible previously. This enabled farmers to operate larger farms and, thus, compensate for lower yields.

A series of major modifications in U.S. land laws increased the acreage that an individual could obtain from the federal government. The Homestead Act of 1863 granted 160 acres (65 hectares) to each settler. The Timber Culture Act several years later allowed settlers to own an additional 160 acres if a portion of that land was planted in trees. It was hoped that this would give farmers the additional farmland they needed and would also introduce wood production to the nearly treeless environment.

Finally, crops that were better adapted to the growing conditions of the region were introduced into the agricultural system, and farmers began to improve their understanding of how to use the Plains environment. Hard winter wheat is perhaps the best example. First brought to the United States by Mennonite immigrants from the Ukraine, it was far better adapted to the dry growing conditions of the Great Plains than were the wheat strains grown there earlier.

Settling the Canadian Prairies. In 1670 the Hudson's Bay Company was given a royal charter to trade through Hudson Bay. It received a title to Rupert's Land, a vaguely defined territory that came to be regarded as the drainage basin of Hudson Bay. For the next 200 years the

Cattle raised on the grasslands of the Plains are commonly taken to a feedlot, such as this one in Colorado, for fattening on a special mix of feed before they are marketed. These lots, which dot the region, are an important factor in the widespread relocation of the beef slaughter industry from major Midwest cities to small facilities in the towns of the Plains. (Grant Heilman/Grant Heilman Photography)

company was the sole agent of British authority across a broad area that included all of what is today the Canadian prairies. Company policy discouraged settlement. Its economic goal was the collection and trade of furs. Agricultural settlement was viewed as generally incompatible with the harvesting of furs. The company did establish inland trading posts at sites such as Ft. Gray (near present-day Winnipeg) and Ft. Edmonton (near the location of Edmonton) that served as seeds for urban development when Rupert's Land was acquired by the newly independent Dominion of Canada in 1870.

The Canadian prairies were owned by the Hudson's Bay Company until 1870. The Canadian government began to foster settlement of the area in the 1870s after it acquired the prairies. The agriculturally uninviting Canadian Shield between southern Ontario and central Manitoba was a barrier to westward movement. The completion of the Canadian Pacific Railway in 1885 made the move much easier. Still, much of the early agricultural settling of the prairies came from the American Middle West. By 1890 good agricultural land was no longer available for frontier settlement in the United States, but there was land in Canada. By World War I over a half-million Americans had moved to the Canadian prairies. The major settling of the prairies occurred in the period from the mid-1890s to World War I when the Canadian government strongly encouraged migration to the prairies. Not only Americans and eastern Canadians, but also immigrants from throughout central and eastern Europe were encouraged to move to the region.

One important result of this settlement hesitation was that the Canadian government and settlers on the prairies learned from the mistakes of American Plains settlement. Farmers were encouraged to increase their land holdings by purchasing railway land grants or other privately owned land. Ranchland was only leased and could be reacquired by the government if needed for agriculture. The government encouraged (and still encourages) mixed farming in the Park Belt in hopes of reducing the climatic hazard to the agricultural economy.

PLAINS AGRICULTURE

The agriculture of the Great Plains is large scale and machinery intensive. In the United States especially, it is dominated by a few crops particularly suited to the Plains climate. As mentioned, the most important is wheat (see Map Appendix). It combines a large market demand and a tolerance for dry conditions better than any other American agricultural product.

Winter wheat is planted in the fall. Before the winter dormant season sets in, the wheat stands several centimeters tall. Its major growth comes in the spring and early summer, when precipitation is at a maximum and before the onset of the dessicating winds of summer. It is harvested in late May and June. Today winter wheat is grown across much of the United States, but its zone of concentration is the southern Plains from northern Texas to southern Nebraska. Production of winter wheat spread steadily northward in recent years with the development of new strains that can better tolerate cold winter weather. In fact, for Montana it is now more important than spring wheat.

Spring wheat, grown primarily from central South Dakota to the south-central Prairie Provinces of Canada, is planted in early spring and harvested in late summer or fall. It is suited to areas where winters are so severe that germinating winter wheat is killed.

Today the Great Plains remains the continent's premier wheat-producing region. The three Prairie Provinces grow most of Canada's wheat, with Saskatchewan easily outstripping its two neighbors combined. Kansas usually leads in the United States, followed closely by North Dakota, and then, in distant order, by Oklahoma and Montana. Of the top seven wheat states in the United States, only Washington (ranked fifth) is not a Plains state. Thus, it is largely on the abundance of Plains agriculture that the United States and Canada rank as the world's two leading wheat exporters (see Chapter 1 and Map Appendix).

Most grasslands wheat is grown using dry farming techniques, that is, methods that enable a successful harvest without irrigation. The soil is plowed very deeply to break the soil and thus slow evaporation. Most visually obvious, especially in the northern Plains, is the widespread use of fallowing, where the land is plowed and tilled but not planted for a season to preserve moisture. This is often done in alternating strips, with the strips rotated each year. The result, when viewed from the air is a striking pattern of long narrow alternating rectangles of brown and either green or gold.

A special kind of migrant labor has long been important in the harvesting of wheat. Beginning in the south around June 1 with the winter wheat harvest in Texas, custom combining crews gradually follow the harvest northward, finishing some four months later with the harvest of spring wheat in the Prairie Provinces. Unlike migrant farm laborers harvesting other crops in the United States, these people, often in large crews that use many combines and trucks, have traditionally been well-paid agricultural workers. The increasing size of wheat farms—average farm size in most of the Wheat Belt now exceeds 400 hectares (1000 acres)—has meant that more wheat farmers can now afford their own combines, and custom combining is less prevalent today than it was before World War II (Figure 13-6). Still, probably a third of all Great Plains wheat, and a higher share of winter wheat, is harvested by these custom combining crews.

The rapid, continuous increase in farm size in the region has resulted largely from the purchase by other farmers of the farms of those who have chosen to leave the area or the task, leading to an increasing number of noncontiguous farms. This, in turn, has helped foster a growing number of "sidewalk farmers," farmers who live in town and travel to their various acreages. Although demanding when performed, fieldwork on a wheat farm must be done only two or three times a year, so there is little need for the farmer to live on his or her land. Thus Saskatchewan, with its greater focus on wheat, has more of a pattern of "sidewalk farmers" than do either of its neighbors. In fact, some so-called "suitcase farmers" live far from their fields and visit them only occasionally. The corresponding processes of farm fragmentation and part ownership also have become increasingly common as a result of land sales (see Chapter 12).

Besides the freedom to live in town, many see a second advantage in this type of disrupted land ownership. A single hailstorm usually affects only a small area. By dividing the total wheat crop over several widely separate areas, the grower is, in effect, ensuring that in any one year, it is unlikely that an entire crop will be lost to hail or to a particularly severe thunderstorm.

A major problem with profitable wheat production continues to be the necessity, and difficulty, of moving the harvest rapidly to storage in the large grain elevators that dot the Plains. In Canada especially, many a small prairie town is dominated by a series of small, often wooden, grain elevators to which grain is brought by truck and then sent on to larger facilities by rail. The small elevators can hold only a part of the local crop; if all of the wheat is to be successfully removed from the fields before it is subjected to rain and destroyed, many grain hoppers must be made available at the right times by the railroads. Occasionally, the harvest outstrips the railcar supply and wheat is piled in large, temporary mounds near the elevators. Competition from truck hauling and, in parts of the winter wheat region, barge transport [a 3-meter (9-foot)-deep channel exists on the Missouri as far as Sioux City, Iowa, and the Arkansas is now navigable to near Tulsa, Oklahoma] has encouraged the railroads to abandon many small country grain elevators in favor of much larger complexes usually in larger towns. Most Canadian wheat goes first to Winnipeg (as do most products leaving the prairies for eastern markets) and then to Thunder Bay for transshipment on the Great Lakes, although an increasing amount is now moving through Pacific Coast ports, reducing somewhat the role of Thunder Bay. All western wheat presently goes to Vancouver. Some is also shipped northward to Churchill on Hudson Bay, usually for export to Europe. Most U.S. export wheat moves through the Great Lakes or in barges down the inland waterway system and the Mississippi River.

The soft international market for wheat during much of the 1960s led to a definite interest in crop diversification in the Canadian Prairie Provinces, and barley acreage increased rapidly. Today, Alberta has more acreage in barley than in wheat. Total wheat area in the Prairie Provinces, which reached 12 million hectares (30 million acres) in 1967, declined to 7.6 million hectares (18.7 million acres) in 1971, with the government's Low Inventories for Tomorrow program playing an important role in the reduction. Harvested area increased later, primarily in response to a growth in international demand. Only in Saskatchewan does wheat continue to dominate total farm production.

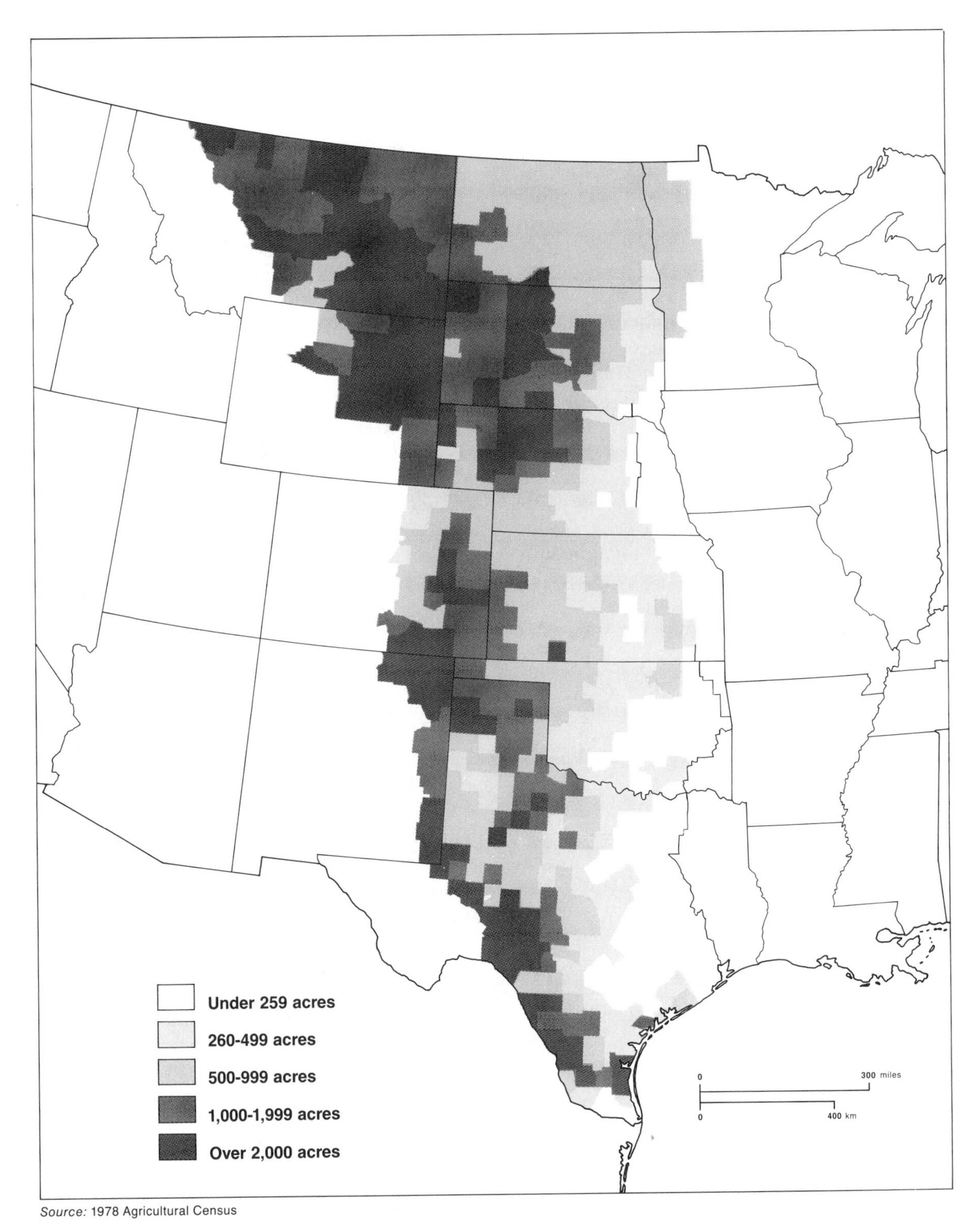

Source: 1978 Agricultural Census

Figure 13-6 Average farm size, 1987. Farmers on the western margins of the Great Plains need larger farms to compensate for lower per-acre economic returns resulting from sparse precipitation.

Sorghum has emerged as a major crop on the southern Plains in recent decades. Able to withstand dry growing conditions, this African grain now equals winter wheat in importance on the hot dry southwest margins of the Plains. Both Texas and Nebraska now have more land in sorghum than in wheat. Most of the grain sorghum crop is used as stock feed. On the northern Plains, barley and oats are major second crops, with most of the continent's barley crop coming from the Lake Agassiz Basin of Manitoba, North Dakota, and Minnesota. Nearly all flaxseed produced in North America also is grown in the northern Plains. Sunflowers are rapidly increasing in importance in the Red River Valley of Minnesota, North Dakota, and Manitoba. Canada is the world's dominant producer of canola, the leading vegetable oil used in Canada and an important protein ingredient in many livestock feeds. Canola production in the United States is also expanding.

The Canadian Park Belt today illustrates the diversity of Plains agriculture. Most farmers in south-central Saskatchewan prefer to raise wheat. Over three-fourths of the cropland in the area is in wheat production. Large farms extend for many miles across the often flat landscape. This is the Canadian stereotype for Plains agriculture. Southern Manitoba's farms are far more diversified. Crops such as barley, flaxseed, canola, rye, and oats; dairying for the Winnipeg market; and feedlots where livestock are fattened before being shipped to meatpacking plants in Ontario and Quebec have long been more important than wheat in the local farm economy. While wheat remains more important in central Alberta, beef cattle and feedlots are more common here than in what is usually thought of as the ranch country of the dry belt of the southern part of the province.

WATER CONTROL AND IRRIGATION

Irrigation in the United States and Canada is usually associated with the dry region of the far west. This is understandable, for crop agriculture is made possible in much of this area only by irrigation. Yet, the benefit derived from irrigation may be higher in many semihumid or even humid areas—in terms of the level of increased production per dollar invested—because irrigation water may be used either as a supplement in dry times to maximize yields for crops already grown in the area or to grow crops for which the available moisture is not quite sufficient. Evaporation rates from irrigation systems are often lower in more humid areas than in drier regions, and it is not as necessary to flood large quantities of water onto the land to provide an off-flow of water sufficient to remove the salts that might otherwise build up in the soil. Generally, increased financial security for the Plains farmer—through an assured crop in drought years—has been the most important reason for irrigation, not the production of crops that would fail to survive there on rainfall alone.

There are a number of Great Plains areas where large-scale irrigation developments are important. Perhaps the most noteworthy of these is on the High Plains from Colorado and Nebraska to Texas. The area is underlain by the Oglala aquifer, a vast underground geologic reservoir under 250,000 square kilometers (100,000 square miles) of the area that contains an estimated 2 billion acre-feet of water. This is "fossil" water, much of it deposited more than 1 million years ago. About a quarter of the aquifer's area is irrigated, almost entirely with Oglala water. The High Plains is a major agricultural region, providing, for example, two-fifths of the country's sorghum, one-sixth of its wheat, and one-fourth of its cotton. Irrigated lands here produce 45 percent more wheat, 70 percent more sorghum, and 135 percent more cotton than do neighboring nonirrigated areas. Groundwater withdrawals have more than tripled since 1950 to more than 20 million acre-feet annually. Recharge replenishes only a small percentage of this amount. Groundwater withdrawal rates are also quite uneven.

Early in this century, the area centered on

Large, expensive center-pivot sprinkler irrigation systems have become common throughout the Great Plains and West, where subsurface water is used for irrigation. Each circle is ½ mile (840 meters) in diameter. A problem for the farmer is how to use the remaining nonirrigated areas. (U.S.D.A./Soil Conservation Service)

Lubbock, Texas, became a significant region of cotton production. Irrigated farming, using water from wells drilled into the water-bearing sands that underlie much of the southern High Plains, gradually replaced the early dry-farming approach. Today the region is the most important area of cotton production both in Texas and in the United States. More than 50,000 wells currently supply irrigation water in the area. One negative result of this massive mining of the groundwater has been a rapid depletion of subsurface water. The average well depth is currently more than 50 meters (155 feet) and is increasing. This represents a serious problem for the area because deeper wells are much more expensive. A recent estimate suggests that wells will be so deep that irrigation will be too expensive on half of the Texas High Plains by 2000. And the resource can be exhausted, meaning there may soon not be enough water available to meet the demand. Even if use is reduced, replenishment of groundwater reserves is a slow process in this subhumid environment.

Laws of New Mexico concerning the conservation of underground water are stricter than those in Texas. Thus, the areas in New Mexico adjacent to Texas have little irrigated land, and most of this area in the more western state is still composed of ranches. The effect of this legal difference is clearly visible from the air; the subdivided farms of Texas stand in sharp contrast to the extensive, unbroken landscape of New Mexico's ranches. It is also reflected in the higher water table on the New Mexico side of the boundary.

The second major irrigated area on the Plains is in northeastern Colorado, with sugar beets the primary specialty crop. The area has long been irrigated from wells and from the waters of the South Platte River in both private and federal irrigation districts. The National Reclamation Act of 1902, passed largely at the instigation of Plains and western farmers and politicians, provided for the development of federal irrigation districts in these areas. The federal government covers the cost of construction, and those who use irrigation pay for water at what many nonfarmers argue are uneconomically low rates. This act has provided one basis for a complex pattern of federal dams and intricate water movement systems scattered widely over the West and constructed at a cost that has reached billions of dollars. Even so, these waters are no longer adequate to meet demands. In recognition of this, the government has funded the recently completed Big Thompson River project, which is designed to carry water from the west slope of the Front Range of the Rocky Mountains to the east slope and the irrigated lands beyond. The most striking technological feature of this project is a 33-kilometer (13-mile) tunnel, lying 1200 meters (3950 feet) below the continental divide in Rocky Mountain National Park.

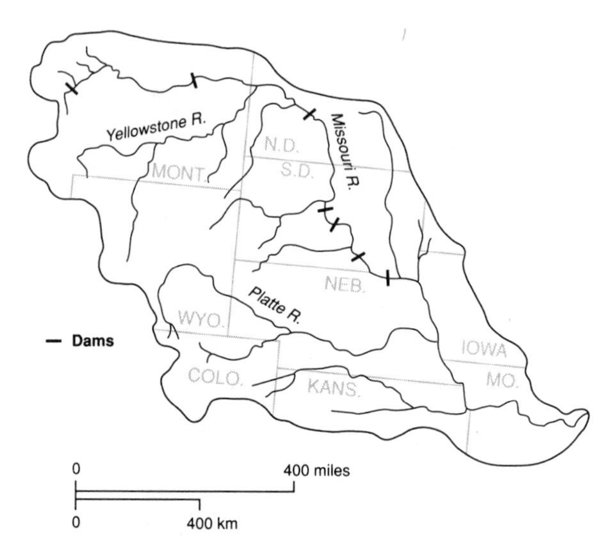

Figure 13-7 The Missouri Valley Project. This project—to meet the need for irrigation water in the western portion of the Missouri River basin and flood reduction downstream—is an example of large-scale federal intervention to help address regional problems.

The largest of the water impoundment projects on the Plains, however, is the Missouri Valley project, originally called the Pick-Sloan Plan (Figure 13-7). The plan was an outgrowth of two different sets of needs of the residents of the Missouri River Valley. People living at the lower end of the valley, including those in Kansas City and St. Louis, desired above all to have an effective system of flood control. An annual average of about 100 centimeters (40 inches) of precipitation falls on this area, and its residents are usually more concerned with an overabundance of water than with drought. In contrast, people in the upper Missouri Valley, especially the Dakotas and Montana, hoped for a system that would provide ample water for irrigation. The resultant multibillion-dollar system, composed of a series of large earth-fill dams on the upper Missouri as well as numerous dams on many of the tributaries of the river, is meant to satisfy both hopes. Irrigation has expanded considerably in many upstream areas, whereas residents in the lower valley hope (with some justification) that major flooding is less likely now. An added result has been a vast boom in water recreation, particularly power boating, throughout an area that previously had almost no standing water. Many argue that the costs, both monetary and in terms of damage to the natural environment, have been too great, but benefits of the plan have been considerable.

These and many other, smaller irrigation projects and individual wells have allowed a great expansion in the diversity of Plains agriculture. Throughout the central and northern Plains, and especially on the western Canadian prairies, alfalfa, the premier hay crop of the West, claims the largest irrigated acreage. Sugar beets are important in the Arkansas River Valley of eastern Colorado and western Kansas, and along the South Platte in northeastern Colorado. Arkansas valley growers also take great pride in the quality of their cantaloupes. Corn, usually irrigated from wells, is a major crop in south-central Nebraska. The cropping diversity of the

Park Belt in central Alberta results partly from expanded irrigation.

OTHER NATURAL RESOURCES OF THE PLAINS

Energy. The sediments of the Plains contain major reserves of energy resources—petroleum, natural gas, and coal (see Figure 2-6). To the south, major petroleum and natural gas fields are traditionally among the country's leading suppliers of these products. The Panhandle field, encompassing western portions of Texas, Oklahoma, and Kansas, is the world's leading supplier of natural gas. The same three states are also major petroleum producers, and recent developments have added Wyoming to the list of leading U.S. petroleum suppliers. Alberta has long been the predominant producer of both petroleum and natural gas in Canada. Coal has been mined for many years at scattered locations in the region, from southern Alberta and Saskatchewan to Texas.

Despite this past energy predominance, it is the future of energy production that excites a mixture of anticipation and concern in many Plains residents. Increases in energy cost and demand have focused attention on the region's rich coal and petroleum resources. Alberta leads the Canadian provinces in coal production. About 40 percent of the total national output is mined there. North Dakota can boast of sizable energy resources, mostly in the form of soft coal. Wyoming is now the leading coal-producing state in the United States. In 1988 Wyoming mines provided 17 percent of the total U.S. coal output, or 164 million tons. Coal accounts for nearly 70 percent of the state's tax base and provides over 90 percent of all railroad freight originating in Wyoming.

Denver has become a focus of considerable petroleum-based wealth. Alliance, Nebraska, nearly doubled in size between 1975 and 1980 because of its location on the Burlington and Northern rail line, which carries coal eastward from the Wyoming fields. Gillette, the largest town in the center of Wyoming mining activity in the Powder River Basin, saw its population increase by a factor of five in a decade. Emerging from years of economic stagnation, scores of other communities are experiencing similar growth spurts.

This energy boom did not happen earlier because of a combination of factors: distance from major markets, technological problems, and abundant alternative energy sources. With substantial energy price increases in recent years, the economic barriers to exploitation have diminished. Although Plains coal is often found in thick, easily mined seams, it is expensive to ship to Eastern markets. Some had considered shipping the coal eastward by pipeline, first transforming the coal into "slurry," a fine coal powder mixed with water, then pumping it inexpensively like liquid petroleum. However, the great demand this would place on the region's limited water supply doomed the plan. The passage of the Clean Air Act in the early 1970s provided an important boost for the West's generally low-sulphur coal. Nearly all the coal mined in the West is used for electrical power generation.

Technological problems have also limited petroleum production from the Athabasca tar sands in Alberta. Located about 400 kilometers (250 miles) north of Edmonton, it is estimated that these sands contain at least 300 billion barrels of petroleum, an amount that approximates the entire world's current known reserves of conventional oil—and some estimates amount to more than twice that. (Although the actual tar sand area is beyond the margins of the Great Plains and Prairie region, its development focuses on Edmonton; hence, it is included here.) Unfortunately, because this petroleum saturates the sands in a solid form, it must be mined somewhat like coal, and technology for its refining is still in the development stage. As energy prices rise and the applicable technology improves, this reserve will be of tremendous importance to Canada.

It is difficult to anticipate the scale of these developments and their possible consequences. At least 100 billion tons of low-sulfur subbituminous coal that meets strict antipollution laws can be found near the surface in the northern Plains of the United States, an amount equivalent to that needed for 125 years at current levels of national consumption. Within 2000 meters (6000 feet) of the surface, the total is perhaps 1.5 trillion tons. Already the structure of the regional economy is shifting, with agriculture and ranching declining in importance. Despite strict regulations, environmentalists fear that future exploitation will result in soil erosion, water and air pollution, and rapid population growth. Whatever form the future development takes, the northern Plains have entered a period of rapid change.

Potash. Central and eastern Saskatchewan contain the world's largest concentration of potash, a fertilizer component. Many small agricultural towns of the area thus have mining and processing facilities in addition to their service center functions. While the scale is much smaller, potash, therefore, has had a local urbanization impact similar to petroleum in Alberta.

POPULATION PATTERNS

Population decline, or at best stagnation, has become the accepted standard across much of the Great Plains during the past 50 years. The Canadian provinces, notably Alberta, have fared somewhat better than have the U.S. Plains states during this period. As recently as 1950 the Canadian prairie population was split fairly evenly between the regions three provinces. Today Alberta's total is greater than the other two combined. Where substantial population increases are found, the areas are almost invariably associated with significant recreational or natural resource developments, or with urbanization. These three factors reflect a pattern found generally in both countries. The difference is that this region has a decided lack of urban centers, major recreational potential is minimal, and until recently there were few important natural resource developments. Regional population growth is concentrated in the larger cities near the margins of the Plains while most smaller communities and rural areas experienced outmigration and often population decline. For example, the Edmonton and Calgary metro areas accounted for three-fourths of Alberta's population increase between 1941 and 1986. Their share of the provinces's population grew from 23 to 61 percent.

One recent work has suggested that most forms of agricultural activity should be abandoned in a broad swath of territory between the 98th meridian and the Rocky Mountains from Texas to Montana and North Dakota. The authors suggest that the best use of most of the area is as a Buffalo Commons, where large herds of the animals are again allowed to roam freely. They argue that U.S. settlement of the area is the longest running environmental mistake in the country's history. They identified over 100 counties that had lost more than 50 percent of their population since 1930 and over 10 percent in the 1980s. The populations of these sparsely settled counties (they have fewer than 11 people per square kilometer (4 per square mile)) have a high median age, a result of ongoing outmigration of young adults seeking opportunity elsewhere. Twenty percent or more of all residents have incomes under the poverty level, and annual new construction per county was miniscule, valued at less than $50. The authors suggest that if current trends continue, two-thirds of the area's farms and towns will vanish by 2020.

The idea of the Buffalo Commons was met on the Plains with a combination of anger, disdain, and sincere expressions of regional pride. There is probably little likelihood that such a drastic proposal will be accepted. Still, the problems of

[1] Popper, Deborah E., and Frank J. Popper. "The Fate of the Plains," in Ed Marston (ed), *Reopening the Western Frontier*. Washington, D.C.: Island Press, 1989, pp. 98–113.

poverty and sparse (and declining) population, with resultant inadequate public services and limited opportunities for young people, persist.

Much of the American portion of the region is served by major urban centers that are found somewhat beyond the peripheries of the Plains. Chief among these are Kansas City and Minneapolis–St. Paul. Denver, Dallas–Fort Worth, and San Antonio, the largest American cities on the Plains, are all peripheral. Denver is a major regional center for much of the central Plains, but it also plays the same role for much of the Interior West. It is a regional office center as well as the focus of financial activity for energy resource development on the northern Plains and in the Interior West. The Dallas–Fort Worth urbanized area, with a population approaching 3 million (easily the region's largest) seems to straddle the margin between the Plains and the South. Overall, oil, the general effects of the Texas boom economy, and this urban area's emergence as a dominant regional office center for the Southwest mark its growth. Still, Dallas seems more of a city of the humid east, whereas the smaller Fort Worth—50 kilometers (30 miles) to the west—is a ranching and stockyard center that is clearly part of the Plains. San Antonio, third in size on the Plains, is the largest commercial center in south Texas plus the home of several major military bases. The city's population is over 50 percent Hispanic, placing it clearly in the Southwest Border Area as well (see Chapter 14).

Many of the somewhat smaller, but still quite important, centers serving the area are also peripheral—cities such as Tulsa and Omaha. The service areas of the cities grouped around the edges of the Plains tend to be elongated east-west zones that cover the region. Within the United States, the eastern peripheral centers are located along the far more densely settled lands found from about the 98th meridian eastward.

In Canada, by comparison, all of the major cities of the Plains are separated from other heavily populated areas by broad expanses of only sparsely settled territory. Winnipeg, for example is nearly 1600 kilometers (1000 miles) from southern Ontario. One result is that the Canadian prairie cities, although peripheral to the region in location, all clearly focus on it. Winnipeg, especially, is also a complex manufacturing center that produces goods for the region and processes agricultural products before they are shipped eastward. Winnipeg was early the largest prairie city. Its location at the southeastern apex of the region, and thus at the focus of transport between the prairies and eastern Canada gave it a considerable locational advantage. The prairies can be thought of as a giant funnel with the narrow end at Winnipeg. Most of the bulky products leaving the Canadian grasslands flow through that narrow outlet. Today, however both Calgary and Edmonton are larger than Winnipeg. Calgary has prospered from Alberta's energy boom. Some call it the most American of Canadian cities, at least in part a result of the impact of American energy firms. Edmonton also shares in the energy development but maintains a more complex commercial and governmental function.

Most towns on the Plains began as transportation centers, commonly strung out along the railroads. Those that have prospered maintain some transport service function, but they have also become established regional market centers. Some are also supported by special local conditions—Oklahoma City and Tulsa, for example, are important petroleum centers. Wichita is a manufacturing center for small aircraft. And, as mentioned, many cities are now growing because of recent energy developments.

The beef processing industry has expanded into many smaller Plains communities in the United States during the last three decades. Formerly, the industry had been concentrated in the large cities of the Midwest east of the Plains, where facilities were large and complex. Changing technology in the slaughter industry, the growth of feedlots on the Plains, and more diversified marketing patterns gradually made smaller plants located near the new feedlots more economical. The vast stockyards of midwestern cities such as Chicago are now gone,

Sited along the Red River near the eastern edge of the Canadian prairies, Winnipeg, Manitoba, possesses strong focality in the northern grasslands. (Courtesy Manitoba Government Travel, Department of Tourism and Recreation)

replaced by production from scores of smaller Plains towns.

Transportation routeways on the Plains were originally built to cross the area rather than to serve it (Figure 13-8). Thus, most major highways and railroads in both countries pass east-west across the Plains, with few lines running north-south. Of the seven interstate highways passing through substantial parts of the region in the United States, six are east-west routes. In Canada, the only substantial north-south highway follows the western edge of the prairies in Alberta. The same overwhelming predominance of east-west movement is also true for railroads in both countries.

Whatever the cause and justification of this pattern, one important result for American residents of the region has been a continuation of the economic and perceptual orientation toward areas off the Plains instead of toward areas within it. North and South Dakotans, for example, focus strongly on the Minneapolis–St. Paul area. The financial and trade institutions of the Twin Cities play a major role in these two states: the principal Twin Cities newspaper is read widely across the region, the activities of the

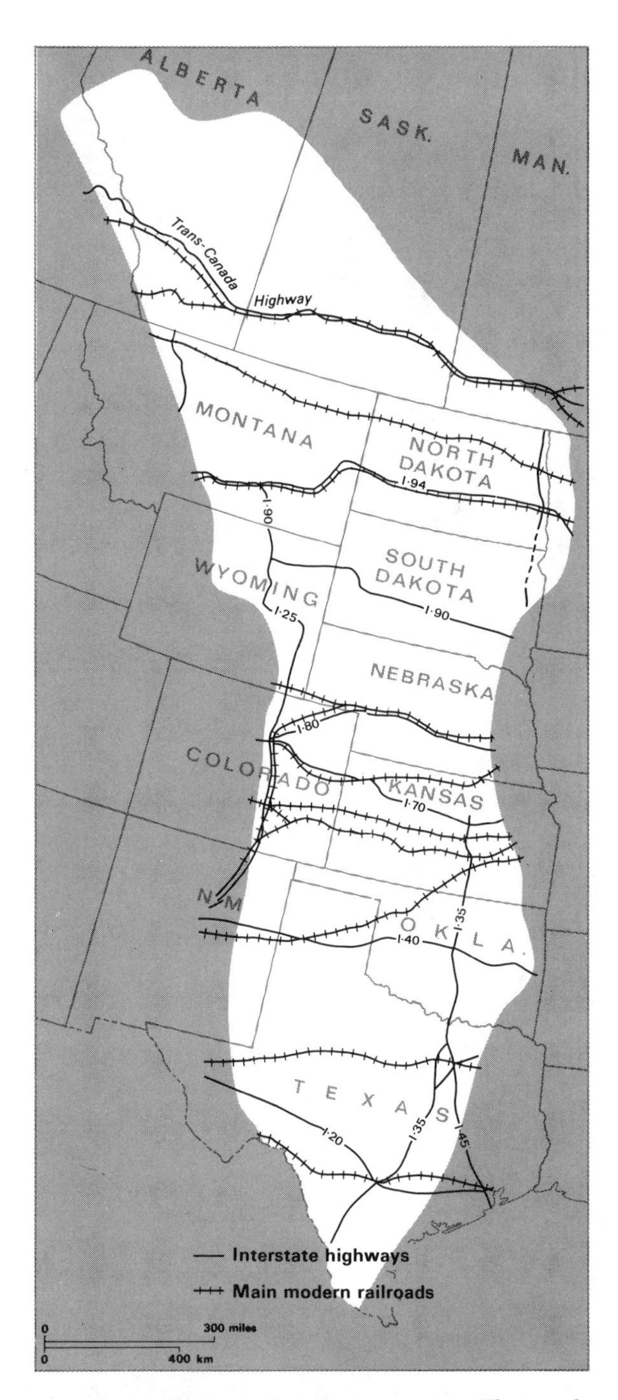

Figure 13-8 Major rail and auto routes. The north-south and, especially, east-west linearity of major overland transportation routes on the Great Plains can be frustrating for anyone hoping to travel diagonally across the region.

various professional sports teams there are followed fervently, and the joys of that urban area become, vicariously at least, those of the Dakotans as well. Typical residents of the Dakotas are likely to be more aware of the Twin Cities than of the bordering Plains areas to the north and south, and the same pattern is repeated elsewhere across the Plains. Again, this is far less the case in Canada.

IDEAS OF THE PLAINS

Interestingly, the concept of the Great Plains as a region is a relatively recent addition to the popular terminology. Even today, most residents of the Plains, if asked to locate themselves in terms of the names of the physical region in which they live are unlikely to answer with the term Great Plains. Instead, a more localized term will usually be given. Thus, a Texan may live on the Staked Plains, a Kansan in the Gypsum Hills, and a Nebraskan in the Platte Valley or Sand Hills. These and several score other distinctive areas are the basic reference regions of the Plains population. The Great Plains as a broader region is, to a substantial extent, an academic invention of the twentieth century. The widespread concern over the problems of the region, both environmental and economic, especially as they developed during the depression years of the 1930s, may have been the major contribution to this conceptual development.

One major and, in a sense, negative result of the development of the Great Plains idea has been the continuation of a belief that the Plains environment is a monotonous one of flat uniformity. This stereotype is often encouraged when travelers cross portions of the Plains by car. Two of the most traveled U.S. routes, I-40 and I-80, pass through the Staked Plains and the Platte River Valley, respectively. These areas are flat and, by some measures, the least attractive portions of the region. In fact, the physical environment of the Plains possesses substantial variation. The Sand Hills of Nebraska, the Badlands of South Dakota, the Alberta uplands, the Mani-

The dark billows of a dust storm are approaching a small Plains town in this photograph from the mid-1930s. Improved farming practices have lessened the likelihood of such storms, although smaller dust storms still develop, especially during the region's cyclic drought periods. (Library of Congress)

toba lowland—these areas are certainly different in both geologic structure and physical appearance.

One other misconception held by many is that the population of the Great Plains, like the physical environment, is ethnically homogeneous. This is not the case. In fact, many parts of the region have striking diversity as the norm, especially in Canada where recent immigration has played an important part in twentieth-century population growth. The Scandinavian settlement of the northern United States is well known. East European settlement, less widely recognized, was important in many areas, as was illustrated by Willa Cather's fine novels of pioneer settlement in Nebraska. Mexican-Americans are a major component of the population of the Colorado Piedmont. Winnipeg is today one of Canada's more ethnically diverse cities; restaurants there offer food from at least a dozen countries. Large numbers of American Indians are also located on the Plains, especially in South Dakota and Oklahoma. In spite of its unifying features, the Great Plains is a region of great diversity in physical and cultural landscapes.

GRASSLANDS LITERATURE

Yes, we buried him there on the lone prairie,
Where the owl all night hoots mournfully,
And the blizzard beats and the winds blow free
O'er his lonely grave on the lone prairie.

(From the folk ballad, "The Ballad of the Dying Cowboy")

It is in the songs, folktales, and literature of a region that we often find the truest measure of the appearance and nature of a place. The regional novelist and balladeer often recognizes the environment as an ever-changing stage for his or her characters. The landscape is, therefore, important as backdrop, as scenery. However, its role is usually not passive. It can influence and mold the thoughts and actions of the characters, and in turn, it can be influenced by those actions. The landscape, thus, becomes a vibrant, important, and even a dominant element in good regional fiction and semifiction.

Like the Deep South, the Plains has produced an uncommonly large volume of such works. Through most of them run a few common threads that seem almost to bind them to the grasslands environment. Hamlin Garland, raised in South Dakota in the late nineteenth century, introduced one of these themes in his novel *Main-Traveled Roads* (1891):

> How poor and dull and sleepy and squalid it seemed! The one main street ended at the hillside at his left and stretched away to the north, between two rows of the usual village stores, unrelieved by a tree or a touch of beauty.

This barrenness, much the result of treelessness, is a theme repeated in Plains novels. Beret, the wife in O. E. Rolvaag's magnificent tale of Norwegian frontier settlement on the Minnesota-Dakota border, *Giants in the Earth,* would look at the treeless horizon of her new home and realize "the full extent of her loneliness, the dreadful nature of the fate that had overtaken her." She wondered: "How will human beings be able to endure this place? Why, there isn't even a thing one can hide behind." In the story, her husband eventually dies in a blizzard, and the isolation and barrenness of her life drives Beret insane.

In *The Grapes of Wrath* (1939), John Steinbeck viewed an Oklahoma wracked by a decade of economic depression and years of severe drought.

> The right of way was fenced, two strands of barbed wire on willow poles. The poles were crooked and badly trimmed. Beyond the fence, the corn lay beaten down by wind and heat and drought, and the cups where leaf joined stalk were filled with dust.

To emptiness, then, is added the theme of a mercurial, often devastating climate.

Marie Sandoz developed this idea of climatic hazard in *Old Jules,* a biography of her father in the Sand Hills of Nebraska published in 1935. Her father was proud of an orchard he had planted; then a hailstorm struck.

> . . . suddenly the hail was upon them, a deafening pounding against the shingles and the side of the house. . . . One window after another crashed inward. . . . Jules came in from the garden and sat hunched over a box before the house. All the trees were stripped, bark on the west and south, gone. Where her garden had been Mary planted radishes and peas and turnips as though it were spring. The corn and wheat were pounded into the ground, the orchard gone, but still they must eat.

Despite this frequent malevolence, the Plains also offers beauty in its starkness and opportunity in its emptiness. Per Hansa, Beret's husband, rushed to his new field to begin breaking sod with a "pleasant buoyancy" and felt "the thrill of joy that surged over him as he sank the plow into his own land for the first time." Willa Cather also focused on this love for the farmland in her *O Pioneers* and *My Antonia.*

A long-term resident, born and raised on the Great Plains, when asked to comment on the beauty of the mountains of central Pennsylvania, responded, "Someone needs to cut down the trees first so I can see them (the mountains)." His response was a bit tongue-in-cheek, but its point was important. The openness of the Great Plains, the feeling of distance that distressed many people of the woodlands when they moved onto the grasslands, is now perceived as a positive aspect of the environment by many of its current residents. This, they proudly state, is "Big Sky" country.

ADDITIONAL READINGS

Alwin, John A. "Jordan Country: A Golden Anniversary Look." *Annals of the Association of American Geographers,* 71 (1981), pp. 479–498.

Blouet, Brian, and Frederick Luebke (eds.). *The Great Plains: Environment and Culture.* Lincoln: University of Nebraska Press, 1979.

Carlyle, William J. "Farm Population in the Canadian Parkland," *Geographical Review,* 79 (1989), pp. 13–35.

Frazier, Ian. *The Great Plains.* New York: Farrar, Straus & Giroux, 1989.

Hudson, John C. *Plains Country Towns.* Minneapolis: University of Minnesota Press, 1985.

Jordan, Terry G. *Trails to Texas: Southern Roots of Western Cattle Ranching.* Lincoln: University of Nebraska Press, 1981.

Lawson, Merlin L., and Maurice E. Baker (eds.). *The Great Plains: Perspectives and Prospects.* Lincoln: University of Nebraska Press, 1981.

McMurtry, Larry. *Lonesome Dove.* New York: Pocket Books, 1965.

Morgan, Dan. *Merchants of Grain.* New York: Viking Press, 1979.

Pearson, Jeff, and Jessica Pearson. *No Time But Place: A Prairie Pastoral.* New York: McGraw-Hill, 1980.

Rees, Ronald. *New and Naked Land: Making the Prairies Home.* Saskatoon, Sask.: Western Producer Prairie Books, 1988.

Shortridge, James R. *The Middle West: Its Meaning in American Culture.* Lawrence: University of Kansas Press, 1989.

Weir, Thomas, R. *Manitoba Atlas.* Winnipeg: Department of Natural Resources, Province of Manitoba, 1984.

CHAPTER

14

The Empty Interior

Stretching from the eastern slopes of the Rocky Mountains westward to the Sierra Nevada of California, the Cascade Range of the Pacific Northwest, and the Coast Mountains of British Columbia is found the largest area of sparse population south of the Arctic and subarctic in North America (Figure 14-1). This low average population density is the key identifying feature of this region. Indeed, there is much variation in other elements of the territory's geography. Portions, notably the Rocky Mountains in the east, the Sierra Nevada–Cascades–Coast Range–Coast Mountains systems along the western margins, and the Alaska and Brooks ranges in the north, have some of the most rugged terrain in the two countries. Between these mountain chains lies a series of plateaus, many of which contain extensive flat areas. Annual precipitation ranges from more than 125 centimeters (50 inches) in the Bitteroot Range of northern Idaho to less than 25 centimeters (10 inches) in the plateau country. The population of the region is mostly of northern European origin, although Hispanic Americans and American Indians are in significant proportions in the south (see Chapter 15). Irrigated agriculture is important in many areas, as is ranching, whereas in other areas, lumbering, tourism, and mining are dominant.

This massive expanse of land extends nearly 1600 kilometers (1000 miles) from east to west in its wider southern sections and some 600 kilometers (375 miles) in parts of British Columbia. The region contains some of the most strikingly scenic portions of the continent. This vast scenic resource and its variable character have created the dominant perception of "the West" held by most Americans. The impact of humans on the region, although locally important, has been overshadowed to a great degree by the varied splendors of the natural environment.

Crisp, clear air. Bold, angular mountains, their peaks capped with snow and bases covered with trees. Few people. This is many Americans' perception of the Empty Interior. (Colorado Office of Tourism)

Major Metropolitan Area Populations in the Empty Interior, 1990	
Phoenix, Ariz.	2,122,101
Salt Lake City-Ogden, Utah	1,072,227
Las Vegas, Nev.	741,459
Tucson, Ariz.	666,880
Albuquerque, N.Mex.	486,977
Spokane, Wash.	361,364
Provo-Orem, Utah	263,590
Reno, Nev.	254,667
Boise, Idaho	205,775

A DIFFICULT ENVIRONMENT

The Mountains

As discussed in the chapter on the Great Plains and prairies, the impressions that any individual or group has about a place are heavily influenced by the past experiences of the people involved. Most Americans, products of the eastern United States and northwestern Europe, often found the grasslands of the midsection of the continent alien. Many of the early evaluations of the Plains are biased by this perception. The reaction of most Americans to the Empty Interior of the West can be viewed in much the same light. Eastern Americans were (and are) products of the eastern woodlands; they are accustomed to an undulating terrain where variations in elevation are seldom dramatic. Where mountains occur, their size, certainly when compared

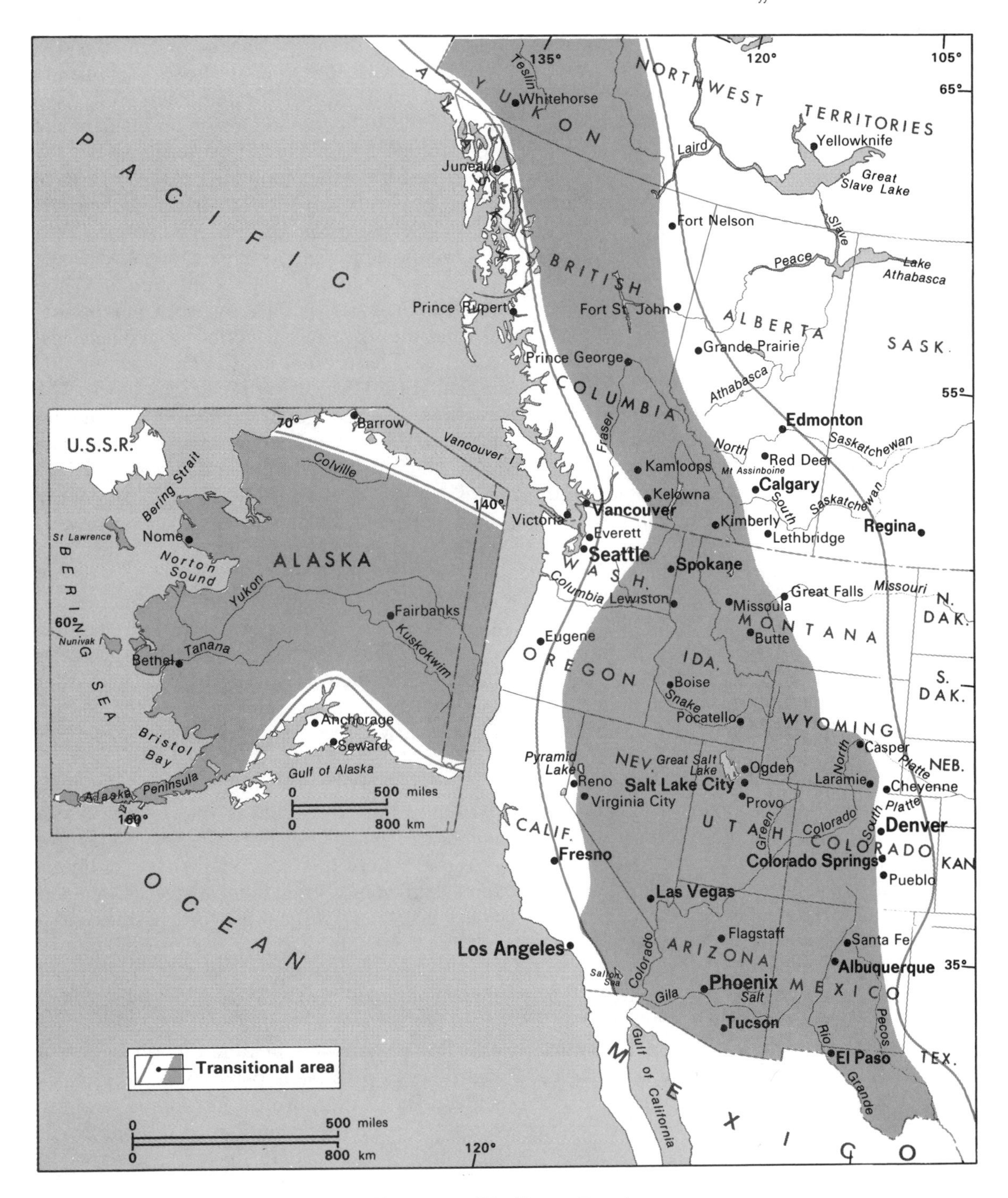

Figure 14-1 The Empty Interior.

to those of the West, is often moderate. Most of the mountains of eastern North America, with the possible exception of a few in the Presidential Range of New Hampshire, do not contain an elevation change from the mountain's base to its top that exceeds 1000 meters (3000 feet), and even this degree of terrain variation is rare. By comparison, dramatic changes of 1000 meters or more are common in the Interior West. The scale of altitudinal variation is vastly different from that found in the East (see Chapters 8 and 9).

Variation in local relief is only one aspect of the difference between the East and the Interior West. A second somewhat related element of the physical geography of the region is its ruggedness. Most of the mountains of the eastern United States appear rounded and molded; the ranges of the West present abrupt, almost vertical slopes, and the peaks frequently appear as jagged edges pointing skyward. This difference is due partly to age. Most of the western mountains, although by no means all of them, are substantially younger than the eastern ranges. Thus, erosion has been active for a much shorter time. One result of long-term erosion is an eventual smoothing of the land surface and the elimination of many of the jagged edges that may have been characteristic at an earlier time.

A drive across the mountains in Colorado's Trail Ridge Road, for example, affords a striking scene of the effects of erosion. The road bisects the alpine country of the higher elevations of Rocky Mountain National Park. Much of the region was arched upward during several different geologic periods; interspersed periods of limited uplift allowed erosion eventually to "smooth" the surface. Today a trip along the road begins with dramatic increases in elevation, scenic vistas of deep canyons and jagged ridges, and a steeply rising, almost frightening drive to the crest. Then, for nearly 18 kilometers (12 miles) along the highest parts of the road, with elevations above 4000 meters (12,500 feet), the terrain is gently rolling; if you can ignore the rugged mountains that are visible in almost every direction, the alpine vegetation, and the low-lying clouds below you, the road is similar to many country drives in the East. You have passed through areas of erosion on the flanks of the uplifted area and into its core. Here, far away from the base of the mountains, is the last area to receive the substantial effects of the geologically new cycle of erosion.

Climatic conditions, both historical and current, have been important contributors to the rugged character of these western mountains. During the most recent period of geologic history, the Pleistocene, substantial parts of this region were affected by the actions of ice. The massive ice sheets that constituted continental glaciation were confined in this region to portions of British Columbia and northward into Alaska, but smaller areas of ice, commonly called *alpine glaciation,* existed as far southward as the mountains of northern New Mexico. The carving done by these mountain glaciers did much to form the topography of the Interior West. Remnants of the glaciers can still be found in parts of the region. Most widespread in the Pacific mountains of southern Alaska, smaller glaciers are found as far south as the central Rockies in Colorado and the Sierra Nevada of California.

Alpine glaciers form in higher elevations and gradually flow downhill as the volume of ice increases. The moving ice is a powerful agent for erosion. Where this erosion pattern continues for a sufficiently long time, a deep U-shaped valley is created with almost vertical sides and a relatively flat bottom. If two glaciers flow side by side, a narrow ridge line is formed, characterized by jagged small peaks called *arêtes.* Where several glaciers formed in the same general locale and flowed outward in different directions, a sharp-faced pyramidal peak called a *horn* resulted. The Matterhorn (in Europe) is certainly the best known such mountain, but Mount Assiniboine in the Canadian Rockies is also a fine example of this feature.

Within North America, much nomenclature for various physical features is borrowed from

Trail Ridge Road in Rocky Mountain National Park, Colorado. The elevation here is 4000 meters (12,500 feet), well above tree line. Glaciers have carved deep valleys into this rolling upland, creating a pattern of gentle slopes dramatically dissected by steep cliffs. (Rocky Mountain National Park Photo, National Park Service)

sections of Europe where the feature is widespread. Thus, most features of alpine glaciation have French names, reflective of the French Alps. Many of the names for arid landscape features, such as *canyon* and *arroyo*, are from the Spanish language and reflect the earlier Spanish and Mexican presence in the arid lands of the U.S. Southwest.

The general result of alpine glaciation, then, is a rugged landscape. Yosemite Valley in the Sierra Nevada in California, an almost classic glacially carved valley nearly a mile deep, is perhaps the region's most photographed example of alpine glaciation. The consequences of such glaciation are visible in the higher ranges throughout the entire region.

The Plateau Country

Most of the Empty Interior is occupied by plateaus rather than mountains. With some exceptions, these plateaus are the basic feature of all the area between the Rocky Mountains and the Sierra Nevada–Cascades–Coast Ranges and Coast Mountains complex from Mexico through Alaska. The term "plateau" is, however, somewhat misleading, because parts of this section have great local relief.

Probably the most scenically dramatic portion of this section is the Colorado Plateau along the middle Colorado River in Utah and Arizona. Although there are some large structural changes in relief, most of the area is underlain by gently dipping sedimentary rocks. The major landscape features result from erosion by the exotic streams (so called because they carry water, something otherwise unknown or exotic, into this arid environment) that cross the plateau, most notably the Colorado River and its tributaries. In this arid environment, streams are easily the predominant erosive influence. Thus, when accompanied by recent substantial geologic uplift over much of the plateau, great downward

Monument Valley, in northern Arizona, was made famous by countless "Cowboy-and-Indian" movies of decades past. These narrow buttes and broader mesas are erosional remnants of a sedimentary rock bed now largely removed by water and wind action. (Santa Fe Railroad photo by Don Erb)

erosion has resulted and been concentrated primarily in the immediate vicinity of the streams. The canyonlands that have been produced, again particularly along the Colorado, are some of the best known examples of America's natural scenic resources. One recent major travel survey conducted in the United States identified the Grand Canyon of the Colorado River in Arizona as the most widely recognized natural scenic attraction in the country. River erosion deeper than the Grand Canyon occurs within parts of the Snake River Canyon. The greater attractiveness of the Colorado canyon, however, particularly in its more fully developed form in Grand Canyon National Park, is partly a result of its many deep tributary canyons. A canyon system has been created that is at places more than 16 kilometers (10 miles) wide. In addition, the variable resistance of strong and weak rocks in these sedimentary formations has created an angular pattern of scarps and benches that is especially characteristic of the area.

Filling the plateau country from the Colorado Plateau to the south across south New Mexico and Arizona, west into the Death Valley and Mojave Desert in California and as far north as Oregon and Idaho is the Basin and Range region. This wide area is composed of a series of more than 200 north-south trending linear mountain ranges that are usually no more than 120 kilometers (75 miles) long and typically rise 1000 to 1600 meters (3100 to 5000 feet) from their base within a collection of some 80 broad, flat basins. North and west of the Colorado River Basin, most of the Basin and Range area has interior drainage; that is, streams begin and end within the region, with no outlet to the sea. One result is that much of this area has received vast quantities of alluvia eroded from the surrounding mountains.

During the late Pleistocene, substantial parts of this section of the region were covered by lakes that resulted from a wetter climate and the melt of alpine glaciers (Figure 14-2). The largest, Lake Bonneville, covered 25,000 square kilometers (9500 square miles) in northern Utah. Most of these lakes are now gone or are greatly diminished in size because stream flow is now dependent on a lower annual precipitation; many of the lakes that remain, such as Pyramid Lake in Nevada or Utah's Great Salt Lake, are heavily saline. Flowing water always picks up small quantities of dissolvable salts; normally, these salts make a minor contribution to the salinity of the world's oceans. But because they lack an outlet to the ocean, lakes in the Basin and Range area have increased their salt concentration. The famous speed-racing course at Bonneville Salt Flats is on the former lakebed of Lake

Figure 14-2 Pleistocene lakes. Large sections of the Empty Interior were filled by lakes during the Ice Age. The Great Salt Lake is a remnant of the largest of these lakes, and the even surface of Bonneville Salt Flats in northwestern Utah is former lakebed.

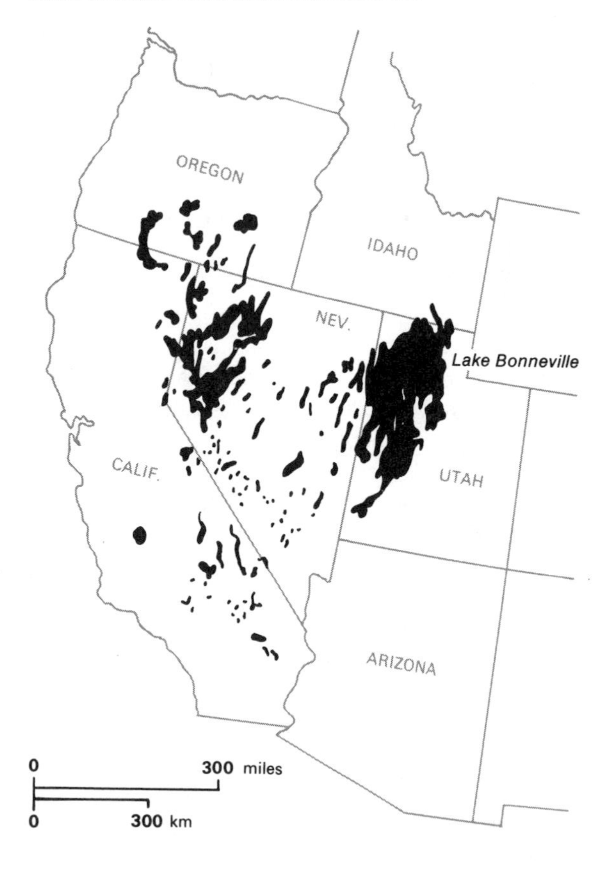

Bonneville. The Great Salt Lake, covering about 5000 square kilometers (1900 square miles), is the remnant of Lake Bonneville and today has a salt content much higher than that of the oceans.

North of the Basin and Range region, the Columbia Plateau is the result of a gradual buildup of lava flows. Contained by the surrounding mountains, these repeated flows, each averaging 3 to 6 meters (10 to 20 feet) thick, have accumulated to a depth of 650 meters (2000 feet) in some areas. A few small volcanoes or cinder cones dot the area, but the prime features of volcanic activity here are the vast flows of formerly molten material. Here, too, streams have eroded deep, steep-sided canyons. The canyons lack the stepped appearance of the Grand Canyon, however, because the lava beds present a more homogenous erosive material. With some gaps, the pattern of eroded plateaus continues northward into the Yukon Territory in the area between the Rocky Mountains and the Pacific Ranges. In central Alaska, the drainage basin of the Yukon River occupies the territory from the Alaska Range to the Brooks Range. Surface materials are mostly sedimentary rocks. Large portions of the middle and lower Yukon Valley are quite flat, and, thus, marshy in summer.

Climate and Vegetation

Much of the Empty Interior has a precipitation regime that is arid or semiarid. This is increasingly true toward the region's southern zones. In the north, especially in northern British Columbia, the Yukon Territory, and Alaska, the long cold season is the dominant climatic theme. Nearly all of the area in the United States that can be classified as desert is found in this region or in the southwestern borderlands. This aridity is of fundamental importance to the region; the water deficiency has played a greater role in the limited population growth of the area than has the ruggedness of the terrain.

There is a strong direct association between precipitation and elevation throughout the Interior West. In fact, a precipitation map of the region would also serve as a reasonably accurate topographic map. Low-lying areas throughout the region are generally dry. Heaviest precipitation amounts are usually found on the midslopes of mountains. The entire region is almost totally dependent for surface water on the exotic streams that flow down and outward from the mountains.

The association between topography, temperature, and precipitation results in a marked altitudinal zonation of vegetation throughout the Empty Interior. The lowest elevations are generally covered with desert shrub vegetation. The dominant low-elevation species almost everywhere in the region is sagebrush. Indeed, one subspecies, big sagebrush, which can grow 5 meters (15 feet) high and live 100 years, may be America's most abundant shrub. In the far south, there is a modest late summer increase in precipitation, locally called the monsoon season, that allows a sagebrush/grasslands combination. Elsewhere, this combination is found at elevations above the desert shrub.

Upslope from the sagebrush is a tree line. Above this lower tree line, precipitation is sufficient to support tree growth. The forests are at first a transitional mix of grass and small trees, like pinon pine and juniper. At higher elevations these blend into more extensive forests of larger trees, such as ponderosa pine, lodgepole pine, or Douglas fir. If the mountains are high enough, smaller tree such as subalpine fir varieties and then a second tree line are encountered. Above this upper tree line, a combination of high winds and a short, cool growing season render tree growth impossible, and the trees are replaced by tundra. The actual elevation at which these vegetation changes occur varies by latitude. Both tree lines are found at substantially lower elevations in the north. The lower tree line may also have a lower elevation on west-facing slopes, which receive higher amounts of precipitation from the normal eastward movement of weather systems, and north-facing slopes where shade lowers temperatures and thus retards water loss due to evapotranspiration. The boundary between forest and tundra is above 3700 meters (11,500 feet) in New Mexico, whereas in southern British Columbia, it occurs at an elevation of less than 2000 meters (6000 feet).

Wildlife

Little mention is made in this text about the wildlife population of North America. Some should be, and the Empty Interior, with much of its territory publicly owned and thus available to support a wildlife population, is an appropriate place to begin.

Much has been written about wildlife decline and species reduction through expanding human activity. Indeed, in some areas the devastation has been staggering. In the United States Hawaii has probably suffered the greatest loss (see Chapter 18). However, the numbers of many large mammals and some birds have increased greatly in recent decades. The bison (buffalo) population has grown from little more than 10,000 in 1935 to 65,000 today. Over the same period the North American elk population grew from 225,000 to 500,000, that of the pronghorn antelope from 40,000 to 750,000, white-tailed deer from 5 to 15 million, and wild turkeys from 30,000 to 2.5 million. The overwhelming reason for this increase has been a dramatic change in public attitudes and a resultant improvement in management and conservation practices.

The states of the west have shared in this wildlife population explosion. Colorado and Idaho each have over 100,000 elk; Montana and Wyoming nearly match that number. California, Idaho, Montana, and Oregon each have over 10,000 wild bear. Each state has a substantial deer population. These wild animals attract large numbers of tourists, as anyone caught in an elk-viewing traffic jam in Yellowstone will attest to. Big game hunting has also become big business. In 1988 hunters spent over $100 million just on hunting licenses in the West. Most of this money is used in state conservation efforts.

THE HUMAN IMPRINT

Public Land Ownership

The human use of the Interior West is strongly influenced by the federal and provincial governments' extensive land control in the region. Most of the land in the Empty Interior in both the United States and Canada remains in governmental hands (Figure 14-3). In Nevada, for example, the various branches of the U.S. government control almost 90 percent of all land. Although the percentages are lower elsewhere, the basic pattern of governmental predominance is universal in the region. About 34 percent of all land in the United States is owned by the federal government.

It is not particularly surprising that so much of the land in the Interior West remains in governmental hands. In both countries, this area and the Far North were the final regions to be occupied by any substantial number of people. And in both cases, federal programs of land distribution, designed to encourage agricultural use, were not relevant because little of the region held any real agricultural promise. In the United States, the Bureau of the Census proclaimed the end of the settlement frontier in the country in 1890, a time when much of the Interior West still remained unsettled. Although significant settlement expansion continued on the prairies of Canada for several more decades, there, too, most of the Interior West was ignored. Also, by the time other interests, such as lumbering or mining, began to push for greater private land ownership, the federal governments were seriously reevaluating their earlier programs in which they distributed land almost for free. Again focusing on the United States, Theodore Roosevelt's creation of the U.S. Forest Service in the first years of this century was largely meant to counter a threatened transfer of much of this land into private ownership. This act was a landmark in the beginnings of an effective conservation movement in the United States. Because Canadian settlement of the Interior West tended

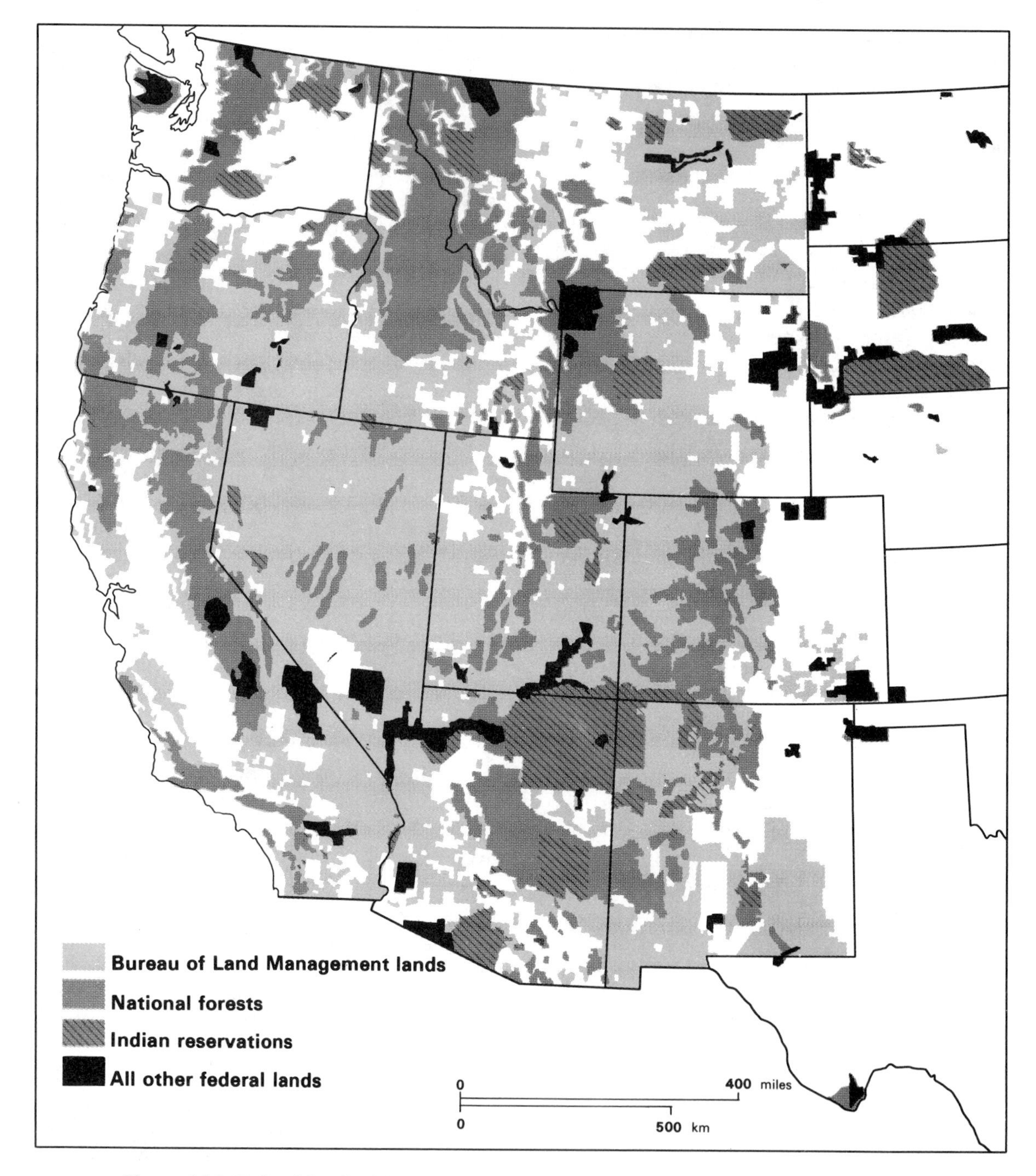

Figure 14-3 Federal lands. Many Westerners question the wisdom of such widespread federal land control. They argue that it lessens the tax base, damages opportunities, and often does not even result in environmentally desirable land use practices.

to lag somewhat behind that of the United States, an even larger portion of the nonurban, nonagricultural land in Canada remains in the hands of the federal and provincial governments.

These public lands have been put to a wide variety of uses in both countries. In each, but especially in the United States, a substantial part of the total national park system is found in the Interior West (Figure 14-4). The U.S. and Canadian national park systems have as their basic task the preservation of the unique, or at least, highly unusual, sections of their respective country's natural environment, and much of the physical landscape that fits this requirement is found in the Empty Interior. Some of the continent's most famous national parks, such as Yellowstone, Glacier, Banff, and Grand Canyon, are found there. Yellowstone, generally accepted as the world's first national park, was set aside in 1872 as the first nature preserve in the United States.

The national parks represent the best known use of this vast resource of governmental lands, but they are only a small portion of the total public land area in the region. The largest share of these lands in the United States is held by the Bureau of Land Management (BLM), a part of the Department of Interior, which puts this land to many uses, grazing being easily the most important. The Bureau has also been the main agent in the construction of irrigation and hydroelectric dams in the area. The U.S. Forest Service, based in the Department of Agriculture, is the second largest of the U.S. federal landholders. Much of the Forest Service's lands in the West were set aside during a few frantic days of the administration of Theodore Roosevelt, when the president feared that impending congressional action would release to private interests most federal lands that had not been specifically set aside for other purposes. The Forest Service today controls a major portion of the forested lands in the Interior West. The service has traditionally emphasized logging and grazing under its multiple-use charge. In recent years, it has also increased the quality and quantity of recreational uses of the land. This has been, at least partly, in response to increased pressures on the limited facilities available in the national parks as tourist numbers have boomed.

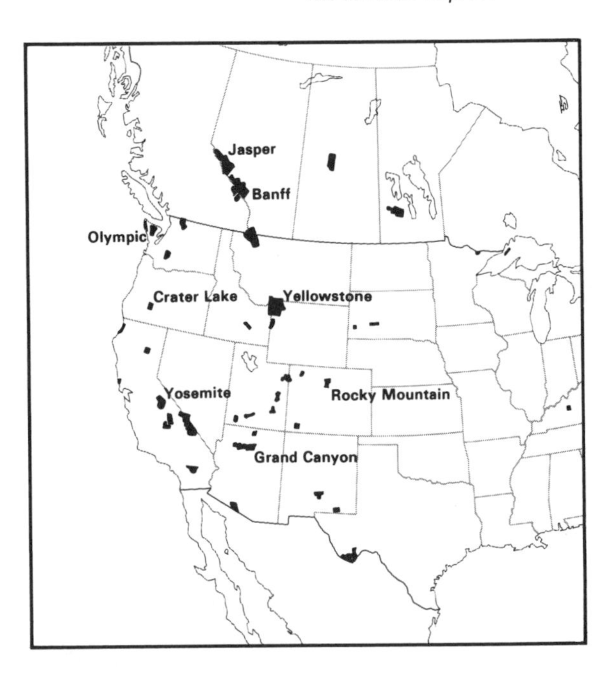

Figure 14-4 Major national parks. The interior West, with its frequently dramatic rugged topography, is the location of many national parks in Canada and the United States.

In the Canadian portion of the region, most governmental land not in federal national parks is controlled by the provincial governments. Most of the land is forested. Alberta and especially British Columbia have used this land resource to control and expand lumber production, and the latter has also established a number of provincial parks.

Many Westerners have questioned this pattern of extensive governmental land ownership. There is a variety of reasons for this concern. Some believe that governmental ownership has a negative impact on regional economic growth.

Governmental decisions on land use often do not satisfy local interests. No property tax is paid on the governmental land, and many feel that federal payments fail to cover that tax loss. Other critics are concerned about what they see as frequently poor management of governmental lands. A recent study indicates that misuse has damaged much of the federally owned grasslands in the West, damage that is difficult to reverse in arid and semiarid regions.

The development of the many U.S. national parks in the region plus a growing number of areas set aside under the Wilderness Act of 1964 (which basically states that humans must honor a hands-off approach to all areas designated as wilderness) has eliminated most kinds of economically productive use from those areas. Many of the region's rural residents object to this expansion of protected land. As wildlife populations of the nature reserves increase, so does their depredation of nearby ranch and farmland. Preservation also denies the opportunity for multiple use of the land for such activities as grazing and hunting. The result can be an income loss for a people who already often find it difficult to make a living.

Perhaps the greatest concern of some Westerners over this question focuses on their inability to plan and control their own region's future. Long-range economic and environmental plans must depend in large part on decisions made by the federal government, and local and national authorities often disagree on planning goals (see the Alaska discussion in Chapter 17). In 1970 the state governments in Nevada and Arizona proposed that the federal government donate 2.4 million hectares (6 million acres) of federal land to each state. That proposal was rejected. In 1978 Nevada then passed a law claiming that all BLM and National Forest lands within its borders belonged to the state. These efforts, and the movement that supported them, were called the Sagebrush Rebellion. Opponents argued that it was unconstitutional for states to claim federal land. The movement died out in the late 1980s, partly because people accepted the constitutional argument and partly because many doubted that the states would be able to do a better job of management. The quality of management of the federal lands has improved. The Forest and Rangeland Renewable Resource and Planning Act of 1974 mandated that a use plan for each national forest be created and updated every five years with public participation. The Federal Land Policy and Management Act of 1976 requires a similar planning program on all BLM lands.

Two other uses of parts of the Empty Interior say much about the region's past and about America's attitude toward the land's quality and usefulness. First, some of the largest American Indian reservations are found here, especially in northern Arizona and New Mexico. The social and economic implications of this are discussed in Chapter 15, but the existence of these reservations is partly a measure of the low regard Americans had for much of the region during the settlement era. In general, if territory controlled by native Americans was perceived as valuable, it was seized from them. Also, some of the country's largest bombing and gunnery ranges, as well as its only atomic bomb testing facility, are found in the region. The population is sparse, and alternative demands on the land are not great.

The Mormon Influence

The region's paucity of population is a principal feature of the Interior West. The combination of rugged terrain and widespread aridity limits the possible scope of agriculture. As the agricultural frontier moved westward in the late decades of the nineteenth century, it largely swept past the Interior West. In fact, were it not for minerals, transportation, and the Latter Day Saints, few people would have chosen the region until well into this century.

Settlement patterns within the Empty Interior have been strongly affected by members of the Church of Jesus Christ of Latter Day Saints, or more commonly, the Mormons. Established in upstate New York in 1830, the church and its

The Great Salt Lake Basin, Utah, along the western edge of the Wasatch Mountains, is a very dry area. Careful use of the meltwater from the mountains has enabled the Basin's farmers to create a major mixed-farming agricultural region. (Georg Gerster/Comstock)

followers were attacked repeatedly, both verbally and physically, for what were considered their "unusual" beliefs. The Mormons moved several times, each time searching for a place to practice their religion in peace. In 1844 a mob killed the church founder, Joseph Smith, in the Mormon town of Nauvoo, Illinois (which at the time was vying with Chicago for the honor of being the largest city in the state). Over the next few years, many Mormons, often on foot, pushed into the West to settle far from American settlement where they hoped to create an independent Mormon state.

The locale they selected for their intial western settlements was the Wasatch Valley, tucked between the Wasatch Mountains and the Great Salt Lake in northern Utah, a location that would later become Salt Lake City. It must have seemed an unlikely spot to begin an agricultural settlement. The climate was dry, the lake saline and useless, and the landscape barren. Nevertheless, the Mormons quickly began their agricultural operations; their settlements expanded as new arrivals came, for the Mormons were, and are, fervent missionaries. A high birth rate also pushed their population numbers upward. They dreamed of founding a country to be called Deseret, stretching northward into what is now Oregon and Idaho and southward to Los Angeles, and Mormons communities were established at greater and greater distances from Salt Lake City.

The Mormons ultimately failed in their hopes for creating Deseret. American expansion moved through and beyond the Mormon area, and the Mormons again found themselves under the will of the United States. Interestingly, the initial site that the Mormons had found so satisfying in its isolation lay directly astride what would become the California Trail, and the discovery of gold in California in 1849 initiated a large movement of non-Mormons through the Mormon settlements within a few years of their arrival. Deseret was divided eventually among a half-dozen different states.

This failure did little to lessen the impact of the Mormons on the settlement of the region. They were the first white people to face the problems of life in the Interior West, and they solved the majority of them. Many of their solutions were later adopted freely by non-Mormons. None was more important than their innovations in irrigation techniques. Americans had previously had no need for extensive irrigation. The techniques and careful central control necessary to collect and move water to a large number of agricultural users were almost unknown in North America. The Mormons, with their theocratic government providing the strong central organization, constructed a large number of storage dams on the western slopes of the Wasatch Range; many miles of canals moved the water to the users in the valley below. The results of these efforts today cover much of the valley with agricultural crops, trees, and green lawns. Even a casual observer flying over the area in early summer is struck by the greens of the irrigated fields contrasting dramatically against the browns of unirrigated portions of the valley. These early efforts at large-scale irrigation were the beginning of an irrigation boom in the Interior West, and today most of the agricultural crops here are produced on irrigated lands.

Mormons continue to have a substantial impact on the Interior West. Of the roughly 11 million persons in this region, more than 1.5 million are Mormons. They constitute three-fourths of the population of Utah, a majority of that of south-central Idaho, and substantial populations in Nevada, northern Arizona, and western Wyoming. Mormon beliefs are definitely procreative, and Mormons have by far the highest birth rate of any major religious denomination in the United States. Their strong belief in cooperation and support within the church has made this religion a central influence in business as well as family life in its region. For example, one of the two principal department stores in Salt Lake City is owned by the church. The Mormons have created a dynamic, optimistic society, although one that some argue is closed to most non-Mormons.

The regional boundaries and variable nature of this distinctive culture region have been defined by Donald Meinig (Figure 14-5).[1] He defines the core, with its focus on the populated Wasatch Valley, as the center of Mormondom. The domain, covering most of the rest of Utah plus southern Idaho, is dominated by Mormons, but it lacks the complexity of the core. The sphere is that area where substantial numbers of Mormons live in a largely non-Mormon society.

DISPERSED ECONOMIC STRUCTURE

Irrigation and Agriculture

As mentioned, irrigated crops dominate the overall agricultural production of the Interior West (Figure 14-6). Without irrigation, no stable agricultural economy would be possible in this dry region. Today much of the flow of several of the more important rivers is diverted for various uses, with irrigation claiming the lion's share. This massive diversion would be impossible under the water laws in effect in most parts of the United States, which usually require that usage

---

[1] Donald W. Meinig, "The Mormon Culture Region," *Annals of the Association of American Geographers*, Vol. 55 (1965), pp. 191–220.

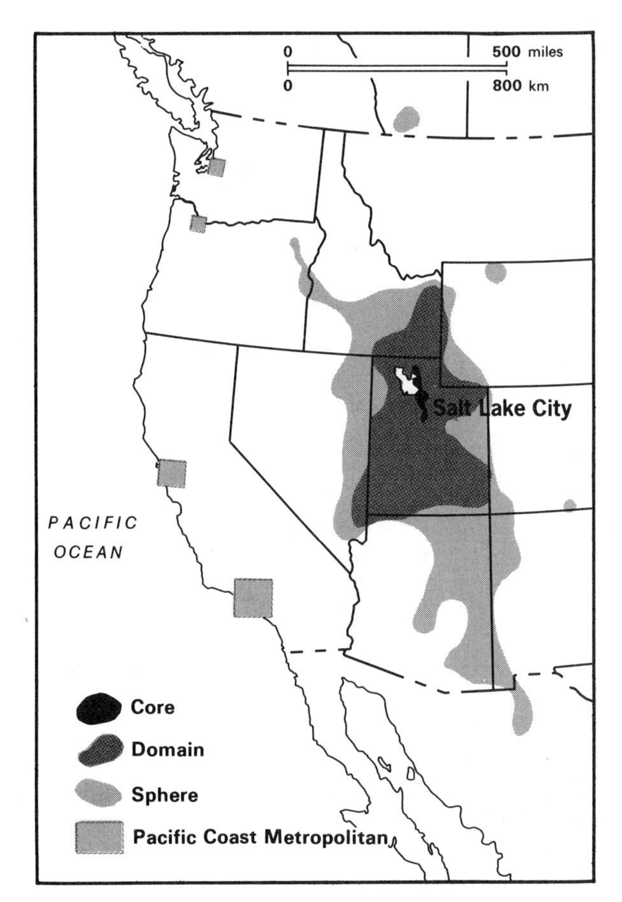

Figure 14-5 The distribution of Mormondom. (Reproduced by permission from *The Annals of the Association of American Geographers,* **55** (1965), Donald W. Meinig.) More than 90 percent of the population of some rural counties of southern Utah are Mormon. By comparison, Mormondom's core in Salt Lake City has a large non-Mormon population.

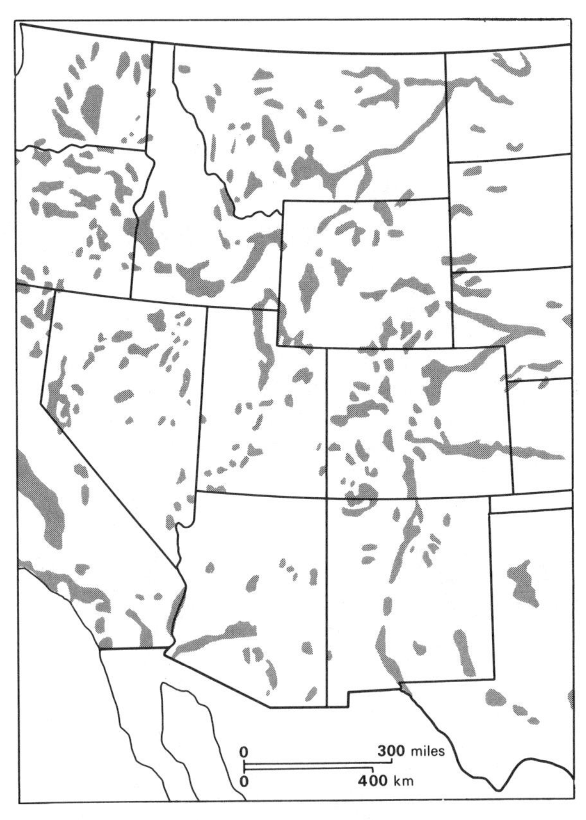

Figure 14-6 Irrigated areas. In the interior West, these areas are important as settlement foci in a region that is generally sparsely populated, as well as being the location of most of the region's agricultural production. Most of these areas are found along larger rivers.

not noticeably diminish stream flow. The unique water problems of the Empty Interior resulted in the development of a new set of standards, the Doctrine of Prior Appropriation, that permits total diversion of a stream.

The Reclamation Act of 1902, passed with strong Western backing, provided for federal support for the construction of dams, canals, and eventually hydroelectric systems for the 17 states (excluding Alaska and Hawaii) west of the 100th meridian. Today over 80 percent of the water from these federally supported projects is used for irrigation, much to the consternation of many urban water users. An area of over 4 million hectares (10 million acres) is irrigated with this water. While most of this irrigated land is in California, large irrigation projects are nevertheless scattered throughout the region. Agricultural water users have been required to pay only a small portion of the real cost of supplying irrigation water. Urban users pay far higher rates, although much of the real cost of development is simply borne by the federal government.

The use (some would argue misuse) of the

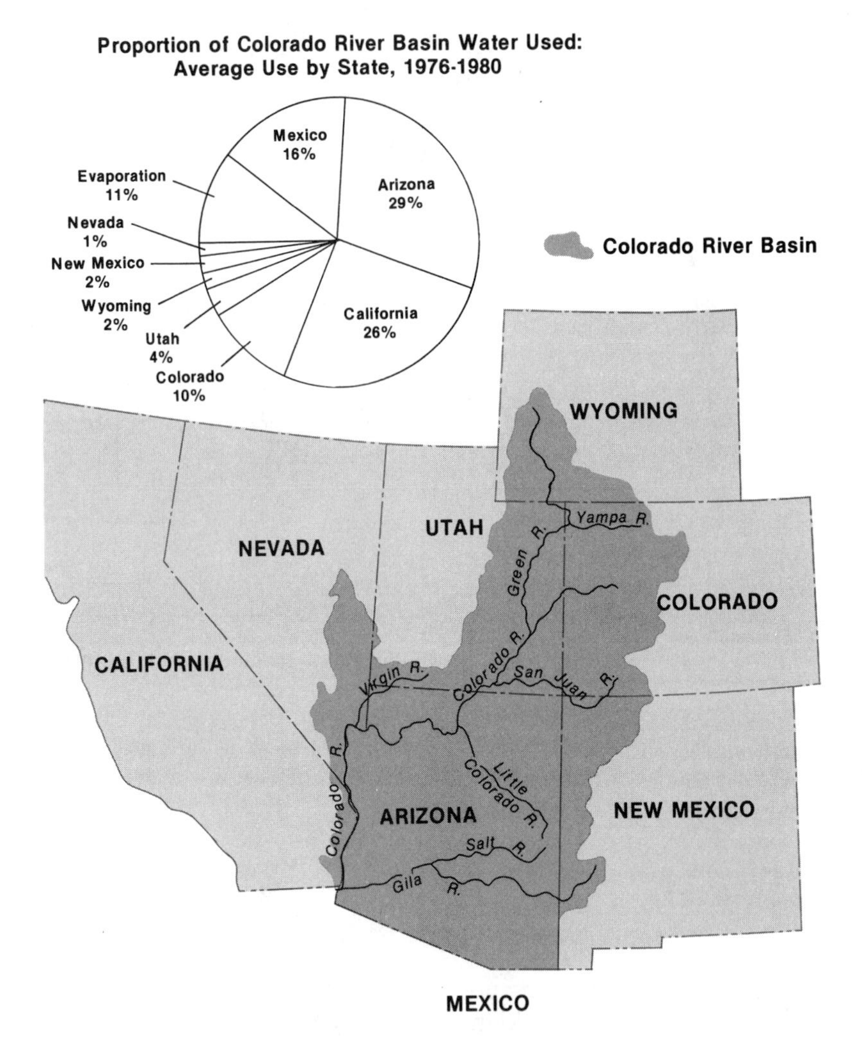

Figure 14-7 Use of water from the Colorado River. Competition for water from the river is fierce. It is a scarce, vital resource in this generally dry region.

longest river in the region, the Colorado, offers an interesting example of several aspects of the region's attitude toward water ownership (Figure 14-7). By 1915, nearly the entire flow of the river had been allocated, mostly at downstream locations in Arizona and California. This meant that the upper basin states of Wyoming, Colorado, Utah, and New Mexico had little right to the stream flow that originated largely in their mountains. In 1929, the Colorado River Compact reallocated these water rights, giving half of the river's total allocated flow of 15 million acre-feet to the upper basin and half to the lower basin states of Nevada, Arizona, and California.

Most stream flow in the region results from the spring and summer melt of winter snows. The amount of winter snow accumulation is, thus, quite important and quite variable. The winters of 1982–83 and 1983–84 were two of the snowiest in history. Water spilled over the top of Boulder Dam, Salt Lake City turned one of its main streets into a temporary river, and the Great Salt Lake, historically shrinking because of irrigation, reached its largest areal extent in over a century. The lake's water, six times saltier than the ocean's in the summer of 1982, had only three times the ocean's salt content by the summer of 1984. In contrast, winters in the late 1980s were dry in the Sierra Nevada and central Rocky Mountains, with resultant water deficiencies the following summers.

Competition among the states for the Colorado River's water is intense. The states of the upper Colorado have, until recently, used relatively little of their allotted share. The lower states, California and Arizona, responded by increasing their withdrawals until they had included most of the water "surrendered" by their northern neighbors. Arizona, confronted with water supplies inadequate to support its continuing population and economic booms, argued that California's share is too large and Arizona's too small. Arizona has gone to the courts several times to gain increases in its share of the Colorado River water. In 1973 it began construction of the Central Arizona project, designed to carry water from Parker Dam on the Colorado to Phoenix and Tucson. Water first reached Phoenix through the project in 1985. California had been using much of the unused portion of Arizona's allotment. That is being transferred to Arizona as the system is developed. The northern states, especially Colorado, have no desire to relinquish permanently any of their allotted share. The Big Thompson River project, discussed in the previous chapter, represents one part of Colorado's move to increase its usage.

The original allotments of Colorado River water have been modified in response to these conflicting interests. The total amount of water distributed will be reduced. Mexico's share will be increased substantially. Water available to the upper basin states will be reduced, whereas that for the lower basin states (primarily California and Arizona) will be increased somewhat. No decision, however, can possibly satisfy everyone, and additional reallocations are likely in the future.

The Colorado River projects are simply the most notable of a number of examples of intensive irrigation development in the intermountain region. Diversion by ranchers of water from the principal stream flowing into Pyramid Lake in western Nevada led residents of the Indian reservation to demand a halt to the practice before their lake disappeared. Irrigation projects along the Gila and Salt rivers in south-central Arizona were using virtually the entire flow of the Gila by the early 1930s. Farmers and the growing urban population in the area then turned to wells for additional supplies. Today Arizona depends on these receding subsurface waters to supply several times as much water as the Gila-Salt river system. In short, the pressures on a deficient water supply are growing more acute throughout the drier southern half of the region.

Future water resources are a matter of pressing concern throughout the dry west. Politics and economics have discouraged the federal government from proposing major new impoundment and diversion programs. Urban and industrial water users demand a larger share of the water. By some estimates, Arizona could support 20 million people with its current water supplies instead of its presnt 3.5 million people if none of its water went to agriculture.

As an example of an innovative response to the problems of water deficits, Arizona in 1980 passed a comprehensive water planning law. All large wells are now metered and withdrawal fees charged. Water rights can now be sold, with the expectation that they will be obtained by urban and industrial users. If that does not happen, the state will eventually begin buying farmland with water rights and retiring it from agriculture. The goal is to reduce the state's 526,000 hectares (1.3 million acres) of irrigated land to 324,000 hectares (800,000 acres) by 2020.

Although irrigated lands are scattered across much of the Empty Interior, a few areas are sufficiently large to merit special mention. The 1 million hectares (2.5 million acres) irrigated in the Snake River Plain makes Idaho the region's leader in terms of the number of acres irrigated. This enables the state to be among the country's leaders in potato and sugar beet production; alfalfa and cattle are also important. The Columbia Valley Reclamation project, supplied by the bountiful waters of the Columbia River, impounded behind Grand Coulee Dam in central Washington, contains well over 400,000 hectares (1 million acres), producing such crops as alfalfa, sugar beets, and potatoes. Irrigation along the Wasatch Valley has expanded little since the first decades of Mormon settlement. About 400,000 irrigated hectares there are devoted primarily to sugar beets and alfalfa. The Grand Valley, along the Colorado River in west-central Colorado, produces alfalfa and potatoes as principal crops, although tree fruits, especially peaches, are also important. In Washington, a series of tributaries of the Columbia River flowing eastward out of the Cascades, notably the Yakima and Wenatchee in Washington and the Okanagan in British Columbia, supply water for both country's most famous apple-producing regions. The Okanagan Valley (spelled Okanogan in Washington) has been irrigated since the 1860s. It was one of the first major irrigated areas in the Interior West. A combination of an attractive product and extremely effective marketing have enabled this dry area with an average annual precipitation of less than 25 centimeters (10 inches) to grow a third of each nation's apple crop.

It should be noted that each of these areas produces basically a limited set of crops. The short growing season precludes production of most long-season crops. Local demand is limited, minimizing the need for dairying or many fresh vegetables. Livestock, sugar from beets, apples, and potatoes, however, can be grown locally and shipped profitably to distant markets.

The more southerly irrigated districts, although particularly affected by irrigation-water shortages, nevertheless have one major advantage over their northern counterparts—a far longer growing season. The Imperial Valley, with a frost-free period in excess of 300 days, is currently one of the premier truck-crop-producing areas in the country. Much of the winter head-lettuce supply in the country comes from here, as do grapes, cotton, and alfalfa for fattening beef. The cattle population of the valley is over 250,000 head. A newly constructed electric-power-generating facility there uses the locally abundant cattle manure for fuel. The growing season is long enough to support double-cropping (growing two or more crops annually on the same area of land), and of course this increases overall productivity. The Coachella Valley north of the Salton Sea produces such crops as dates, grapes, and grapefruit. The Yuma Valley along the lower Colorado River supplies principal crops of cotton, sugar beets, and oranges. In the Salt River Valley, near Phoenix, winter lettuce, oranges, and cotton are the major crops. These southern crops, unlike those grown farther north, face little direct competition from the agricultural centers in the major market areas of the eastern United States.

Hoover Dam and Lake Mead on the Colorado River. The canyonlands of the Colorado River are an efficient area for water storage—the lakes can be very deep, minimizing water loss due to evaporation. The several lakes along the Colorado and its tributaries have created a major power-boat recreation industry in the arid Southwest. (Bureau of Reclamation/D.O.E.)

Transportation Services

Transportation has also played a major role in the population size and economic development of the Empty Interior. With the growth of population along the West Coast, the Interior West constituted a broad barrier to transportation between more densely populated areas in both countries. Because little traffic is generated within the area, a prime goal of transport developers has been to permit movement across the region as speedily and inexpensively as possible. Consequently, most major highway and rail routes pass through the region east-west, from the urban centers of the Midwest to those of the West Coast, as directly as the topography will allow.

Despite these requirements of transport design, the great width of the region demands development of many service facilities for the traffic passing through. Many of the smaller towns of the region, as well as a fair number of the larger ones, began as rail centers established to service and administer the railroads. The centers were founded wherever railroad personnel were needed, whether the region was populated or unpopulated. The arrival of automobile traffic in this century provided a demand for gas stations, car-repair facilities, motels, and restaurants for travelers. Although fewer local railroad workers were needed as technological innovations were implemented after World War II, this population reduction was more than compensated for by the growing need for people to serve truck and automobile traffic. Few regions today have as large a share of the total work force involved in transportation services as the Empty Interior.

Although the provision of transport services was the principal early influence on the growth of urban centers in the region, cities that have become the largest have usually been aided by some additional attribute. Spokane, Washington, with a population of 357,000, has, for example, become the principal center for the "Inland Empire" of Washington. That area, geographically defined and half encircled by the sweep of the Columbia River across central Washington State, has long been a region with substantial agricultural production. Albuquerque, New Mexico, with a Metropolitan Statistical Area population of 475,000, has gained a role similar to Spokane's through its centrality and accessibility in that state. Phoenix, Arizona, grew initially as an agricultural center and then boomed as Americans flocked to its warm, dry environment. Its 1.9 million people rank it twentieth among the country's urban regions. It has become a retirement center as well as a focus of manufacturing activities. Industries that produce a small, high-value product, such as the electronics industry, have been particularly important in the city's growth. The higher transportation costs engendered by locating a great distance away from most of their markets are not of major significance to these "light industries". The amenity factor of outdoor recreation opportunities, which attracts both labor and management, far outweighs the slight economic locational disadvantage. Locational considerations that largely ignore a traditional factor, such as accessibilty to either resources or a major market, have grown more common as technological advances place a greater share of all manufacturing in such industries (see Chapter 6).

Ogden, Utah, is one community that exists as a major rail center and was early among the most important such places in the region, but it has not become a major urban place. It is about 55 kilometers (35 miles) north of Salt Lake City, a city whose continued dominance derived from its early development and its key functions as the capital of Mormondom.

Tourism

Any mention of automobiles and travelers in the Empty Interior brings to mind the importance of tourism to the regional economy. The great variety and appeal of the region's scenic wonders attracts millions annually. Most of the major attractions are controlled by the federal government, but this has not slowed the growth of tourist-oriented industries. Visitors to most of the major parks must first pass a long, garish strip of motels, snack bars, gift shops, and other sources of ersatz local color. In addition, distances between attractions are usually great, and thus services are needed in countless locations. When Nevada's gambling operations are included as part of the area's tourist industry, the overall regional impact of tourism becomes even greater.

In 1931 Nevada passed laws legalizing many forms of gambling and simplifying divorce proceedings. The state soon became the gambling and divorce capital of the nation. Liberalized divorce laws in other states have eliminated part of this claim, but the gaming industry remains the cornerstone of the state's economy. State gambling taxes account for 45 percent of Nevada's revenues. The industry first centered in the Reno/Lake Tahoe area, but Las Vegas in the south now dominates. It has the advantages of relatively cheap electricity from nearby Hoover Dam to power the amazing display of lights that advertise the city's gambling and entertainment activities and proximity to urban southern California. A direct interstate highway and frequent, inexpensive air travel connect Las Vegas with its most important source of customers.

The total volume of tourism has grown so much in this thinly populated region that many of the tourist attractions have become overburdened. The U.S. National Park Service, faced with substantial local environmental damage from overuse of its facilities, has chosen to cut facility availability in some parks. In Yellowstone, where the summer morning lineup to obtain emptying campsites is as much a part of the landscape as the roadside bears once were, the service has reduced the number of campsites. Part of the road along the south rim of Grand Canyon National Park has been closed to private vehicles, as have sections of the road system in Yosemite National Park in California. One must have a reservation to stay in many of the more populated national park campgrounds during the busy summer season. This tremendous tourist growth has exerted pressure on the U.S. For-

Yosemite Valley in Yosemite National Park, California. Eroded primarily by a series of alpine glaciers, this is one of America's premier natural features. The valley is so heavily used in summer that the National Park Service has been forced to limit camping and to close a portion of the valley to private automobiles. (National Park Service)

Overgrazing in the late 1800s destroyed much of the sparse grassland vegetation that formerly covered large sections of the interior West. This area in Arizona has been protected from grazing for 50 years; still, little grass has returned because of competition from mesquite trees. (U.S. Forest Service)

est Service and the BLM to provide recreational facilities on its lands to decrease the pressure on the national parks. Current Forest Service and BLM policy encourages such multiple-use actvities on their land. Many also hope that substantial portions of the forestlands will be selected under the Wilderness Act of 1964, which would restrict human use of areas designated as wilderness.

Lumbering and Ranching

Ranching, lumbering, and, to a lesser extent mining depend on governmental land for many of their basic materials: The holdings of the U.S Forest Service and the Bureau of Land Management are open to grazing, and most lumbering here is carried out on Forest Service lands in the United States and Canada. Levels of production per acre for both ranching and lumbering products are relatively low, particularly in the United States. This is especially true when a comparison is made with land held in private hands.

There are several reasons for this apparent inefficiency. A greater interest in conservation on the part of governmental agencies is probably not one of these reasons, although the Taylor Grazing Act of 1934 did mandate the preservation of public range lands by regulating grazing. A more likely explanation is the limited quality of the land, particularly that controlled by the Bureau of Land Management. In many drier areas, 40 hectares (100 acres) of land per head are needed for satisfactory cattle grazing. The great seasonal climatic variations found in much of the region make this one of the few areas in North America where transhumance is practiced. This is a seasonal movement of animal flocks by those who tend them from the lowlands in winter to mountain pastures and meadows in summer. It is especially important in the sheep-ranching economy. Many Basques, expert shepherds from the Pyrenees of Spain and Portugal, came to this area as contract laborers to manage the herds. Today, descendants of the Basques are a substantial part of the population of several states, especially Nevada. Many who oppose the government's land-use policies contend that poor management has resulted in widespread overgrazing of federal lands and that the land's carrying capacity has declined as a result.

The wood products industry in British Columbia remains the cornerstone of the provincial economy. Wood products and paper and allied products are its two most important manufactured goods. Over 90 percent of the forest is Crown Land—held largely by the provincial government. The government has actively encouraged both cutting and reforestation.

Mining

The other element of the region's economic resource base that has supported the growth of substantial population nodes is mining. The Klondike gold discovery in 1898 was important in the early settlement of northern British Columbia, the Yukon Territory, and the Yukon Valley of Alaska. Miners followed shortly on the heels of the Mormons to become the second largest group of settlers in the region from the United States. The discovery of the Comstock Lode in Nevada gave rise to Virginia City, which grew into a city of 20,000 during its heyday around 1870 before nearly disappearing with the decline of high-quality ore. The boom in gold and silver mining in the years immediately following the Comstock discovery created rapid population growth in Nevada. This culminated in the admission of the state into the union in 1864, long before most of its neighbors. The depletion of much of this mineral resource by the late nineteenth century resulted in a widespread population decline in Nevada, one from which it did not fully recover until well into the twentieth century. Today, the mining economy is of little importance to the state, although some of the abandoned mining centers, notably Virginia City, are important tourist attractions. In fact, gold, a foundation of the mining boom over much of the region during the last century, is today of little economic importance in any part of the Interior West. Continued high gold and silver prices have encouraged some renewed mining, but most of this new interest remains at a small scale of operation.

Leading any list of mineral contributions to the region's economy is copper, with production today concentrated in Arizona and Utah. The vast open pit of the Bingham mine outside Salt Lake City, said to be the largest human-made excavation in the world, has yielded some 8 million tons of copper. Among the several score major and minor copper-mining centers in Arizona, the most important is at Morenci in the eastern part of the state. Other important mines are at San Manuel, Globe, and Bisbee, all in southern Arizona.

Most of the copper ore mined in the Empty Interior is low grade, with a metal content of under 5 percent. Consequently, most mines have a smelting or concentrating facility located nearby to lessen shipment costs by greatly reducing the weight of the material being shipped (see Chapter 6 for a discussion of the role of bulk or weight reduction in manufacuting location). Refining is, thus, a major manufacturing industry in the region, employing approximately 10 percent of the total manufacturing labor force in Arizona.

Lead and zinc follow copper in regional importance in the United States but are of greater importance in the Canadian portion of the region. The vast mine near Kimberley, British Columbia, dominates Canadian production of both metals. Production is more widespread in the

Copper mined in the Empty Interior is usually dug from vast open "pits," such as the Bingham Canyon Mine in Utah. The mineral content of the region's ore is low and the cost of production high; in consequence, many of the mines have been forced to close during the last decade. (Don Green/Kennecott Copper Corporation)

United States, with the two often joined by several other metals mined at the same location. The Butte Hill mine in Montana, for example, long was a significant producer of lead and zinc as well as copper. The Coeur d'Alene district in northern Idaho produces gold, silver, lead, and zinc; the Leadville district in Colorado has those four plus molybdenum, used in the manufacture of steel products. The last metal now ranks above silver in regional importance owing primarily to work at the Climax mine north of Leadville. Some three quarters of the world's supply of molybdenum comes from the Leadville district. Uranium exploration has also been widespread in the region, and today Utah and Colorado are the principal producing states. The northern interior of British Columbia is one of Canada's main resource frontiers. Large coal deposits have been developed in the Rockies in southwestern British Columbia. Most of the approximately 25 million tons of coal mined annually is exported to Japan.

The economic recession of the early 1980s was accompanied by increasingly effective foreign competition and, in some cases, by concern over the environmental and health consequences of

Lake City, New Mexico, once a thriving mining community, is now nearly abandoned. "Ghost" towns are common in the Empty Interior. Developed to serve mining, they lost their economic function entirely when the mineral resource was gone. New Mexico alone has 2000 such ghost towns. (Courtesy New Mexico Tourism and Travel Div. Commerce and Industry Department)

mining and smelting. This was devastating to the Interior West's mineral economy. The massive Butte Hill mine in Montana is closed, its underground mines flooded because the constant pumping needed to keep the mines dry has been stopped. The economies of Butte and the nearby former smelting center of Anaconda have been severely affected. The large zinc smelter in Kellogg, Idaho, is also closed. Many other mines have cut production.

One potential resource of the region must be emphasized. Spread across thousands of square miles of the area where Utah, Colorado, and Wyoming meet are the vast oil shale deposits of the Green River Formation (Figure 14-8). Locked in these rocks are as much as a trillion barrels of oil, vastly more than the entire proven alternative oil reserves of the world, including those of the Athabasca tar sands of Alberta (see Chapter 13). The technology needed for economically feasible extraction, however, is just developing. Processing the shale requires crushing and heating the material, so that the oil turns to gas and separates from the rest of the rock. The gas is then cooled, forming a liquid oil-like substance. The process is currently very expensive. Unfortunately, much of the shale is covered by overburden that is several hundred feet thick in many places. It has been suggested that the shale-converting process may leave behind enough waste to level the topography of the entire production area if most of the petroleum is removed. Also, most of the extraction process requires large amounts of water, a problem in this dry area where most available water is already used for agriculture. As a result of all these problems, efforts to develop this industry are largely on hold.

There has been little sustained or substantial urban growth based on these mineral resources.

As with all such resources, they are ephemeral, with depletion a standard segment of the extraction process. Cities, other than temporary boom towns, seldom develop on such a basis. In this region, copper provides the only real exception. Even in this, the results are modest. Butte, with a population of 34,000 in 1990, is perhaps the region's largest city developed with mining as the main base of the economy, yet it has long been an important processing center for agricultural products as well.

In summary, this is a region of sparse settlement, vast open spaces, and a quality of ruggedness and a sense of the outdoors that greatly appeals to its residents. The population is concentrated largely in isolated nodes. They know theirs is a region to which many travel thousands of miles for vacations, and they are generally firm in their belief that life in their region is superior. The Empty Interior's isolation and its lack of a broad resource base are clearly important here, making it for many, then, a delightful place to live, but a difficult one in which to earn a living.

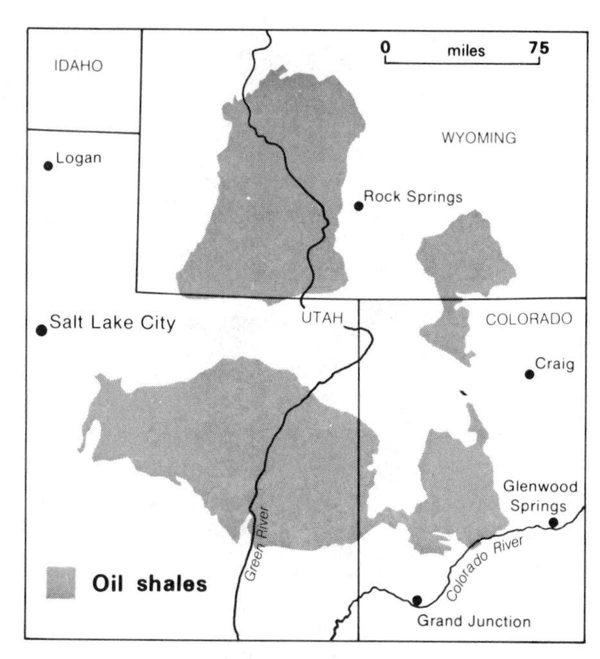

Figure 14-8 Oil shale deposits. Few people presently live in this isolated area. How much local population growth would a major development of this resource create?

NATURAL HAZARDS CASE STUDY

THE YELLOWSTONE FIRES

The summer of 1988 started like any other in Yellowstone National Park. The park, like much of the northern Rockies, was in a regime of long-term drought that was interrupted by periods heavy precipitation. May was a wet month. Natural fires that started in the park through much of June burned themselves out. Human-caused fires were quickly put out, as policy dictated.

Until 1972, National Park Service policy had been to extinguish all fires within a national park as quickly as possible. In that year the service adopted a "let-it-burn" policy for natural fires under the philosophy that such burns are necessary to maintain vegetative diversity and thus a diverse animal life. The fire management plan for Yellowstone gave park officials latitude to monitor and manage such fires, but they would be allowed to burn if they did not threaten life or property. Still, by 1988 most of the park's forest was littered with 100 years' accumulation of forest deadfall.

The rainy May gave way to a summer of no precipitation, very low humidity, high winds, and high temperatures. By midsummer much of the deadfall was actually drier than kiln-dried lumber. By July, one of the worst forest fire seasons in western U.S. history was

The rich green of new grass quickly began to replace the trees killed by the Yellowstone fire of 1988. This photo was taken in the summer of 1989. (Tom & Pat Leeson/Photo Researchers)

underway. Before it was over, some 70,000 fires would burn 1.5 million hectares (3.75 million acres) in the West. The cost of fighting those fires would exceed $500 million.

The first fires that continued to burn in Yellowstone started in late June and early July at scattered locations throughout the park. Most were caused by lightning, although one of largest, the North Fork fire, was probably started accidentally by a lumberman working a power saw near the western border of the park. By mid-August, more than 40 separate fires had been identified. On July 21 park officials decided to attempt to suppress all fires. The situation was clearly more critical than had been expected in early June. At the time, fires had burned some 6000 hectares (15,000 acres). Still no rain fell. On August 20 alone, 25,000 hectares (62,000 acres) of park land burned. On September 7, flames nearly surrounded the Old Faithful geyser area, and it was feared that the historic inn there would be lost. It survived, and finally, on September 10, the rain arrived. The worst was over, although isolated fires would continue into November. Almost 400,000 hectares (1 million acres), 45 percent of the park's land, had burned.

The fire generated a storm of controversy over the National Park Service's let-it-burn policy. Many in the tourist industry peripheral to the park feared a loss of business if fewer people chose to visit Yellowstone. Loggers and ranchers who worked near the park argued that fire did not respect park boundaries and that the policy thus threatened their livelihood unnecessarily. Others were simply saddened over the disappearance of a portion of Yellowstone's beauty. Many ecologists countered that fire was necessary to maintain a balanced environment and that the century-long absence of major fires in Yellowstone had resulted in loss of meadowlands and in an old forest that supported less animal and bird life. The fire forced the service to modify its policy to encourage a more rapid response to major fire threats if they might later be uncontrollable or if they threatened to move beyond park boundaries.

The park itself began its recovery with amazing speed. Many large animals had died, but no more than are lost during a difficult winter. Grasses were greening in some locations by late September. The park provided its most spectacular wildflower display in memory the next spring and early summer. One type of the cone on the lodgepole pine, a dominant species in the park, opens only when heated. By the end of the summer of 1989, tree seedlings from the lodgepole and other species common to the park seemed to be sprouting everywhere. And the tourists returned in record numbers, anxious to see the new wonder of Yellowstone.

SETTLEMENT CASE STUDY

MOAB, UTAH*

The "Invincible Three" were jittery after robbing the San Miguel Bank at Telluride, Colorado, on a June morning in 1889. Seeing a posse in every dustcloud, Butch Cassidy and his sidekicks, Tom McCarty and Matt Warner, spurred their horses toward a favorite northeastern Utah hideout called Brown's Hole. From Telluride, they galloped through Mancos and Monticello in a relentless ride to Moab, a small, isolated town situated at the only sensible crossing of the Colorado River for miles in either direction. There, they grabbed fresh horses and ferried across the river to find refuge in rugged Utah terrain.

Indeed, outlaws passing through southeastern Utah late in the last century must have had an instinctive appreciation for Moab's geographic attributes. Located astride a natural transportation route amid a region of irregular terrain and harsh climate, Moab was remote from other populated areas. These same attributes, appealing to someone like Butch Cassidy a century ago, have kept Moab small and its growth sporadic.

Moab exists because it is located at what, until modern times, was the region's only reasonable place to cross the Colorado River. Typically, the Colorado is deeply entrenched in the Colorado Plateau across southeastern Utah as well as across much of Arizona. Few sidestreams and tributary rivers enter the river along this portion of its length, and almost all of those that do have cut steep-walled canyons that join the main river's larger canyon.

Mormon missionaries made an early, unsuccessful attempt at permanent settlement in 1855. The site appealed to them because not only was the river crossing a natural transportation control point, it is also in the Moab Valley, a *graben* that lies perpendicular to the course of the Colorado. There also were modest, but apparently promising grazing lands in the valley south of the river. And access to river water meant that agriculture was possible in a section of the river's floodplain where it cut across the valley.

* Written by Margo L. Price.

Most American settlement at this time followed east-west trails many miles to the north or south. Moab's site was too isolated to be worth sustained attention when the 1855 mission was closed following an attack by Ute Indians. Decades passed before permanent settlement was achieved, even though prospectors, trappers, cattlemen, and outlaws continued to use the river crossing.

By 1902, Moab had accumulated sufficient population and organization to become incorporated as an independent community, even though most of the land for many miles around was of little economic appeal. Minerals could be sought, but with difficulty because of the harsh environment. Cattle could be grazed, but only at low densities and at some risk because of the limited availability of water. The region's average annual precipitation, well below 25 centimeters (10 inches), also was inadequate for agriculture. An enthusiastic but fruitless search for oil in the area stimulated Moab's population growth briefly in the 1920s. A more successful search for uranium tripled the town's population in the early 1950s, but this increase, too, only lasted about three decades, and the population growth stalled once again.

During the first 75 years of the present century, a succession of good years stimulated growth in Moab several times; then the town shrank when the good years ended. In this boom-and-bust cycle, Moab's experience is typical of Empty Interior communities, but the town had sufficient geographic resources to survive even during the difficult times. Some of the limited agricultural land on the floodplain between the town and the river was planted in fruit trees. Water was reliable, not only from the Colorado River but also from snowmelt streams flowing off the nearby LaSal Mountains. Local mineral resources, while spotty, were somewhat diverse. Potash mining continues, even though the larger uranium mining operation of the 1950s, 1960s, and 1970s has closed, and the earlier oil exploration efforts never did meet expectations. Most important, local roads were improved, and

The canyon country around Moab is stark and extremely rugged, but its beauty attracts visitors by the thousands each year. The Colorado River is almost lost in this scene, but is visible just to the right of the photo's center. (Stephen S. Birdsall)

Moab became less isolated from the major east-west traffic patterns.

Ironically, the harsh and broken landscape that provided a haven for outlaws a century ago and that kept most land ownership in federal hands is now seen as the magnet needed to attract economic activities to Moab. Moab's scenic setting is spectacular. Red, soft orange, rust, and mahogany cliffs, buttes, and spires form fantastic landform arrangements that are exposed to full view in this desert climate. Arches National Park and Canyonlands National Park lie on either side of Moab, and cooler, more moist, and mountainous national forestland is less than an hour's drive up into the LaSal Mountains.

As this landscape became more widely known and more accessible, summer tourists planned their travels so that several days could be spent in the area, based in Moab. Even as the uranium mining operations went bust, businesses oriented toward tourism boomed. Whitewater rafting on the lowest undammed portion of the Colorado River offered families a day of respite (or three or four days) between repeated short hikes and continuing windshield vistas. Representatives of the motion picture industry regularly take advantage of the picturesque landscapes of the region. And the dry climate and attractive scenery have stimulated a small retirement community on the outskirts of town.

The latest wave of economic growth is a mixed blessing. The desert landscape is very fragile and easily damaged. As the number of tourists increases, the number who are careless of the land also increases. Motels, jewelry stores, T-shirt shops, and gas stations line the main highway in town. And the loss of solitude is also decried by some who remember the 1950s and early 1960s. Moab's economy is somewhat more stable than earlier, however, and most who chose to live in this beautiful but unforgiving landscape believe this stability is worth the cost.

ADDITIONAL READINGS

Ballard, Stephen C. *Water Policy and Western Energy.* Boulder, Colo.: Westview Press, 1982.

Brown, Robert H. *Wyoming: A Geography,* Boulder, Colo.: Westview Press, 1980.

Clawson, Marion (ed.). *The Federal Lands Revisited.* Baltimore: Johns Hopkins University Press, 1983.

Cole, David B., and John L. Dietz. "The Changing Rocky Mountain Region." *Focus,* 34 (1984).

Francaviglia, Richard V. *The Mormon Landscape.* New York: AMS Press, 1978.

Franklin, Philip L. *A River No More: The Colorado River and the West.* New York: Alfred A. Knopf, 1981.

Franklin, Philip L. *Sagebrush Country: Land and the American West.* New York: Alfred A. Knopf, 1989.

Griffiths, Mel, and Lynnell Rubright. *Colorado: A Geography,* Boulder, Colo.: Westview Press, 1983.

McPhee, John. *Basin and Range.* New York: Farrar, Straus & Giroux, 1981.

Malone, Michael P., and Richard W. Etulain. *The American West: A Twentieth-Century History,* Lincoln: University of Nebraska Press, 1989.

Rogers, Garry F. *Then and Now: A Photographic History of Vegetation Change in the Central Great Basin Desert.* Salt Lake City: University of Utah Press, 1982.

Vale, Thomas R., and Geraldine R. Vale. *Western Images, Western Landscapes: Travels Along Route 89.* Tuscon: The University of Arizona Press, 1989.

CHAPTER

15

The Southwest Border Area: Tricultural Development

The Southwest is a distinctive place to the American mind but a somewhat blurred place on American maps which is to say that there is a Southwest but there is little agreement to just where it is.

Donald W. Meinig

The area known generally as the Southwest is one of the most widely recognized yet also one of the most transitionary of American regions (Figure 15-1). Many Americans, when asked to regionalize the United States, will create some place that they call the Southwest. Its extremities may be vague, but its existence is clear. However, many geographers, when asked to do the same task, create no place that even vaguely resembles the Southwest. Of the existing textbooks on the geography of the United States and Canada, probably no more than half choose to identify the Southwest as a separate place. We do feel that it should be treated as an identifiable region, but at the same time, we include almost all of its territory in four other regions. No other region shares portions of its territory with as many other regions as the tricultural border area.

This combination of a clear regional identity for most Americans but a rather ambiguous one for geographers is unusual, although its roots are understandable. As mentioned in Chapter 1, regions of the kind used in most textbooks are mental constructs, somewhat dependent on the whims of their creators. Even so, identifiable regionalizations follow some logic. For geographers, this logic frequently depends heavily on a distinctive combination of physical and economic conditions. The Southwest, however, is a region of substantial economic diversity. Moreover, its apparent physical uniformity can be attributed primarily to its aridity, but much of the Empty Interior is dry also. Topographically, the region includes the broad flatlands of the lower Rio Grande Valley; the plateaus of New Mexico; the dramatic mesas, buttes, and deserts of Arizona; and the Sangre de Cristo Mountains of New Mexico.

Most Americans, we feel, would call the Southwest a region because of its clear, dry climate. To us it is a culture region; it is made distinctive by the coexistence of Spanish-American, American Indian, and Northwest European–American (Anglo) cultures. Its special character results from this cultural complexity; although somewhat unusual, the physical environment is almost like a stage that serves to emphasize aspects of each culture and further sets apart the region from the rest of the country. The American Indian and Spanish populations coexisted in much of the region for 250 years after Spanish arrival at the very end of the sixteenth century before Anglos began immigrating in the middle of the nineteenth century. The result was a long period of *aculturation,* of cultural borrowing and sharing between the two groups. One consequence is that today the significance of both groups to the region is enhanced because their cultures often reinforce one another.

The tricultural border region is preponderantly non-Spanish and non-Indian. Perhaps 1 person in 4 has a Spanish surname, and little more than 1 in 100 is American Indian. The expectation, then, might be that these minority populations would be engulfed by the much larger and relatively homogeneous Anglo population. This has not been the case. Both minority groups have had a major sustained impact on the region. In the United States, only Hawaii is as clearly tied to non-Anglo ethnic patterns. The importance of the cultural impact of both non-Anglo groups is obvious on the regional land-

---

The United States of America and los Estados Unidos Mexicanos are linked by a common border over 1000 miles (1600 kilometers) long, much shared culture, and a tradition of peaceful relations. Amistad ("Friendship") Dam is on the Rio Grande near Del Rio, Texas. (Texas Highways Magazine)

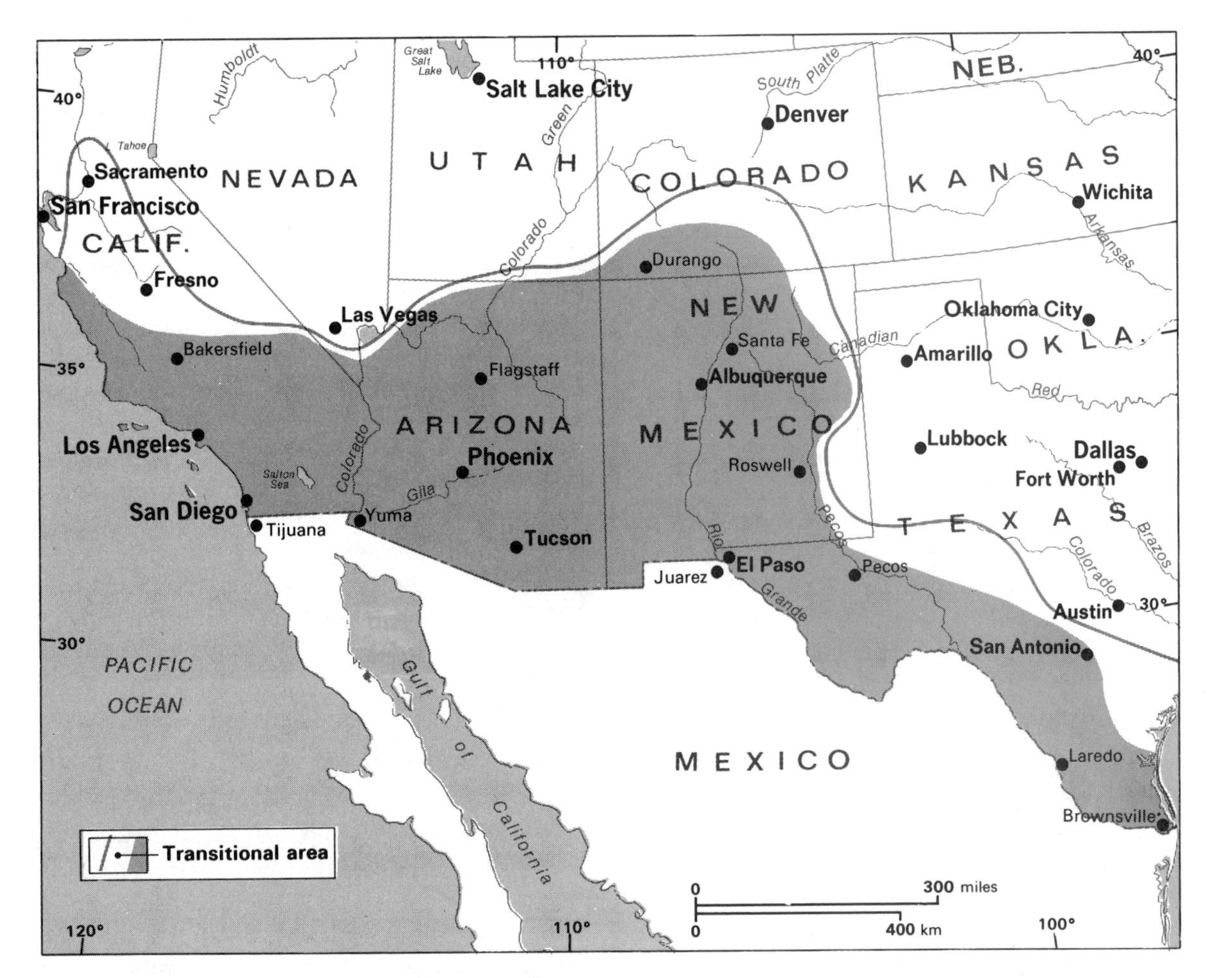

Figure 15-1 The Southwest Border area.

scape. Spanish place-names abound, especially along the Rio Grande and in coastal California. American Indian place-names are locally important, especially on the Navajo, Hopi, and Papago reservations in Arizona. Hispanic neighborhoods are sometimes identified by the use of adobe, but more often by the use of bright colors in house painting and outside ornamentation, and by yards encircled with bold fences. The distinctive conical hogan can still be found on the Navajo reservation, and the pueblos of New Mexico are a striking element of that state's architecture. Northern Mexican foods, rich in hot chiles, corn products (especially tortillas), and pinto beans, are a mainstay of regional fare. Catholicism, the religion of the Hispanics as well as many American Indians, dominates the region. The Southwest Border Area's outer limits are directly defined by the distribution of the two minority groups.

The canyonlands of northern Arizona and southern Utah provided an effective barrier to Spanish expansion northward from Mexico. The Spanish moved up the Rio Grande to the broad expanse of the Rocky Mountains, north of which little Hispanic settlement developed. In Texas,

Major Metropolitan Area Populations in the Southwest Border Region, 1990	
Los Angeles–Long Beach, Calif.	8,863,164
Riverside–San Bernardino, Calif.	2,588,793
San Diego, Calif.	2,498,016
Anaheim–Santa Ana, Calif.	2,410,556
Phoenix, Ariz.	2,122,101
San Antonio, Tex.	1,302,098
Las Vegas, Nev.	741,459
Oxnard–Ventura, Calif.	669,016
Tucson, Ariz.	666,880
El Paso, Tex.	591,610
Bakersfield, Calif.	543,477
Albuquerque, N.Mex.	486,577

most settlement remained concentrated along the Rio Grande and the Nueces Rivers. The extensive cattle-grazing industry that the Spanish introduced into south Texas was not well suited to the moist forested lands of the eastern portion of the state. That area, left as a frontier, was largely unsettled by the Spaniards. Most migration by Spanish Americans beyond this original settlement area has been to urban places, where their cultural impact has been far more limited. Thus, the region was most clearly established by the extent of Spanish settlement, and its basic outline was manifested well before the arrival of the westward migrating American population.

The aridity of Arizona, New Mexico, and bordering areas in Utah and Colorado discouraged large-scale Anglo agricultural settlement in the nineteenth century. The slow pace of Anglo penetration ensured that substantial numbers of American Indians remained in those states, although Indian strength was as important as Anglo hesitancy. The Pueblo of the upper Rio Grande Valley had developed the technologically most advanced pre-European Indian civilization in what was to become the United States, and they remain important in New Mexico. The Navajo, Hopi, and Apache, all primarily in Arizona, also survived the European wave better than had most eastern tribes.

ETHNIC DIVERSITY

Most Americans take pride in the concept of "America, the melting pot," in which numerous and diverse peoples have formed a population now united by common goals and sharing a basically common culture. There is much truth in the melting pot idea: 45 million Europeans moved to the United States, and many of their progeny continue to hold dear the customs of their European homelands. In such divergent characteristics as eating and drinking habits, religious preferences, and musical tastes, Americans demonstrate their heterogeneous ethnic heritage. As we have seen in Chapter 6, they sometimes choose to live with others of the same background in closely knit neighborhoods. Yet nearly all of these people are Americans who have far more in common with other Americans than with the present residents of their European homelands.

American Indians

Although there are American Indians in every part of the United States, today most are found in areas that white settlers had deemed undesirable or in areas that were not part of the earlier settlement frontiers (Figure 15-2). The tribes of the East were much reduced, succumbing more often to disease than to the bullet; others were moved westward, out of the way of European settlement. The Sioux and other Plains tribes and the tribes of the Interior West (particularly in the Southwest) fared better numerically. Their lands were not easily suited to crop agriculture and, within a few decades after their first substantial contact with the United States, the wanton governmental disregard of Indian rights was replaced with a less overtly hostile approach. As a result, the great majority of all American Indian reservations as well as most Indians are in the Plains and Western states (Figure 15-3).

The region's American Indian population is itself culturally diverse. The largest tribes are the

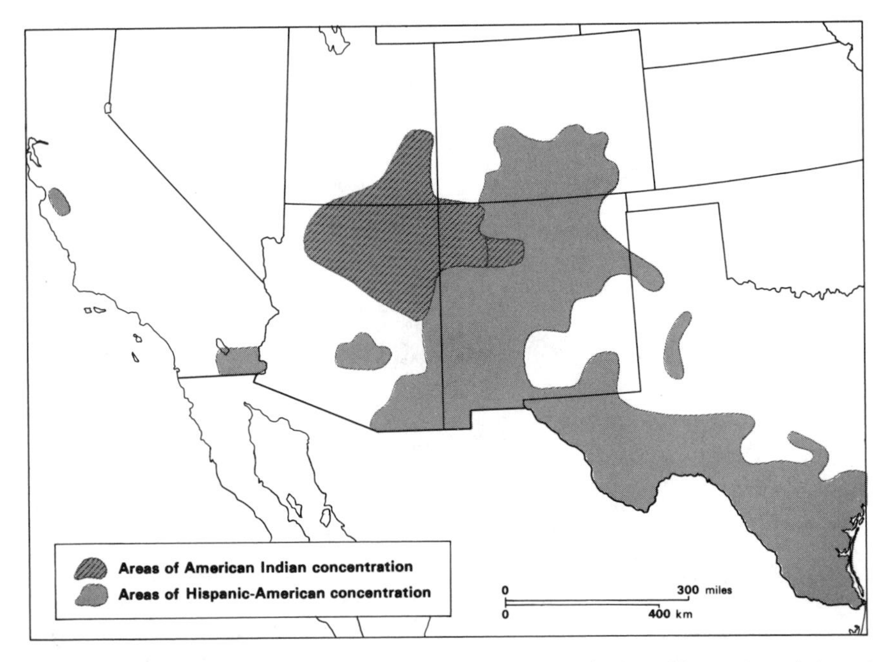

Figure 15-2 Areas of Hispanic and American Indian population. Hispanic and American Indian populations numerically dominate much of rural New Mexico, south Texas, and northern Arizona.

Navajo in the Four Corners area where the states of Colorado, Utah, Arizona, and New Mexico meet; several Apache tribes in Arizona and New Mexico; the various Pueblo groups in New Mexico; the Papago in southern Arizona; the Hopi in northwestern Arizona; and the Utes in southwestern Colorado. The American Indian population is by no means evenly distributed across the region. Most American Indians are found in the major reservation areas, especially those centered on the Four Corners area and in the cities of California. Arizona and New Mexico together are the home of over 300,000 American Indians. Arizona ranks a close second to Oklahoma in the size of its American Indian population. California ranks third, and New Mexico fourth. The Navajo reservation of the Four Corners area has 10 times the population of any other reservation. The Los Angeles–Long Beach SMSA has far more Indians than any other urban area.

Hispanic-Americans

What is in a name? Sometimes a great deal. For the Spanish-surnamed population of the Southwest, the issue of what to call themselves has been important but difficult to agree on.[1] Many younger militants, angered over injustices suffered at the hands of Anglo society, call themselves *la raza* (the race) or *Chicano.* Recent migrants from Mexico also often prefer the latter term. Traditional Spanish-surnamed people

---

[1] For an interesting discussion of this problem of nomenclature, see Arthur L. Campa, *Hispanic Culture in the Southwest* (Norman: University of Oklahoma Press, 1979),pp. 6–10.

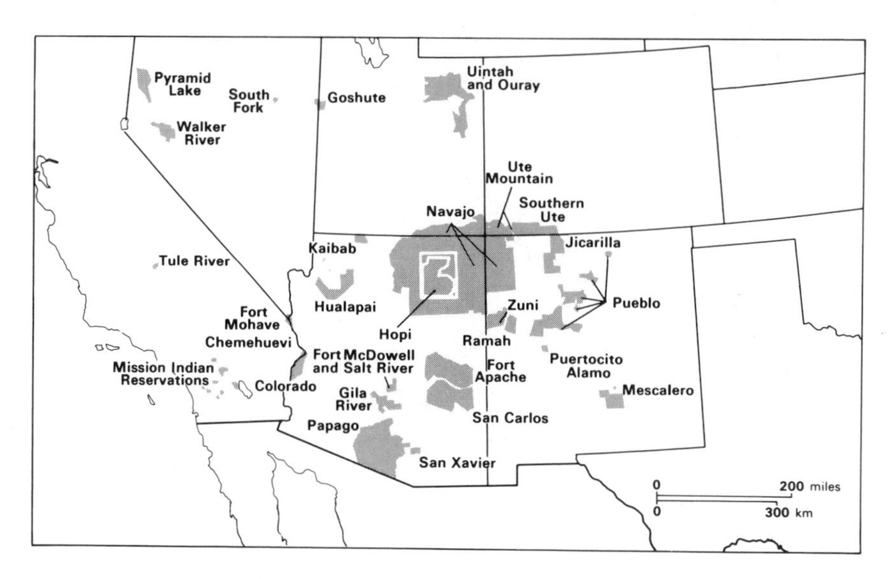

Figure 15-3 American Indian reservations. Ignored by Anglo settlers in the nineteenth century as being of little economic worth, some reservation lands are now valued for their resources, notably the coal deposits that underlie portions of the Navajo Reservation. Most, however, remain isolated and outside the main pattern of U.S. economic growth.

from the Southwest prefer *Hispano.* Many in Texas call themselves *Texanos. Mexican* is usually reserved for people born in Mexico, although some would also use the term when referring to the older, rural, conservative population. Somewhat arbitrarily, but based on widespread usage, we have adopted *Hispanic.*

History texts in the United States commonly cite the early coastal settlements, such as the Raleigh colony in North Carolina or the Pilgrim landing in Massachusetts, as the start of European occupation of the country. In fact, Spanish settlements in Florida and the Southwest easily predated the English arrivals. St. Augustine, Florida, founded in 1565, validly claims to be the oldest European city in the United States. Some southwestern Hispanic communities were founded more than 200 years before Anglo settlement of the region began during the early nineteenth century.

All of what is today the southwestern United States was incorporated into the Spanish Empire during the early years of the sixteenth century. By 1550, the Spanish had explored widely across the region. They found little of the gold and silver they so eagerly sought, but their travels did serve to establish a firm Spanish claim to the territory. The lack of any identified, easily extractable riches coupled with the great distance to the core of Spanish development in Mexico (near present-day Mexico City) minimized Spanish concern for their northern territory. Before 1700, the only permanent Spanish settlements north of the present U.S.-Mexico border were along the valley of the upper Rio Grande in New Mexico. Santa Fe was founded in 1610, and other pueblos (civilian communities that could loosely be viewed as small towns), notably, Taos and Albuquerque, soon followed. This area was to remain the most important part of northern Mexico.

A tenative Spanish occupation of Arizona began in 1700. The Apache Indians were a constant threat, repeatedly raiding Spanish settlements in the area; Arizona, consequently, remained a small settlement outlier. Colonization of Texas

Taos Pueblo, New Mexico, is a frequently-visited tourist attraction. It is also an Indian village, one that has been inhabited for hundreds of years, and is the best known and perhaps most picturesque of several score pueblos that continue to exist in the state. (U.S.D.A.)

began at about the same time, with long-range results that were considerably more successful. Nacogdoches was founded in 1716, followed two years later by San Antonio. During the middle 1700s, the lower Rio Grande valley was settled by the Spanish. Still, by the early nineteenth century these and other centers of Hispanic settlement were viewed by the authorities as a small, inadequate occupation when compared with the large number of Americans then pushing westward toward Texas. Thus, foreigners, mostly Americans, were allowed to develop settlements there during the 1820s and 1830s; it was hoped that these settlers would be loyal to Mexico (independent from Spain after 1821) and form an effective bulwark to American expansionist designs on Texas. History was to prove this hope a false one.

California, the most distant of Spain's northern territories, was the last to be settled. A mission and presidio (military post) were not established at San Diego until 1769. During the next

The village of Chimayo, north of Santa Fe in New Mexico, is one of a small number of examples of a Spanish influence on town form in the region. Spanish towns were usually small and often unplanned, and their pattern has usually been lost with the urban growth of the last 100 years. (Dick Kent/FPG)

two decades, a string of missions, with a few presidios and pueblos as well, was established along the coast as far as Sonoma, north of San Francisco. This thin band of coastal occupation was encouraged partly in response to a growing British and Russian interest in the west coast of the continent.

After U.S. acquisition of Texas in 1845 and the Mexican Cession, which ended the Mexican-American War in 1848, the estimated Mexican population of this broad territory was 82,500. Of this number, roughly 60,000 were in New Mexico, 14,000 in Texas, 7500 in California, and 1000 in Arizona.[2]

By 1850, the Mexican population of Texas and California represented less than 10 percent of the two states' total populations. There were good

[2] For a presentation of the settlement geography of the Hispanic population to 1960 see Richard L. Nostrand, "The Hispanic-American Borderland: Delimitation of an American Culture Region," *Annals of the Association of American Geographers*, no. 60 (December 1970),pp. 638–661.

reasons for the rapid increase in the non-Spanish population. East Texas was the new western frontier for Southern settlements, and the discovery of gold in California in 1848 was largely responsible for the influx of non-Spanish peoples into central and northern portions of the state. Only in New Mexico, southern California, and Texas south of San Antonio did the Hispanic population continue to dominate for a few more decades.

This original Hispanic population of the Southwest has been greatly increased by substantial immigration, especially during the twentieth century. However difficult the existence of many recent Hispanic immigrants, their access to economic opportunities is an improvement over what is available in northern Mexico. Between 1900 and 1990, there were roughly 2.9 million legal Mexican immigrants to the United States; California, Texas, and Arizona easily ranked as the leading destinations for these immigrants. Annual Mexican immigration into the United States in the late 1980s averaged between 75,000 and 90,000, or about 15 percent of total immigration to the United States. Although an increasing number of Mexican people are moving into areas quite distant from the Southwest, most remain in the border states of the region.

Because of substantial illegal immigration, it is difficult to estimate exactly how many Mexicans enter the United States each year. Although legal migration into the United States has been considerable, general quota restrictions and numerous denials of individual Mexicans' requests to immigrate have meant that far more have wanted to enter than have been admitted. The illegal wetback movement (so named because many entered the United States by swimming or wading across the Rio Grande into Texas) has existed along the Mexican-American border for decades. Many of these illegal migrants are caught by immigration authorities and returned to Mexico. Many simply renew their efforts until they are ultimately successful.

Although an accurate figure for such illegal migration is impossible, recent estimates have ranged from 2 to 12 million, a variation that suggests the lack of reliable data. Perhaps half of the illegal migrant population is from Mexico. In recent years, more than 1 million illegal aliens have been arrested annually, in the states from Texas to California. Most were citizens of Mexico, and many had been arrested multiple times previously by U.S. authorities for illegal immigration into the country. A high rate of population growth in Mexico, coupled with widespread unemployment and poverty in the north, provide strong reasons for this migration. The international border, although more closely regulated than that between the United States and Canada, is generally marked by only a fence and can be crossed with relative ease. Despite increased efforts by border officials to control access and despite attempts by Mexican and U.S. officials to negotiate a migration agreement that is enforceable, there is little likelihood of a significant decline in this flow of people in the near future.

Much of this illegal movement is a temporary labor migration; workers who enter the United States often work in the agricultural harvest in the Southwest for a few months, then return to Mexico with the profits of their labor. Until the early 1960s, the United States admitted large numbers of such temporary workers into the country legally. Then the federal government decided that this influx of low-wage labor had a detrimental impact on living and working conditions of migrant farm laborers. Organized labor, attempting to unionize the migrant agricultural labor force, also opposed a large temporary labor migration. Labor leaders argued that it held down wages by providing a plentiful and inexpensive labor force. The number of such temporary workers admitted into the United States was, therefore, greatly restricted. It is debatable whether this decision had a positive effect on the working conditions of U.S. migrant workers. It certainly has been a factor in the rapid mechanization of the harvesting of crops that had formerly been picked by hand. Tomatoes, iceberg lettuce, peaches, asparagus—these and other types of farm produce can now be picked by machine. What is known is that many workers like those who had formerly entered the country legally decided to enter illegally.

Many residents of the region have a more ambivalent attitude toward illegal migrants. For many, they represent a cheap, dependable labor supply. Such migrants play a major role in the regional economy. They often work for years undisturbed by immigration officials.

In 1990, persons of Spanish surname represented 18.8 percent of the population of Arizona; in California, the proportion was 26 percent; in Colorado, 13 percent; in New Mexico, 38 percent; and in Texas, 26 percent. In 1990 the Bureau of the Census counted the Hispanic population of the United States at 20.8 million, an increase of 34 percent over 1980. One American in 12 is now Hispanic. Over 60 percent of the Hispanic population is Mexican-American. Hispanic percentages have increased in every state in the tricultural border region. Hispanics now represent about a quarter of the populations of each of the region's two most populous states, California and Texas. Almost a third of California's schoolchildren are Hispanic. In short, the proportion of the Spanish-surname population in the states that lie wholly or partly in this tricultural border region increased, even though all these states, except New Mexico, had a substantial immigration of non-Spanish-surname people, as well. This Spanish-surname population increase is due both to immigration and to a higher fertility rate as compared to that for the total population. Completed family size for Spanish-surname families averages about one more child than that for the non-Hispanic population of the region. Part of the explanation for this higher fertility rate is economic. On the average, the Hispanic population is poorer than the Anglo population, and in the U.S. poorer people have larger families than their wealthier neighbors. However, culture is also an important part of the reason for larger Hispanic families. In much of the region, the family, especially a large, extended one (to include grandparents, aunts, uncles, cousins, etc.) forms the heart of Hispanic life. Even in the larger cities, family events from weddings to graduations bring the family together to celebrate and share kinship.

The present Hispanic population is concentrated in those areas where most were found in 1850 (see Figure 15-2). Nearly every county in the Rio Grande Valley is populated predominately by people with Spanish surnames. Major clusters exist along the lower Rio Grande in south Texas and in northern New Mexico and southern Colorado. In some south Texas counties, 90 percent of the population is Hispanic. The Hispanic population is also overwhelmingly urban. More than 2.5 million Spanish-origin people reside in the Los Angeles–Long Beach MSA, a population that represents roughly 1 out of every 7 Hispanics in the country. Greater Los Angeles is the largest Spanish culture community north of Mexico City. Of the 10 MSAs with the largest Hispanic populations, 7 are in the Southwest. The popular image of the Hispanic as a stoop-shouldered farm worker stoically following the harvest is no more valid than is that of a Southern black picking cotton or a midwestern white riding atop an expensive corn-picking machine. Today all are far more likely to be urban residents with urban jobs and urban attitudes.

Socioeconomic Disparities

The cultural and economic gap between the Anglo population, on the one hand, and the Hispanic and American Indian groups, on the other hand, is immense in many instances. Most of the cities of the region are marked by patterns of segregation similar to those found in other American cities shared by blacks and whites. One result is that in most of the region's cities, neighborhoods with low-income populations, and large families living in poor-quality housing are likely to be Hispanic. Across the region there is a close association between a high percentage of the population living below the poverty level and a high percentage of the population being Hispanic or American Indian (Figure 15-4).

The cultural differences among Anglos, His-

The adobe brick house of a Mexican-American family in southwest Texas. The Spanish/Mexican cultural influence and dry climate of the region have encouraged the use of easily eroded adobe brick in construction. (Baldwin—Watriss/Woodfin Camp)

panics, and American Indians are important in much of the Southwest, especially in rural areas. Here the American melting pot has worked to only a limited extent, and a wide cultural gap remains between these various groups. The Hispanics, residents of the region for nearly two centuries before the Southwest was annexed by the United States, suddenly became part of a country with a different language, a different legal system, and a different economic structure. The Anglos, quickly and clearly in control of the politics and economics of the entire region, also grew rapidly in numbers to predominate over most of the region. As a group, Anglos viewed the Hispanic with a feeling of superiority; equality of opportunity seldom existed. For example, as of this writing no Hispanic has even been elected to a county office in Los Angeles.

Placing most American Indians on reservations effectively maintained the distinctions between Indian and other cultures. Reservations were located almost by definition in out-of-the-way areas that offered little obvious economic opportunity and little contact with Anglos, let alone integration into Anglo culture. Until recently, reservation educational opportunities were minimal, and local opportunities to use an education were almost nonexistent. Even in economic hardship, most Indians did not choose to leave. Ties to the land and to family, plus the problems of coping successfully within an Anglo culture, lessened what on economic grounds

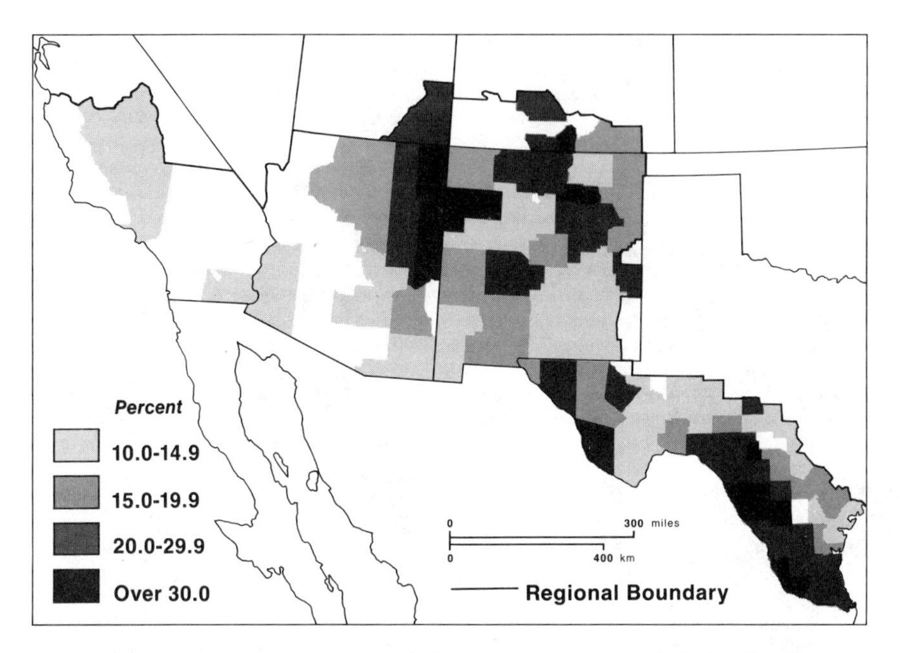

Figure 15-4 Percentage below the poverty level, 1979. In general, the higher the percentage Hispanic and American Indian, the greater the portion of the population living below the poverty level.

might have been expected to be a wholesale movement off the reservations. The large reservations of the Southwest, such as the 62,000-square-kilometer (24,000-square-mile) Navajo reservation centered on northern Arizona or the Hopi or Papago or Apache reservations, provided with their size and isolation a separate, although poverty-stricken existence.

In many small, but nevertheless meaningful, ways, conditions for the two minorities have changed for the better in recent years. Developments on the Navajo reservation, although not entirely typical, are indicative of altered reservation conditions. Final authority remains, much to many Navajos' dislike, with the Bureau of Indian Affairs, but an elected tribal council now makes most economic decisions for the reservation. Since the passage of the Rehabilitation Act of 1950, appropriations to the reservation have increased dramatically. All-weather roads now cross the reservation, greatly reducing isolation. Health and educational facilities have been improved. Huge reserves of fossil fuel, particularly coal, have been found on the Navajo land. Several large power plants located on the reservation serve southern California, which has banned such plants from its own state because of the atmospheric pollution they produce. Additional plants are under construction or planned. Many Navajo oppose this disruption of their tribal lands; others feel they are receiving far too little monetary return for the disruption. The large amounts of water used by power plants are also a concern, because readily available water is in limited supply and its diversion to power production may restrict the development of irrigated agriculture or other forms of industrialization on the reservation. This controversy has split the Navajo leadership at times. Nevertheless, the power companies annually move millions of dollars into the reservation economy. The reservation also has expanded greatly its

This occupied Navajo house, or hogan, is on the Navajo reservation in Arizona next to the Black Mesa Peabody Power Company plant (note top of crane operating beyond the hill). (Ilka Hartmann/Jeroboam)

tourist industry and has attracted a number of new industries with its large, available, and now better educated labor force.

Still, family incomes on the reservation remain less than half of that for white rural Americans, and the reservation fertility rate is twice the national average. Few Navajos enter college; few of those who do, graduate.

THE CROSS-BORDER ECONOMY

Perhaps nowhere else in the world do the First World and Third World meet in such close geographic proximity as along the U.S.–Republic of Mexico border. The economic disparities north-south across the border are striking to even the casual observer. It also has long been a relatively open boundary, with movement and exchange easily accomplished and often encouraged. During World War I and the economic boom times of the early 1920s, large numbers of Mexicans moved across the border to fill labor needs in the United States. Again in the 1940s the United States had a warfare-generated labor shortage. In 1942, the two countries negotiated the Mexican Labor Program, commonly called the Bracero Program. Under its auspices Mexican laborers could legally enter the United States and work as seasonal laborers in the agricultural sector. That program lasted until 1964.

In 1965 Mexico started the Border Industrialization Program. Its goal was to attract U.S. labor-intensive manufacturing industries to border communities in northern Mexico. The law allowed foreign companies, called *maquiladoras,* to import equipment and material duty free into Mexico if the manufactured products were then exported from Mexico. In 1989 that regulation was eased, and now maquiladoras can sell 50 percent of their total product in Mexico. At first maquiladora locations were restricted to a zone within 20 kilometers (8 miles) of the U.S. border to attract foreign investment and employment opportunities to an area far from the center of Mexican economic activity around the large cities to the south. That restriction has also been relaxed, and today maquiladoras can locate almost anywhere in the country outside Mexico City.

For Mexico, the program offered the possibility of jobs for its people. The attraction for U.S. firms in this relationship was the opportunity to use low-cost labor at locations near the U.S. marketplace and sources of supply where transportation costs could be minimized. The result could be a significant reduction in the overall cost of production for many labor-intensive products. Many firms were attracted by this cost-saving opportunity during the decade after the inception of the program. There was little additional expansion until the late 1980s, when wage rates in most Third World areas began increasing while the continued devaluation of the Mexican peso meant that wages there, measured in U.S. dollars, were actually declining. The result was the addition of several hundred new firms each year to the maquiladora list. By late 1990, an estimated 1800 maquiladoras employed over 500,000 Mexican laborers.

There is also a massive daily movement across the border. U.S. citizens travel south for recreation, for an easy opportunity to experience a different culture, and to buy goods at a lower cost. Mexicans travel northward to find products not easily available back home. Hispanics in large numbers travel back and forth to visit family and friends on the other side.

POPULATION GROWTH TODAY

Along with its distinctive cultural diversity, the principal identifying features of the Southwest in the minds of most Americans are sunshine and aridity. This is a reasonably accurate mental image. This is the sunniest and driest of all the regions considered in this textbook. Throughout the area the characteristic vegetation is bunch grass, mesquite, and—although far less common than most believe—cactus. Temperature conditions across the region vary widely. Southern California and Arizona and south Texas normally have hot summers and only a short mild winter; the hot summers of the upper Rio Grande Valley in New Mexico are balanced by winter conditions that can drive temperatures far below freezing. The common image of the Southwest as a warm area will be discounted easily by anyone who has been in northern New Mexico or Arizona—where elevations are generally much higher than in the south—during a midwinter storm, when cold, wind, and snow can combine in chilling intensity.

Despite this variety, the sunny climate of the Southwest, especially in mild winter areas, has proven to be a powerful attraction for many Americans. The spectacular growth of southern California's population during the late nineteenth century and the emergence of its vast metropolitan areas in the twentieth century will be discussed in Chapter 16. This population and economic growth has spread well beyond California. Arizona was the third most rapidly growing state on a percentage basis in the 1980s, following only Nevada and Alaska. Only three states—California, Florida, and Texas—added more to their populations than Arizona. In fact, all of the region's states during this period grew at a rate well above the national average. The city of Phoenix doubled in size between 1950 and 1960, added an additional 50 percent to its population between 1960 and 1970, then added yet another 60 percent between 1970 and 1980, and another 30 percent in the 1980s. It is now a booming urban area, the 18th largest in the country, with a frequent cover of atmospheric pollution to attest to its status. Greater Tucson grew from 266,000 in 1960 to 667,000 in 1990. These low-density urban areas now roll for miles across large expanses of former desert.

Some of the early attraction of the Southwest stemmed from the healthful effects of the dry environment for people with respiratory ailments. The warmer parts of the region today attract many thousands of retired Americans. Large retirement communities have grown in the desert. One, Lake Havasu City, Arizona, gained national recognition when its developer purchased London Bridge and carefully transported it to Arizona, where it now arches over a newly created and artifically maintained waterway.

Still, Arizona's growth cannot be attributed only to retirement migration. Many industries and corporate offices have been attracted. An aircraft industry developed in Phoenix during World War II, taking advantage of its proximity to the large aircraft complex in southern California, plus the promise of good flying weather. Many employers have located in southern Arizona because the environment has a strong appeal for a work force (and for the employers themselves). The relative isolation of the state from most of the major national markets, once possibly significant to Arizona's growth, has lost much of its impact with the emergence of high-value, low-weight manufactured goods, especially electronics.

Other MSAs such as El Paso, Texas, and Albuquerque, New Mexico, roughly doubled in size between 1950 and 1970 and since then have continued to grow rapidly. Both cities, and San Antonio as well, have benefited from the presence of large military bases, although they also share in the diversified growth of light industry common to the entire region's cities.

Elsewhere, in New Mexico and in that part of Texas included in this region, population growth has been far more spotty. Many rural counties of the lower Rio Grande Valley and most in southern Colorado and eastern New Mexico have lost population during the last few

The street system of a large land development project near Albuquerque, New Mexico, cuts across the landscape. Plans such as this one have placed several hundred thousand potential second home and retirement lots in rural parts of the state. This boom, a criterion of the American movement toward the "Sun Belt," has resulted in a supply of building lots that currently far exceeds the demand for such lots. (George Gerster/Comstock)

decades, sharing the fate of most other strongly rural areas in America. Today, New Mexico remains one of the least industrialized states in the country.

PERSISTENCE OF A PLURAL SOCIETY

The Hispanic and American Indian cultures give portions of this region a unique place in the American landscape. The rural highlands of central and northern New Mexico, the principal core of Spanish settlement in the United States, continue to show characteristics remarkably unaffected by the Anglo tide that has engulfed Albuquerque and southern Arizona. Hispanics make up perhaps 70 percent of the highland population of northern New Mexico and comprise the entire population of many small towns. American Indians are a much smaller but highly visible element of the region's rural non-Anglo culture.

This 1970s street scene is from a small town in New Mexico. The Spanish influence, and the poverty, are obvious. (Danny Lyon/Magnum)

Traveling the back roads north of Santa Fe, one can get the impression of leaving twentieth-century Anglo society entirely behind. Old adobe villages, poverty-stricken agriculture, and public signs in Spanish dominate the cultural landscape. Along the highway near the dominantly Anglo city of Albuquerque, as throughout the north central part of the state, are several-centuries-old apartmentlike Indian villages called pueblos; their ancient appearance is in striking contrast to the low, sprawling modern city. The pueblos each control substantial areas of surrounding lands that insulate them from the Anglo community. Pueblo society and traditions are not merely surviving remnants of a past period; they are vibrant and thriving. New Mexico's capital city of Santa Fe, in the heart of Hispanic country, retains a Spanish flavor with its adobe architecture, open central square, and restaurants and stores offering the food and goods of northern Mexico, that gives it a pleasant distinctiveness, despite its rapid growth and increasing Anglo influence. Hispanics, in many cases poor and cut off from the economic advancement offered by the Anglo economy, have become an important political force in New Mexico and frequently hold statewide elective offices.

The Winter Garden area of the lower Rio Grand valley in Texas, also overwhelmingly Hispanic, is a major area of irrigated agriculture. The average growing season is longer than 280 days and supports such crops as oranges, grapefruit, and winter lettuce and tomatoes. The Hispanic population has long provided stoop labor for this agriculture. Here, too, politics has become an important Hispanic activity.

In Los Angeles, Hispanic enclaves may contain hundreds of thousands of inhabitants. Despite far more acculturation into Anglo society than has occurred in the upper or lower Rio Grande Valley, Hispanic traditions remain important; Spanish-language radio stations and newspapers abound, and major Mexican-American festive occasions attract huge throngs.

In the Southwest border area, then, the impact of the Hispanic and the American Indian cultures remains strong. The abundant evidence of a flourishing plural society continues to distinguish this region from others in the western United States.

URBAN CASE STUDY

SAN ANTONIO, THE MEXICAN-AMERICAN CULTURAL CAPITAL[3]

An area of springs at the southern edge of the Texas Hill Country was long popular for American Indian settlement. The Spanish showed an interest in the area for future development in 1691, and in 1718 established mission San Antonio de Bexar (the Alamo) at the present site of downtown San Antonio. Four other San Antonio missions, dating from 1720 to 1731, were soon established along the San Antonio River by Franciscan friars. That last year also saw the creation of a villa (civilian community) at San Antonio, the first real Spanish effort to implement colonization in Texas. That concentration of different

[3] Daniel D. Arreola, "The Mexican American Cultural Capital," *Geographical Review*, no. 77 (1987), p. 18. Many of the arguments presented for San Antonio as Mexican-American cultural capital are Arreola's.

San Antonio's El Mercado (Market Square), location of several of the city's Hispanic heritage celebrations, lies near the heart of the modern city. (Texas Highways Magazine)

institutions at one general location made San Antonio (usually called Bexar during the Spanish period) unique in the Spanish borderlands.

San Antonio remained the chief Spanish, then Mexican, community in Texas until the Texas Revolution. It remained overwhelmingly Spanish/Mexican until the 1840s, when German, American, and Irish migrants began entering the city. The beautiful, rolling countryside to the north and northwest of the city is still often called the "German Hill Country" in recognition of the large number of Germans who settled that area in the middle of the last century and whose successors are still there. By 1850 Mexicans accounted for less than half San Antonio's 3500 residents.

San Antonio and Galveston contested for the role of largest city in Texas between 1850 and 1900. During that period San Antonio established good railroad connections with Mexico, enabling the city to become a major focus for exchange between the United States and Mexico and a center for Mexicans entering this country. The city attracted several tens of thousands of Mexican political refugees in the early part of this century. They maintained close ties with their homeland.

Los Angeles first surpassed San Antonio in the size of its Mexican-American population in 1930. Today the California city is home to four times the number of Mexican-Americans living in San Antonio. Still the "City of Missions" lays strong claim to the title of Mexican-American cultural capital. It is the only major U.S. city that is predominantly Mexican-American. San Antonians of Mexican ethnicity cling to the use of Spanish. They are likely to marry within their ethnic group. San Antonians have long played a central role in many Mexican-American political organizations and activity.

San Antonio today presents a striking blend of landscapes and influences. Its four U.S. air force bases make it seem like home to many in the air force. The Paseo del Rio, or River Walk, offers the visitor a pleasant meander for several miles along the San Antonio River through the heart of the city. A German influence, with accordions and a strong polka base to many songs, gives a distinctive sound to local Texas–Mexican music, called *conjunto.* Unlike most American cities, awnings cover most central-city sidewalks, reminding some visitors of the hot climate and others of a frontier community. Still, a walk into the west side barrio brings home to visitors the city's Mexican heritage. Signs and conversations are in Spanish. The annual Fiesta San Antonio in April celebrates that Mexican heritage.

CULTURAL GEOGRAPHY CASE STUDY

THE HOPI RESERVATION

Land, and their attachment to it, is of central importance to many people in the world. Some have argued that the European population of the United States and Canada, with its great willingness to relocate and leave the land, does not share fully in this attachment. That is not the case with many American Indian tribes in both countries.

The Hopi, a tribe of perhaps 8500, are direct descendants of the Anasazi, the people who built the magnificant cliff dwellings of such places as Mesa Verde in Colorado and Hovenweep in Utah plus equally impressive pueblos like Chaco, New Mexico. The Anasazi dominated a large area of the Southwest for centuries. A succession of droughts in the 1200s apparently forced the abandonment of their great urban centers and irrigated agriculture. Some moved southeastward to contribute to the emerging pueblo culture along the Rio Grande. Others, after a period of wandering, settled around a series of three mesas in what is now northeastern Arizona. These mesas form the core of the current Hopi Reservation (Figure 15-5).

The Hopi lived peacefully around their mesas for several centuries. They settled into a series of compact villages (pueblos) in the Anasazi fashion. They continued (and still continue) an agricultural economy focused on such crops as corn and beans. In the late 1600s, they removed their pueblos to the mesa tops to avoid anticipated Spanish reprisals after the Pueblo revolt against those Europeans settling along the Rio Grande. At about the same time, the Navajo began raiding and then settling in Hopi land. The aggressive, expansive Navajo soon surrounded the Hopi, forcing them from much of their ancestral land. The Hopi even invited Tewa Indians from the pueblo country of New Mexico to join them in their battles. The Tewa still live in a village on the Hopi reservation and maintain a seperate cultural identity.

The U.S. government ignored this ongoing conflict in the decades after the Mexican War of 1848. The Hopi were finally granted a reservation in 1882. A portion of the order establishing the reservation said it might also be used by "such other Indians as the Secretary of the Interior might see fit to settle thereon." The relatively small Hopi population used only a small part of the reservation for their agriculture. The numerically dominant Navajo had adopted a pastoral economy based on sheep. Their need for grazing land for their animals was great in the dry environment of northern Arizona. Thus, the Navajo had soon moved onto the Hopi reservation so that only a relatively small portion remained exclusively Hopi. A federal panel recognized this situation by proclaiming all of the reservation except the Hopi-dominated section a Joint-Use Area. The Navajo were the larger group, however. In practice, the Navajo occupied virtually the entire Joint-Use Area. Hopi attempts to claim

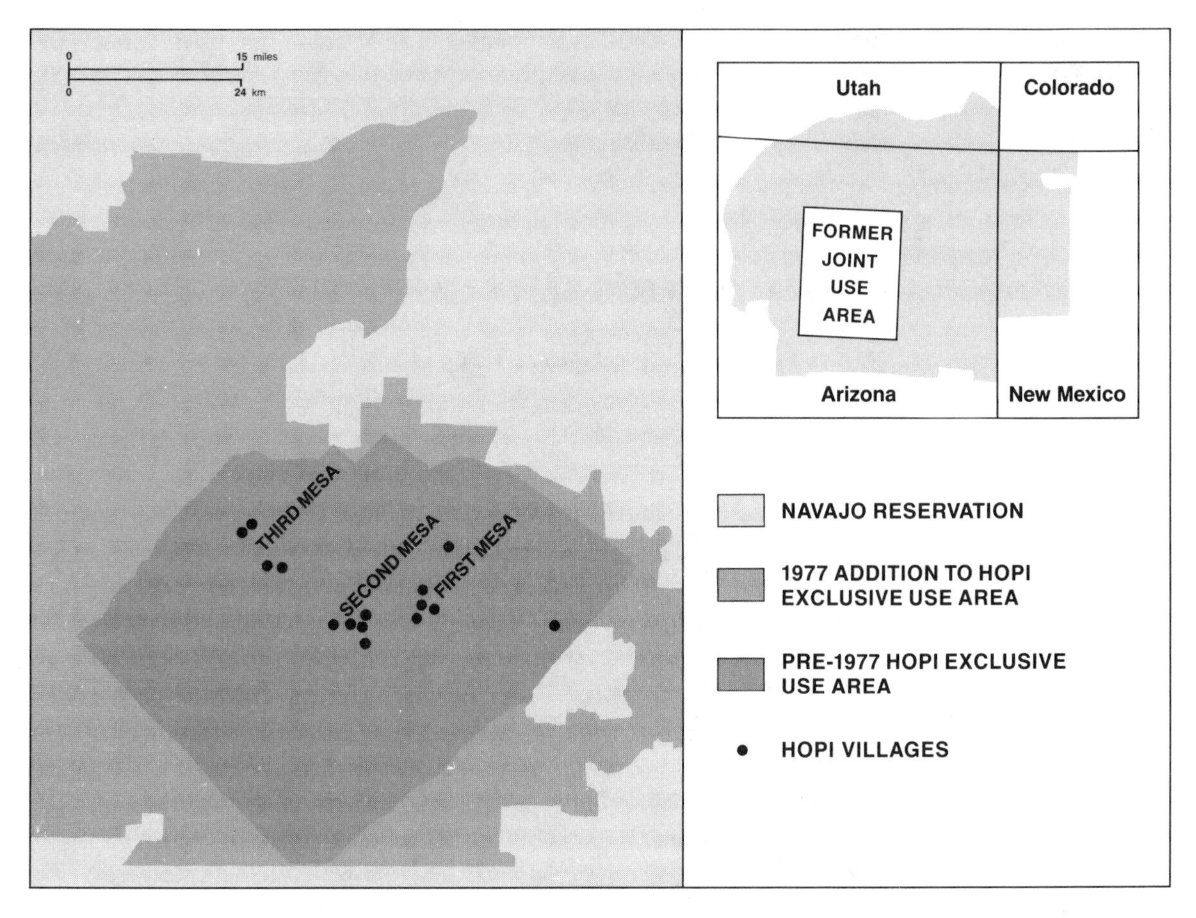

Figure 15-5 The Hopi reservation. Hopi communities are clustered atop the reservation's mesas. Navajo settlement is generally more widely distributed.

what they considered to be their share of the area were rejected.

In 1974, Congress mandated a division of the Joint-Use Area. The two tribes attempted and failed to negotiate a boundary. Finally, in 1977, a federal district court divided the area equally between the two tribes. The decision requires the relocation of 60 Hopi and over 3000 Navajo. Despite government promises of substantial funding of all moves, many Navajo bitterly oppose being forced from their homes.

The Hopi are a conservative people. They cling to their traditional ways of life. The land is very important to them, and they feel a strong sense of responsibility to it. The Hopi were very disturbed at the persistent overgrazing of what they considered Hopi land by the sheep herds of the pastoralist Navajo. They see the 1977 settlement as just. The Navajo, for their part, have lived on what is now Hopi land for a century and also feel an attachment to the land. They consider their overgrazing problem a result of too little land for a large and rapidly increasingly population. The Hopi, they believe, do not use Hopi land effectively. To the Navajo, the settlement is an unfair limitation of their right to earn a living by traditional pastoral means.

ADDITIONAL READINGS

Comeaux, Malcolm R. *Arizona: A Geography*, Boulder, Colo.: Westview Press, 1982.

Davis, Cary, et al. "U.S. Hispanics: Changing the Face of America" (*Population Bulletin* #38). Washington, D.C.: Population Reference Bureau, 1983.

Gann, L. H., and Peter J. Dunigan. *The Hispanics in the United States: A History*. Boulder, Colo.: Westview Press, 1986.

Goodman, James O. *The Navajo Atlas: Environments, Resources, People and the History of the 'Dine Bikeyah,'"* Norman: University of Oklahoma Press, 1981.

Hansen, Niles S. *The Border Economy: Regional Development in the Southwest*. Austin: University of Texas Press, 1981.

Martinez, Oscar. *Troublesome Border*. Tucson: University of Arizona Press, 1988.

Meinig, Donald W. *Imperial Texas: An Interpretive Essay in Cultural Geography*. Austin: University of Texas Press, 1971.

Meinig, Donald W. *Southwest: Three Peoples in Geographic Change, 1600–1970*. New York: Oxford University Press, 1971.

Nichols, John. *The Milargo Beanfield War*. New York: Random House, 1974.

Norwood, Vera, and Janice Monk (eds.). *The Desert Is No Lady: Southwestern Landscapes in Women's Writings and Art*. New Haven, Conn.: Yale University Press, 1987.

Romo, Ricardo. *East Los Angeles: History of a Barrio*. Austin: University of Texas Press, 1983.

South, Robert B. "Transnational 'Maquiladora' Location." *Annals of the Association of American Geographers*, 80 (1990), pp. 549–570.

Stoddard, Ellwyn L., et al. (eds.). *Borderlands Sourcebook: A Guide to the Literature on Northern Mexico and the American Southwest*. Norman: University of Oklahoma Press, 1983.

CHAPTER

16

California

Perhaps no American region excites such a diversity of perception among people living outside the region as California. Many see it as the desirable ideal of the modern, outdoor-oriented American life-style. At the opposite extreme, others see in it a decadent example of what can go wrong with the country. Whatever viewpoint one chooses, California, home to more than 10 percent of all Americans, is a central element in the American cultural fabric.

Our identification of California as a separate region calls for some discussion (Figure 16-1). By most of the criteria used in the definition of regions, California is not a single unit. Culturally, the agricultural population of the Imperial Valley in the southeast is quite different from the urban population of San Francisco. The striking flatness of the San Joaquin Valley is in sharp contrast to the ruggedness of the Sierra Nevada. In the same region are found broad areas of desert in the southern interior and heavily forested slopes along the coastal north, where average annual precipitation exceeds 200 centimeters (80 inches). The lowest and highest elevations in the coterminous United States, Death Valley and Mount Whitney, respectively, are almost within sight of each other.

California's dramatic and, certainly, varied physical environment has played a strong role in the state's human settlement. The variety and value of agricultural crops grown there is unsurpassed by any other state. Most of the state's population today is crowded into a small part of its territory, constricted by expanses of rugged topography and a widespread lack of water. Californians have invested billions of dollars in attempts to overcome these two problems. Their success is partial, at best, and in many areas it is fragile. It is a surprising contradiction that this mecca for America's worship of the outdoor life, this migration destination for hundreds of thousands seeking to free themselves from the indoor enslavement of the colder, physically less attractive midsection of the country, should also surpass every other state in its level of urbanization. Yet it does. No state has a higher percentage of its population classified as urban by the Bureau of the Census. Several factors account for this, but the restrictive aspects of the physical environment are certainly an important element.

The beach, sunshine, physical activity—and crowds. These participants and spectators at a Santa Barbara beach volleyball tournament are part of a classic California lifestyle. (Steve Malone/Jeroboam)

Major Metropolitan Area Populations in California, 1990	
Los Angeles–Long Beach	8,863,164
Riverside–San Bernardino	2,588,793
San Diego	2,498,016
Anaheim–Santa Ana	2,410,556
Oakland	2,082,914
San Francisco	1,603,678
San Jose	1,497,577
Sacramento	1,481,102
Fresno	667,490
Oxnard–Ventura	661,106
Bakersfield	543,477
Stockton	480,628
Vallejo–Fairfield–Napa	451,186

THE PHYSICAL ENVIRONMENT

The coast of California is lined by a series of long, linear mountain ranges that trend in a generally northwesterly direction. They are collectively called the Coast Ranges. Most are not particularly high—summits are between 1000 and 1600

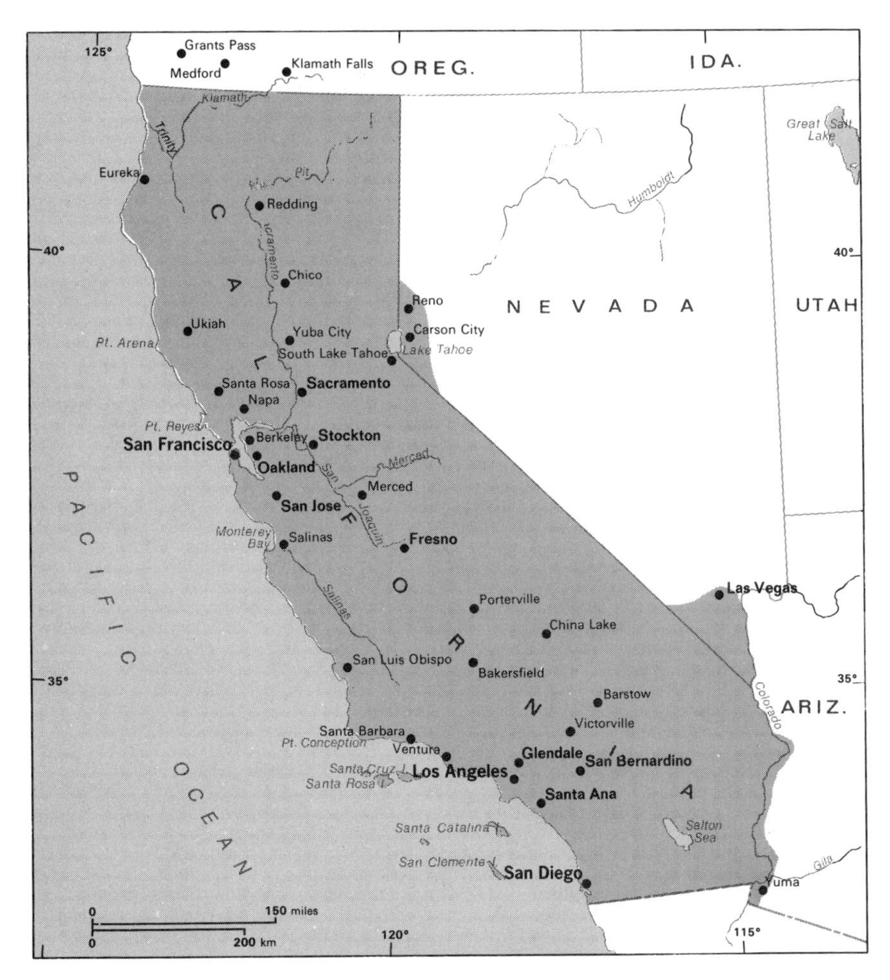

Figure 16-1 California.

meters (3000 and 5000 feet). They are heavily folded and faulted as a result of the pressures of plate contact just to the west. The California faults follow the same northwesterly trend as the Coast Ranges (Figure 16-2). The most famous of these faults, the San Andreas, extends from the Gulf of California through the Imperial Valley to Point Arena well north of San Francisco, where it extends into the Pacific Ocean. Lateral earth movement (i.e., horizontal shifting as opposed to vertical displacement) was as much as 6 meters (20 feet) along the fault at the time of the 1906 San Francisco earthquake.

Small earthquakes are common across large sections of the region, especially from the San Francisco Bay area southward to near Bakersfield and from the Los Angeles area southeastward through the Imperial Valley. Seismologists (scientists who study earthquake activity) believe the regions of major continuous stress, such as the Coast Ranges area, will periodically experience large earthquakes as a result of this stress, which gradually accumulates during periods of less violent activity. Many believe that this fault zone, which has had no major earthquake since 1906, is due for a large one. The 1989

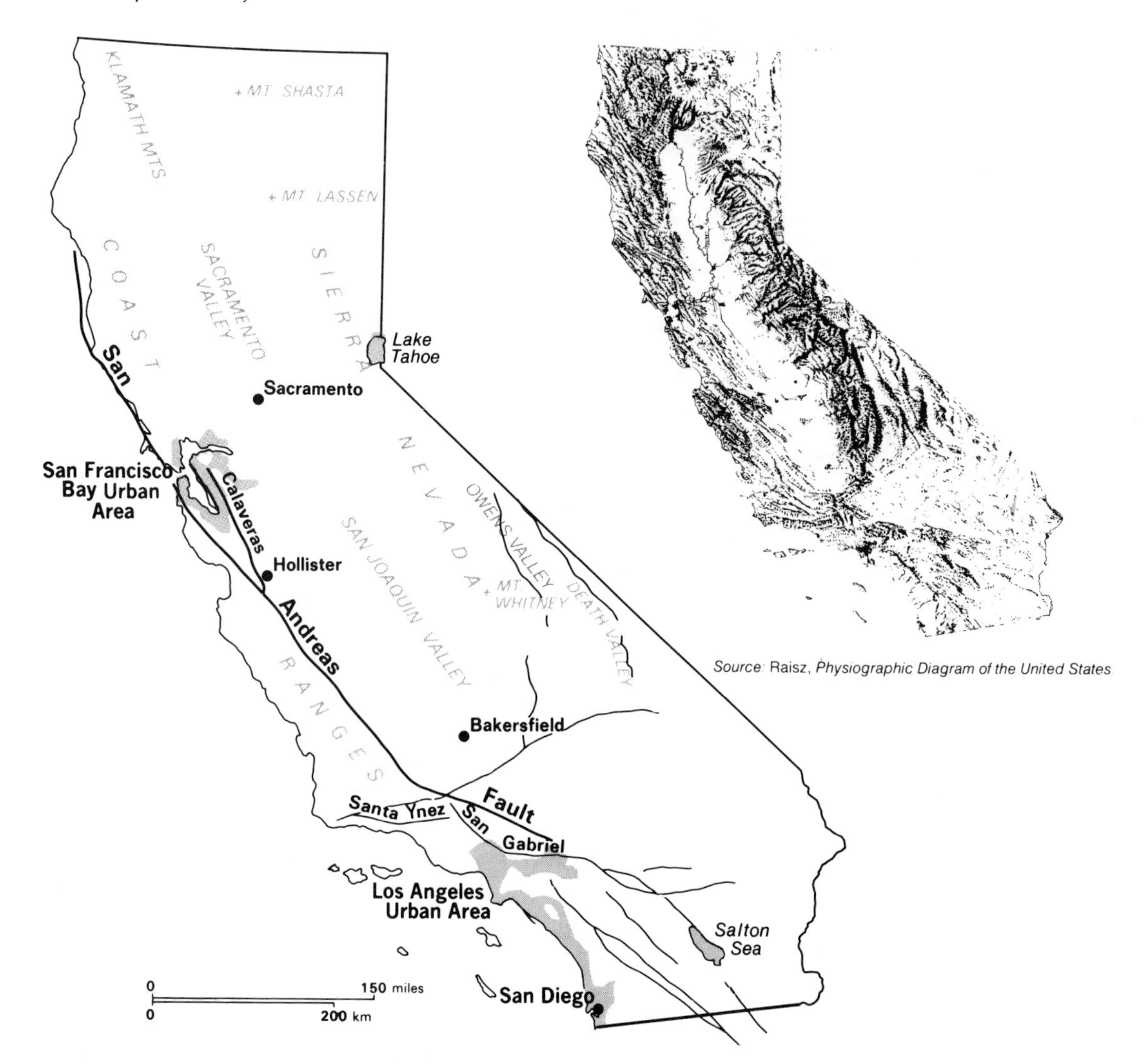

Figure 16-2 Earthquake fault zones. The topographic linearity of coastal California is largely the result of land movement along these fault zones.

San Francisco earthquake, while large, was not that awaited major earthquake. The area near Los Angeles is considered especially vulnerable, because it has not experienced a major earthquake from movement along the San Andreas fault for 120 years.

Earthquake activity represents a substantial hazard for many Californians. The two major foci of population growth in the state, the San Francisco Bay Area and the Los Angeles Basin, are also principal centers of seismic activity. Thus, the majority of all Californians must live with the threat of an earthquake. Major highways, commercial and industrial concerns, and thousands of homes are located on or near known fault zones. For many, earthquake insurance is prohibitively expensive; insurers understandably have fears of suffering catastrophic financial losses in the event of a major earthquake.

Perhaps the most visible impact of the earthquake threat has been on the skylines of the state's larger cities. Until recently, local "earthquake laws" limited the height to which buildings could be constructed as a protection against great loss of life during an earthquake. The result for many years was a low, even skyline, with no skyscrapers in San Francisco and only City Hall rising more than a few stories in Los Angeles. These pleasantly homogeneous skylines (especially that of San Francisco) set these cities apart visually from their counterparts to the east. Substantial innovations in structural materials and construction techniques have led to modification of many of these laws; now a growing number of tall buildings punctuate the skylines in both cities. Many native Californians find these substantial changes in the state's urban landscapes visually jarring, especially in the case of San Francisco.

To the east of the Coast Ranges for much of their distance lies the Central Valley. This valley is extremely flat, extends 650 kilometers (400 miles) from north to south, and is nearly 150 kilometers (95 miles) wide in places. The Central Valley was originally a massive extension of the Pacific Ocean. A recipient of the erosion material carried off the western slope of the Sierra Nevada, this former section of the sea has been filled in with sediments. The result is a low-relief

The modern skyline of San Francisco, built after the repeal of the "earthquake laws" that formerly limited building height in the city, towers over the older buildings that extend westward toward the Golden Gate. (San Francisco Examiner)

landscape rich in potential for large-scale agricultural pursuits.

To the east of the Central Valley, the Sierra Nevada rise gradually and have been heavily eroded. In contrast, the eastern face of the mountains offers a dramatic change in elevation. These are *fault-block mountains,* large rock masses that rose as a whole unit, and the eastern side was lifted far more sharply than the western face. Because they reach higher elevations and contain few passes, the Sierra Nevada have proven a major barrier to movement between middle and northern California and areas to the east. Today the tragic tale of the Donner Party or the exploits of Chinese laborers during construction of the first transcontinental railroad are partly matched by the problems, if not the drama, of highway construction in the "Sierras."

Two other landscape areas, less identifiably Californian, complete the state's topography. In the north, the mountain-valley-mountain pattern breaks down, and much of the northern tier of the state is generally mountainous. The central plateau portion, directly north of the Central Valley, contains two of the state's major volcanic peaks, Mount Lassen and Mount Shasta. This is the southern extension of the Cascades of Oregon and Washington. To the southeast of the Central Valley, the Great Basin of the interior extends well into California; topography there is composed of low-lying mountains (at least by comparison with the Sierra Nevada) interspersed with large areas of fairly flat land.

Climate and Vegetation

Variations in the state's climate and vegetation are nearly as great as the diversity of its topography (Figure 16-3). Much of the weather that spreads over the state is initiated offshore, along the eastern margins of the Pacific Ocean. Of greatest importance for precipitation is the frequent movement of moisture-laden air southeastward out of the northeast Pacific. This is a large source area of moist maritime air. The distance southward that storm systems from this air mass are able to move is influenced substantially by the occurrence of a stable high-pressure center usually found somewhere off the west coast of Mexico. Such a stationary center blocks the southward movement of storms emerging from the maritime air mass and forces the moisture-producing system eastward onto shore. This blocking high tends to drift northward during summer and southward during winter in conjunction with the latitudinal movements of most climatic zones during the year.

A definite north-south gradient in average annual precipitation occurs in California as a result of the interaction of these two major air masses; the north is much more moist on the average than the south (see Figure 2-2). Furthermore, summers are characteristically drier than winters, especially in the south. In summer, southern California frequently experiences long periods during which there is no rainfall. Consequently, the wooded mountain slopes grow dry as tinder. Forest fires, another of the region's recurring environmental problems, are most frequent in late summer and fall, toward the end of the long dry period. A climate with moderate precipitation concentrated almost entirely in the winter months, and one that has mild winters and hot summers, is called a Mediterranean climate, so named because it is found in much of the Mediterranean Basin lands of Europe, Africa, and Asia. In California, the entire coast from San Diego northward past San Francisco Bay, all of the northern Central Valley, and the western margins of the southern valley represent the only zone of Mediterranean climate found in North America.

Northward from San Francisco along the coast, average annual precipitation increases greatly, and its seasonality partially disappears. The northern coasts of the state have a climate similar to that of the Pacific Northwest—mild temperatures with relatively little seasonal variation, plentiful year-round precipitation, and frequent periods of overcast skies. Such a climate is called Marine West Coast, an indication of its customary location along the western margins

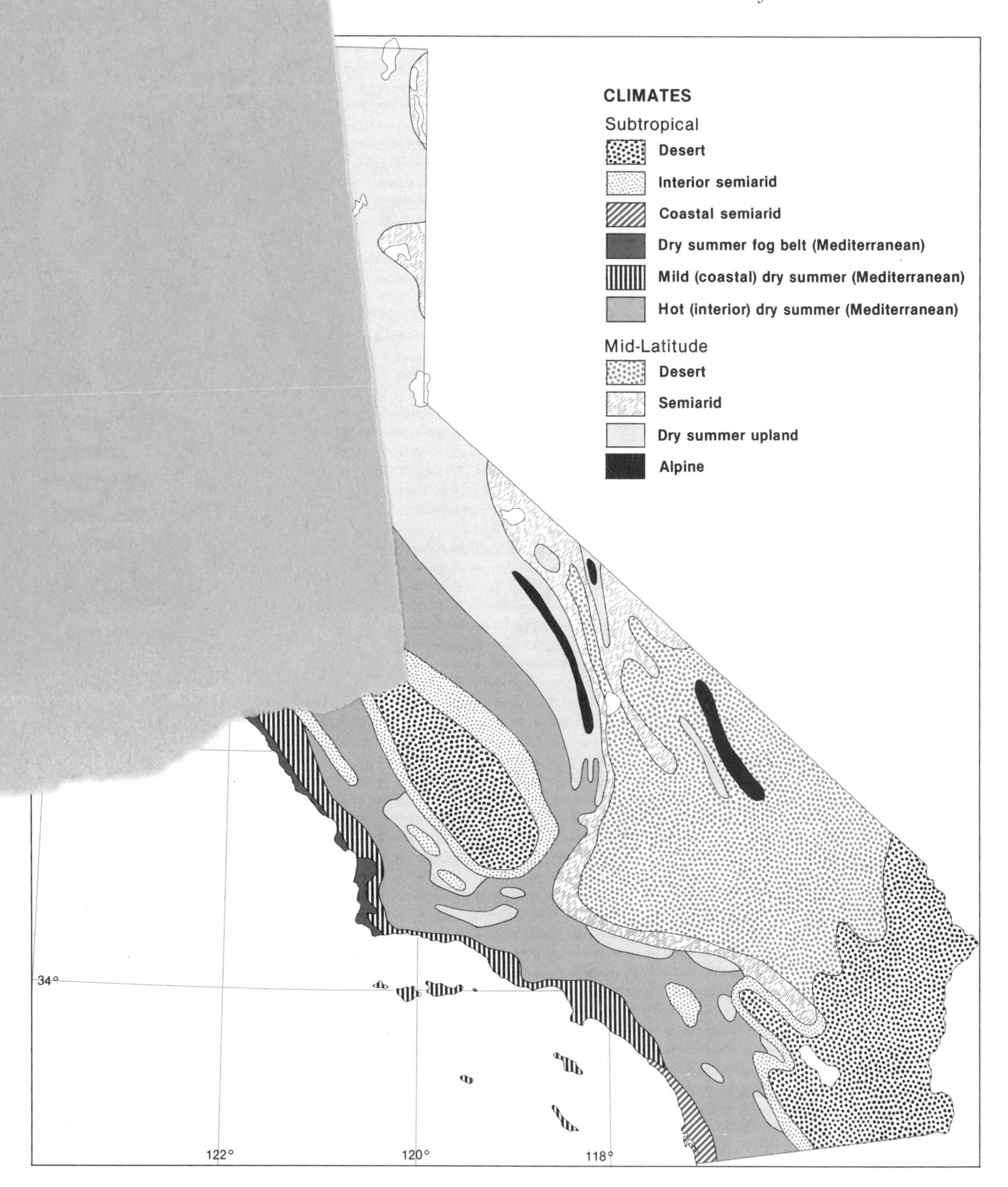

Figure 16-3 California's climate. (Adapted from *California: The Pacific Connection* by David W. Lantis, Rodney Steiner, and Arthur E. Karinen. Chico, CA: Creekside Press, 1989.) Topography plays a substantial role in influencing the climate of much of the state.

of continents in the middle latitudes (see Chapter 17).

The Central Valley is much drier than the coastal margins of the state. As air masses pass over the Coast Ranges, they rise, cool, and lose much of their moisture. When those air masses descend into the valley, they warm. Little moisture is added to the air and its relative humidity (the percentage of the absolute humidity represented by the moisture actually in the air) declines. The result is a much lower average annual precipitation. Annual precipitation in the Central Valley is usually less than half that found at a similar latitude on the western slopes of the Coast Ranges. This dryness is especially pronounced in the rain shadow along the west side of the valley. For example, Mendocino, on the coast north of San Francisco averages 92 centimeters (37 inches) of precipitation annually, while Yuba City directly east of Mendocino in the heart of the Sacramento Valley averages only 52 centimeters (21 inches); southward, coastal San Luis Obispo averages 52 centimeters (21 inches) while inland, Bakersfield must make do with only 15 centimeters (6 inches) annually.

Summer temperature differences between coastal and inland points at a similar latitude are equally dramatic. San Luis Obispo's average July temperature is 18°C (64°F); the average for Bakersfield is nearly 12°C (20°F) higher. Daytime high temperatures in San Francisco in late summer are usually under 27°C (80°F) while Stockton, 100 kilometers (60 miles) east is baking in temperatures over 38°C (100°F). Much of the explanation for this difference is based on the moderating effect of the cold ocean current offshore and the usual pattern of afternoon and evening fog in summer along the coast from Point Conception northward (again see Figure 16-3).

To the interior of the Coast Ranges and the Sierra Nevada, in California's southeast, is a broad region of dry steppe or desert environment. The limited precipitation that reaches southern California is reduced further as the air masses move eastward over the coastal mountains. In addition, the entire area frequently comes under the influence of very dry air generated from the south and east. During the late summer months, these dry winds occasionally press westward to the coast, bringing extremely low humidity and temperatures that may register more than 40°C (104°F), even along the coast. The southeastern interior area of California receives on the average less than 20 centimeters (8 inches) of precipitation annually.

Vegetation patterns closely parallel variations in climate in the state. Nearly all of lowland southern California and the area east of the Sierra Nevada-Cascade ranges is covered with sage, creosote bush, chaparral, and other characteristic desert and semidesert growth. The Central Valley and the valleys of the southern Coast Range are somewhat better watered than are areas farther south; they are steppe grasslands. These grasses are green during the moist period from early winter through spring. They turn light-brown during the dry season. Wrapping around the Central Valley and following the coast from Santa Barbara to Monterey Bay are mixed open forests of live oaks and pines. The coast from Monterey Bay northward is the home of the redwoods, the world's tallest trees. At higher elevations in the Coast Ranges and Sierra Nevada are mixed forests of pine and fir, and high in the Sierra Nevada are found subalpine hemlock-fir forests, including those of the sequoia.

CLIMATE AND CALIFORNIA'S GROWTH

California's climate, which provides much of the unquestionable attraction of the state, is also at the basis of some of its greatest problems. Viewed from a purely economic standpoint, it is almost inconceivable that fully 1 out of every 9 Americans should live in California. Geographers, and many others, have long viewed the population growth of a place as a result of several factors; two of the most important are the natural resources of that place and the strategic advantages of its location. [For a discussion of relative location (situation), see Chapters 3, 5.]

Perhaps California's greatest disadvantage, at least until recently, is its location at the far western periphery of the country. The United States was settled primarily from east to west. The major population concentrations and most areas of principal economic importance remain near or east of the Mississippi River. Thus, California is located as much as 3500 kilometers (2200 miles) from the most important areas of economic demand and supply in the country. This relative isolation is compounded by the nature of most of the land that lies between the Sierra Nevada and the South and Midwest, a broad section of the country that generates little local freight (except at a few sites). Therefore, nearly the entire cost of movement between California and the cities east of the Mississippi River was absorbed by through traffic. Normal economic considerations would suggest that high transportation costs would limit economic growth in California.

California is certainly not without its natural resources; San Francisco Bay, for example, is one of the world's premier harbors. Still, these are not nearly sufficient to account for the state's population size. More than anything else, climate was the key in overcoming the state's apparent locational disadvantage; climate has been important to both the state's settlement history and its agricultural development.

Settlement History

The pre-European native American population in California were hunters and gatherers. For food, they depended on either seafood or grains and nuts that could be collected in the wild and ground into flour. There were few large tribes. Instead, most groups were organized into small units of perhaps 10 to 20 families. They usually spent their lives in a very small areas, never traveling far from home. One result of this geographic isolation was a pattern of substantial cultural variation within the California Indian population. The region also supported a very large Indian population. At the time of first European arrival in the Americas, perhaps 1 native American in 10 in what was to become the United States and Canada lived in California.

Although Spanish explorers brushed the edge of California in the mid-1500s and claimed it as a part of Spain's large North American holdings, they basically ignored California for the next two centuries. The region was at the extremity of Spain's North American empire (see Chapter 15). Not until concern arose over British and Russian expansion in western North America, just before the American Revolution, did Spanish missionaries establish a string of missions from San Diego to Sonoma, near San Francisco. These mission settlements (now some of the most sentimentally revered artifacts of the state's past) were joined by presidios (forts) and a few pueblos (towns). Most of the native American population disappeared during the early period of Spanish settlement, the victims of mistreatment and European disease. During the next few decades, the Spanish and Mexican governments granted a series of large landholdings (ranchos) to immigrants. Still, the area remained peripheral; the towns were small and ramshackle, and hides and tallow were the ranchos' most important exports. Indeed, the decision by the Spaniards to settle California was the result of a desire to provide a buffer against the Russians, who at the time were extending their settlements down the coast from Alaska, and not because of any real interest in California.

Following on the heels of the U.S. seizure of California in 1846, the great gold strike in the foothills of the central Sierra Nevada in 1848 brought the first significant change in the region's settlement fortunes. Within a year, 40,000 people had come to the gold fields by sea, passing through San Francisco harbor. Perhaps as many more came overland. By 1850, California was a state. The frantic period of the gold rush lasted only a few years, but it succeeded in breaking the state's isolation from the rest of the country. It also stimulated the growth of San

A San Francisco street scene. Chinese have been important in the life of the city since the gold rush days of the 1850s. (San Francisco Convention Visitors Bureau)

Francisco, a growth that easily weathered the end of the gold boom. San Francisco remained the largest U.S. city on the West Coast until World War I.

Southern California, the center of Spanish occupation in the state, did not share in the early population expansion. The completion of the Southern Pacific Railroad to Los Angeles in 1881 and soon thereafter the arrival of the Atchison, Topeka, and Santa Fe Railroad to the same destination brought a sudden end to the area's quiet existence. In an effort to create a demand for their facilities, the railroads advertised widely for settlers, aided new arrivals in finding housing and jobs, and lowered fares. For one day in 1887, rail fares on the Southern Pacific from Kansas City to Los Angeles were $1. During the first southern California land boom, between 1881 and 1887, the population of Los Angeles grew from 10,000 to 70,000. Land speculators offered literally hundreds of thousands of town lots for sale in communities throughout the Los Angeles Basin, and many paper millionaires were created almost overnight.

Although this first land investment boom burst in 1888 as a result of wildly exaggerated development fueled by land speculation, the growth of Los Angeles, like that of San Francisco after the gold rush, continued. Promotional literature propagandized the delightful and healthful climate, appealing to sufferers from tuberculosis and asthma. As this propaganda continued to spread, thousands of the ailing flowed into the area.

A large number of crops were also introduced into southern California during this period, including the navel orange (1873), lemon (1874), Valencia orange (1880), avocado (1910), and date (1912). They were in demand in eastern markets and, at that time, in the United States only southern California could provide them in large quantities. Several technical improvements in food handling, notably artificial dehydration (1870) and refrigerated freight cars (1880), meant that these crops could reach eastern markets with only a minimal loss owing to spoilage. Agriculture was to remain the backbone of southern California's economy until after World War I.

California's Agriculture Today

California, by some measures the country's most urbanized state, is at the same time its most agricultural state in terms of total farm income. In 1988, the total market value of agricultural products sold in California was $16.6 billion. Its closest competitors were Texas ($10.3 billion) and Iowa ($9.1 billion). In no other state was the total value of agricultural products as much as half that of California. The state produces more than half of all the country's production of many agricultural products, and ranks high for many others (Table 16-1). California's agriculture, although typified by its many specialty products,

Table 16-1 California Leads the Country in the Production of These Agricultural Products

Crop	Percent of 1987 U.S. Production	Crop	Percent of 1987 U.S. Production
Alfalfa seed	50	Nectarines	97
Almonds	100	Nursery plants	28
Apricots	91	Olives	100
Artichokes	100	Onions	29
Avocados	85	Oriental vegetables	25
Broccoli	97	Peaches	61
Brussels sprouts	97	Pears (Bartlett)	40
Cantaloupes	52	Persimmons	83
Carrots	56	Pigeons and squabs	33
Cauliflower	79	Pistachios	100
Celery	72	Plums	92
Chinchillas	20	Pomegranates	100
Cut flowers	29	Potted plants	22
Dates	100	Prunes	100
Eggs	11	Rabbits	12
Figs	100	Safflower	76
Garlic	87	Spinach	24
Green lima beans	40	Strawberries	77
Grapes	91	Sudan grass	83
Honey	14	Tangerines	49
Honeydew melons	69	Tomatoes for processing	88
Kiwifruit	100	Vegetable and flower seeds	41
Ladino clover seed	97	Walnuts	99
Lemons	82	Worms	18
Lettuce	72		

is, in fact, broadly based. The wide variety of climatic regions and the market demand of the state's own large population are the principal contributors to this diversity.

The importance of the many specialty crops in California does much to explain why the state's farmers have had such success in penetrating the markets of the distant East. These are crops that can be grown, or at least grown on a large scale, in only a few parts of the country. Most require long growing seasons. Thus, there is no local competition in the demand areas. Southern California, especially the Imperial Valley, is also able to provide vegetable crops, such as tomatoes or lettuce, during the winter season when competition is minimal and sale prices are at a maximum. Cooperatives such as Sunkist have been organized to facilitate marketing and to guarantee low shipping costs.

Although the state as a whole produces many agricultural products, local areas within the state tend to specialize in the growth of one or a few local crops. There are several reasons for this pattern of local specialization. One factor is reflected in the general trend toward agricultural specialization that exists across the country. The image of a typical farm as producing a variety of agricultural goods and being at least partly self-sufficient in terms of food is no longer accurate in American agriculture. Often, the local market is structured in a way that makes it difficult to sell all but a few products. As the required technology and knowledge for production has increased, many farmers have found it too difficult or too expensive to handle more than a few basic products. For all these reasons, the national trend has definitely been toward increased specialization on both a regional and an individual farm basis (see Chapter 12).

A second important factor is the major role played by large agricultural operations in the state. Some of the specialty crops produced in California are grown by only a handful of farmers. In the San Joaquin Valley (the southern half of the Central Valley) some landholdings extend over many thousands of acres. Such large-scale growers generally consider carefully the profitability of their operations and have been leaders in the move toward further specialization.

Finally, the diversity of climatic and physiographic niches within the state is a major influence for regional specialization. Many of the specialty crops are particularly sensitive to what may seem minor variations in climate or soil type. The Coastal Range valleys that open onto the Pacific are frequently foggy and have moderate temperatures. Vegetables, such as artichokes, lettuce, broccoli, or Brussels sprouts, grow well under such conditions. The Salinas Valley "Salad Bowl" is a prime example of such a locational advantage. The varietal grapes that people often feel produce the best of American wines need a mild, sunny climate such as that found in the inland Coastal Range valleys around San Francisco Bay. The grapes of the San Joaquin Valley or of southern California, where summer temperatures are much higher, are used for table grapes, raisins, or for what most (although by no means all) wine drinkers consider less distinguished wines. Most flowers grown for seed in the country are planted in the Lompoc Valley west of Santa Barbara. Navel oranges and particularly lemons are grown almost exclusively along the coast and the interior surrounding the Los Angeles Basin. It is unlikely that early spring frosts would occur there to damage the emerging flowers or that heavy winter frosts would kill the trees themselves.

Such local concentrations, although economically advantageous, mean that moderate environmental changes in a relatively small area of California may have a major impact on the overall national production of specialty crops if they are grown in only a few places across the country. Because of this, the rapid expansion of California's urban areas represents an acute problem for many agricultural areas in the state. For example, the Napa Valley, containing many of the finest vineyards in the state, is being "invaded" by waves of residential construction. Although most vintners are relocating elsewhere (total wine grape acreage in the state has increased in recent years), these new locations are less suitable for growth of quality wine grapes. South of San Francisco, some of the coastal valleys, such as the Santa Clara, are urbanizing rapidly. Production of such crops as apricots, prune plums, and pears are threatened by this expansion. California's orange and lemon groves were concentrated in the Los Angeles Basin through the 1940s. Urbanization and smog damage (and competition from the Florida citrus industry, see Chapter 11) have forced almost all production from the Basin.

Whenever land-use competition develops between urban and agricultural users, it is the latter who almost always lose. Urban users, both commercial and residential, are willing and able to pay far more for land than are their agricultural counterparts. In addition, rural land has traditionally been taxed in the United States on the basis of its appraisal sale value and not on its value as a strictly agricultural unit. Thus, the twin pressures of increasing taxes and higher and higher offers from people who wish to buy land, both normal consequences of urban expansion, may convince even the most dedicated farmers to sell. (See the discussion of this point in Chapters 4, 5, and 7.)

The consequences of this changing land use, estimated to result in the annual changeover of 800,000 hectares (2 million acres) from rural to urban uses in the United States, have been of little national concern in the past. It appeared to many that enough rural land remained to provide adequate agricultural production and to of-

California's orange groves, squeezed out of the Los Angeles basin by urban expansion, are concentrated in irrigated areas of the southern San Joaquin valley. Wind machines and smudge pots are needed to protect trees from frost damage, especially during the early spring blossoming period. (Courtesy Sunkist)

fer the aesthetic reward many find in a trip to the country. But this lack of concern has begun to change. Many cities are located in the midst of good farmland, and because urban land users tend to prefer the same flat land that is most valuable to farmers, the farmland lost to urban expansion tends to be better than average. Also, in an increasing number of areas, residents have come to feel that in losing local farmland they are losing a valued part of the landscape.

Attempts to preserve farmland in an urbanizing environment have followed several tactics. Most common is a decision to grant special tax breaks to farmers; this is usually accomplished through the valuation of their lands for tax purposes as farmland rather than as potential urban land. One problem is that urban expansion, already quite jumbled in its spatial pattern, may become even more so as some farmers choose not to sell their land until the surrounding land is urbanized. Another, more negative problem is that urban services (water, sewer, streets, etc.) tend to become less orderly; cost increases are often the result. Within the Los Angeles region, several dairying areas attempted to protect themselves by incorporating, thereby creating their own tax structure. This system broke down when individual farmers within the "town" decided to sell to residential developers. These new residents soon outnumbered the farmers, and the urbanites' demands for improved services began to push taxes upward. Other areas have reacted with substantial zoning restrictions, such as a 2- or 4-hectare (5- or 10-acre) minimum for residential plots. Often used to keep "undesirable elements" out of exclusive areas, it may also keep land in farms. However, the future of such restrictions are in doubt because some courts have found this type of zoning to be illegal.

Today nearly every state has laws that provide some property tax relief to farmers. California requires owners to enter into a long-term agreement not to develop their land as a condition for eligibility for reduced property taxes. Some 300 counties nationally have some type of agricultural zoning to preserve farmland. A few locales have programs to purchase development rights to ensure that the land remains in farming.

Water Supply

California's agriculture, more than its manufacturing or urbanization, has created a massive demand for water across much of the state. California has more irrigated farmland than does any other state, about 3.5 million hectares (8.6 million acres). Only Texas has as much as half as many irrigated acres as California. Nearly all of California's cotton, sugar beets, vegetables, rice, fruits, flowers, and nuts are grown on irrigated farmland. The state's farmers use more than one-fourth of all the irrigation water used in the United States. On the average, irrigated land in the state receives about 1 meter (40 inches) of "artificial" water annually.

Crop selection at a location is dependent upon water availability plus other factors such as soils, drainage, terrain, and growing season. But the potential for irrigation is usually critical. A transect across the San Jaoquin Valley finds livestock grazing in the Sierra foothills; dry farming for grains in the flatter, but still too high for irrigation, land below; irrigated fruit trees and vine crops in the better drained soils near the valley floor; and irrigated field crops such as cotton, vegetables, and sugar beets on the flat valley floor.

Some 70 percent of the state's precipitation falls in the northern mountains and valleys and in the Sierra Nevada where few farms or urban places exist, whereas 80 percent is used in the drier south. Of the state's major farming regions, only areas north of San Francisco and a few coastal valleys to the south receive as much as 50 centimeters (20 inches) of precipitation in an average year; two of the most important farming areas—the southern end of the Central Valley and the Imperial-Coachella area in the southeast—average annually less than 25 centimeters (10 inches).

Farms are the principal users of water, but it was the cities that initiated the development of the state's tremendous water movement complex. At the beginning of this century, Los Angeles outgrew its local groundwater supplies and looked for a supplementary source in the Owens Valley, east of the Sierra Nevada and some 300 kilometers (185 miles) north of the city. By 1913, the Los Angeles Aqueduct was carrying water to Los Angeles, much to the dismay of Owens Valley farmers who lost virtually their entire water supply (Figure 16-4). This aqueduct still provides half of the city's needs. In 1928, Los Angeles and 10 other southern California cities formed the Metropolitan Water District to de-

Figure 16-4 Water movement systems. Chief beneficiaries of this massive system are the farms of the San Joaquin and Imperial valleys, and the cities of the Los Angeles basin.

velop an adequate water supply for their entire area. By 1939, the group had completed the Colorado River Aqueduct, which carried water from Parker Dam on the Colorado to coastal cities from San Diego to the Los Angeles Basin. Today the Metropolitan Water District, which serves 6 counties, over 130 cities, and half of California's population, is one of the state's most powerful political bodies.

Farmers were by no means ignored, however. Perhaps the most spectacular episode in the state's water history occurred in 1905 in the Imperial Valley. In 1901, private groups constructed canals to carry water from the Colorado River into the Imperial Valley; the result was an immediate agricultural land boom. Then, in February 1905, the Colorado River flooded, broke out of its channels, and flowed into the irrigation ditches. Before a massive effort returned the river to its channel in the fall of 1906, 1100 square kilometers (400 square miles) of the valley had been filled by the Salton Sea, a body that still

The automobile, which greatly increased the spatial flexibility of the American population, encouraged a very rapid expansion of urban space. By 1965 (the year this photo was taken), the San Francisco urban area was spilling southward along the east side of the Bay toward Palo Alto. The San Andreas fault is marked here by the line of reservoirs toward the top of the photo. (U.S.G.S)

exists, its size maintained by the ongoing drainage from irrigated farmland in the valley. Most of the valley's irrigation water is provided today from the Colorado River by the All-America Canal, built by the federal government in the 1930s.

In the 1940s, the federal Bureau of Reclamation began the Central Valley project. The aim of the project was simple—to improve the local availability of irrigation waters in the Central Valley. Most of the best potential irrigation land was in the southern San Joaquin Valley, but most of the water was in the northern Sacramento River. Today the project is fully functional. Water, which is removed from the Sacramento River by the Delta-Mendota Canal, flows southward along the west side of the San Joaquin Valley to Mendota, where it is put into the San Joaquin River. This transferred water meets the irrigation needs of the San Joaquin Valley from Mendota northward. Thus, most of the normal flow of the San Joaquin River can be used for irrigation in the southern part of the valley. The southern valley is the leading agricultural region in the state, supplying nearly 40 percent of the state's total agricultural output. Of the country's 10 leading agricultural counties, measured in terms of value of farm products sold, 5 are in the San Joaquin Valley. The value of farm products sold in Fresno County alone is greater than that for 17 states. Only 7 states, other than California, exceed the combined value of agricultural output for these 5 counties.

In 1957, California brought all of the existing water projects and all schemes for new ones under the California Water Plan, which was administered by the state. The plan developed the California Water Project, designed to obtain water in California where available and to transfer that water to where it was needed. The project, easily the largest water movement program in American, perhaps in world, history, focuses on a massive system to move water from the north to the south in the state. The California Aqueduct, also starting from the delta area, carries water southward; this water is used partly in the western San Joaquin Valley and partly to supply the ever-growing water needs of the southern coast. Most of the cost of the southern extension is borne by users in the Metropolitan Water District. The total cost of the system continues to increase, with the estimates for improvements now exceeding $20 billion.

It is predicted that even this monumental system will satisfy the state's increased demands for only a few decades, with critical water shortages perhaps by the year 2020. Some hope that desalination of seawater will provide a major new source of freshwater by then. However, existing processes are expensive and are heavy users of energy. Others look northward to the Columbia River, whose flow dwarfs that of the Colorado, for a future supply. As early as 1949, the federal

Bureau of Reclamation suggested the possibility of transferring Columbia River water to California. Apart from the huge costs of such an undertaking, people outside of California in the Northwest show no inclination to accept such a suggestion because they believe they will eventually need all of their own water supplies.

Several central points emerge from a study of water use in California. One is fairly obvious—given adequate financing, we have a tremendous technical ability to move water from one place to another. This, in turn, has led to a rapid increase in the demand for water. In recent decades, California's consumption of water has increased even more rapidly than has its population size. Suburban swimming pools dot southern California and a growing industrial demand has also been important, but farm irrigation has been the major user. Between 1964 and 1987, the total acreage of California's farms dropped by more than 20 percent, but both the number of acres irrigated and the volume of water used per acre increased. The main reason is that irrigated lands in this generally dry environment produce very high yields. California's cotton production per acre, for example, is double the national average. As mentioned earlier, governmental agencies tend to absorb most of the costs involved in irrigation projects, thus providing water at an artificially low price to farmers, and this encourages expanded use. Finally, there is a clear tendency to reach across ever-increasing distances to meet existing needs. The water supply of the state and of the entire Southwest is limited. The region is quite close to using all there is.

Three problems in connection with the water supply in California are worth mentioning. First, much of the Imperial Valley and Coachella Valley face an impending problem because of a gradual buildup of salts in the soil. Suspended salts are carried in large concentrations in the Colorado River. Once in the soil, they can be removed only through the application of large quantities of water to flush them out. If left, the salts will eventually destroy the soil's productivity. As discussed earlier (Chapter 14), much damage is already apparent in the irrigated lands just south of the California border in Mexico.

Second, many northern Californians are not happy about the transfer southward of much of what they view as their water supply. They too feel that their part of the state will eventually need all the water it can get. California's center of population is in the mountains just north of Los Angeles. Hence, in any statewide political decision the south has far more potential voting power than the north. Northerners see the water decision as one proof of that. Antagonisms have long existed between the two sections of the state. Angelenos denigrate what San Franciscans like to view as "The City" by calling it "Frisco." For their part, San Franciscans feel a certain condescending pity for those who they feel are the uncultured people of the south. Mention is even made occasionally, usually by northerners, of splitting the state in half. A bill proposing such a split reached Congress in 1940. Although a separation is more an interesting speculation than a likelihood, the differences between the two areas are real; political conflict generated by the state water movement program has been a significant reflection of these differences.

Third, many Californians question the economic and environmental costs of water transfer projects. In 1980, the state legislature proposed construction of a peripheral canal to carry water from the Sacramento River southward east of the delta to increase the volume of flow to the dry south. The state estimated that the project would cost $5.4 billion; opponents argued that $20 billion was closer to correct. Environmentalists were concerned that the loss of freshwater would damage the rich aquatic and waterfowl populations located in the intricate lacework of waterways of the delta. Delta farmers feared loss of their rich vegetable-producing lands. San Franciscans again worried over a water loss to the south. A coalition of these diverse interest groups defeated the canal in a 1982 referendum.

The California Aqueduct, and the Delta-Mendota Canal in the background, carry water southward to irrigate the San Joaquin Valley and for metropolitan Los Angeles. Note the intensity of land use in the irrigated area. (U.S. Department of Water Resources)

URBAN CALIFORNIA

Despite the lengthy treatment of California's agriculture—necessary in the light of the state's agricultural importance nationally, its unique character, and the fact that agriculture's influence on the state far outstrips its share of the total population—its population is overwhelmingly urban, and still increasing. Over 90 percent of the people live in one or another of the state's 23 MSAs. California had 14 of the country's 50 fastest growing MSAs in the 1980s. Los Angeles has passed Chicago to become the nation's second largest city. Most of the state's population live in one of its two major urban regions, one centered on Los Angeles, the other on San Francisco.

The Southern Metropolis

"A city in search of a center!" This is the title that many give to the loose assemblage of cities normally lumped under the general name Los Angeles (Figure 16-5). The land boom of the 1880s led to the establishment of several score of cities scattered across the Los Angeles Basin and southern California coastlands. As their populations increased, these communities squeezed out the intervening rural lands that formerly separated them spatially. At the same time, the communities maintained their jurisdictional independence. The result is a jumbled complex of politically independent but economically intertwined units. The cities have surrendered some of their individual decision making and taxing jurisdiction in recognition of common regional need that can only be met cooperatively. The Metropolitan Water District is the best example of such a regional agreement. Still, the political complexity results in many problems in taxation and planning (see Chapter 5).

Most of the 300-kilometer (185-mile) stretch of coastline from Santa Barbara to San Diego is now occupied by one long megalopolis. The home of about 15 million Californians, the vast spread of this urban landscape is a reflection of a growth process shared by other cities in North America and, at the same time, a strikingly unique metropolis.

The entire complex is basically a creation of the twentieth century. In 1900, its population was less than one-tenth of its present size, and, as mentioned earlier, it was not until after World War I that the area began to shift from an essentially agricultural to an urban region. Thus, many of the elements of eastern cities, elements placed on the landscape largely during the nineteenth century and the first two decades of the twentieth century, are not present in Los Angeles. Among these missing features we might list the four- or five-story walk-up apartment buildings, warehouses of about the same elevation, fixed-rail elevated (or underground for that matter) public transportation lines, and a strong nodality centered on the central business district. Knowledgeable Angelenos could undoubtedly expand this list greatly, but it is representative.

The southern metropolis's major identifying characteristics are obviously of the twentieth century. The most important stimulus has been the family automobile. Without it, the region would surely strangle. Automobiles offered a degree of spatial flexibility unknown previously. People could live, work, shop, and recreate where they chose, and the bonds of structured travel patterns were removed from urban residents. If we measure distance by time rather than by miles (a far more relevant method of measurement to most travelers), the maximum one hour's distance that a worker might be willing to place between residence and place of work increased during the present century from a 5-kilometer (3-mile) walk to a 90-kilometer (55-mile) drive. Furthermore, the linear inflexibility of public transport was no longer an important factor in Los Angeles; this made it possible to live far from work and not be bound to the limited residential opportunities located along streetcar lines.

Today fully half of the central portion of Los Angeles has been surrendered to the automobile, either for roadways or for parking. The urban area's dense system of freeways makes possible fairly high-speed movement across much of the metropolitan region. The Los Angeles area has more cars per capita than any other part of the country. It also has only a minimal public transportation system. Although neither fact should be surprising given the diffuse pattern of housing, employment, and commercial facilities, nevertheless, it means that the private car carries the overwhelming percentage of the region's daily population circulation. That easy mobility by automobile is central to the area's economy and life-style.

Increasingly heavy use has led to a growing problem of congestion on these arteries. Los Angeles county has four of North America's five busiest freeways. On most mornings during

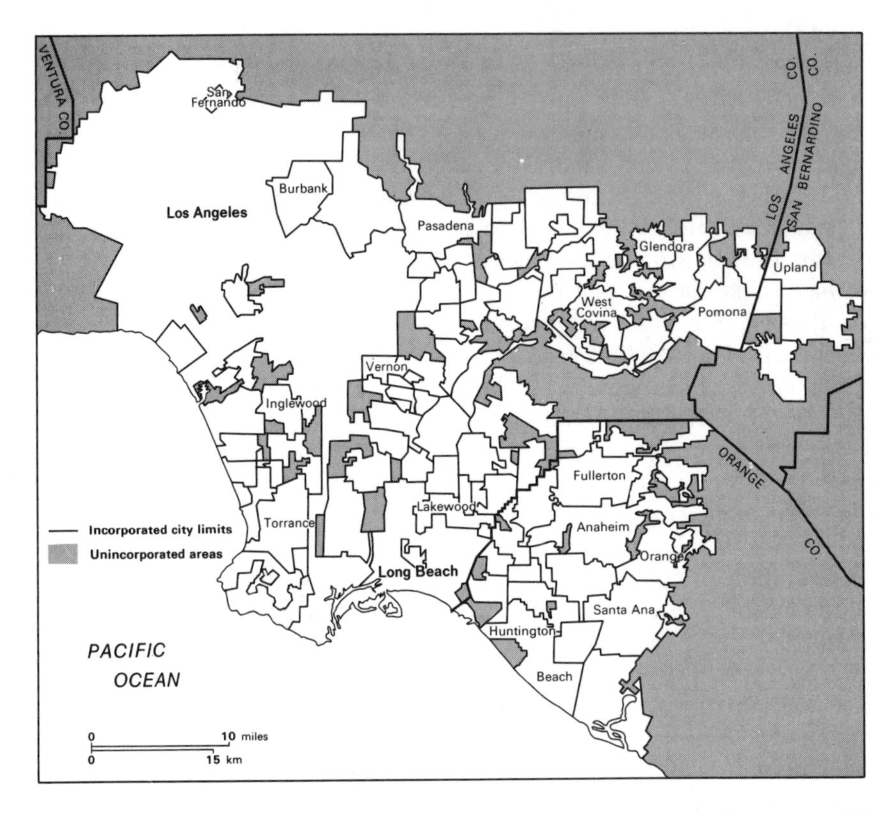

Figure 16-5 Incorporated cities in the Los Angeles area. Los Angeles annexed large portions of the San Fernando Valley northwest of the early focus of the city to avoid being entirely surrounded by other independent urban places.

rush hour, about 40 percent of the region's freeways are clogged with commuter traffic moving no more than 25 kilometers (15 miles) per hour. Until a 1990 yes vote on a major highway funding program, California was spending less per capita on highway construction than any of the other 49 states, and Los Angeles ranked only 17 among American urban areas in miles of freeway per capita. The rapid geographic expansion of greater Los Angeles, especially eastward toward San Bernadino and the Mojave Desert, will only increase commuting distances and make the problem worse.

Much about southern California's urban landscape reflects the impact of the automobile. The population density of Los Angeles is 17,000 per square kilometer (6100 per square mile); in San Diego it is about half that. By comparison, Philadelphia has an average density of 38,000 per square kilometer (14,200 per square mile). Housing units throughout the area are predominantly of the detached, single-family type. Even low-income areas tend to be composed of such housing, a feature that often leads the casual observer to conclude mistakenly that slum or near-slum conditions do not exist in Los Angeles.

Finally, as suggested earlier, Los Angeles is a city without a center. The traditional singular central business district (CBD), as a focus of urban activity, barely exists here. Los Angeles is really many cities that have grown together as they increased in size; 14 of these cities currently have populations of more than 100,000. The absence of a single dominant CBD results in part

The Los Angeles River, usually a trickle but a major stream after winter rains, is paved for much of its distance across the Los Angeles basin. The basin's freeway system connects its many cities. (Georg Gerster/Comstock)

from the continued existence of independent centers for each of these communities. The metropolitan area seems to be coalescing into about 18 "urban village" cores, which are business and retail focal points within the generally low-density urban landscape. Los Angeles itself is, of course, the largest. Some, such as Pasadena or Westwood, are older communities that have grown into regional hubs. Others, like Ontario and Costa Mesa/Irvine/Newport Beach seemed to have simply emerged out of the expanding suburban sprawl. Some in southern California talk about the area as a series of "constellations" forming a metropolitan "galaxy." The terms city center and suburb, commonly used elsewhere to describe the geography of an urban area, have little meaning in Los Angeles.

The automobile is also partly responsible for one of Los Angeles's most infamous features—smog. Ringed by mountains to the north and east with hot dry deserts beyond and bordered by the cool waters of the Pacific to the south and west, the Los Angeles Basin frequently is subjected to temperature inversions. Such a situation occurs when a body of warm air lies above cooler air. This warmer air forms an atmospheric "lid," and pollutants in the cooler air are unable to rise into the higher, warmer air. Onshore winds push the pollutants inland, but a combination of the inversion lid and the surrounding mountains keep much of this pollution within the Basin. Palm Springs, just east of the Basin, has grown up as a high-status community because it is beyond the smog's reach. In reality, Los Angeles produces only moderate amounts of pollution, for it has little heavy industry, burning is carefully controlled, and there are strict controls on automobile emissions. The Basin's automobile population has more than doubled since the mid-1950s, but their total emissions have actually been reduced over the period. The city has made a continuous effort to lessen smog problems since World War II. Overall smog levels in the Basin today are lower than they were a quarter of a century ago. The South Coast Air Quality Management District is responsible for developing far-reaching plans to deal with the smog problem. The area's peculiar physical geography, a pattern that led early Spanish explorers to comment on the buildup of smoke from Indian fires, is the remaining, probably unbeatable, problem. Some have suggested, semiseriously, that a large cut be carved through the mountains and fans set up to blow the smog out. People living in areas beyond the mountains, already hit by some effects of the smog that drifts over the mountains, are understandably opposed to such suggestions.

If it is difficult to explain the existence of almost 30 million Americans in California using the normal logic of location theory, it is nearly impossible to explain the presence of nearly 15 million Americans in southern California. Although the area is not without resources, their overall importance is not overwhelming. We have already looked at agriculture. Petroleum production is also important; three of the country's major fields are in southern California. Offshore development began in 1965. Long Beach has reaped a large financial harvest from the sale of leases and from a tariff on oil pumped. Many find the offshore pumping stations, built as islands complete with trees and waterfalls, visually less offensive than they had expected. The heavy demand for petroleum products, especially gasoline, results in the local consumption of virtually all of southern California's production.

Southern California is known worldwide as the location of Hollywood, long the center of the American motion picture industry. In the early days of filmmaking, outdoor settings and natural light were the norm. The area's cloudless skies and short cold-temperature periods made its streets and fields a fine home for countless motion pictures. Los Angeles remains one of the centers of American filmmaking and television, both of which have provided invaluable national and international advertisement for California as well. The silent movies introduced California's palm-lined streets, and contemporary television continues to glamorize the region. However, today the industry plays only a small role in the Los Angeles economy. Only about 75,000 people in Los Angeles county are employed in motion picture and television production. That is less than 2 percent of the metropolitan area's total employment.

The Mediterranean climate has also supported the recreation industry. That, and the varied scenery, especially along the coast, early made southern California one of the country's centers of outdoor recreation. Towns and railroads emphasized the area's fine climate and scenery for recreation late in the nineteenth century. Pasadena's Rose Parade was partly developed as an advertisement for the area's mild winters. Today, these natural advantages have been supplemented by some of the country's largest and best developed recreation facilities. Balboa Park in San Diego, with its excellent zoo, Knott's Berry Farm, and Marineland are major attractions. Disneyland has become an American phenomenon and the main destination of countless tourists.

Still, all of this does not add up to resource support for 15 million people. Heavy resources, such as coal or iron ore, are virtually nonexistent. San Diego has a good harbor, but Los Angeles's harbor, entirely human-made, is only average. To understand this population concentration better, three additional elements must be included: general national affluence, the growth of a postindustrial economy, and the role of federal government spending.

Southern California has profited from government spending far more than have most other areas of the country. California now receives about 20 percent af all Department of Defense spending and nearly half of that of the National Aeronautics and Space Administration. San Diego is the West Coast home of the U.S. navy, and the navy is easily the city's principal employer. San Diego has relatively little manufacturing employment for an urbanized area of over 2 million, surely an important part of its justifiable claim to be one of the country's most livable cities. Southern California has been the country's leading aircraft-manufacturing center for nearly 40 years, with much of the business generated by the purchase of military aircraft. This means that a portion of the regional economy is cyclical, with the fluctuations dependent on government defense spending.

The declining share of total employment held by manufacturing and the corresponding growth of the tertiary and quaternary sectors has tended to free American workers from even the tenuous locational impact that the physical resource base has on manufacturing. This shift is essentially a transition into a postindustrial economy. The nature of manufacturing itself has also shifted; an increasing share of all workers are in light industries, where the value added by manufacture is high and worker skills particularly important. Electronics is such an industry, and it is important in southern California manufacturing. In this postindustrial period, new factors, such as the availability of an adequate secretarial pool or the presence of major universities with their diverse intellectual and research resources, become more meaningful locational considerations. The question of where the workers and more particularly, perhaps, the management want to live becomes highly significant (see Chapter 11). Southern California, with its mountains, seashore, mild climate, and outdoor image, has proved to be a most attractive location for many. As never before, we have the ability to choose where to live; studies show that southern California offers the kind of environment that many Americans find especially desirable.

Southern California is also the major destination for Latin American and Asian migrants entering the United States. Today only a minority of Los Angeles county's 7 million people are Anglos. Greater Los Angeles may be the largest Mexican metropolitan area outside Mexico; the second largest Chinese metropolitan area outside China; the second largest Japanese metropolitan area outside Japan; and the largest Korean, Filipino, and Vietnamese metropolitan areas outside those countries. Over one schoolchild in four in the Los Angeles school district speaks one of 104 different languages better than English. Los Angeles is close to Asia and Latin America, its climate is moderate, and in their numbers, its diverse ethnic population offers some security to new arrivals. It is estimated that by 2010 the metropolitan area will be 40 percent Anglo, 40 percent Hispanic, 10 percent Asian, and 10 percent non-Hispanic black.

The cityscape of Los Angeles reflects this ethnic diversity. Recent migrants especially often settle in ethnic neighborhoods. Little Tokyo, long a part of the city, has a renewed vibrance. Monterey Park in the San Gabriel Valley is 50 percent Asian, making it the most heavily Asian city in the mainland United States. A rich diversity of ethnic restaurants, many found within ethnic enclaves, dot the city.

In 1950, 7 of the world's 12 largest cities were European or North American. By 2000, only 2—New York City and Los Angeles—will be. Of the two, only Los Angeles is still growing rapidly. It has a higher dollar volume of retail sales than does the New York metropolitan area, and the value of its manufactured goods is also higher. A decade ago the city passed San Francisco as the West Coast's financial center; in 1986 it surpassed Chicago in banking deposits, and now ranks second nationally to New York City. The twin ports of Los Angeles and Long Beach together form the fastest-growing major cargo center in the world. The dollar value of their import-export oceanborne cargo now easily surpasses that of the Port of New York and New Jersey. U.S. trade with Pacific Rim countries is now 30 percent greater than U.S. trade with Atlantic Basin countries, and 60 percent of that trade passes through Los Angeles's two ports and its international airport.

San Francisco Bay

By comparison with their upstart competitors in the south, San Franciscans choose to view their city as old, cosmopolitan, and civilized. There is something to be said for this attitude. The city was the northern core of Spanish and Mexican interest in California. It served as the supply center for the California gold rush. By 1850, it was the largest city on the Pacific Coast, a ranking it maintained until 1920. The completion of the first transcontinental railroad in 1869 coupled with the city's size and its excellent harbor made it not only the focus of western U.S. growth, but also the key location for U.S. commerce with the Pacific as well. Into the city came large numbers of immigrants from Asia, especially Chinese, plus substantial numbers of other foreigners. They created a cosmopolitan ethnic mixture that today remains a readily apparent aspect of the city's character. Much of the early history of the city was more bombastic than civilized, but San Francisco is certainly the oldest West Coast city.

The romantic flair of this early history is one piece of the mosaic that makes San Francisco among the most popular of American cities. Its physical geography provides a splendid setting for a city, steep slopes that offer dramatic views of the Pacific Ocean or of the Bay coupled with a mild climate that escapes the sometimes staggering summer heat of southern California. Recent quantitative evaluations of the quality of American cities, which tend to emphasize readily available socioeconomic statistics, have generally given the city a good ranking. Most high rankings tend to go to the newer cities of the West; hence, San Francisco's position is neither surprising nor exceptional. Even so, it is another component of the pattern that places San Francisco at or near the top of a list of "desirable" places in studies of urban perception. One consequence of this perception has been a continued migration of newcomers into the city that has led to unemployment levels generally well above the national average.

In some ways, this attitude is based on a false perception of the area. The city of San Francisco is currently home for fewer than one-eighth of the Bay Area's 5.4 million people. Hemmed into its small peninsula, the city is actually losing population while the entire urban area grows.

Most of this growth has been diffused, random, and very much like that found in most large metropolitan regions (Figure 16-6). The exotic appearance of Chinatown, or the vaguely romantic seediness of the Mission District, both in San Francisco, today must be matched against the stark ghettos of Oakland, the refineries of Richmond, or the almost endless sprawl of San Jose. A folk song lamenting "houses made of ticky-tacky," where they "all look just the same," refers to the San Francisco MSA, not Chicago or New York. The sweep of brown hills and rich vineyards around the region, so long a part of its special nature, pass daily under the bulldozers of the metropolitan area's leapfrogging urban growth.

The Bay Area today is really composed of several different areas, each with its own special character. The East Bay is the most varied, with a mix of blacks, Berkeley students, large tracts occupied by middle-class residents, and most of the port facilities and heavy industry of the region. Some have suggested that it is not very different from northern New Jersey in its landscape character. The San Jose–South Bay area is upper middle class, white, with new houses, fine yards, and major regional shopping centers. Along the bay north of San Jose is Silicon Valley, so named because of its concentration of businesses engaged in chemical and electronics research associated with production of computer components. North of the Golden Gate Bridge, the cities are smaller, there is little manufacturing, and the conflict between agricultural and

The San Francisco Bay area is unusually hilly for a major urban center. In newer sections, the housing often appears like ropes coiled over the hills. (Peter Menzel)

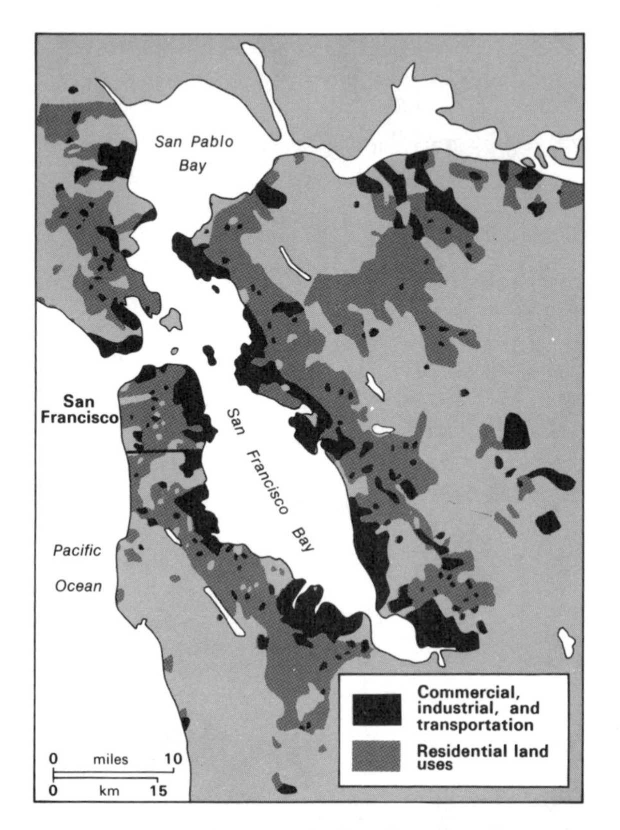

Figure 16-6 Land use in the San Francisco Bay area. Industrial and large-scale transportation activities hug the margins of the bay. Accessibility provided by this superb site has been critical to the area's economic growth.

urban land use is sometimes obvious. Here are found the well-to-do urbanites searching for a place in the country. The city of San Francisco itself, with its grid pattern of streets incongruously placed on a hilly terrain, its closely spaced late nineteenth- or early twentieth-century housing, and its ethnic diversity maintains a special appeal, despite the recent appearance of skyscrapers along the skyline. San Francisco has adopted zoning ordinances aimed at maintaining the special character of the city. Buildings over 18 meters (60 feet) high are banned from many areas, architectural appropriateness is encouraged in public housing, private redevelopment plans are encouraged, and the city, in general, takes an uncommon interest in preserving what it sees as its special character.

San Francisco is far less the focus of the area than it once was. Oakland is the principal center of the East Bay; San Jose is the hub of Santa Clara county and is now more populous than San Francisco. The bay, so important to the quality of San Francisco's site, is in some ways also a problem. Movement between different sections of the area has been difficult and expensive, with few bridges crossing the bay. As the urbanized area grew, water barriers made a continued regional focus on San Francisco difficult. The Bay Area Rapid Transit (BART) system, one of the few new mass transit systems in the country, was designed in an attempt to alleviate part of this problem. Only three of the area's nine counties chose to join the system, however. It serves only the East Bay and San Francisco, thus doing little to ease circulation in the entire region. Still, BART is an innovative attempt to meet the difficulties of urban movement.

Unlike Los Angeles, the existence of a major urban center in the Bay Area is not surprising. Its excellent harbor and good climate are important site factors. In fact, access to the bay has been so important and sizable tracts of flat land for construction so limited that large portions of the bay have been filled to create new land. Communities around the bay have created a regional government authority to restrict such landfill development, spurred by the fear of huge economic and ecological losses that might result from excessive filling; little reshaping of the bay has been allowed by the authority in recent years.

Furthermore, the relative location, or situation, of the bay region also remains excellent. By volume, it is the major Pacific port in North America. Its rail and highway ties to the East at least equal those for any other West Coast city. Just as Megalopolis is America's hinge with Europe, San Francisco is its hinge with Asia.

This air photo of a portion of San Francisco Bay is a dramatic indication of human encroachment into the Bay. The irregularly shaped areas surrounding much of the Bay are areas for salt collection. They are repeatedly flooded, then allowed to dry. The salt that remains can then be scooped up and sold. (NASA)

Cities of the Valley

Beyond the southern coastline and the San Francisco Bay region, the largest cities in California are in the Central Valley. The most populous is Sacramento, the state capital, with an MSA population of 1.5 million. In common with all of the valley cities, it is a major agricultural processing center. Much of the processing of fruit and vegtables for canned or frozen products is accomplished near the source. It also has a substantial aerospace industry. Two large nearby air force bases are additional important contributors to the regional economy. Fresno (MSA population, 667,000) and Stockton (population, 481,000) are regional service and agricultural processing points for the central and northern San Joaquin Valley. Bakersfield (population 543,000) serves a similar role at the southern end of the valley. It has also been influenced by local petroleum extraction. Bakersfield is the home of country-western music in California.

A STATE OF IMMIGRANTS

"California is a state of immigrants, mostly from the Middle West." "A native Californian is a rarity." These are commonly held beliefs about Californians. As with most such statements, they contain a substantial kernel of truth, but they also can be misleading. Although more than two-thirds of all native-born Americans live in their state of birth, it is true that fewer than half of all Californians were born in the state. California has been an important destination for U.S. internal migration in nearly every decade since 1850 and has become the major state of residence, other than the home state, for people born in an area extending from California eastward and northeastward through the Midwest. The political climate, especially in San Diego and Los Angeles, often reflects the conservatism of the area's southern and midwestern migrants. Between 1940 and 1990, California's share of the total population of the United States grew from 5 percent to over 10 percent. California, for decades, had attracted more migrants than any other state until Florida gained that distinction for the 1970s. California regained the number-one position during the 1980s.

California's Future

California has also already established itself as a dominant political and economic component of American life. A presidential candidate is well advised to focus upon winning the state's nearly 20 percent of the electoral votes needed for victory. The victor in the Golden State needs only three-eighths of the remaining electoral votes of the other 49 states. Los Angeles alone produces enough goods and services each year to rank the

Recreational vehicles, such as these dune buggies, can cause significant damage to the environment. (Virginia Blaisdell/Stock, Boston)

metropolitan area among the world's 12 wealthiest countries. As the countries of the Pacific Rim continue their economic development, and as the U.S. trade with Rim countries increases, California benefits. Many have suggested that California has emerged as the country's cultural core, replacing New England and the Middle Atlantic states. It has taken a leadership role in such varied efforts as local tax reduction, environmental legislation, term limits for elected officials, and gay and lesbian rights. Certainly it represents what Americans view as the good life. One author has written that southern California may be "the super-American region or the outpost of a rapidly approaching postindustrial future." It is certainly a new kind of America, one that is copied if not totally admired.

California's problems are as great as its possibilities, with growth the main creator of both. The state's physical environment, so admired and desired, is a fragile one; the desert and the alpine ecosystems are easily damaged. The outdoor recreation demands of 30 million Americans are immense and could easily outstrip the state's resources. Only so much seacoast exists, and much of it is already unavailable for public use. Although the state has some of the country's most rigorous, well-enforced environmental laws, the magnitude of the task makes these laws barely adequate.

Perhaps the two greatest problems California faces are related to urban expansion and water supplies. Although cities are expanding everywhere, the loss of farmland in this Mediterranean climate, a climate found nowhere else in the country, poses a serious threat to the domestic production of certain crops. The greatest losses of land, centered around Los Angeles and San Francisco, have been offset largely by expansion elsewhere, particularly in the San Joaquin Valley. Most of the easily available new land is now in use. Additional expansion will require irrigation on a massive scale, and this will bring the need for maintaining agricultural production levels into conflict with the state's limited water supplies; if present trends continue, water demands will outstrip California's available supply early in the next century. The chances of obtaining water from the Columbia River are problematic, at best.

It may well be that, in the long run, irrigation is not the most economical use for the state's water. The monetary return for water invested in crop growth is relatively small when compared with alternative industrial uses. The repeated development of new sources of supply has thus far allowed the state to sidestep this difficult question. Soon it may no longer be able to do so.

The problem of water sufficiency makes the future of California's agriculture somewhat difficult to predict. Although California does lead the nation in the value of its agricultural production, the relative importance of agriculture within the state's total economy is much less than it is for states like Iowa or Nebraska. Also, many of the crops grown are not a basic part of the American diet. Demand for such crops may fall if their prices are pushed up by more expensive irrigation water. Finally, these expensive irrigation projects help relatively few farm operators, who are often corporations rather than individual farmers. Thus, despite the considerable political clout these few operators may have, the decision might eventually be made to shift water from agricultural land to urban areas where the vast majority of Californians live and where most of the state's income is generated; this might be done in spite of the loss it would represent to corporate farmers, to the state, and to the nation.

PHYSICAL CASE STUDY

PLATE TECTONICS AND THE RING OF FIRE

California is associated with what geologists have called the "Ring of Fire," a belt of intense earthquake and volcanic activity that encircles much of the Pacific Ocean. In what almost amounts to a landslide of intellectual activity during the last 25 years, earth scientists have improved their understanding of the reasons for such a pattern of concentrated, violent earth movement. Their findings are important to an understanding of both the physical and human geography of California and, indeed, of most of the west coast of North America.

Most geologists now agree that the crust of the earth, the *lithosphere*, is composed of a large number of large, solid zones called plates that move slowly on the molten rock which lies beneath them. In general, oceanic plates are rather thin, averaging about 50 kilometers (32 miles) in thickness, but are dense and heavy, while continental plates are thicker, averaging 150 kilometers (95 miles) thick, but are brittle and more boyant. In midocean areas these plates may pull apart forming a deep trench bordered by ridges. A mid-Atlantic ridge, stretching nearly the entire north-south extent of that ocean, is a prominent example of such a feature. In other areas plates may be pushing together and colliding. Where that collision is between oceanic and continental plates the denser oceanic plate usually plunges deep beneath the continental plate in a process called *subduction*. Deep ocean trenches may be created along the zone of surface contact between the two plates. The crust of the oceanic plate may melt as it is subducted. The resulting *magma,* or molten rock, can move to the surface, creating a line of volcanos well away from the surface contact zone. A third possibility is that the two plates are in contact but one is merely sliding along the other, with no movement toward or away from each other. The results are usually *faults* that run along the line of contact between the two plates. Faults are fractures in the earth's crust where relative displacement has occurred. Mountain building activty is often found where two plates come into contact. The general theory given to explain these associated activities is called *plate tectonics.*

The entire west coast of the United States and Canada is a zone of active plate contact. Most of the continent is dominated by a single plate, called the North American plate. An even larger Pacific plate borders it along the west coast. In California, from the Imperial Valley northwestward to Cape Mendocino, the Pacific plate extends inland to meet the North American plate roughly along the line of the San Andreas fault (Figure 16-2). Coastal California north of the Transverse Ranges that form the northern edge of the Los Angeles Basin is sliding northwestward along the North America plate at a steady rate of 2 centimeters (1 inch) or more a year. The two plates are colliding along the Transverse Ranges. This entire zone of contact is an area of frequent earthquake activity.

From northern California northward to British Columbia a much smaller oceanic plate, the Juan de Fuca plate, is being subducted by the North American plate as the latter follows a generally westward drift. One dramatic consequence is the line of volcanic peaks from northern California to Washington that mark the zone of crustal melt for the subducted oceanic plate. This is a zone of active volcanic activity, as the 1980 eruption of Mt. St. Helens in southwest Oregon reminded us (see Chapter 17).

Compare maps of the coastal margins of the east and west coasts of the United States (see, for example, Figure 2-1). The Atlantic Coast is indented and irregular. It is a classic example of a coastline of submergence, where the land is sinking relative to sea level. The coastal plain itself is broad and flat, especially in the South. Mountain building associated with plate contact has generated a very different appearance along the Pacific Coast. The coastline is one of emergence. It is very regular, with few embayments and no estuaries of significance. Offshore, the broad continental shelf that marks the Atlantic Coast is replaced on the Pacific coast by deep trenches near the coast. There is no coastal plain. Instead the land is generally hilly or mountainous to the water's edge. Rivers seem to just suddenly meet the ocean, with none of the gradual transition from river to estuary to ocean so common on the east and Gulf coasts. Few large west coast cities, and none north of San Francisco, are at the ocean's edge.

ADDITIONAL READINGS

Furseth, Owen J., and John F. Pierce. *Agricultural Land in an Urban Society.* Washington, D.C.: Association of American Geographers, 1982.

Hartman, Chester W. *The Transformation of San Francisco.* Totowa, N.J.: Rowman and Allenheld, 1987.

Kahrl, William L., et al. *The California Water Atlas.* Sacramento: Governor's Office of Planning and Research and the California Department of Water Resources, 1979.

Lantis, David W., et al. *California: The Pacific Connection.* Chico, Calif.: Creekside Press, 1989.

Liebman, Ellen (ed.). *California Farmland: A History of*

Large Agricultural Holdings. Totowa, N.J.: Rowman and Allanheld, 1983.

McWilliams, Carey. *Southern California: An Island in the Sun*. Santa Barbara, Calif.: Peregrine Smith, 1973.

Miller, Crane S., and Richard S. Hyslop. *California: The Geography of Diversity*. Palo Alto, Calif.: Mayfield, 1983.

Nelson, Howard. *The Los Angeles Metropolis*. Dubuque, Iowa: Kendall/Hunt, 1983.

Norris, Frank. *The Octopus: A Story of California*. Garden City, N.J.: Doubleday, 1901.

Starr, Kevin. *Inventing the Dream: California Through the Progressive Era*. New York: Oxford University Press, 1985.

Steiner, Rodney. "Large Private Landholdings in California." *Geographical Review*, 72 (1982), pp. 315–326.

West, Nathaniel. *The Day of the Locust*. New York: Random House, 1939.

CHAPTER 17

The North Pacific Coast

Come to visit us again and again. . . . But for Heaven's sake, don't come here to live.

Tom McCall, former governor of Oregon

Cold, clear mountain streams tumble down rock-strewn courses. The destination: a rugged, unused coastline where precipitous, fog-enshrouded cliffs rise out of pounding surf. Mountains are visible in the distance, lofty, majestic, covered with snow. Tall needleleaf evergreens—pines, spruces, and firs—cover the land between with a mantle of green. Cities, where they exist, give the impression that they are new. They fit within the general landscape, clean-looking and somehow better than most other urban places. The people here are satisfied and friendly. They ask for little but the beauty of their land and a life filled with the North American virtues of independence and self-reliance. This is North America's North Pacific Coast, or more popularly, the Pacific Northwest (Figure 17-1).

At least that is how many Canadians and Americans (both from within the region and from outside) think of the coastal zone that stretches from northern California to southern Alaska. The appeal of the region's natural environment and the variety of outdoor activities it supports led one writer to label this region "Ecotopia."[1] It is an image that has induced many to seek their fortunes there. A traditional western belief (in both countries) is that regional progress can be measured in the ability to attract people to fill an empty land. However, a change of attitude occurred in the southern half of the region in the 1960s as residents of Oregon and Washington began to realize that they had something good and the continued influx of newcomers threatened to destroy it. Their fear was given-nationwide attention by the plea—quoted at the beginning of this chapter—made on national television by Tom McCall, then (1976) the governor of Oregon. That attitude, too, is now a firm part of the region's image even though it is not fully shared by all of the region's residents.

Another important element of its regional character is the North Pacific Coast's relative isolation from the rest of North America. Populated sections of the region are separated from the other principal population centers of the two countries by substantial distances of arid or mountainous terrain. Residents of the region often view this isolation as positive, a characteristic that provides a geographic buffer against the less desirable rest of the world. Economically, however, it is a hindrance. High transportation costs inflate the price of Pacific Northwest products in distant eastern markets and also discourage some manufacturers from locating in the region. It is the sense of regional independence, and the mix of characteristics that support or diminish this independence, that is the theme adopted for this portion of North America.

Like many images, that of the Pacific Northwest is made up of equal parts reality and myth. Whether the image of the region is more true or more false depends partly on alternative interpretations of its boundaries. Oregon, Washington, and Idaho, together with the province of British Columbia, are sometimes labeled "The Northwest." Others would not include Idaho and might find the regional title "The North Pacific Coast" more acceptable. Identification of the region by using state and provincial boundaries also creates a territory with great internal

[1] Joel Garreau, *The Nine Nations of North America* (Boston: Houghton Mifflin, 1981).

Natural landscapes in the North Pacific Coast have a spectacular beauty that draws the region's residents to recreate in the outdoors as nowhere else in North America. (Roy Bishop/Stock, Boston)

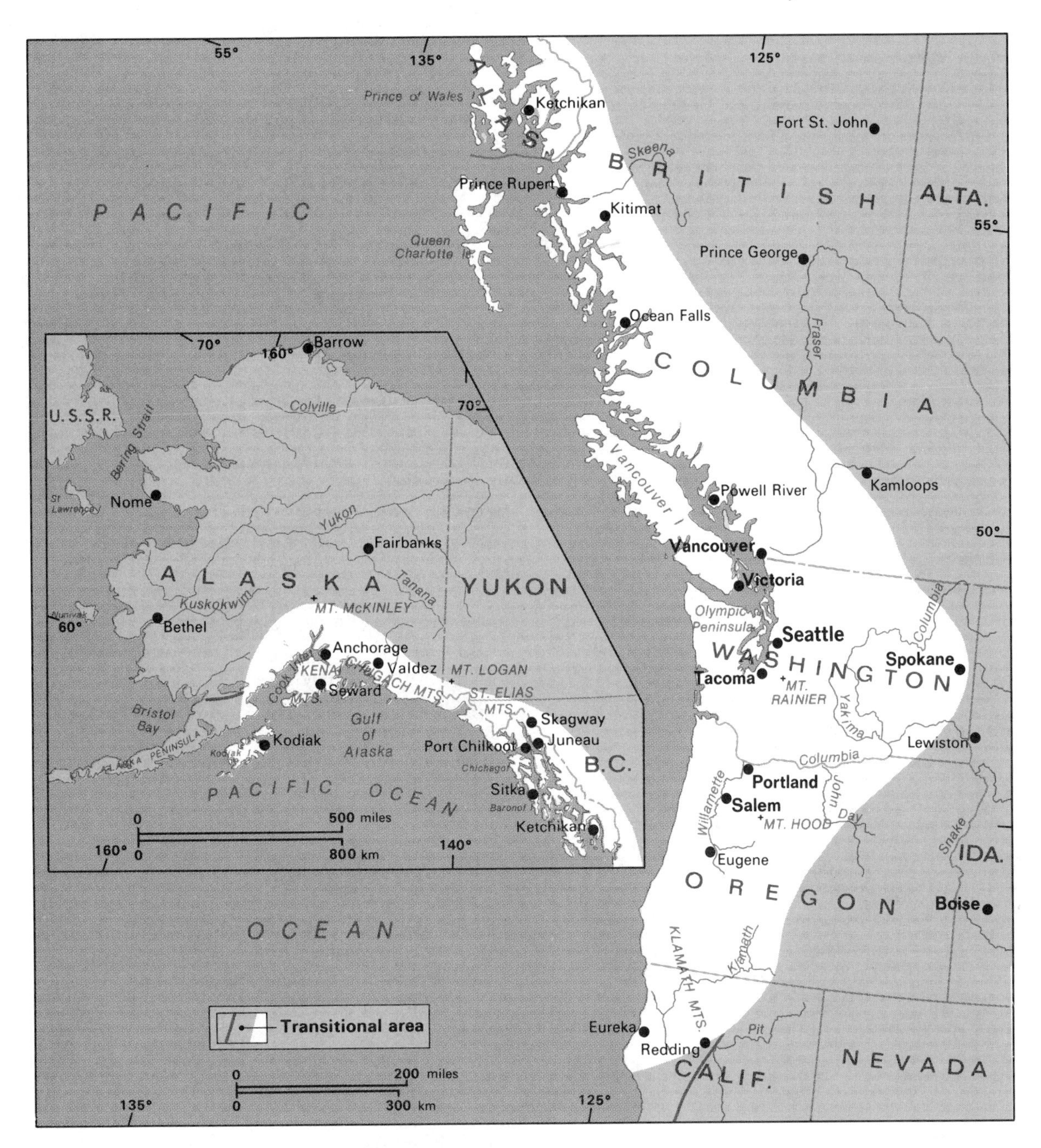

Figure 17-1 The North Pacific coast.

Major Metropolitan Area Populations in the North Pacific Coast, 1990	
Seattle–Everett, Wash.	1,972,961
Vancouver, B.C.	1,380,729[a]
Portland, Oreg.–Wash.	1,239,842
Tacoma, Wash.	586,203
Spokane, Wash.	361,364
Salem, Oreg.	278,024
Victoria, B.C.	255,547[a]
Anchorage, Alaska	226,338
Bremerton, Wash.	189,731

[a] Indicates 1986 population.

diversity. Broad expanses in the eastern two-thirds of Oregon are desert or a semidesert environment, much of it virtually treeless. This comes as a shock to the uninitiated visitor from the East who expects an environment like the well-watered Pacific margins of the state. The climate of southeastern Washington has led to a land of large wheat farms that consistently place the state among the country's half-dozen leading producers of that grain—a group that also contains Kansas, North Dakota, Oklahoma, Montana, and Texas.

THE PHYSICAL ENVIRONMENT

The North Pacific Coast is, with one or two notable exceptions, defined primarily on the basis of the physical environment. Stated very simply, it is a region strongly subject to maritime influence and rugged terrain. Precipitation is high, and vegetation associated with heavy moisture is located near the coast but with marked variation over short distances because of the influence of surrounding mountains on the region's climate.

Climate

The mountain is out today.

Any discussion of the nature of the geography of the Pacific Northwest begins with its climate. The greatest average annual precipitation on the continent is found here; averages above 190 centimeters (75 inches) are common, and averages are double that amount on the western slopes of the Olympic Mountains in northwestern Washington and in the coastal mountains of British Columbia (Figure 17-2). The 600 centimeters (230 inches) recorded on the northwest end of Vancouver Island easily surpasses even these high amounts and is more than twice the amount received anywhere else in the United States (with the exception of Hawaii) or Canada. During the winter, the cloud cover is so constant that residents of Seattle tell each other when the slopes of nearby Mount Rainier are visible. The heavy precipitation on the Olympic Peninsula supports rainforests where ferns and mosses grow in profusion, and trees (including western hemlock, red cedar, Sitka spruce, and the world's largest Douglas fir) are frequently more than 60 meters (200 feet) tall. Naturalist Roger Torey Peterson suggests that the Olympic rainforest contains more weight of living matter per unit of land than anywhere else in the world.

The northern Pacific Ocean is a spawning ground for great masses of moisture-laden air. As these air masses move, they are pushed south and east by the prevailing winds onto the Pacific shores of North America. A high-pressure system located off the coast of California in summer and off northwestern Mexico in winter prevents many of these maritime air masses from drifting farther southward and ensures that most of their moisture will fall over the North Pacific Coast. The seasonal movement of weather belts has a noticeable impact on precipitation in the Pacific Northwest. Winter precipitation amounts are everywhere higher than summer levels, but the seasonal difference is more marked on the southern margin of the region. The coast of southern Oregon and northern California receives less than 10 centimeters (4 inches) of precipitation during the summer months of July and August, only a tenth of the amount that falls there between December and February.

It is tempting to suggest that this area has

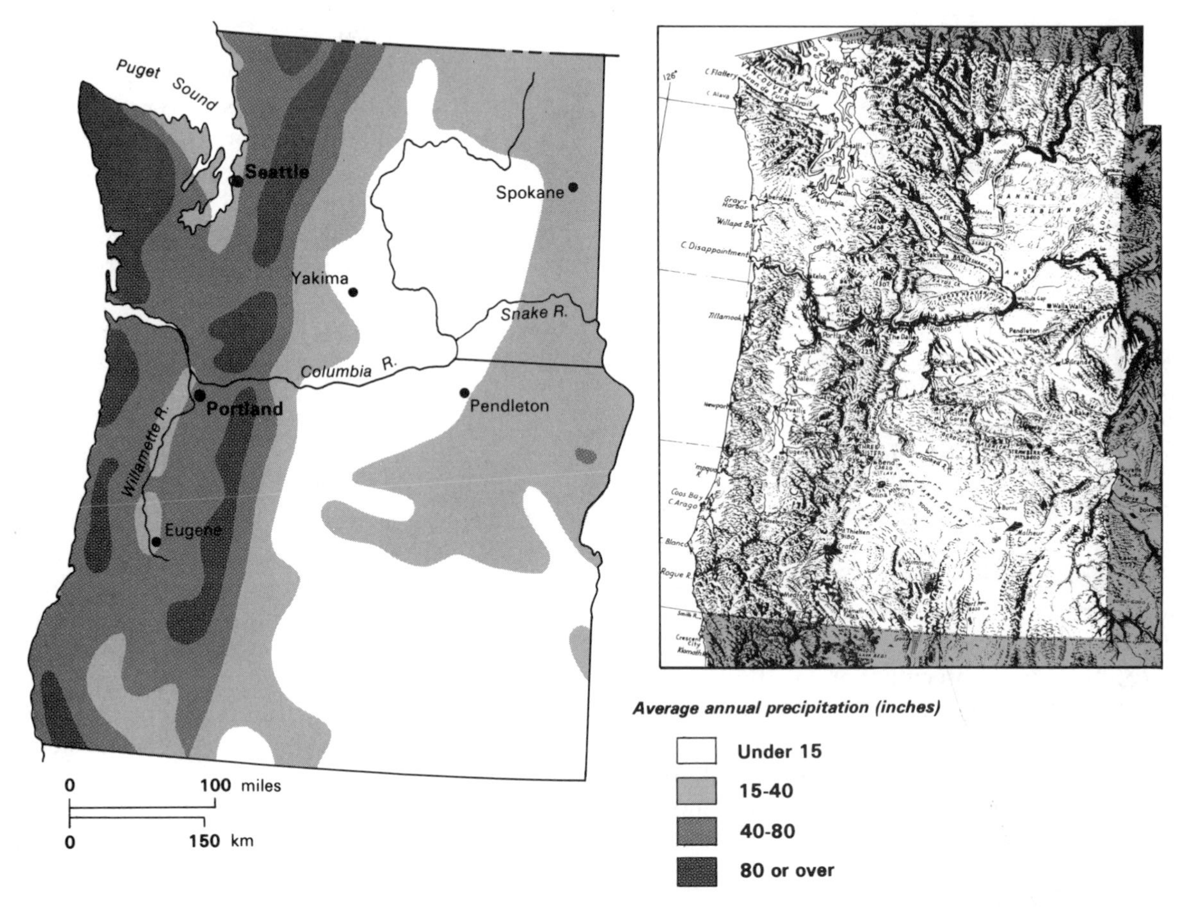

Figure 17-2 Precipitation and topography. (Topographic map from Raisz, *Physiographic Diagram of the United States.*) Notice the association between areas of rugged topography (right map) and higher precipitation (left map).

generally high precipitation, and then leave the subject. That would be an error; considerable portions of the region are semiarid. Parts of the borderlands of Puget Sound in Washington receive only about 60 centimeters (25 inches) annually, no more than the amount that falls on Wichita, Kansas, in an average year. Portland's average, like Seattle's, is similar to that for Boston or New York and is far less than the precipitation of a southeastern U.S. city such as Atlanta. In part, these areas of the Northwest seem wetter than they are because of the high amount of cloud cover during the midfall to midwinter season. Precipitation seldom falls in the form of heavy thundershowers; more typical is a gentle, light, frequent rain that feels more like a heavy mist. One consequence of this precipitation form is that runoff, so normal in heavy rains, is lessened, and vegetation is able to make maximum use of the moisture.

The region's mountains are the main reason for both the high precipitation along the coast and the substantial climatic variations that exist in close proximity. As a Pacific air mass passes over land, moving eastward and southeastward, it strikes the mountain ranges that line the North

Pacific Coast and is forced to rise. As the air rises, it cools and its moisture-carrying capacity is reduced, resulting in precipitation. Such precipitation is called *orographic* (mountain-induced) precipitation, and it is common on the windward side of most extensive mountain chains.

Along a belt extending from south-central Oregon to southwestern British Columbia, the Coast Ranges in the United States and the Insular Mountains in Canada are backed by a trough of low-lying land, including the Willamette Valley in Oregon, the Puget Sound lowland in Washington, and the Fraser River lowland in British Columbia. As the east-moving air descends into these lowlands it warms, and its moisture-carrying capacity is increased. Because additional moisture is not introduced into the air, little precipitation will occur. Thus, the same air masses that provide abundant coastal moisture and heavy precipitation on the western slopes of the coastal ranges provide substantially less moisture along the lowland trough.

To the east of the lowland is a second north-south trending range of mountains. The moving air must again rise, and orographic precipitation, although not quite as heavy as along the coast, results again. These inland mountains, called the Cascades in the United States and the Coast Mountains in British Columbia, are high; Mount Rainier in Washington has an elevation of 4390 meters (14,410 feet), and many peaks are between 2750 and 3650 meters (9000 to 12,000 feet) high. Winter precipitation in the mountains falls in the form of snow, making this the snowiest portion of the continent.

Finally, in the eastern extension of the region beyond the Cascades in interior Washington, air masses again descend and warm. The little moisture that remains in the air is retained, and most of eastern Washington averages less than 30 centimeters (12 inches) of precipitation yearly. It is only through the use of irrigation and dry-farming techniques that most of the area's important crop agriculture is possible.

South and north of this mountain-valley-mountain system, the mountain ranges merge and the separating valley disappears. Heaviest precipitation amounts are concentrated in a single band along the northern coast, including the central and northern coast of British Columbia and Alaska's adjacent "panhandle." This coastline receives substantial moisture throughout the year, and average cloud cover percentages are among the continent's highest. The climate of Alaska's panhandle, especially, is dominated by moisture and by cloudiness.

Average precipitation levels drop sharply along the coast in Alaska north and west of the panhandle; most of the southward-facing coast of central Alaska averages 100 to 200 centimeters (40 to 80 inches) annually. As mentioned, many of the weather systems created off southern Alaska's shores move south and east, often only brushing the shoreline of central Alaska. Many residents in Anchorage look with disdain at the damp climate of the southern peninsula, favoring their own clearer, if somewhat colder, weather.

In addition to bringing rain, the region's maritime location provides a moderate temperature regime. Summers are cool. Winters are surprisingly warm, although the dampness means that the air can feel raw and less comfortable outdoors than the thermometer might suggest. Still, Seattle is warmer in winter than St. Louis, and Juneau, Alaska, is no colder than Washington, D.C. Snow is uncommon along the coast south of Vancouver, but heavy snow can occur on the coast of northern British Columbia and the Alaskan panhandle.

A third consequence of the seasonal movement of air masses is frequent periods of high winds along the region's coastal margin, especially in Oregon, northern California, and southwestern Washington. The large storms built up during passage across the North Pacific can strike the coast with tremendous ferocity. These storms are not hurricanes, a tropical storm, but it is not uncommon during the long winter months for winds to exceed 125 kilometers (80 miles) per hour during the stormier periods. Although the coastal mountains provide some protection and winds are generally lower in the summer, high winds can reach the eastern portions of the region even in the summer. When they do, the danger of fire is worsened.

Topography

The Pacific Northwest is a land of mountains; few places in the region do not offer a view of neighboring peaks when the weather is clear. Mount McKinley at the region's northern extremity is, at 6200 meters (20,300 feet), the highest in North America. The St. Elias Mountains in Canada are the world's highest coastal mountains; Mt. Logan is just short of 6000 meters (19,700 feet). Elevations are generally lower farther south, but it remains a land of high mountains and rugged terrain.

The peaks of the Coast Range of Oregon and Washington are generally lower than other mountains in the region. The ranges are fairly continuous in Oregon, with elevations reaching about 1200 meters (4000 feet). In Washington they are discontinuous, with several rivers, notably the Columbia and the Chehalis, cutting pathways across them. Coast Range elevations in Washington are seldom above 300 meters (1000 feet).

The Klamath Mountains of northern California and southern Oregon offer a jumbled topography, one that has been heavily eroded by Pleistocene glaciation and stream erosion. Little pattern is apparent in the terrain, in marked contrast to the parallel ridge and valley structure of the Coast Range. This is a wild, rugged, empty area.

The lowlands of western British Columbia and Oregon are part of a structural trough that was created when that area sank at the same time that the Cascades to the east were elevated. This trough extends northward in the form of straits separating Vancouver Island from the rest of British Columbia, then passes through the complex of islands that line the Alaska panhandle and provides the Inside Passage north as far as Juneau.

Farther inland, the Cascades extend from the Klamath Mountains northward into southern British Columbia. The southern section of these mountains appears as a high, eroded plateau topped by a line of volcanic peaks. Between Mount Lassen in California (the only volcano in the United States outside Hawaii to have been active in historic time until the 1980 eruption of Mount St. Helens) and Mount Hood in Oregon, these peaks are especially splendid in their isolation above the surrounding plateaus. This line of peaks marks the zone along which the Pacific plate is passing beneath the North American plate (see Chapter 16). The northern Cascades are more rugged and have long proved a difficult barrier to movement eastward from the populous Puget Sound lowland. Here again, extinct volcanoes, most notably Mount Rainier, provide the highest elevations and best defined peaks.

In British Columbia, the Cascades—known there as the Coast Mountains—merge with the Insular Mountains and push against the ocean. This is a land of dramatic coastal mountains cut by glacially eroded fiords and islands. Land transportation along the coast is nearly impossible, and coastal shipping provides the principal transportation connections. Between Vancouver and the Alaska panhandle, overland transportation lines connect the coast with the interior at only two points: one at Prince Rupert in British Columbia and the other at Port Chilcoot and Skagway at the northern end of the Inside Passage in Alaska.

Beyond the Alaska panhandle and the massive, glacier-covered St. Elias Mountains, the mountains again divide in southern Alaska. The Coast Ranges, notably the Chugach and Kenai Mountains, decline in elevation from east to west. The interior mountains, the Alaska Range, are much higher and more continuous. A large lowland at the head of Cook Inlet is south

Mt. Rainier, Washington, is one of a series of volcanic peaks that punctuate the southern Cascades. (George Grant/U.S. Department of the Interior)

of a gap through the Alaska Range, and here Anchorage, easily the largest city in Alaska (est. population 229,627 in 1989), is located around its good harbor and with easy connection to the interior.

Juneau, the capital of Alaska, is located on a narrow coastal lowland on the panhandle; its only transportation connections to the rest of the state are by air or water. The farthest that one can drive from town is about 15 kilometers (10 miles). This location for the capital was reasonable when Alaska's wealth was in the panhandle's forests and salmon fisheries and when access to the Yukon gold fields through Skagway was a consideration. As the state's economy changed and other resources became more important, the panhandle languished. Fairbanks (est. population 32,293 in 1989) in central Alaska and especially Anchorage, which is accessible to the southern portion of the state, far outstripped Juneau in population growth; the capital city still had an estimated population of fewer than 29,000 in 1989. Alaska's voters and political leaders, dissatisfied with the inaccessibility of Juneau, voted in 1974 to move the capital to a point some 80 kilometers (50 miles) north of Anchorage. The cost of constructing a new capital city, now estimated at over $3 billion, has caused many Alaskans to reconsider that decision. In 1978 and again in 1982, the voters refused to allocate money for the move, so despite its location, Juneau may yet remain the capital.

Trees and More Trees

There are magnificent redwood stands in the Klamath Mountains; Douglas fir, hemlock, and red cedar in Washington and Oregon; red cedar, hemlock, and Douglas fir in British Columbia; and Sitka spruce on the Alaska peninsula. This is a land not just of forest, but of beautiful expanses of tall trees reaching straight for the sky, trees that are among the largest on earth, trees that encourage people to stand and stare in awe or admiration . . . and to cut for profit.

Except for the drier lowlands, where the normal vegetation of the Willamette Valley is prairie grass and that of the land east of the Cascades a mix of grass and desert shrub, and except for the tundra above the treeline, all of the Pacific Northwest is, or was, covered by forest. Tree growth is encouraged by plentiful moisture and moderate winter temperatures. Forest products were long the economic mainstay of the region. Even today, while the southeastern United States produces more wood for pulp and paper products, no other part of Canada or the United States provides as much lumber as does the North Pacific Coast.

PATTERNS OF HUMAN OCCUPATION

No other coastline, except for the polar areas, was explored by Europeans as late as was the North Pacific Coast. Vitus Bering had claimed the Alaska coast for Russia by 1740, but it was not until 1778 that Captain Cook sailed the coast from Oregon to southeastern Alaska. By that date, several million colonists at the other side of the continent were fighting a war for independence against the very country for which Cook was sailing. By the time Lewis and Clark worked their way across the Cascades to the mouth of the Columbia River in 1805, Philadelphia and New York City, each with about 75,000 people, were vying for the title of the country's largest city. By the mid-1840s, when American settlers began traveling the Oregon Trail to the Willamette Valley in substantial numbers, New York's population was fast approaching 500,000, and its city administration was already wrestling with problems of urban transportation and waste disposal. Since the date of substantial European influence at a place was related to its distance from Europe, this late exploration and settlement of the region is understandable. In the days of sail and slow overland transportation, no nonpolar coast in the world was as far from Europe at this one.

The American Indians

The pre-European population of the region was relatively large. The moderate environment provided a plentiful supply of food throughout the year. Deer, berries, roots, shellfish, and especially salmon represented a natural bonanza of food that seemed without limit. The American Indians responded to this with a hunting and gathering economy and no food crop cultivation. Concentrated along the coast, they were divided into many distinct ethnic groups, each occupying their own, often small, coastal valley. They constructed large, impressive houses of red cedar planks and went to sea in dugout canoes made of the same wood. They shared many cultural features, notably the *potlatch* (a ritual during which property is given away in a large celebration) and the carving of totem poles (the process of recording on an upright log a series of key incidents in an individual's life).

Along most of this coast, the Indians seemed simply to melt away when Europeans arrived. Because their extreme isolation made organized opposition impossible, each small tribe succumbed quietly, making little impact on European settlement; their numbers dwindled, unnoticed by the incoming European settlers. Today, few Indians remain in the south, although one small tribe called the Puyallups recently won in the courts the right (for all Indians) to half of the total salmon catch in Washington. This decision, based on an early agreement in which the Indians surrendered much of their land in return for certain vaguely stated rights, resulted in restrictions on salmon fishing that created a great clamor among the state's many non-Indian sport fishermen. Farther north, the Indian population is proportionately larger and remains a substantial ethnic group in northern British Columbia and the panhandle of Alaska.

Early Canadian and American Settlement

Much of the early European settlement history of this region reads like a collection of pulp novels, with prospectors, lumbermen, land barons, fur trappers, settlers, sober-minded preachers, railroad magnates, and imported brides trooping across the pages in the best Bret Harte fashion. It is the stuff of good historical fiction, and it is a wonder that the region's writers haven't done better with it. Several events in this history are so important to the region's development that they must be mentioned here.

Russians were the first Europeans to establish permanent settlements along the coast. They came late in the eighteenth century, motivated by the predictable European search for easily extracted riches. In the North Pacific Coast, these riches proved to be furs, and the Russians established a series of trading posts and missions that were concentrated in southeastern Alaska but that extended as far south as northern California. The outposts never became self-sufficient in foodstuffs, and the cost of maintaining these scattered, distant sites usually exceeded the income from fur sales. Following several earlier Russian attempts to sell the colony to the United States, a $7.2 million sale price was finally agreed upon in 1867. Many Americans considered the price far too high, and the U.S. government was roundly criticized for the purchase. Evidence of Russian occupation still exists in some parts of Alaska in the architecture of wooden onion-domed Eastern Orthodox churches, in cemeteries, and in a few pockets of cultural survival from early Russian settlers.

The Hudson's Bay Company moved its fur-trading operations into the Columbia River basin early in the nineteenth century. It was the dominant influence between northern Oregon and British Columbia until the late 1830s when American missionaries and settlers began the long journey across the Oregon Trail from Missouri. Most new American settlers moved into the Willamette Valley, but their numbers quickly outweighed the small British population of the entire Northwest. The cry of these American settlers, and of Westerners in general, "Fifty-four forty or fight," claimed all of the land as far north as 54°40′ N and was a significant part of the presidential election of 1846. In that year, the British and American governments agreed to establish the boundary as it exists today at 49° N. However acceptable this may have been as a political decision, it disrupted the normal north-south movement patterns along the Puget Sound and Columbia River transportation corridors. Victoria, on Vancouver Island, was for decades the main urban focus for the British-held portion of the region. As the first permanent European settlement established in 1843, it remained the largest town in British Columbia until 1886. In that year, the settlement of Vancouver, located on the mainland adjacent to the mouth of the Fraser River, became the western terminus of Canada's first transcontinental railroad. While Victoria continued to be important as the province's administrative and commercial center, Vancouver's eventual position as Canada's main port of entry on the Pacific was ensured once the railroad was completed.

The railroads were of key importance in the eventual growth of Oregon and Washington as well. In 1870, these two states combined had fewer people than did the city of San Francisco. In 1883, the Northern Pacific Railroad was completed to Seattle and was followed a decade later by the Great Northern. This ended the region's overwhelming dependence on oceanic shipment, which sailed via the southern tip of South America to the eastern United States and European markets. The region's bulky resources could now be more quickly and easily shipped in volume. For a period after completion of the railroads, twice as many people moved to Washington each year as moved to California.

The immigration that flowed into Washington and Oregon between the 1880s and the beginning of World War I created an interesting difference in the two states' ethnic composition. In 1880, Oregon had 175,000 people, and Washington, only 75,000. By 1910, Washington had outpaced Oregon's growth so that the northern state's population was nearly double the southern state's, with 1,140,000 persons in Washington compared to 672,000 in Oregon. While most of Oregon's population was composed of immigrants from other parts of the United States, booming Washington attracted large numbers of European immigrants as well, especially Scandinavians. Even today ties to that ethnic heritage can be important in Washington, and many of the state's politicians are of Scandinavian extraction. Oregon, by comparison, has a population with a strong New England heritage, the result of early overland settlement.

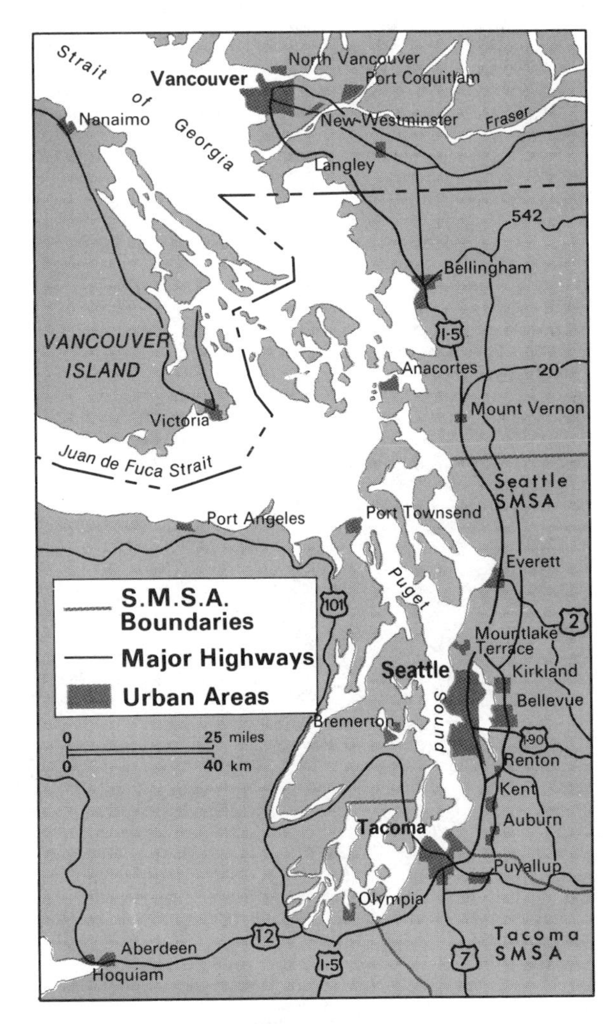

Figure 17-3 The Puget Sound-Fraser River lowland. This area, home to a majority of the region's residents, benefits from excellent water transportation, adequate precipitation, and a desirable recreational environment.

Population Distribution Today

More than 6 million people live in the Pacific Northwest. Their numbers are increasing at a more rapid rate than the national averages for either Canada or the United States. Most live in a long cluster of cities and towns along the lowland extending from the Fraser River in southwestern British Columbia through the Willamette Valley of Oregon (Figure 17-3). Three of these cities, Vancouver, Seattle, and Portland, have metropolitan area populations of more than 1 million people each. This land of the great outdoors is, like nearly every other part of the two countries, peopled by an urban population. The cities, in turn, are very much a part of the region.

Vancouver. Now Canada's third largest city, this bustling metropolitan area of more than 1.2 million people has grown greatly in size and cosmopolitan character since World War II. Its urban area now extends southward to the U.S. border, and local planners expect the city to double in population during the next 30 years. In support of this growth, Vancouver's transportation function has become more important as the western outlet for interior Canada. Most products of the Prairie Provinces, formerly shipped almost entirely to and through the cities of east-

Vancouver, British Columbia. Excellent harbor facilities and easy access into the interior up the Fraser River Valley help make this handsome city Canada's major western urban center. (David Watson/National Film Board—Photothèque)

ern Canada, are now funneled westward to Vancouver. The development of Asian markets for western Canada's wood products and wheat encouraged the growth of the city's port facilities. With an annual cargo volume of nearly 50 million metric tons (55 million tons), the city is easily Canada's busiest port. Frequent overcrowding of the port facilities and of the eastern rail lines has prompted Canadian officials to encourage greater use of the alternative port at Prince Rupert, which the government is considering as the site for construction of a major grain transshipment facility.

Vancouver is the western headquarters for nearly all of Canada's major businesses. Most of British Columbia's manufacturing, with wood processing easily most important, is in or near the city. Only the fact that the provincial capital is across the Strait of Georgia, at Victoria on Vancouver Island, prevents Vancouver from totally dominating British Columbia.

Seattle. Seattle has been the largest city along the North Pacific Coast since the boom era of the late nineteenth century, although it is now close to losing that distinction to Vancouver. Founded as a logging center, Seattle began to achieve regional dominance when it was linked to the northern U.S. transcontinental railroads.

The city has been the home of Boeing Aircraft

since World War I, and it has been called the world's largest company town. A mainstay of the construction of bomber aircraft in World War II, the company's fame spread after the war as it introduced a series of highly successful passenger jets. Employment at Boeing has fluctuated wildly as airplane orders responded to changes in market demand and the general condition of the world economy. From a high of 103,000 in the late 1960s, employment at Boeing fell to about 50,000 during the early 1970s. The economy of the city, and to an extent that of all Washington, suffered as a consequence; population decline followed the heady growth period of the 1960s, and only nine states grew less rapidly than Washington during the economic slowdown. Comparative employment stability at Boeing, which had recovered to more than 106,000 by the late 1980s, and local efforts at economic diversification have enabled the area's economy to recover, and today the metropolitan area is again experiencing substantial immigration.

Seattle's urban core is tucked onto a narrow isthmus bordered by Puget Sound on the west and Lake Washington on the east. Like Vancouver, it has a beautiful site, with views of mountains and water offered to the residents of its many hills. It is a middle-class city, with pleasant, scattered neighborhoods of tree-lined streets. Evaluations of the quality of life in the largest cities in the United States during the mid 1970s and again in 1989, two of many that have been conducted over the years, concluded that Seattle was the "best" city in the country in which to live.

When the weather is bright and the air clear, those who live in Seattle cannot help but be aware of the towering cone of Mt. Rainier to the east. (Seattle-King Convention and Visitor's Bureau)

Portland. Two other studies of urban quality done at about the same time as the earlier one cited above disagreed with its conclusion. A study by the Federal Environmental Protection Agency called Portland the "most livable city" in the country. Another, produced by Midwest Research Institute in Kansas City, evaluated more than 100 variables before agreeing that Portland was indeed the best of the country's 67 largest cities. (Eugene, Oregon, in the Willamette Valley, was judged the best medium-sized city in both studies.) Studies like these have been criticized for overemphasizing new cities and cities with small ethnic minorities and for discounting less measurable characteristics.[2] Nevertheless, they say much about the general livability of Oregon's principal city.

Residents of Portland greeted both laudatory studies with mixed emotions. They were largely inclined to agree, but they feared that such publicity would bring waves of newcomers who by their numbers would destroy the qualities of the city that attracted them. An old city by the standards of the region (new by most others), Portland residents consider their city more conservative and stable than Seattle, and they hope to maintain that atmosphere. Its economy is more diversified than Seattle's, and its relations with the region's interior are closer because of the lowland routeway to the east provided by the valley of the Columbia River. Portland is a major transshipment point for the grain from eastern Washington, for example, and wood products and food processing are principal activities of the local manufacturing economy. Portland, inland about 160 kilometers (100 miles) from the coast, nevertheless rivals Seattle as an ocean port because the lower Columbia River is navigable.

---

[2] Susan L. Cutter, *Rating Places: A Geographer's View on Quality of Life* (Washington, D.C.: The Association of American Geographers, 1985).

THE REGIONAL ECONOMY

In many ways, the economic structure of the North Pacific Coast is dominated by the production of staple products (see Chapter 7) and by its isolation from the major markets of both countries. This latter factor is especially applicable for the region's U. S. portion. Less than 3 percent of the American population lives here, although about 10 percent of all Canadians are in British Columbia. Most major markets of both countries are more than 3500 kilometers (2200 miles) away. British Columbia is the exclusive western terminus for national economic activity in Canada, but goods and people passing in or out of the United States through Pacific Coast ports can go through California, bypassing Oregon and Washington completely. Freight-rate structures placed many of the region's products at a competitive disadvantage in the major markets of the two countries for many years. The region has always contained a number of high-demand products, notably lumber and foodstuffs. However, movement costs reduced the ability of producers to get their products to market at reasonable cost; that is, the great distance has limited the *transferability* of these products. Consequently, market areas turned to other sources of supply that were nearer and less expensive to obtain. The higher freight rate charged for finished wood products in Canada, for example, has encouraged manufacturing in British Columbia to remain heavily involved in the initial processing stages only.

Agriculture

The region's agriculture provides a good example of the impact that transferability problems can have on a regional economy. The kinds of crops grown in the North Pacific Coast are duplicated for the most part closer to the areas of major consumption in the East (Figure 17-4). Washington apples, for example, are in direct competition in the Eastern markets with those

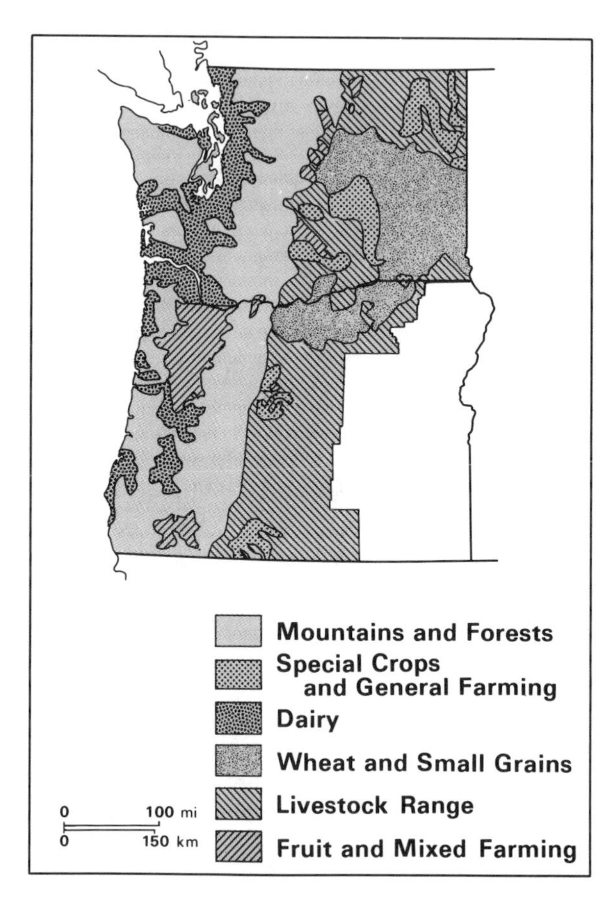

Figure 17-4 Major crop-producing areas. (From Richard Highsmith, *Atlas of the Pacific Northwest: Resources and Development.* Corvallis: Oregon State University, 1968.) The Williamette Valley and Puget Sound lowland is clearly identifiable. Areas east of the Cascades either emphasize dry-area crops like wheat or depend on irrigation.

from Wisconsin, Pennsylvania, and Virginia. The same is true for a variety of other fruit and vegetable crops of the region. This is quite different from the situation in central and southern California, where the Mediterranean climate enables growers to produce many crops that simply cannot be grown in the colder north and to provide a greater volume of common fruits and vegetables fresh during the winter season. Thus, for decades California cornered the market for many of the North Pacific Coast's agricultural products (see Chapter 16). *Intervening opportunities* (other sources of supply) were not available, and distance to market had little negative impact on California's agricultural economy. The result of all of this is that much of the North Pacific Coast region's agricultural products are grown for the local market, not for export. Aggresive advertising campaigns and cost advantages accompanying innovations in air cargo transport have increased the East Coast market share for Washington apples and pears and Oregon strawberries, but the relative impact of food exports on the region's economy remains modest.

The broad Willamette Valley in Oregon is easily the largest agricultural area near the region's coast. The land has been cultivated for more than a century, and its farms are prosperous and well established. Much farmland is in forage crops, and many farmers still follow the practice of burning these fields in the fall—with the result that for a period of several weeks, large sections of the valley are covered by a layer of smoke.

Dairy products, also generated mostly for the local markets, are of greatest importance to the agriculture of the Willamette Valley; strawberries are perhaps the most important specialty crop. A variety of other specialty crops also thrive in the valley's climate, including hops, grass for turf seed, cherries, and spearmint. Even grape production, supporting a local wine industry, has increased in recent years (although it remains tiny when compared to viticulture in California). The total value of these crops, however, does not match the area's income from dairy production.

The Puget Sound lowland in Washington is another important dairy area, as is the Fraser River Delta in British Columbia. Again, a variety of specialty crops are also grown, with peas leading the pack. Quick frozen and then shipped to markets throughout North America, this cool-weather crop is particularly well adapted to the local climate.

The area east of the Cascades in Washington presents a very different kind of agricultural landscape. Most of this area is semiarid, and grasses and desert shrubs replace the majestic evergreens of the coast and mountains. Although called the Columbia Plateau, the area has little of the characteristic flatness one expects from the term plateau. Much of the area consists of rolling hills. Elsewhere, in central Washington, the landscape has been cut by a series of steep-sided dry canyons, called *coulees.* That section—properly called "the channeled scablands" because lava pockets dot the surface with what looklike scablike knolls—is covered by a deep blanket of lava flows eroded by the floods from glacial melt during the Pleistocene ice retreat.

We have placed most of the Columbia Plateau in a different region, the western Empty Interior, because of its aridity, nodal settlement pattern, and economic focus on minerals, grazing, and transportation services. An exception is found along the Oregon-Washington border and across much of eastern Washington. This is a substantial farming area, easily the most important in the Pacific Northwest. In many ways separate from the coast, residents of the area—who call it the Inland Empire—nevertheless feel an association with the coast equal to that felt with other parts of the interior. Their crops are sold down the Columbia River, they share a common state government with the coast, and they look to Seattle and Portland for many goods and services. For all of these reasons, the Inland Empire is a part of the North Pacific Coast region, although an atypical one.

The agriculture of the Inland Empire is quite varied. The hilly country of east-central Washington, called the Palouse, averages between 35 and 65 centimeters (15 and 25 inches) of precipitation annually, somewhat more than other parts of the interior. It is today one of the most important dry-farming areas in the country. Dry farming is usually confined to subhumid climate areas and emphasizes water conservation more than other types of farming. Wheat is the primary crop of the area, with both spring and winter varieties grown. In the Palouse, wheat is normally planted on a given field every other year. In alternate years the land is left fallow; that is, it is plowed, but nothing is planted. This practice retards evapotranspiration and allows soil moisture to increase. The large wheat farms of the Palouse are heavily mechanized and highly productive. Most of the product is exported through Portland to Asia.

Irrigation has played a major role in the Inland Empire's agriculture in recent decades. Two major irrigated areas have been developed. The water from a number of streams flowing eastward out of the Cascades is used to irrigate their relatively narrow valleys. The result is one of the country's most famous apple-producing areas. As mentioned earlier, effective national advertising has helped convince Americans that these apples from the Yakima, Wenatchee, and neighboring valleys are the best available. More recently, pears from this same area have been given equivalent national marketing attention and distribution.

The Columbia River, at Grand Coulee northwest of Spokane, was dammed primarily to provide hydroelectric power. It also made available large amounts of irrigation water to the Big Bend area of south-central Washington. After these waters became available in the late 1950s, crop acreage in the area expanded considerably in response. Major agricultural products include sugar beets, potatoes, alfalfa, and dry beans.

Forest Products

British Columbia is easily the largest lumber producer in Canada, supplying approximately 45 percent of all timber harvested in that country. In the United States, Washington, California, and Oregon together provide more than half of all timber cut, and Washington and Georgia vie for the lead in pulp and paper production. Although forestry was the first major industry in

The rolling wheatfields of eastern Washington give the area an appearance that is more like portions of the Great Plains than the usual perception of the Pacific Northwest. (Grant Heilman)

the region, the rich forest in the North Pacific Coast did not become nationally important until well into the twentieth century when improved transportation facilities, coupled with the destructive overcutting of many eastern forests, opened the region's woods to lumbering.

Douglas fir (of prime importance as structural supports for houses and for flooring, doors, and plywood) is easily the major lumber tree of the region although each section of the region has its own mix of trees to harvest. In British Columbia, for example, spruces, hemlock, and balsam are most important. In northern California, redwoods remain locally important, although their numbers are dwindling; the western red cedar is also widely cut in the area from Oregon northward.

The large size of many of the trees plus the distances to market tends to encourage large-scale logging operations. Weyerhauser Lumber Company, for example, owns some 690,000 hectares (1.7 million acres) of land in Washington and is easily the state's largest private landowner. A substantial part of Washington and a majority of all land in British Columbia, Oregon, and northern California is government owned. Private logging on government land plays a large role in overall production. Effective marketing systems have enabled lumber products of the region to penetrate all market areas of both countries, even the southeastern United States, America's other major forestry region. In recent years the Japanese have become major buyers, especially from British Columbia and Alaska;

most of Alaska's lumber is now shipped to Japan.

Changes in the practices of the forest products industry are visible in this region's landscape. Formerly, loggers cut everything in their path and moved on, always sure that more trees could be found elsewhere. Today, a new tree is planted for every one cut, although wide swaths are still clear-cut during the logging process. Governments of both countries, as well as private companies, carry on extensive research programs to develop stronger, straighter, healthier, faster-growing trees; and young trees are now being fertilized from airplanes and helicopters. Computers now control the process by which trees are cut into lumber, with each tree cut so that the maximum amount of lumber is obtained. New stands now reach maturity in 40 to 75 years, and it is arguable whether or not yields of most varieties of trees can be sustained indefinitely under these methods of operation (see Chapter 4).

Because lumber and pulp production are so important to the regional economy, any issue that might affect the availability of trees for cutting is strongly felt and widely discussed. In recent years two major controversies have been subject to debate, especially in the United States. The first involves the redwoods of northern California, the world's largest trees and among its oldest. The majority of redwoods, the most important lumber tree in California, are on private land. A major environmental drive developed to keep these trees from being cut and led to the creation of Redwoods National Park in the late 1960s. Conservationists contend that the park is too small and that some lumber companies are hastily cutting trees on land that may soon be set aside as parks. Loggers respond that some people would have every tree saved, creating ruin for the lumber industry and large increases in the price of forest products.

The second controversy involves the practice of *clear-cutting,* in which all trees in a given area are cut at one time. Loggers argue that this is the most efficient way to remove trees and to ease reforestation. Many conservationist groups consider clear-cutting to be a major contributor to soil erosion and stream pollution, destructive of forest ecosystems, and an ugly blotch on the landscape. In 1973, the Izaak Walton League sued to halt clear-cutting in the Monongahela National Forest in West Virginia, contending that the Organic Act of 1897 provided only for the cutting of "dead, matured, or large growth" trees. They won. In 1975, cutting was halted in Alaska's Tongas National Forest on the same basis. A wave of concern swept the lumber industry of the North Pacific Coast. Although the eventual solution to the problem will come from the U.S. Congress, the conflict is indicative of continuing disagreements between conservation groups and the forest-products industry.

Power and Dams

The plentiful precipitation and rugged topography of this region provide a hydroelectric potential unmatched on the continent—40 percent of the entire U.S. potential is contained in Oregon and Washington alone. The Columbia River, in particular, with a flow volume larger than that of the Mississippi River and a drop of nearly 300 meters (1000 feet) during its 1200-kilometer (750-mile) course from the Canadian-U.S. border to the sea, is a power developer's delight.

Begun in 1933 and still the region's largest, Grand Coulee Dam was the first of many dams constructed on the Columbia River. It was followed by no fewer than 10 dams downstream. The Dalles, between Oregon and Washington, where Lewis and Clark portaged around dangerous falls, is now a lake. British Columbia and the United States agreed to the construction in Canada of 3 additional dams that would store water during periods of heavy flow and then release the water when flow was low to guarantee consistent power generation. The government of British Columbia was given a check for $273 million as part of the agreement. This guaranteed flow has resulted in the addition of new

Clear, or block, cutting, such as is seen here in Washington, is a controversial approach today. Loggers argue that it is the most efficient method, but some environmentalist groups fear that it may increase the ecological damage. (U.S. Forest Service)

generators at Grand Coulee Dam that will triple its capacity, making it the world's largest single power producer. The treaty also led to expansion of power production elsewhere in British Columbia for provincial use, most notably in the Peace River area.

These developments have provided inexpensive electricity for the North Pacific Coast. Inexpensive electrical power, in turn, has attracted manufacturers that are heavy power consumers; most notable is the aluminum-smelting industry. Kitimat, isolated on the northern coast of British Columbia, is the site of one of the world's largest aluminum refineries. It is located there because a nearby dam provides the very large quantities of cheap electricity needed for aluminum production and the coastal location permits cheap transportation of bauxite, a bulky ore.

Just east of the region, Hell's Canyon—the world's deepest gorge—has been eyed for its potential as a dam site and source of further electric power generation. Cut by the Snake River along the Oregon-Idaho border, Hell's

Canyon has been at the center of controversy between power producers and environmentalist groups for decades. In 1975, the U.S. Congress named much of this stretch of the Snake a "wild and scenic river," presumably stating that no dams will ever be allowed there.

Fishing

Forestry and fishing at one time formed the backbone of the region's economy. The cold waters of the North Pacific were, and to an extent still are, rich fishing grounds. Large numbers of whaling vessels were attracted to these waters during the late eighteenth century and first half of the nineteenth century. Heavy overharvesting has reduced the North Pacific's whale population to a small fraction of former levels; most nations have agreed to an international ban on whale hunting, and the few remaining whalers have turned to the waters off Antarctica for most of their catch.

Salmon has been the most important fish of the North Pacific Coast for a very long time. It contributed a major part of the foodstuffs of the coastal tribes before the arrival of Europeans and remains the principal fish caught throughout the region. Salmon are anadromous; that is, they migrate upstream from the ocean to spawn in freshwater. Years ago their spawning runs filled the rivers, and massive catches were easily available to people on the banks. The Salmon Chief of the Colvilles, near Grand Coulee, reportedly could once catch an average of 400 fish a day, each weighing perhaps 14 kilograms (30 pounds), simply by scooping them up and throwing them onto shore. Most of the tribes revered the salmon, making it an important part in their religious rituals.

The size of the salmon catch has declined greatly over the past five decades, and today it is less than half its former level. Most salmon are caught off the Alaska coast, with British Columbia following in second place. When the region's streams were dammed, access to many traditional spawning grounds was blocked, especially on the upper Columbia and its tributaries. *Fish ladders*—a series of gradual water-carrying steps that allow fish to jump from level to level and thus bypass a dam—have been constructed around some of the smaller dams, but they do not work on the larger ones. As a consequence, nearly all of the Snake River and its tributaries plus all of the branches of the Columbia above Grand Coulee are closed to salmon. This has led to continued controversy between hydroelectric power developers and fishermen; the two seem totally incompatible. The salmon and those who depend on them have been the chief losers in dam construction.

Overfishing has also had a substantial impact on the North Pacific's fish population. As they fish these waters, Canadian and U.S. commercial fishermen are joined by many others, most notably Japanese. Because of this foreign competition, many U.S. fishermen in the Northwest supported the federal government's 1976 decision to extend exclusive offshore fishing rights to 320 kilometers (200 miles) for a wide variety of fish, including the salmon. The decision on whether or not to favor this extension has continued to be debated vigorously by West Coast fishermen. It cuts foreign fish harvest in these extended United States waters and thus provides larger catches for domestic fishermen, but it also encourages other countries to enforce similar limits already established in their areas or to enact new 320-kilometer limits and force U.S. fishermen to observe them. This is especially bothersome to tuna fishermen, based primarily in California. Much of their tuna catch originates off the shores of Ecuador and Peru, whose 320-kilometer limit these U.S. fishermen have largely ignored, claiming it was illegal.

In addition, controls on fishing exerted by Canadian governmental bodies have not always been matched by equivalent conservation measures on the U.S. side of the border. One result has been occasional contention betweeen U.S. and British Columbian fishing interests, similar in nature to the disagreements between the two countries in Atlantic fishing areas.

ALASKA—A POLITICAL ISLAND

Coastal southern Alaska is clearly a part of the North Pacific Coast, but it must be viewed as somewhat separate from the rest of the region. No railroad connects Alaska with the more populated parts of the continent, and only a single, long highway, part of which remains unpaved, connects coastal southern Alaska through interior Canada to the rest of the United States. People in the panhandle of southeastern Alaska are crowded by coastal mountains onto a narrow shoreline rarely more than a few hundred meters wide. All parts of this portion of the region look to air and sea transportation for connection with the rest of the world. This leads to an even greater sense of detachment than might be typical of the rest of the region, a greater sense of separation from the activities of the "lower forty-eight," and an economy of high prices resulting from scarcity and high transportation costs. Anchorage is the most expensive city in the United States in which to live.

Many believe that the Alaskan economy is based heavily on minerals, lumbering, and fishing. In fact, the federal government, primarily the Department of Defense, is the dominant employer in the state. Even the petroleum development boom on the state's North Slope has only modified, not eliminated, this orientation. Anchorage, home of one out of three Alaskans, is dependent on the military as the cornerstone of its economy.

Alaska is almost as close to Japan as to the coterminous United States, and Japan is the major market for many Alaskan products. As indicated, most lumber cut in Alaska is destined for Japan. The federal government defended its strong support of the construction of the Trans-Alaska Pipeline system as necessary to bring North Slope petroleum to the rest of the United States in order to meet growing energy demands and to cut dependence on foreign suppliers. Some people have suggested that a bigger profit could be gained if a substantial portion of the output were sold to energy-hungry Japan.

LAND OF CONFLICTING OPPORTUNITIES

> The Pacific Northwest, where the unspoiled is not yet spoiled, attracts a different kind of person. The weather is too cool, or at times too uncomfortable, to attract the sun-seeking elderly. Economic opportunities rarely provide a quick kill. The Northwest is for people more interested in being than achieving. . . . And in the Northwest it is not necessary to own a great deal to share in a great wealth, the outdoors all around.[3]

This self-serving statement, written by a lifelong resident of the Northwest, says much about both its people and environment. It is a place where many want to live, and most who are there seem relatively satisfied with their lot, but it is not the home of a booming economy. Unemployment is frequently high, particularly when there is a decline in demand for one of the region's few staple products, as happened with Boeing Aircraft in Seattle in the early 1970s. Although the growth of a diversified manufacturing economy in the northern Willamette Valley is lessening the role of the lumber industry, unemployment in Oregon approached 13 percent in early 1982 as a national decline in home building suppressed the local lumber market. Limited economic opportunity is surely the major reason why far more people have not moved here.

Nearly 200,000 pleasure boats crowd Puget Sound every year. Washington has more than 500,000 licensed fishermen. Literally dozens of magnificent state parks and campgrounds dot Oregon's majestic 650-kilometer (400-mile) coast. To reduce litter, Oregon was the first state to pass a law requiring that beer and soda be sold only in returnable containers. Much of British Columbia remains true mountain wilderness in spite of development projects undertaken in the last several decades across the province. This is a

---

[3] Thomas Griffith, "The Pacific Northwest," *The Atlantic Monthly*, April 1976, p. 51.

Juneau, the capital of Alaska, is cut off from land communication with the rest of the state by the sea and encircling mountains. (Air Photo Tech, Inc., Anchorage)

land of the great outdoors. The environment seems to call people to go out and enjoy it. Most do. That so much of the land is publicly owned simplifies their search. The land is the people's, and they use it. All of this sounds a little mushy, almost quaint. Still, as Thomas Griffith puts it, "Northwesterners are a people by nature possessed."[4]

A LAND USE CASE STUDY

ALASKAN LAND: WHO IS IT FOR?

When the United States acquired Alaska in 1867, nearly all of the territory's 150 million hectares (375 million acres) fell under the control of the federal government. The Alaska Organic Act of 1884 recognized vaguely that Indians and Eskimos in the territory "shall not be disturbed in possession of lands actually in their use or occupation or now claimed by them." Otherwise, most Alaska remained in the public domain.

Historically, all public land initially controlled by the federal government in a newly acquired area was opened to private ownership through one of a series of land transfer laws, such as the Homestead Act of 1863 or the Mining Law of 1872. In Alaska, however, the Department of the Interior imposed a ban on private preemption or settlement that remained in force for nearly a century. Thus began a spirited struggle for Alaskan land that continues to the present, a struggle that has involved argument between the U.S. federal government and Alaska state government, between native groups and each government, between settler and native groups, and between developers and environmentalists. The opponents in these arguments have occasionally shifted in their alliance with others involved in the issue, but the scope of the struggle has grown dramatically and regularly.

The federal government chose to maintain its hold on Alaskan land until 1958, the year of statehood. On the eve of statehood, only 200,000 hectares (500,000 acres) were owned privately, with most private land located around Anchorage and Fairbanks—and most of it patented recently under the Homestead Act. The federal government had reserved a number of plots for military purposes and a large area north of the Yukon River for a Naval Petroleum Reserve.

The Alaska Statehood Act of 1958 ignored Indian and Eskimo claims and granted the state the right to select 42 million hectares (104 million acres) for state use by 1983. Most of what has been selected is in the Alaska Range and the Yukon Valley, although the state also controls a large area on the North Slope.

In 1968, petroleum was discovered at Prudhoe Bay on the North Slope. At least partly to dispose of native claims to Prudhoe Bay and to accelerate petroleum development generally, Congress passed in 1971 the Alaska Native Land Claims Settlement Act. This gave the state's Eskimos, Aleuts, and Indians nearly $1 billion and the rights to 18 million hectares (44 million acres) of land. Most of this land was given to profit-making regional development corporations, although about 40 percent went to serve 200 villages who maintain control of surface rights.

The transfer of land to the state and to native groups proceeded slowly amid a tangle of conflicting claims. It was during this period

[4] Ibid., p. 47.

that environmentalists became concerned at what they saw as the potential loss of a rich natural resource, the vast wilderness areas of Alaska. In 1978, the U.S. House of Representatives passed the Alaska National Interest Lands Conservation Act, which withdrew 58 million hectares (144 million acres)—more than a third of the state—primarily for national parks and wildlife refuges. A threatened filibuster by Alaska Senator Gravel kept the Senate from acting on the bill, but President Carter used a little-known 1906 Antiquities Act to create 22 million hectares (54 million acres) of Alaskan national monuments. Furthermore, he proposed another 22 million hectares for wildlife refuges and 4.5 million hectares (11 million acres) to be exempted from mining.

Outraged at these moves to set aside this land from possible development, many Alaskans argued that land withdrawal at the scale proposed would strangle the state's economy. The governor at the time, Jay Hammond, spoke for this view when he complained that "Alaska is being called upon at once to be the oil barrel for America and the national park for the world."

Congress passed a compromise Conservation Act in 1980. This act designated 42 million hectares (104 million acres) for parks and refuges, including 23 million hectares (57 million acres) specified as wilderness (Figure 17-5). It also expedited the conveyance of land to native groups and to the state, and it extended the transfer deadline to 1994. Less than 20 percent of the state's total to be transferred from the federal government had actually been shifted when transfers were halted because of conflicting claims. The 1980 commitment became further clouded a decade

Figure 17-5 Alaska's national parks and wildlife refuges. The state has nearly one-third of the United States' entire national park and refuge acreage.

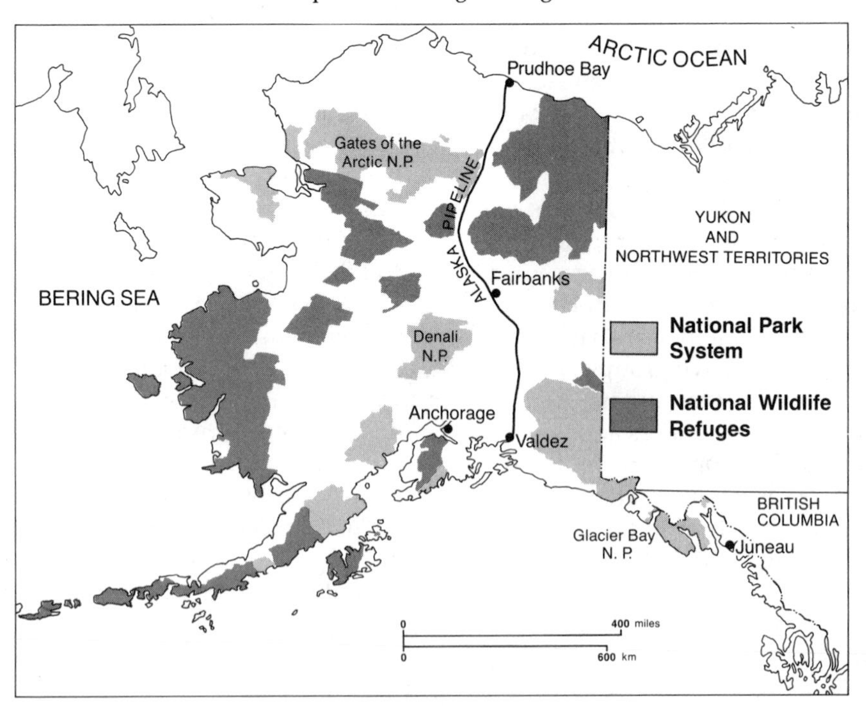

later as discussions within the federal government raised the possibility of opening parts of the wildlife refuge along the Arctic coast to oil exploration.

The struggle for control of Alaskan lands is not likely to be over, because conflicting goals remain:

> Fundamentally, the issues have been ones of priority. The environmental movement argues that Alaska's lands belong primarily to all the people of the United States and require legal protection to prevent undue development, particularly of mineral resources. The native corporations established by the Alaskan Native Claims Settlement Act of 1971 argue that their rights to land should take first priority because they represent aboriginal rights recognized in the Alaska Organic Act of 1884 and implicitly confirmed in the Alaska Statehood Act of 1958. The state, while recognizing the paramount legal rights of the federal government and the rights of the natives, argues that the recent massive federal withdrawals of land from individuals have seriously impinged upon the rights confirmed to Alaska by the Statehood Act.[5]

It seems reasonable to expect that even the extension of the transfer deadline to 1994 will be insufficient to permit resolution of the conflicting claims for use of Alaska's lands.

---

[5] Donald F. Lynch, David W. Lantis, and Roger W. Pearson, "Alaska: Land and Resource Issues," *Focus,* Vol. 31 (January–February 1981), p. 2.

ADDITIONAL READINGS

Baird, David M. "The Mighty Fraser Untamed Still." *Canadian Geographic,* 109 (October/November 1989), pp. 18–29.

Booth, C. W. *The Northwestern United States.* New York: Van Nostrand, 1971.

Dicken, Samuel N., and Emily F. Dicken. *Oregon Divided: A Regional Geography.* Portland: Oregon Historical Society, 1982.

Farley, A. L. *Atlas of British Columbia: People, Environment, and Resource Use.* Vancouver: University of British Columbia, 1979.

Gastil, Raymond. "The Pacific Northwest as a Cultural Region." *Pacific Northwest Quarterly,* 64 (1973), pp. 147–162.

Hull, Raymond, et al. *Vancouver's Past.* Seattle: University of Washington Press, 1974.

Ley, David. "Styles of the Times: Liberal and Neoconservative Landscapes in Inner Vancouver, 1968–1986." *Journal of Historical Geography,* 13 (1987), pp. 40–56.

Meinig, Donald W. *The Great Columbia Plain: A Historical Geography, 1805–1910.* Seattle: University of Washington Press, 1968.

Robinson, J. Lewis. "Sorting Out All the Mountains in British Columbia." *Canadian Geographic,* 107 (February/March 1987), pp. 42–53.

Towle, Jerry C. "The Pacific Salmon and the Process of Domestication." *Geographical Review,* 73 (1983), pp. 287–300.

Warf, Barney. "Regional Transformation, Everyday Life, and Pacific Northwest Lumber Production." *Annals of the Association of American Geographers,* 78 (1988), pp. 326–346.

CHAPTER 18

The Northlands

The United States and Canada are in many ways creations of a frontier experience. The push westward remains part of recent history. Many still live who remember the days of early settlement, of the often heroic struggle with the land that is such a large part of our settlement history.

The American historian Frederick Jackson Turner, and many other students of American attitudes and customs who agreed with him, have argued that because this frontier experience is still recent, it continues to have a noticeable impact on our society. The high value we attach to land ownership, by comparison with most Europeans, stems in part from the widespread availability of land and from the drive to possess land that helped push settlement westward. More land always seemed available over the next hill or beyond the next river. Those who lacked land believed that, through perseverance and diligence, land could be acquired; those who owned land could always reach for more.

The Canadian-American frontier is largely gone today. One can still find ads promising "government lands at $5 an acre," and some families do still move off from settled society and attempt to begin anew in areas that were heretofore unoccupied and unused. The ads are basically flimflam, and the infrequency of frontier settlement can be seen in the wide reporting it receives when it does occur. Although humans presumably have the technology to live anywhere on the earth's surface, those areas of Canada and the United States that can be occupied with moderate physical and economic effort are already staked out.

One vast region of North America remains sparsely settled; many tens of thousands of square miles are totally unoccupied. This is the North American Northlands (Figure 18-1). Extending as far south as the northern Great Lakes states and including the great majority of interior and northern Canada and Alaska, it is easily the largest of North America's regions. The size of this region is sometimes difficult for most of us to comprehend. A few comparisons might help. If it were a separate country, its area would make it the world's sixth largest. Hudson's Bay is larger than the Sea of Japan. Canada's Arctic archipelago, 20 major islands and some 18,000 smaller ones, is the largest in the world. Generally difficult for human settlement, it is the inhospitable nature of the physical environment plus the consequent thinness of settlement that gives the Northlands its special character.

This giant replica of a Canadian nickel at Sudbury, Ontario, celebrates the productivity of the Northland's mineral industry. The tall smokestack in the background puts many tons of sulphur into the air daily. (Kirby Harrison/The Image Works)

A HARSH ENVIRONMENT

A brief identification of the key features of the Northlands environment must include mention of its cold temperatures, long winters, thin soils, poor drainage, and low precipitation. It is the combination of these basic elements that limit broad human use of this region.

Climate

If Americans were asked to give a one-word definition of the nature of the Northlands, "cold" would probably be the most commonly used adjective. This would certainly be a valid judgment. Average January temperatures in this region range from a high of about −7°C (20°F)

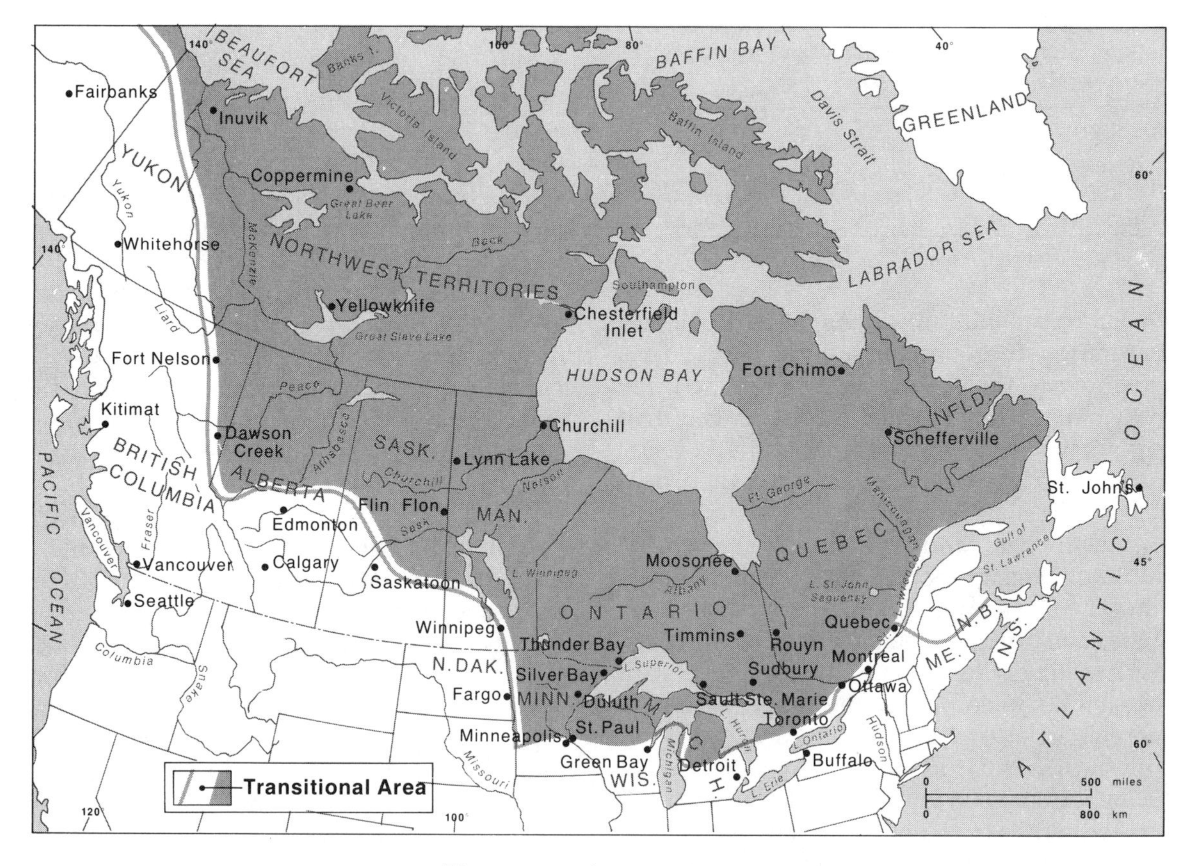

Figure 18-1 The Northlands.

along its southern Great Lakes margin to a full −40°C (−39°F) in parts of arctic Canada. Temperatures can reach −60°C (−76°F).

Not only are winter temperatures quite low across most of the region, but winters are also long. The frost-free period (the average time between the last frost in the spring and the first in the fall) is roughly 135 days at the southern margins of the area but little more than 14 days in length along parts of the Arctic Ocean borders of the two countries. Across most of the region, the frost-free period is less than 90 days. Because virtually all major food crops need a growing season of longer than 90 days, they can be grown in only a few small areas along the southern margins of the Northlands.

Summers, generally short and cool, can have surprisingly warm days. The region's climate is largely continental. Maritime moderation is significant only along the peripheries, mainly in the east and west. Thus, along with the low winter temperatures come brief but warm summers. An old joke among residents of the Yukon and the Northwest Territories is that, when asked by visitors what they do in summer, they reply, "Well, if it falls on a weekend, we go on a picnic."

This dramatic seasonal variation in temperature results from the great shifts in length of day and the angle of incidence of the sun's rays. The earth's axis, that straight line connecting the two poles, tilts at an angle of 23½° from the plane of

Major Metropolitan Area Populations in the Northlands, 1986	
Sudbury, Ont.	148,887
Thunder Bay, Ont.	122,217
Duluth–Minn.–Wis.	93,745[a]
Fairbanks, Alaska	30,843[a]

[a] Indicates 1990 population.

revolution around the sun. As the earth follows its annual path around the sun, the north pole is tilted toward the sun in our summer and away from it in our winter. Thus, everywhere north of $66\frac{1}{2}$°N latitude, the Arctic Circle, is in darkness for at least one day at midwinter and experiences at least a 24-hour period without the sun setting at midsummer. Moreover, during the winter, the sun, when it rises, remains low on the horizon. The sun's rays, striking the earth at a low angle, must cover a relatively large area. Much less heating results from this in comparison with the summer period, when the sun's rays are far more direct.

In midwinter, Fairbanks must make do with the light from a sun hanging just above the horizon for perhaps two hours. In midsummer, the sun sets briefly around midnight. Even southern sections of the region receive only six or seven daylight hours during the winter. For people accustomed to sleeping in darkness and being active during daylight, this seasonal fluctuation demands substantial psychological adjustments.

Precipitation amounts vary widely across the Northlands. Highest levels are found in the far southeast where both winter and summer storm systems move along the coast and combine to dump more than 100 centimeters (40 inches) of annual precipitation along the southern shore of Labrador. Precipitation levels drop markedly toward the interior and north. Most of the Northwest Territories in Canada receive less than 25 centimeters (10 inches) annually; parts of the Arctic islands average less than 15 centimeters (6 inches). The cold air masses usually found over this area hold little moisture and, therefore, produce little precipitation. Also, the Arctic Ocean, covered by ice for all or much of the year, supplies little or no evaporation from the sea. The northern Arctic Islands are the most arid parts of Canada and the United States. Snowfall amounts in the Northlands are consequently much lower than might be expected. Whereas parts of Labrador and interior Quebec may average more than 150 centimeters (60 inches) of snow a year, most of the Northwest Territories receive less than 60 centimeters (24 inches). The Arctic Islands get only about 25 centimeters (10 inches) of snow annually, roughly the same amount as Washington, D.C.

Terrain and Soils

Despite the paucity of precipitation, little of the Northlands provides the appearance of a dry environment. In the summer, in fact, much of the region is covered with standing water, and the mosquito stories of the Northlands match those generated anywhere on the continent. This is due, in part, to the low levels of evaporation and evapotranspiration found in this cold climate. In the northern portions of the region, standing water is also supported by the widespread existence of *permafrost* (Figure 18-2). This is a subsurface layer of permanently frozen ground, which may be only a few meters thick, but which is commonly about 100 meters (330 feet) thick and sometimes extends downward for more than 300 meters (1000 feet). In warmer areas the permafrost is discontinuous with areas of frozen ground interspersed with unfrozen soil. As the surface layer thaws during the short summer to a depth of perhaps 1 meter (3 feet), water is held on the surface by the frozen layer underneath instead of penetrating the soil, creating an extremely boggy, shifting surface. Construction in permafrost is difficult. Buildings must be placed on piles sunk deeply into the permafrost for stability, and roads must be repaired extensively each year to maintain any resemblance of an even roadbed. Heat given off by buildings or absorbed from the sun's rays by road surfaces causes additional melt of the per-

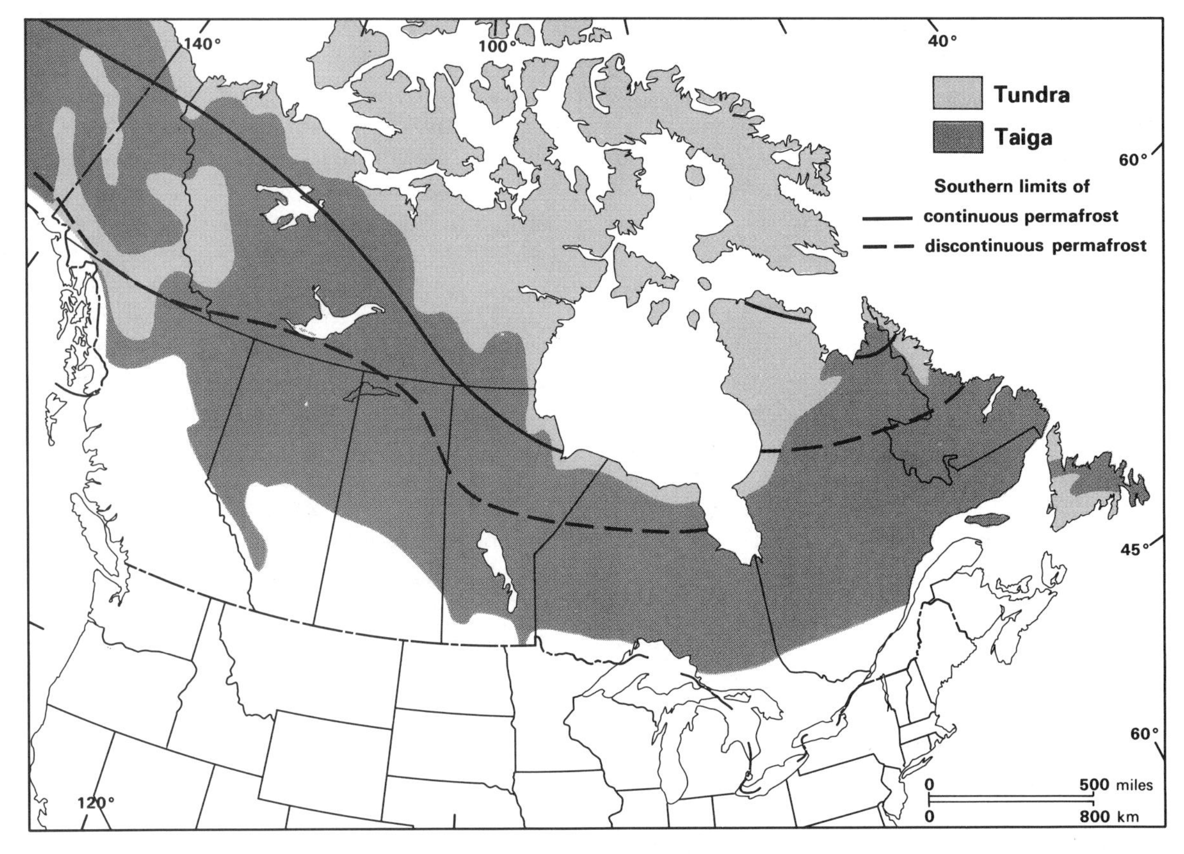

Figure 18-2 Distribution of tundra, taiga, and permafrost. A harsh winter and a short growing season dominate much of interior and northern Canada. Most Canadians live south of the taiga.

mafrost under the construction. Unless care is taken, a building constructed on permafrost can sink gradually into the ground, plunging deeper and deeper as heat given off by the building continues to melt the permafrost directly under it. Roughly half of Canada and most of Alaska are underlain by continuous or discontinuous permafrost.

Although considerable local terrain variation is found, much of the Northlands topogragphy is either flat or gently rolling. The north slope of Alaska and much of the central Canadian Arctic is a broad, flat coastal plain, as are the broad valleys of the Mackenzie River and the Yukon River in central Alaska.

Almost all of the area to the south and east of the Arctic coastal plains is a part of the Canadian Shield, North America's largest physiographic province. The Shield is underlain by very old, heavily metamorphosed rocks. Substantial local relief is found mainly along the southeastern edges of the Shield, especially just north of the St. Lawrence Valley. Nearly all of this 3-million-square-kilometer (1.1-million-square-mile) area was scoured and contoured repeatedly during the past million or so years of the Pleistocene (Ice Age), removing much of the soil from the Shield and depositing it in valleys or to the south. Drainage patterns were greatly disrupted; the landscape is dotted with thousands of lakes and bogs that mark former lakes and is drained by rivers following tortured, twisted pathways to their destinations.

Northland soils are varied, but are generally acidic, poorly drained, and of low agricultural quality. Soils in the southern portion of the region are mostly boralfs or spodosols soils of a cool needleleaf forest environment. To the north, tundra soils, often water saturated and frozen, dominate. Fertile soil is confined to some of the old river valleys and to those lakes that have been filled by sedimentation and decayed vegetation.

The rate of the soil formation in this cold climate is slow. Relatively little organic material is deposited on the surface, a result of slow growth rates for most vegetation. Additionally, the long cold season means that decomposition and humus formation, accomplished during warm periods, is also hindered.

Vegetation

Most of the Northlands can be placed in two large, distinct vegetation areas. Stretching across the entire southern arc of the region is a coniferous forest called the *boreal forest*, or *taiga*. Covering hundreds of thousands of square miles, these closely ranked spruces, firs, pines, and tamaracks appear to blanket the landscape in a dark, almost black mass when seen from the air. Slow growing and never really tall, these trees decrease in number and height from south to north across the taiga. Around the Great Lakes, a mixture of pines and hardwoods predominates.

Passing just south of Hudson's Bay, then angling northwest to the mouth of the Mackenzie River and across the northern edge of Alaska, is the *tree line*. Tree line identifies the boundary between the taiga and tundra in the Northlands

An upland tundra in central Alaska. The vegetation of the tundra is often a rich mixture, with small flowers blooming in profusion in early summer.

(see again Figure 18-2). It should be viewed as a transition zone rather than an actual line. North of it, climatic conditions are too harsh for treelike vegetation. Beyond lies the tundra, a region of lichens, grasses, mosses, and shrubs. At first glance, tundra vegetation appears homogeneous because it is all of the same low height. In fact, considerable variation exists based on temperature, precipitation, and soil differences. Still, to those not sensitive to these differences, the overwhelming visual impact of the tundra is of a broad, usually flat or rolling terrain devoid of any tall vegetation.

Arctic Ice

The great Arctic ice pack, covering some 4.8 million square kilometers (1.8 million square miles) is a thin [usually 3 to 6 meters (10 to 20 feet) thick], rugged sheet of nearly salt-free ice floating on the Arctic Ocean. It holds as much water as all the freshwater lakes in the world. In winter it extends southward to enshroud all of the Arctic coast of Canada and northern Alaska. Summer melt briefly frees all but the northern islands of Canada's Arctic archipelago from its grasp.

The ice limits ocean transport in the Arctic to a brief, often hectic period each summer. An unusually early return of the ice can sometimes block ships (and whales that migrate into the Arctic Ocean during the summer) from escaping to open water to the south. The ice also minimizes any moderating impact by the Arctic Ocean on the Northland's climate.

The Arctic ice pack is slowly melting, apparently in response to gradual global warming. It has lost about 1 percent of its bulk in the last 100

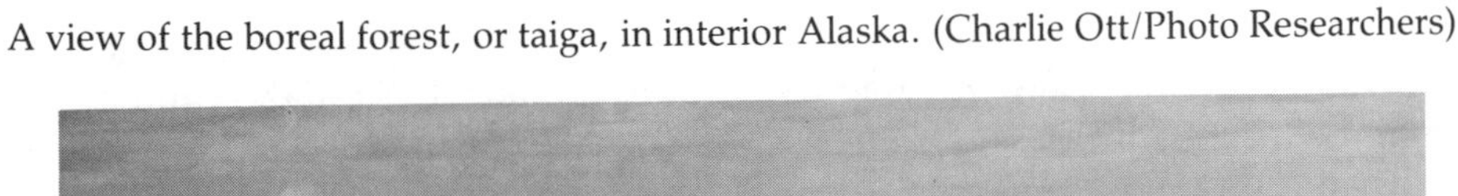

A view of the boreal forest, or taiga, in interior Alaska. (Charlie Ott/Photo Researchers)

years. Because the ice largely floats on Arctic water, this melt has little impact on global sea levels.

HUMAN OCCUPATION

Nearly all parts of the Northlands are sparsely populated with highest densities found along the southern margins. American Indians, Metis, and Inuit (Eskimos) are numerically dominant over much of the region north of the United States. Inuit are the predominant population in most of the Arctic. American Indians are found mainly in the boreal forest area. In pre-European days, the Indians were hunters and fishers; their characteristically widespread population was generally associated with hunting societies. The Metis are the result of intermarriage between American Indian women and whites during the early fur trading period of European settlement in the taiga. Metis now probably outnumber American Indians in the region. They have a distinct identity, but economically they have much in common with the Indians.

The Inuit culture was, and is, remarkably uniform across its broad distribution, which extends around the Arctic margins of the continent from Siberia to Greenland. An Inuit in Greenland can be understood by one in northern Alaska on the other side of the Arctic Ocean. The most recent arrivals of America's pre-European population, the Inuit settled largely along the water margins of the Arctic. The Inuit population now totals over 85,000. They depended on fishing and hunting for their livelihood and developed an effective, self-sufficient life-style in that harsh environment.

The arrival of Europeans in the Northlands brought an end to much of the American Indian and Inuit traditional native economy. Fur traders early acquired many of their pelts from the Indians of the taiga, and European goods entered the Indian economy as a result. Along with these trade goods came European diseases, liquor, and Christianity, all of which worked against the structure of the Indian culture. Although these changes did not reach many Inuit until comparatively recently, the results have been much the same. Where hunting and fishing continue, the motorboat, rifle, and snowmobile have usually replaced the kayak, bow, and dogsled.

Most Northlands Indians and Inuit no longer exist by hunting and fishing. They have moved in substantial numbers into towns, or the economy of their villages has changed. Many urban places of the Northlands today have large native populations. They generally occupy the bottom rungs of the social and economic ladder, with far more than their share of unemployment, disease, and poverty. Those who are poorly educated and trained are less able to compete for the better paying urban jobs.

Both governments have had difficulty defining their posture in their dealings with the Northlands' natives. The Canadian government never adopted the "remove them" approach that long characterized American attitudes toward Indians. Even in the United States, contact with Alaskan natives came sufficiently late to allow some softening in these harsh atitudes. The native population remains scattered across the state, and the American government has recently approved the return of large blocks of Alaskan land to Indian and Inuit control (see Chapter 17).

Canadian Inuit and Indian group in the Northwest Territories and parts of Quebec claim aboriginal rights to the land, rights not surrendered by treaty as elsewhere in Canada. Canada established a Native Claims procedure in 1974 in which the government agreed to negotiate settlements with native groups whose rights had not been surrendered by treaty. In 1975 Quebec reached an accord with local American Indians on the control and use of land around the massive James Bay hydroelectric project. In 1983 the Indians of the Yukon Territory agreed to surrender most of their land claims in return for a $183 million settlement. Elsewhere, in the Northwest Territories, Canada is negotiating a

A modern Eskimo village at Payne Bay, on the Ungava peninsula in Quebec. Electricity, gas heat, public schools, and prefabricated dwellings are all part of the rapid decline of the pre-European culture of the Eskimo population within the space of one or two generations. (Dunkin Bancroft/National Film Board—Photothèque)

number of claims with Indian and Inuit groups. One native proposal would subdivide the Territories into two areas along the tree line, with an American Indian unit in the taiga and an Inuit unit on the tundra. A referendum in the Territories in the early 1980s approved the concept, and the Canadian government has given its limited approval. Control of natural resource development and the sharing of their revenues is one of several continuing government concerns.

Still, even the most benign government cannot easily solve the problem of how to deal with a population subgroup that is vastly different technologically from the much larger and dominant group. Even if it were possible to create the kind of isolation that would allow the continuance of traditional life-styles, such a policy would refuse to Indians and Inuit the opportunity to integrate into the predominant economy of the country. To encourage wholesale change, on the other hand, would result in the destruction of much that is valuable in the traditional culture. Not surprisingly, both countries follow something of a middle course, providing services such as basic education and health care while also encouraging the maintenance of traditional culture. Although certainly not an unqualified success, this is probably a reasonable approach and has the support of most native leaders.

Early European Settlement and the Northern Economy

The Northlands offered little of interest to most Europeans who came to North America. Early settlers generally selected the far greater agricultural promise of the warmer climate and better soils to the south. Where Northlands' settlement did occur, its focus was usually either extractive or military. French voyageurs, fur trappers and traders in search of animal pelts, pushed their canoes far beyond the agricultural settlements along the lower St. Lawrence as early as the middle of the seventeenth century. They extended French political control across the Great Lakes and northward into the headwaters of streams draining into Hudson's Bay. The Hudson's Bay Company, an early British fur-trading company, established itself on the margins of Hudson's Bay and then pushed south and west, thus blocking further French expansion westward. By the mideighteenth century the Hudson's Bay Company, which had been granted a trade monopoly to the area by the British government, was in control of the entire boreal forest reaching from Hudson's Bay westward to the Rocky Mountains, with further extension of influence into the Arctic as well. This vast extractive empire brought with it only a minimal number of small and widely scattered settlements.

The voyageurs and Hudson's Bay Company relied on the numerous lakes and streams of their area for transportation, and they located small forts at control points along the water routeways. At places where important streams met lakes, where streams ended and an overland portage began, or where rapids or falls were encountered, vessels had to be unloaded and their goods moved and reloaded onto other boats; these break-in-bulk points provided effective control of the entire water system. The sites of many early French forts are occupied today by important urban centers; the continuing break-in-bulk function remains one reason for their existence. Cities, such as Ottawa and Sault Ste. Marie in Canada and Chicago, Detroit, and Pittsburgh in the United States, have grown around the sites of former French forts (Figure 18-3).

Logging

The boreal forest of the southern half of the Northlands contains the largest area of uncut forest remaining in North America. Until recently, the lumbering and the pulp and paper industries only nibbled at the edges of this vast forest. The area of the upper Great Lakes was logged on a massive, devastating scale during the late 1800s and the early twentieth century. Tree removal was so complete that the area became known as the Cutover Region. Because little reforestation was practiced at that time and because the cold climate of the boreal forest slows regrowth, much of this area is only now recovering its previous appearance. Repeated fires, which were fueled by wood debris left from wasteful logging, and a hardpan soil that developed after all vegetation cover had been removed, further slowed the reforestation process. Even today, the Cutover Region is still not an important logging area.

Canada is the world's leading exporter of forest products. Lumber and, particularly, pulp and paper operations now dot the southern margin of the boreal forest from Quebec to Manitoba, and cutting is gradually moving northward. The rich spruce forests south and southeast of Hudson's Bay are the prime source of raw material for paper mills, which are themselves located on some of the plentiful water power sites along the southern margins of the Canadian Shield, many in the St. Lawrence lowland. Quebec is today the regional leader in Canadian pulp and paper production, and Canada produces about 40 percent of the world's newsprint. After motor vehicles, pulp and paper products are Canada's most important export in value, with lumber also ranking high.

Much of the capital for the Canadian lumbering and pulp and paper industries, and also

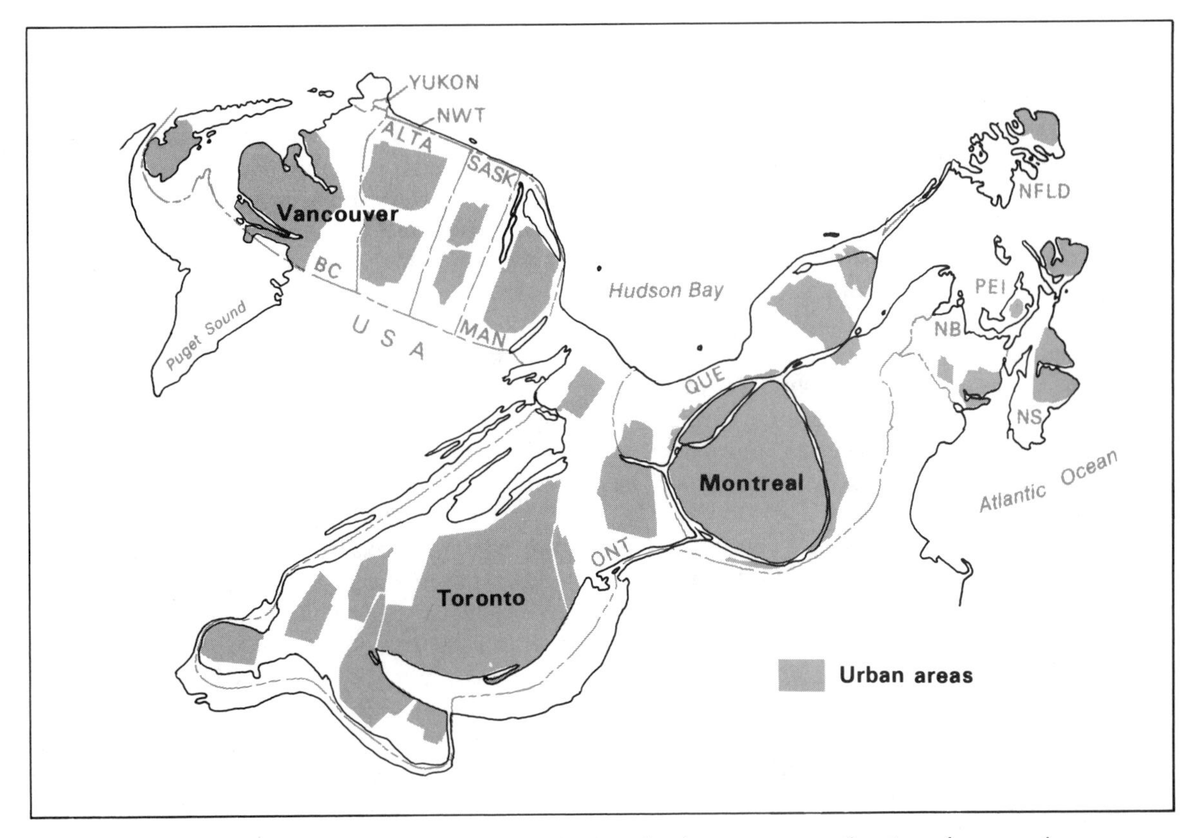

Figure 18-3 Canada's population distribution. In the cartogram, the size of an area is proportional to its population, not its geographic extent. Most Canadians are urbanites, and about 30 percent live in one of the country's three largest metropolitan areas (Toronto, Montreal, and Vancouver), while so few live in the Northlands that it is hardly visible here.

most other Canadian extractive industries, has come from outside Canada. Foreign investors control nearly a quarter of Canada's nonfinancial industries, including 40 percent of all manufacturing. Until the end of World War I, the United Kingdom was the principal supplier of investment capital, but since than an increasing majority has come from the United States (see Chapter 7). Today about 70 percent of all foreign direct investment in Canada comes from the United States. Of course, about 70 percent of Canadian direct investment abroad goes to the United States. Because adequate development capital for large-scale growth was unavailable locally, foreign capital has aided Canada in many positive ways. However, outside ownership frequently has also meant outside control. This rankles many Canadians, who, with justification, have often felt unable to govern their own extraction policies. For example, after 1911, Canadian newsprint was admitted duty free into the United States, and American interests subsequently expanded greatly their Canadian operations. Still, in 1988 the governments of the two countries signed the United States–Canada Free Trade Agreement. This historic step's goal is to make trade and investment across the border less cumbersome and more secure. Under the pact, all goods produced in both countries will be allowed to cross the border duty free in less than 10 years.

The total value of Canadian exports in 1987 was $121 billion (Canadian currency); about 75 percent of that export was sent to the United States. Imports totaled just over $116 billion, with the United States claiming a 70 percent share. The bilateral trade flow is the largest between any two countries in the world. The composition of Canadian exports to the United States has changed markedly. In 1960 over 50 percent of all exports by value were fabricated raw material (lumber, newsprint, refined metals, and the like), 20 percent was crude materials like mineral ores, and 10 percent was foodstuffs. Today, manufactured products account for about half of the total and fabricated materials only 25 percent.

Mining and Petroleum

Canada produces a wide range of metals and other minerals. It is the world's leading producer of nickel, zinc, and asbestos, and a major supplier of potash, copper, lead, and iron ore. The great majority of all production comes from the Northlands (see Figures 2-6, 2-8). Also, the region's U.S. portion in the upper Great Lakes area is that country's leading source of iron ore plus a substantial contributor of copper. Alaskan North Slope petroleum has recently provided a large addition to the American energy supply, presently pumping about 25 percent of the country's total production. In short, mining and petroleum extraction together with lumbering comprise the bulk of economic activity in the Northlands.

As with logging, the accessible peripheries of the Northlands became the region's first important mining districts. The Mesabi Range in northern Minnesota, an area of gently rolling elongated hills, along with neighboring areas in Minnesota, Wisconsin, and Michigan developed into America's chief source of iron ore late in the nineteenth century. The Great Lakes waterway provided an easy route for moving those ores to the industrial centers of the manufacturing core (see Chapter 6). Billions of tons of high-grade ore were moved by rail to Lake Superior ports and were loaded there onto large specialized lake ships that carried the ore to ports in northeastern Ohio, where it was then transferred to railroads for the trip to iron- and steel-processing plants in the Pittsburgh-Youngstown area. Today, most ore goes to the newer integrated iron and steel facilities at the southern end of Lake Michigan. Locks at Sault Ste. Marie that connect Lake Superior with the rest of the Great Lakes are the busiest in the world, largely as a result of this ore traffic.

Most of the high-quality iron ores are now gone from the Lake Superior mining district. Attention has turned to a lower-grade ore called taconite, also found in huge quantities in the district. The iron content of taconite, roughly 30 percent compared to perhaps double that for the richer ores, is so low that to ship the ore to the lower Great Lakes for processing is considered far too expensive. Thus, the bulk of the ore is reduced (and its iron content increased) through a process called beneficiation. The ore is ground into a fine powder, much of the rock removed, and the resulting material pressed into small pellets with a much higher iron content. This greatly lowers the cost of shipping taconite, thereby increasing its transferability. That is, its movement costs are sufficiently low relative to the value of the improved ore to make it profitable to ship. Without this reduction in bulk, the taconite might be ignored in favor of other sources of supply that, although more distant, offer an ore that is actually cheaper to move to market because of its higher ore content.

The cost of shipping low-grade ores is the major factor in choosing to locate many smelting operations near the source of supply instead of near the market. For example, copper, which seldom represents as much as 5 percent of its ore and often less than 1 percent, is nearly always refined near the mine. The smelting and refining of ores is the major form of manufacturing em-

Thunder Bay, Ontario, at the western end of Lake Superior, Canada's second largest port, is a major transshipment center for Canadian grain, ore, and coal. A similar facility for American iron ore is found a short distance southwest at Duluth-Superior. (Lakehead Harbour Commission)

ployment in the Northlands, and the large smokestacks of the refineries are the central element of the skyline of some of the largest cities in the region.

These smokestacks, incidentally, also reflect one of the key problems with this and almost any mining company—environmental disruption. Lake Superior iron ore is mined by the open pit method. Until recently, the land was simply left in its disturbed condition to recover as it might. With the topsoil gone, sterile materials exposed on the surface, and normal drainage patterns destroyed, natural recovery is extrememly slow, and abandoned mines remain as ugly scars on the landscape. Today, in some places, an attempt is being made to aid recovery by recontouring land, planting, and fertilizing. There have been some successes, but such efforts tend to be difficult and expensive.

Other environmental problems developed at the processing plants. A vast quantity of waste material, or slag, is created during ore processing. The topography around many refining centers gradually changes as the waste is piled into great hills. Also, many kinds of refineries produce gaseous and liquid wastes that can be harmful. If one were to visit Sudbury, Ontario, during the mid-1960s, the first indication that this important mining and smelting center was near were the trees covered with brown leaves in midsummer. Substantial improvement is apparent now, but treatment of waste is usually expensive, and often only partially successful.

The mining frontier pushed northward from the Great Lakes early in the 1900s. At first gold and silver were the principal minerals removed; then attention turned to such metals as copper, nickel, lead, and zinc. Sudbury, at the southern end of this great mining concentration along the Ontario-Quebec border, started first. With an urban area population of 150,000 people, it is the largest urban center in the Northlands. To the northeast of Sudbury, the Clay Belt district crosses the provincial border from Timmins to Val d'Or, with copper the leading mineral. The Chibougamau area, 350 kilometers (220 miles) farther northeast, is a newer mining region that produces a variety of metals, including lead, copper, and zinc. Together, these several mining districts comprise the major mining region in Canada.

Of the other metallic mining areas scattered across the Shield, the most notable is the Labrador Trough along the border between Labrador and Quebec. The existence of large bodies of high-grade iron ore in the region was first confirmed early in the twentieth century. Its inaccessibility coupled with the adequate supplies of the Lake Superior district slowed development of the area. The decline in the quality of the latter district's ores plus growing demand in Canada and the United States encouraged the exploitation of the Labrador Trough. A railroad, open in 1954 at a cost of nearly $400 million, connected Schefferville, the early center of production, with Sept Iles, some 550 kilometers (340 miles) south on the St. Lawrence River. Today, Schefferville's mines have been closed, and the community is abandoned. Other mines in the area, such as those at Wabush and Carol Lake, Newfoundland, and Gagnon-Fremont, Quebec, maintain the two provinces as the leaders in Canadian iron ore mining.

Smaller concentrations of mining activity are found in a number of other locations in Canada's Northlands, including mines north of Lake Superior in western Ontario (iron ore), Flin Flon on the Manitoba-Saskatchewan border (copper and zinc), Thompson, some 320 kilometers (200 miles) north of Flin Flon (a leader in nickel production), and Yellowknife on Great Slave Lake (a center of gold mining).

Canada is the world's third leading metal producer, behind the Soviet Union and the United States. On a per capita basis, Canada is surely in first place. As noted, much of that product is exported. One result is that the Canadian mining industry is particularly susceptible to international economic shifts, especially those felt by the American metals industry. For example, the recession in the U.S. metals industry in the early 1980s depressed the sale of Canadian copper by one-third, iron ore by over one-third, and nickel by nearly one-half between 1980 and 1983. The economic results were devastating to many of the single-industry mining communities of the Northlands. Recovery during the 1980s was generally less than complete.

Accessibility remains a major problem for Northlands mineral production; construction of railroads was necessary before large-scale mining could begin in many of these areas. Many other sites, particularly in the far north, still have not been fully explored geologically.

Until recently, little petroleum was produced in the Northlands. Canada's major proven resources of liquid petroleum are on the prairies of Alberta; American demands were met by domestic production elsewhere or by imported supplies. Recent declines in developed reserves elsewhere has encouraged exploitation and development in both countries. In Alberta, a major initiative financed by both public and private funds is underway to develop the Athabasca tar sand petroleum reserves (see Chapter 13). Elsewhere, geologic conditions suggest that the best prospects for petroleum and natural gas discoveries are found along the mountain margins in the west to the Mackenzie River Delta and in a broad band extending from Alaska north of the

This section of the Alaskan petroleum pipeline on the North Slope about 200 kilometers (120 miles) south of Prudhoe Bay has been elevated above ground to help ensure that the heated petroleum does not melt the permafrost. (Alyeska Pipeline Service Company.)

Brooks Range (the North Slope) across Canada's northern Arctic islands. Oil deposits have definitely been indicated in the Mackenzie Delta, and gas has been found on Arctic islands. Exploitation of these northern fields will be gradual, however, slowed by the high cost of developing a resource that is cheaper at other locales and by Canada's growing desire to husband its natural resources.

The United States, in contrast, has moved rapidly to develop its North Slope petroleum fields. Some oil producers paid well over $1 billion just for the right to search for oil in the region. America's fear of an inadequate energy supply resulted in the country's most dramatic technological adventure since the boom days of extraterrestrial exploration in the 1960s.

Transportation of crude petroleum was the principal problem involved in opening the North Slope fields. At first, many assumed that large tankers could carry the petroleum through waterways between Canada's Arctic islands to the Atlantic Ocean, and one specially reinforced ship did make the voyage. Canadian concern for the long-term ecological damage that an oil spill could cause in cold Arctic waters, coupled with the near impossibility of making the trip during any period other than the few warm months, forced abandonment of this plan. The only real alternative was to construct a pipeline. For a time a joint American-Canadian project was considered in which a pipeline extending from the north slope perhaps as far as the lower Great Lakes was to be built. Differences over management and construction timetables plus the huge cost of construction led to its abandonment as well. A pipeline costing $8 billion and crossing central Alaska to the port of Valdez on the Pacific was finally built, and opened in 1977.

Hydroelectricity

About 70 percent of all electrical power in Canada comes from hydroelectric facilities, mostly located along the southern margins of the Shield where streams crossing the Shield's hard rock drop onto the lowlands of southern Ontario and Quebec. Until recently, the Churchill Falls project in Labrador was the largest in the country. Quebec has finished an even larger project on the margin of James Bay. These projects, like so many other developments, have been extended northward onto the Shield. The electricity generated by them is both cheap and abundant. As a result, Quebec is a major center for aluminum smelting, a process that requires large amounts of electricity. The province has also contracted to sell surplus electricity to New York, Ontario, and several power companies in northern New England. Canada now exports to the United States about 10 percent of its generated electric power.

Transportation

In a region where isolation is so pervasive, transportation is quite important. The region's problems in developing transportation result from more than the difficult environment, however important that factor might be. The Northlands' sparse population and especially its lack of cities means that even if transportation routes could be constructed cheaply, they would be used relatively little. The large nodal growth necessary to support a well-integrated transport network is unlikely to develop in the region. Hence, it is difficult to justify high transportation expenditures unless a large deposit of a particularly needed mineral is available.

Today only the Mackenzie River is used to any extent for river transport to the north. Diesel fuel is the most carried freight. Few railroads penetrate far into the Northlands. They include one to the Labrador Trough from Sept Iles, one that carries mostly export wheat from the Prairie Provinces to Churchill on Hudson's Bay, one to the mining developments on Great Slave Lake, and one to James Bay in Quebec. Other rail lines serve the southern margins of the region, such as the mining districts on the Quebec-Ontario border. The Mackenzie Highway also reaches Great Slave Lake, and the Alaska Highway and some of its branches pass through the western margins of the region.

Taken together, these conventional routeways provide access to only a small part of the Northlands. Small ships serve some coastal towns on an occasional basis. For much of the region, however, the light airplane and its bush pilot is the only transportation link available. Close to a score of carriers operate a relatively dense pattern of scheduled routes in the north. Although expensive, some studies suggest that air transport may be only twice as expensive as truck transport on the winter highways, and it cuts development costs tremendously. No region of the world, with the possible exception of the Soviet Union's own northlands and the Australian outback, places greater reliance on the airplane for basic transportation.

Tourism

As North Americans have gained higher and higher incomes with an accompanying growth in leisure time, they have extended their need for recreation space. As nearby areas have become overused, their search has reached farther and farther away from the settled portions of the continent. No major national, provincial, or state park is today so isolated that it does not face the problem of overuse during peak tourist periods. The Northlands, the continent's largest empty space, is felt by many to have almost unlimited tourist potential. This belief is only partly justified.

For many in the large urban centers just south of the region's margins, a summer vacation in the North Woods has long been a standard part of life. Large sections of the southern boreal forest are heavily used by tourists. Northern Minnesota, Wisconsin, and Michigan are an easy day's drive from some of the largest U.S. cities. In Canada, Montrealers flock to the Laurentians a short way north. The jam of cars carrying weekend vacationers north of Toronto can extend for 300 kilometers (185 miles) on a fine Friday afternoon in summer.

To the north, in the central and northern boreal forests and in the tundra, tourist volumes are much smaller. The big game and sport fish found there are justifiably famous, attracting many hunters and fishermen who are willing to fly to isolated lodges in the area. The total number of visitors involved is small, however, and it is likely to remain so. For the average vacationer, these areas offer nothing that cannot be found closer to home with less effort and expense. Also, the fragile environment cannot support anything approaching heavy use. For example, this is an area where vehicle tire marks made on the tundra more than 45 years ago during World War II are still visible, and many people fear that the traffic traveling the road constructed along the Alaska pipeline will have a detrimental impact on the environment.

A Nodal Settlement Pattern

Although the total regional population is not large, the great majority of people in the Northlands live in villages, towns, and cities (Figure

18-4). Agriculture, a major support of dispersed settlement elsewhere, is only locally important. The principal agricultural settlement areas are the Lake St. John–Saguenay Lowland of Quebec, the Clay Belt along the Quebec-Ontario border, and the Peace River District on the British Columbia–Alberta border. (The last two areas were settled in this century and represent the last significant examples of something that was once a key part of North American life—frontier agricultural settlement.) Nearly all of the larger cities are dominated by a single major economic activity and are located in the south. Many, such as Sudbury or Flin Flon, are mining and smelting towns. Others, such as Duluth or Thunder Bay, are primarily transportation centers. Chicoutimi's economy focuses on aluminum smelting. Most smaller towns in the boreal forest are similarly unifunctional.

In the far north, European development has

Figure 18-4 Settlement zones of northern Canada. Most of northern Canada is a zone of recent European settlement, usually the result of resource extraction activities. The Far North remains largely beyond the settlement frontier.

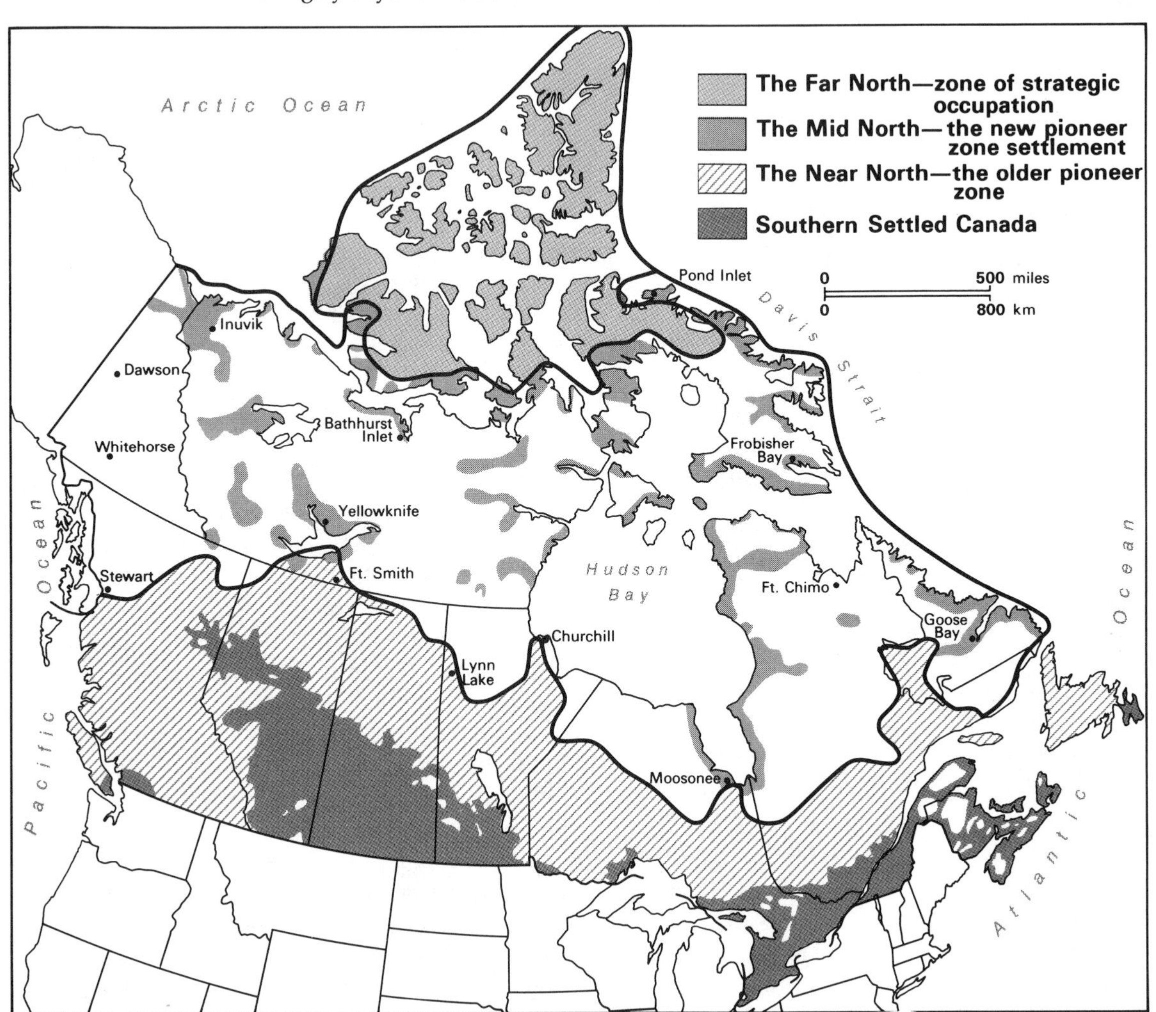

resulted in few permanent settlements. Most whites in the area work, in some capacity or another, for the Canadian or U.S. governments or in resource exploitation. Far northern communities are extremely isolated, often with predominantly male populations, and the labor force frequently spends periods of weeks away from the communities for family visits and recreation. As with communities everywhere that are totally focused on minerals extraction, they are often short-lived, owing to resource depletion.

The Inuit constitute the far north's majority population. Most live in small villages along the coastline, although more and more of them are drifting into the towns of the area. Again, it is a nuclear settlement pattern; few people live away from these settlement nodes.

GOVERNMENT ATTITUDES AND NORTHLANDS DEVELOPMENT

The Northlands have presented both governments with a series of important and often perplexing problems. This is a vast area, largely unoccupied, and only now gradually being exploited. At an earlier time, perhaps neither country would have concerned itself with the integrity of these lands or their people. However, times have changed. Today, the region, still largely frontier, is controlled by industrial, urban economies that share an increasing concern for what their growth has meant to the natural environment.

Canadians, particularly, with the Northlands so much a part of their country, are concerned about the problems of development there. The government and many universities in Canada actively sponsor a wide range of research efforts to help learn more about the Northlands environment. As mentioned earlier, Canadians are tempering their desire for the wealth almost surely available in the region's resources with a concern for what extraction will do to its obviously fragile environment. The United States shares these concerns, although on a much smaller scale.

The native population in the Northlands compounds these problems. Both Indian and Inuit fertility rates are high. Even with a fairly high mortality rate and with some migration toward the margins of the region, rapid growth in numbers still is occurring. Unemployment rates among these groups are already high—often above 50 percent—and good jobs cannot be provided without substantial economic growth. Indians and Inuit are caught in a nearly impossible situation; they have learned to desire many of the comforts and products of a developed economy, but they seldom have the means of attaining those goals.

The necessity of focusing development on mineral resources coupled with the high cost of transportation to widely scattered locations, suggests that even a decision to encourage extractive industries would have only a small and short-lived impact on the economy of most far northern communities. Economic opportunities are restricted and will probably remain so.

Several writers interested in the political geography of places have suggested that it is useful to look at what can be called a country's "effective" national territory. They contend that whatever a country's boundaries might seem to be, it is really in control only of that land actually occupied and actively administered. Large areas of empty land are, by this measure, not firmly a part of a country's territory.

No one is threatening to seize the empty lands of either Canada or the United States, but at least some aspects of this idea can be seen in their approaches to the Northlands. The construction of improved transportation facilities, the development of military bases, and the study of development schemes for the region are in many ways all indicative of this concern. This region will become more and more integrated into the economy and society of the rest of the two countries. It will not be set aside as a massive preserve for things as they once were. In the final analysis, neither country views that as an acceptable alternative.

Many of the towns of the Northlands are mining centers, and their skylines are often dominated by the chimneys of the refineries that reduce the bulk of the product to lessen the cost of shipment. This is a view of Flin Flon, Manitoba. (Manitoba Government Photograph)

ADDITIONAL READINGS

Archer, Clive, and David Scivener (eds.). *Northern Waters*. Towanda, N.J.: Barnes and Noble, 1986.

Bradbury, John, and Isabelle St. Martin. "Winding Down in a Quebec Mining Town: A Case Study of Schefferville." *Canadian Geographer*, 27 (1983), pp. 191–206.

Canada, Department of Energy, Mines, and Resources. *Annual Report*. Ottawa: Department of Energy, Mines, and Resources, 1966/67– .

Coates, Kenneth, and Judith Powell. *The Modern North*. Toronto: James Lorimer, 1989.

Flader, Susan L. *The Great Lakes Forest: An Environmental and Social History*. Minneapolis: University of Minnesota Press, 1983.

Hamelin, Louis-Edmond. *Canadian Nordicity: It's Your North Too*. Montreal: Harvest House, 1979.

McGinnis, Joe. *Going to Extemes*. New York: Alfred A. Knopf, 1980.

McPhee, John. *Coming into the Country*. New York: Farrar, Straus & Giroux, 1977.

Mowat, Farley. *The Great Betrayal: Arctic Canada Now*. Boston: Little, Brown, 1976.

Vanderhill, Burke G. "The Passing of the Frontier Fringe in Western Canada" *The Geographical Review*, 72 (1982), pp. 200–217.

CHAPTER

19

Hawaii

When we initially designed this text, we seriously considered excluding any major discussion of Hawaii. After all, we reasoned, this is a book that attempts an overview of the regional geography of North America, and a group of islands in the middle of the Pacific Ocean is certainly not a part of North America. Other calmer and more reasonable heads prevailed, however. They contended that because Hawaii was a part of the United States and because our North America really meant Canada and the United States, the book would be incomplete without a section on Hawaii.

Those who encouraged us to include Hawaii were correct, of course, and we were likely to have reached the same decision after further reflection. What is most interesting is that we even considered ignoring it. That we, as geographers, considered not discussing Hawaii is a good indication of both the physical separation of Hawaii from the continent and the distinctiveness of its physical and cultural landscapes.

Technically, the Hawaiian archipelago is a string of islands and reefs 3300 kilometers (2050 miles) long that forms a broad arc in the mid-Pacific. The archipelago begins in the east with the island of Hawaii and ends almost at the International Date Line with a small speck in the ocean called Kure Atoll (Figure 19-1). Only the easternmost 650 kilometers (400 miles) of the state contains islands of any size as well as almost all the people. It is to this portion, which is usually considered as the actual "Hawaii," that we will limit our discussion. Among the hundred or so bits of land to the west, Midway Island (which is occupied by a military base), 2500 kilometers (1500 miles) west of Honolulu, is easily the largest.

The eight main islands of Hawaii contain more than 99 percent of the state's land area and all but a handful of its people. The island of Hawaii, at 8150 square kilometers (4021 square miles), itself comprises nearly two-thirds of the state's total area. The smallest of the eight is Kahoolawe, at 125 square kilometers (45 square miles). It is uninhabited and used only as a military bombing and firing range, to the annoyance of many residents of the state.

The rugged Na Pali coast line of Kauai, the Garden Island. The spectacular beauty of Hawaii has captured the imagination, and the tourist visits, of millions of Americans from "The Mainland." (Joe Sohm/The Image Works)

LOCATION AND PHYSICAL SETTING

Mid-Pacific

Hawaii is near the middle of the Pacific Ocean. Honolulu, the state capital, is 3850 kilometers (2400 miles) west of San Francisco, 6500 kilometers (4000 miles) east of Tokyo, and roughly 7300 kilometers (4500 miles) northeast of the Australian coast. The great swath of islands stretching across so much of the south and west Pacific are, with a few small and largely inconsequential exceptions, themselves well over 3500 kilometers (2175 miles) away. This might be viewed as a case of extreme isolation. Until the last couple of centuries, this was probably true, but as countries around the Pacific Basin began to communicate more with one another and to use the ocean's resources, these islands in the middle of the ocean became an important center of interaction. The strategic value of Hawaii's location is great in the affairs of the central and north Pacific.

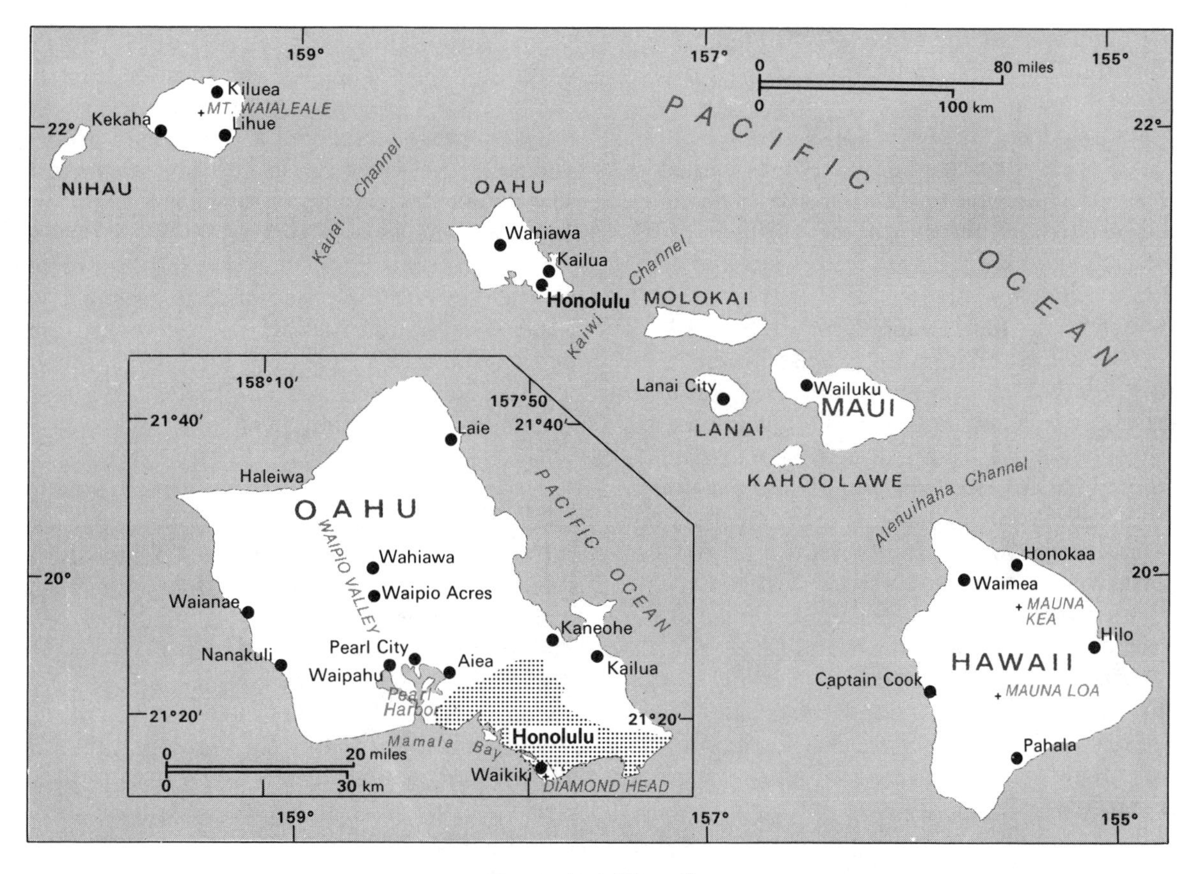

Figure 19-1 Hawaii.

Geology

The Hawaiian chain is merely the visible portion of a series of massive volcanoes. The ocean floor in this area is 4000 to 5000 meters (13,000 to 15,000 feet) below sea level. Hence, for a volcano to break the water's surface requires a mountain already approaching 5 kilometers (3 miles) in height. If one wishes to be a little picky and ignore the oceans in determining the height of mountains, Mauna Kea, on the "Big Island" of Hawaii, may be the world's tallest. It stands 4528 meters (13,784 feet) above sea level. In addition, its base is some 5400 meters (18,000 feet) below sea level, for a total height of nearly 10,000 meters (32,000 feet), almost 1000 meters (3000 feet) taller than Mount Everest.

The kind of volcanic activity that created the islands and that continues there today has, for the most part, not been of the explosive type where large pieces of material are thrown great distances by violent explosions. Instead, the islands have been built up by repeated flows of lava from fissures. Volcanic cones resulting from explosive eruptions do exist on the islands. Diamond Head, the famous Honolulu landmark, is

Major Metropolitan Area Populations in Hawaii, 1990	
Honolulu	880,835
Hilo	37,808

the highest at about 240 meters (810 feet). More common, however, are features formed from a gradual buildup of material as a sequence of lava flows piled one layer on top of another. The usual shape of volcanic mountains formed in this way is domelike, with the main feature being undulating slopes instead of steep cliffs. Such rolling highlands are found in many areas on the islands.

Several of the volcanoes on the Big Island remain active. Mauna Loa pours out lava on the average of once every four years. A 1950 eruption covered some 100 square kilometers (36 square miles). Volcanic activity poses a constant threat to Hilo, the island's largest town. Another volcano, Kilauea, is usually active, but lava actually flows from it about once in every seven years. A 1960 flow from Kilauea covered 10 square kilometers (4 square miles), adding nearly a square mile to the island's size.

Hawaii is characteristically a state of rugged slopes and abrupt changes in elevation. This is the result of the erosion of the volcanic surfaces by moving water. Sea cliffs cut by waves form a spectacular edge to parts of the islands. Such cliffs on the northeast side of Molokai stand as much as 1150 meters (3600 feet) above the water and are among the world's highest; others on Kauai exceed 600 meters (2000 feet). Some small streams on the northeast side of the Big Island drop over such cliffs directly into the sea.

Stream erosion has heavily dissected many of the lava surfaces. Canyons lace many of the domes. The floor of Waimea Canyon, on Kauai, is more than 800 meters (2500 feet) below the surface of the surrounding land. Waterfalls several hundred feet high are common on the islands. Hiilawe Falls, in Waipio Valley on Hawaii, has a drop of about 300 meters (950 feet). The Pali, on Oahu, is a line of cliffs where the headwaters of streams eroding from opposite sides of the island meet. Those flowing east have eroded the ridges separating them to cut a broad lowland; the westward-facing valleys are higher and remain separated by ridges. The result, viewed from the coast, is a jagged cliff that looks like a series of connected V's.

One important result of this intense erosive action is a limited amount of level land on the islands. Kauai is particularly rugged, with the only lowlands formed as a thin coastal fringe, and even that does not exist in the northwest. Maui has a flat, narrow central portion separating mountainous extremities. Molokai is reasonably flat on its western end. Oahu has a broad central valley plus some sizable coastal lowlands. The island of Hawaii has only some limited coastal lava plains. The total area of these level sections is not great; flat land on the islands is limited and often quite valuable.

Climate

Hawaii's oceanic location has, obviously, a substantial impact on its climate. It is the ocean that fills the winds with the water that brush the islands' mountains. The ocean also moderates the islands' temperature extremes—Honolulu's record high of 31°C (88°F) is matched by a record low of only 13°C (57°F). The maritime influence tends to delay seasons on the islands a bit as well. Whereas most places in the United States have their highest temperatures in July and August and their lowest in January and February, highs in Hawaii occur in September and October and the yearly low may be delayed until early March. This lag reflects slower changes in the ocean's water temperatures.

The state has a definite tropical location. The latitude of Honolulu, about 20° N, is the same as Calcutta and Mexico City. As a result, there is little change in the length of daylight or the angle of incidence of the sun's rays from one season to another. This factor plus the state's maritime position means that there is little seasonal variation in temperature. Honolulu in January averages 22°C (72°F); the average July temperature is 26°C (78°F). It is variations in precipitation, not in temperature, that mark the major changes in season on the islands.

It has been suggested that Hawaii has only two seasons, a drier summer between May and October, and a moister winter from October to April. During the summer the islands are under the persistent influence of northeast trade winds. They approach Hawaii over cool waters located to the northeast and create characteristic Hawaiian weather—breezy, sunny with some clouds, warm but not hot. In winter, these trade winds disappear, sometimes for weeks, allowing "invasions" of storms from the north and northwest that bring greater amounts of moisture. These storms may bring the very heavy rainfall expected in tropical locations. Honolulu has received as much as 43 centimeters (17 inches) of rain in a single 24-hour period. Other Hawaiian weather stations have recorded 28 centimeters (11 inches) in an hour and 100 centimeters (40 inches) in a day, both of which rank near world records.

The topography of the islands creates extreme variations in precipitation from one location to another. As air masses are forced to rise up and over the islands' mountains, they cool, their moisture-carrying capacity is reduced, and orographic precipitation results in a process similar to that found along the shores of the North Pacific Coast (see Chapter 17). Most precipitation falls along the north and east side of the mountains as a result of the dominant direction of wind flow. Mount Waialeale, on Kauai, receives 1234 centimeters (486 inches) annually, making it one of the world's wettest spots, and Waimea, also on Kauai, receives about 50 centimeters (20 inches) annually—yet these two sites are only 25 kilometers (15 miles) apart. For another example, within the metropolitan area of Honolulu it is possible to live near the beach in a semiarid climate with less than 50 centimeters (20 inches) of rainfall annually or inland near Pali on the margins of a rainforest drenched by 300 centimeters (120 inches) of precipitation a year. Unlike the Pacific Northwest, the greatest precipitation on the higher mountains in Hawaii occurs at fairly low elevations, usually between 600 and 1200 meters (2000 to 4000 feet). Air masses, tending to take the easiest path, push around these isolated mountains instead of up

A substantial ranching economy is found on many of the drier portions of the islands, especially on the Big Island of Hawaii. Despite the state's small size, the pattern of concentrated private landholding results in the existence of a number of ranches each covering many thousands of acres. (Hawaii Visitors Bureau)

and over them. Hence, the lower northeastern flanks of Mauna Kea average well over 250 centimeters (100 inches) of precipitation yearly, with as much as 750 centimeters (295 inches) received in some areas; the opposite southwestern slopes average less than a tenth that amount. Leeward mountain slopes and leeward locations in general, are much drier than windward slopes, because the descending air warms and increases its moisture-carrying capacity.

Much of the volcanic soil is permeable. This allows water to percolate rapidly, draining beyond the reach of many plants. Thus, many areas of moderate to low precipitation are arid in appearance. Visitors to Hawaii are often surprised at the expanses of desert and semidesert that are present on most of the islands.

Biota

The isolation of the Hawaiian islands coupled with their generally temperate climate and great environmental variation has created a plant and bird community of vast diversity, the great majority of which are unique to the islands. There are several thousand plants native there and found naturally nowhere else, and 66 uniquely Hawaiian land birds have also been identified. Interestingly, there were no land mammals on the islands until humans arrived.

Unfortunately, island ecosystems are especially fragile, as numerous studies have shown. Many island species have developed in small, isolated locations in response to a peculiar ecological framework and to the absence of competition. These small niches can disappear easily under human influence; when they do, plant or animal species disappear with them. Of Hawaii's 66 unique birds, more than one-third are now extinct, and most of the rest are on the list of rare and endangered species. It is harder to determine the impact on the islands' plants. Still, by 1900, most native plant cover below 450 meters (1500 feet) had been destroyed. Much of that at higher elevations is also succumbing to human encroachment.

Simple human destruction may not be the primary way in which these native species are eliminated. Alien plants and animals introduced to the islands often do a far better job. For example, people brought with them rats, dogs, pigs, and goats. Rats took an especially heavy toll of native bird life by eating their eggs. Because few Hawaiian plants had thorns or any other protection from grazing animals, feral animals often did great damage to the vegetation. Also, plants by the hundreds were imported either intentionally or accidentally, and many native plants could not compete with them. Government attempts to protect endangered species have resulted in the creation of a number of sanctuaries on the islands, but the success of these efforts has not been great.

POPULATING THE ISLANDS

The Polynesians

The Polynesian settlement of Hawaii was a segment in one of humankind's most audacious periods of ocean voyaging. These people set out on repeated voyages in open canoes across broad oceanic expanses separating small island clusters. Settlers who came to Hawaii 1000 years ago, for example, are presumed to have come from the Marquesas, 4000 kilometers (2500 miles) to the southwest. There was some kind of pre-Polynesian population on the island, but it was probably absorbed by the newcomers. A second substantial wave of Polynesian migrants arrived 500 or 600 years ago.

The massive effort of these voyages, which matched anything accomplished by the Vikings, apparently became too great. As a result, Hawaii spent several hundred years in isolation after the second migration period. During the isolation, the Hawaiians solidified a complicated social organization in their insular paradise. Hereditary rulers held absolute sway over their populations and owned all of the land. By the late eighteenth century, when Europeans found the islands, the benign environment supported a large population that numbered about 300,000 natives.

Early European Impact

The first European to visit Hawaii—which he dubbed the Sandwich Islands—was Captain James Cook in 1778. Cook himself was killed on the shore of the Big Island, but news of his discovery spread rapidly after reaching Europe and North America; it was quickly recognized that the islands were the best location for a way station to exploit the trade developing between North America and Asia. In the 1820s, the whaling industry moved into the North Pacific and, for the next half-century, the islands became the principal rest and resupply center for whalers. About the same time, conservative Protestant missionaries came to the islands. Like most of the whalers, they were from New England. They were very successful in their missionary work, and for decades had a major influence on the islanders.

The effect of all of this on the Polynesians and their culture was catastrophic. It was an almost classic example of the negative consequences that can occur when a long-isolated culture comes into sudden contact with more broadly based outsiders. As Hawaiians accepted aspects of European culture, their balance with their own environment was disrupted. Famine, heretofore unknown, became a part of life for the native islanders. A whole series of infectious diseases, endemic among the visitors but previously unknown on the islands, came to Hawaii. Measles, leprosy, smallpox, syphilis, tuberculosis, and the like swept the islands. One outbreak of either cholera or bubonic plague reportedly halved the population in 1804.

Within three generations, the economic and social order of the Polynesian Hawaiians was destroyed. Their numbers, falling from 150,000 in 1804 to perhaps 75,000 in 1850, are fewer than 10,000 today (although state of Hawaii statistics indicate that an additional one-sixth of the state's population is part Polynesian).

The Asian Arrival

The first Hawaiian sugar plantation was established in 1837, although the islands did not become a substantial producer until after the middle of the century. Between then and the end of the century, Hawaii grew to the rank of a major world sugar exporter.

This development led to a great need for agricultural laborers. Native Hawaiians were used for a time, but they proved to be unwilling workers. Besides, their declining numbers provided nothing like the labor force needed. The plantation owners, who were Americans and Europeans, turned to Asia for the required work force. The first few hundred contract workers from China arrived in 1852. Japanese contract laborers began arriving in 1868; Filipinos in 1906. Between 1852 and 1930, 400,000 agricultural laborers, mostly Asian, were brought to Hawaii. Most were men, and many chose to leave the plantations at the end of their contract period; hence, it was necessary constantly to replenish the labor force.

In 1852, ethnic Hawaiians represented over 95 percent of the population of the islands. By 1900, they were less than 15 percent of the total, whereas nearly 75 percent were Oriental. Japanese were the largest single ethnic group, comprising more than 50 percent of the total population of just over 150,000.

Current Ethnic Patterns and the Second American Influence

After 1930, the mainland United States became the main source of new residents in Hawaii (Figure 19-2). In 1910, only about one resident of Hawaii in five was of European ancestry (referred to in Hawaii as Caucasian). By 1980, Caucasians outnumbered Japanese 222,000 to 198,000 in the state's nonmilitary population. Nearly 40 percent of the state's total current population is Caucasian or part-Caucasian.

Modifications in U.S. immigration laws during the 1960s resulted in a much larger Asian share of total immigration into the country.

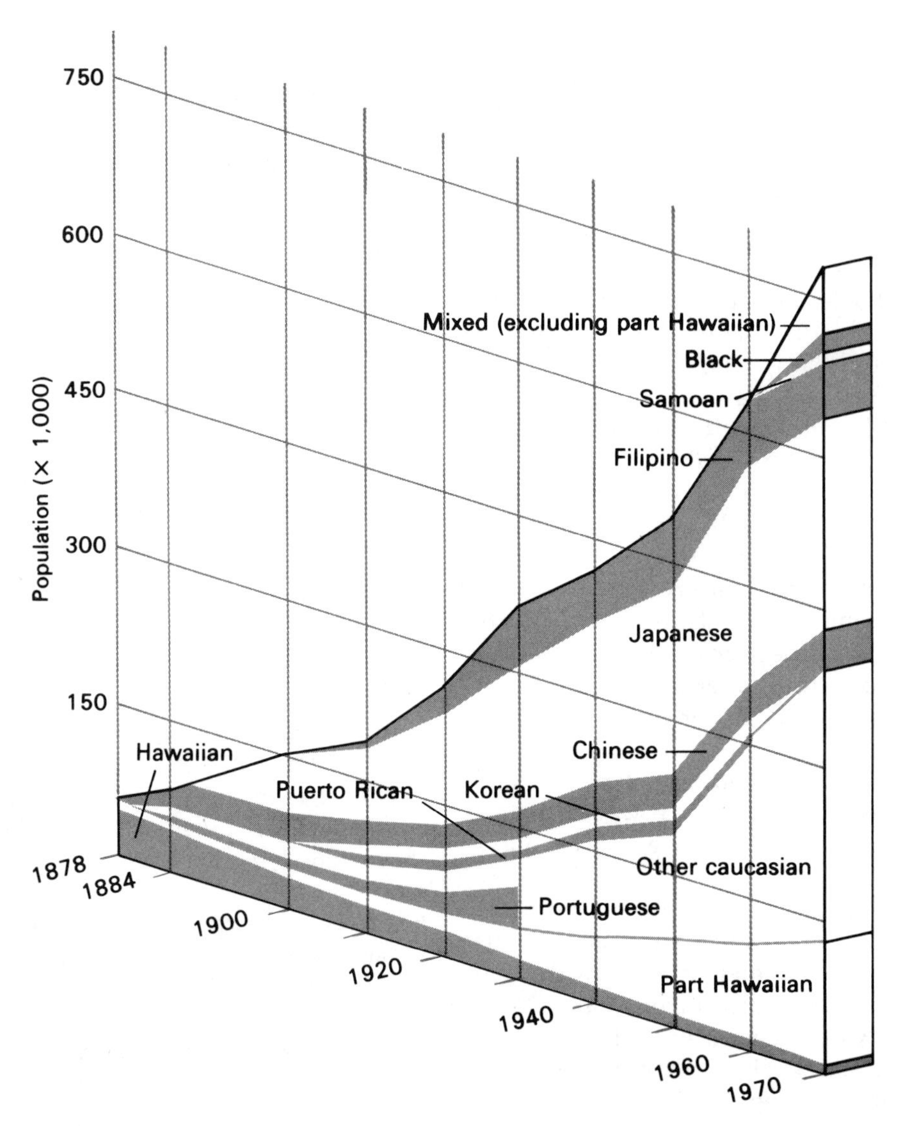

Figure 19-2 (From R. Warwick Armstrong, ed., *Atlas of Hawaii.* Honolulu: University of Hawaii Press, 1973.) Diversity has long been the dominant theme in Hawaiian ethnicity. Migration from the mainland U.S. resulted in a rapid increase in the Caucasian population after 1950.

Hawaii, close to Asia, has been an important destination for these migrants. About half of the new increase in the Hawaiian population resulting from immigration is currently due to Asian migrants, mostly from the Philippines. Net migration accounts for about 40 percent of the state's total annual population increase.

Implications of Ethnicity

Marriage between people of different ethnic groups is increasingly common in Hawaii. About 75 percent of the Caucasians still marry other Caucasians, and 60 percent of the Japanese still marry Japanese. For most of the other ethnic

The ethnic diversity of the Hawaiian population is captured in this crowd scene in Honolulu. (Werner Stoy/Camera Hawaii)

groups, about half or more of all marriages are outside the group. In mainland United States, social acceptance of such marriages has remained uncommon.

The Japanese and particularly the Chinese have done well in the Hawaiian economy. Educational standards among both groups have been high, and average Chinese income levels have surpassed Caucasian averages in recent decades. Newer immigrant groups, notable Filipinos, and native Hawaiians have comparatively low incomes and low-status occupations.

Although by no means a completely integrated society, Hawaiian residential integration easily surpasses that in other parts of the United States. No census tract anywhere in Hawaii is as much as 40 percent Chinese in population, and none is more than 70 percent Japanese. Within Honolulu, where most of the state's residents are found, a clear majority of tracts have a population mixture that includes at least 10 percent of each of the city's three major ethnic groups—Caucasians, Japanese, and Chinese. The poorer ethnic groups are less well integrated.

Population Distribution

The total population of Hawaii fell from its pre-European peak to a low of 54,000 in 1876 before beginning to grow again. By the early 1920s, the state's population had reached pre-European levels. Over 400,000 service personnel, stationed on the islands during World War II, temporarily pushed the total to about 850,000. In 1988, the state had 1.1 million residents. Because of immigration, Hawaii's annual rate of population growth is well above the national average.

The pre-European population was spread across the islands, with the Big Island occupied by the largest number of people. Since European discovery, the islands' population has been concentrated increasingly on Oahu (Table 18-1). Honolulu, with its fine harbor, became the principal port city. The other islands also lost in absolute terms. Their postcontact maximum of 165,000 was well below precontact estimates. That number had declined to 132,000 in 1960, although, by 1990, there had been a substantial increase to 272,000. The trend in outer islands population growth will probably continue as new tourist developments search out less crowded areas of the state.

POLITICAL INCORPORATION INTO THE UNITED STATES

The political history of Hawaii was turbulent during the 120 years after Cook's discovery. The various kingdoms of the islands were eliminated by a strong chief, Kamehameha, between 1785 and 1795. The missionaries' growing influence gradually made a sham of the authority of the Hawaiian rulers, and, during the nineteenth century, competing European political interests moved in to fill the resulting vacuum. From the 1820s to 1850s the French pushed their influence in the region. Much of the Hawaiian monarchy seemed to favor a relationship with Britain, however, and for a brief period in 1843 Britain annexed Hawaii. It was only the balance between the different powers operating in Hawaii that kept the islands independent during much of the nineteenth century.

The increasing role of Americans made it inevitable that, if it was to lose its political independence, Hawaii would be annexed by the United States. As American plantation owners increased in number and influence, their dissatisfaction with the Hawaiian government grew. In 1887, they forced the monarchy to accept an elected government controlled by the planters. The monarchy was overthrown completely in 1893, and the new revolutionary government immediately requested annexation by the United States. Initially refused, they were finally accepted as a territory in 1898.

Unlike earlier acquisitions, no provision was made at the time of annexation for the eventual admission of Hawaii to statehood. Despite repeated requests, Congress long refused to grant the islands the status of a state. The reasons for this steadfast refusal were varied and never really clear, but they likely centered around concerns over the allegiance of the Asian population of Hawaii and the long distance separating the islands from the mainland. Not until 1959, after Alaska was admitted as a state and all possible economic arguments against Hawaiian statehood were gone, did it become the fiftieth state.

THE HAWAIIAN ECONOMY

Landownership

Roughly half of all land in Hawaii is government owned. This in itself is not unusual, be-

Table 19-1 Percentage Distribution of the Hawaiian Population

	Date					
Location	1831	1878	1910	1940	1970	1990
Oahu	23	35	43	61	82	79
Honolulu	10	24	27	42	42	33
Rest of Oahu	13	11	16	19	40	42
Other islands	77	65	57	39	18	21

Source: R. Warwick Armstrong, ed., *Atlas of Hawaii* (Honolulu: University Press of Hawaii, 1973), p. 100; *1990 U.S. Census of Population.*

cause large proportions of all of the Western states and provinces are governmentally owned. What is unusual is that it is the state, not the federal government, that controls 80 percent of that land. Most of it is in the agriculturally less desirable portions of the islands, and the bulk is in forest reserves or conservation districts. Most federal lands are primarily in national parks on the Big Island and Maui or in military holdings on Oahu and Kahoolawe.

It is the unusual concentration of private lands in the hands of a relatively few major landowners that sets the state off from the rest of the country (Figure 19-3). Seven-eighths of all privately owned land in Hawaii is in the hands of only 39 owners; each owns 2000 hectares (5000 acres) or more. Six different landowners each control more than 40,000 hectares (100,000 acres) out of a state total of about 1,040,000 hectares (2,570,000 acres), and the largest, the Bernice P. Bishop estate, holds more than one-sixth of all the privately owned land in the state. Smaller unit ownership of private land is most extensive on Oahu, but even there the larger owners control more than two-thirds of all privately owned land. Two of the islands, Lanai and Niihau, are each nearly entirely controlled by a single owner, and on all of the other islands (except Oahu) major landowners control about 90 percent of all privately held property.

Figure 19-3 Patterns of land ownership on Oahu. (From R. Warwick Armstrong, *Atlas of Hawaii.* Honolulu: University of Hawaii Press, 1973.) Most of the land area is held either by the state or by a very small number of large land owners. One result is that land prices in Hawaii are extremely high.

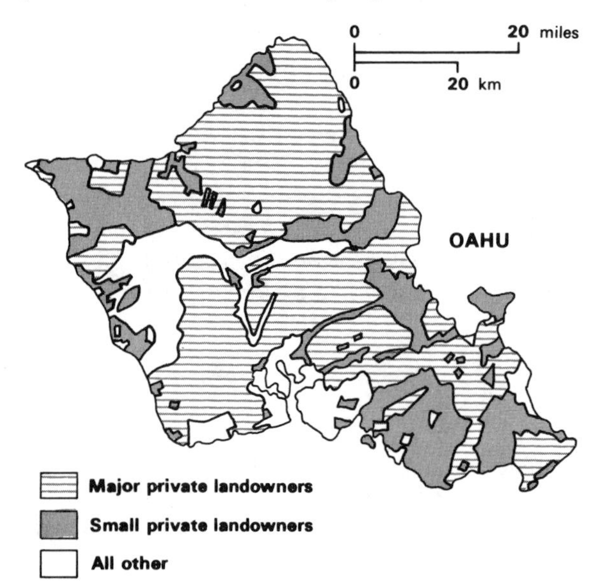

Most of these large landholdings were created during the nineteenth-century period of freewheeling exploitation on the islands. Land had previously been held entirely by the monarchies, not by commoners. This land passed into the hands of non-Hawaiian private owners during the political decline of the monarchy. With the deaths of the early owners, most estates have been given over to trusts to administer rather than passing directly to heirs. This has made it difficult to break up the ownership patterns.

This concentration of landownership represents a major problem for the state, especially in urban areas. The landowners have leased land for urban use. As the leases expire, they can be renewed at a much higher rate, because the urban development provided by the lessors has greatly increased the value of the land. The security of landownership, so long a part of the American psyche, is, thus, effectively denied to many in Hawaii. Where smaller holdings are available and where population growth is rapid, most notably on Oahu, this has led to a tremendous increase in land prices.

In addition to high land values, other sometimes less direct problems result from the landownership pattern. High land costs have helped create pockets of high population density, again particularly on Oahu. Many major investors are unwilling to build on land they do not own.

Hawaii attempted to moderate the impact of large landholdings through the Land Reform Act of 1967. This act extended the state's power of eminent domain by allowing Hawaii's housing authority to condemn individual parcels of land, usually at the time of lease renewal. Title would be transferred to the tenants at a fair market price determined by arbitration or negotiation. The law was declared unconstitutional by lower courts, but in 1984 the U.S. Supreme Court overturned that ruling, thus paving the way for change.

Agriculture

Sugar, and later pineapples, fueled the Hawaiian economy for many decades after the 1860s. The economy remained primarily agricultural until World War II. In recent decades agriculture has continued to show modest gains in income, but its relative importance has declined greatly. Sugar's share of all income generated in the state fell from 20 percent in 1950 to 4 percent in 1985. The income from pineapples dropped from 16.5 to 2 percent of the total. Even diversified agriculture, such as livestock raising on the island of Hawaii, which has been the prime growth component of the agricultural industry, had its income share decline from 4.6 to about 2 percent. Only 1 Hawaiian worker in 30 is currently employed in agriculture. Once the kingpin of the Hawaiian economy, agriculture today ranks no better than fourth or fifth.

In some ways this emphasis on declining agricultural importance is misleading. First, even with the move of some plantations out of production in the face of declining profits, Hawaii continues to provide a substantial share of the world's sugar harvest. Hawaii once controlled most of the world's pineapple production but lost part of its market after World War II to newer producing areas; however, production has recently increased somewhat and is about 650,000 tons annually. Hawaii remains the world's largest supplier of pineapples, although both acreage and employment are less than half those of the peak years of production.

Gross economic statistics for the state overwhelmingly emphasize the position of Oahu, where today more than 80 percent of the state's entire economy is concentrated. The role of agriculture remains great on the other islands. Lanai depends on pineapples for much of its employment and income. Livestock and sugar form the backbone of the economy on the Big Island, as does sugar and pineapples on Maui and Kauai. In short, only on Oahu has agriculture been replaced as the most important sector of the economy.

The Federal Government

As agriculture declined and lost its dominance over the Hawaiian economy, its place was first taken by the federal government. Over the past several decades, governmental expenditures have increased at a rate roughly comparable to the growth of the total economy, maintaining about a one-third share of all expenditures. Most of this has come from the military. The strategic value of Hawaii's location in the mid-Pacific is obvious. It is the headquarters for the U.S. Pacific Command, center of Pacific operations for all branches of the armed forces, and home of a number of major military bases. The military controls almost 25 percent of Oahu, including the land around Pearl Harbor, one of the finest natural harbors in the Pacific. Nearly one Hawaiian worker in four is an employee of the military. Military personnel and their dependents together represent over 10 percent of Hawaii's population. The armed forces are also the largest civilian employer in the state.

The armed forces have long been a substantial part of the Hawaiian community, and they are generally well received. Still, the heavy dependence on the military is often viewed as unfortunate, especially when periods of military cutbacks affect the local economy adversely. The relative importance of the military is likely to decline somewhat in the future, although it will almost surely remain a major presence.

Tourism

For decades, Hawaii represented an exotic, distant tropical paradise to most Americans. Many dreamed of a visit to its balmy shores. The airplane and cheap fares turned that dream into reality. The first regularly scheduled transpacific

Hawaii's apparently urban economy is largely a function of Honolulu's domination of the state's economic structure. Urban functions, other than tourist services, remain relatively unimportant on all the other islands. These pineapple fields are on the island of Maui. (Arnold Nowotny/USDA Soil Conservation Service)

flights began in 1936. As late as 1951, however, nearly half of the 52,000 annual visitors to Hawaii reached there by ship. Then came a period of larger airplanes, lower fares, and a booming national economy. The number of visitors increased sixfold in every decade, to 320,000 in 1965 and 1,819,000 in 1971. Now over 4.5 million people (most of them tourists) visit the state each year. Hawaii's government and businesses encourages these visits through the Hawaii Visitors Bureau, which widely advertises the tropical and exotic image of the islands. Tourism has become the principal growth sector of the economy, increasing its share of total island income from 4 percent in 1950 to 18 percent in 1970 and over 30 percent today.

To many Hawaiians, this influx of tourists is a mixed blessing. In 1952, only 2800 hotel units existed on the islands. Now, there are over 60,000. Over half of the existing units are on Oahu, with most crowded along the famous beaches of the Waikiki-Kahala coast in Honolulu. On a typical day 60,000 tourists flock to the high-rise hotels and, increasingly, rental condo-

These two photographs, both of Waikiki Beach and Diamond Head in Honolulu, were taken approximately 25 years apart, the first in the early 1950s, the second in the mid-1970s. Comparatively inexpensive air flights form the mainland opened Hawaii to the mass of American tourists. For most, Waikiki was the principal destination. (State of Hawaii Public Archives)

minium units along the beach, and another 50,000 fan out across the rest of the state. Diamond Head, long the city's most famous landmark, now has hotels shrouding its lower elevations. The congestion, decline in scenic quality, and possible pollution problems resulting from this crowded development are of great concern to many. Recent expansion has focused on the outer islands, and there is fear that without careful control these problems will spread throughout the state.

A second problem with the tourist economy is its unstable nature. The downturn in the national economy in the early and mid-1970s had a heavy impact on the Hawaiian tourist industry because many Americans chose to economize by cutting travel costs with a vacation closer to home. Also, although Hawaii has striven to de-

velop a year-round tourist flow, the summer months remain the peak period; this results in higher unemployment during the winter.

Most visitors to Hawaii come from the mainland of the United States, especially the West Coast. Perhaps a quarter of all tourists now come from Asia, mostly Japan. These latter visitors are causing some local concern. In an attempt to control the total tour package and, thus, increase profits, Japanese tour operations have purchased a number of hotels and other tourist facilities in Hawaii. Other Japanese firms, simply looking for a good investment, have bought a variety of Hawaiian businesses. Golfing packaged tours are popular with many Japanese tourists. The sudden development of many new golf courses has made Hawaii decide to limit the number of new courses that can be constructed each year. All of this has caused some Hawaiians to urge control of such sales to foreigners, arguing that the state's economy could become Japanese controlled. Perhaps they can appreciate better the native Hawaiian reaction to the American influx of a century ago. Japanese tourists typically spend more than twice as much daily as their American counterparts, and spend far more for gift and souvenir items. Thus, the growth in Asian tourism is especially important to Hawaii.

Transportation

The insular nature of Hawaii creates a transport situation far different from that of the mainland United States. First, it is entirely dependent on air and sea movement; overland transport is of only local importance. Nearly all visitors arrive by air. Hawaii imports much of its food, nearly all of its energy supplies and vehicles, and a host of other commodities. Its agricultural exports still amount to over 10 million tons annually.

HONOLULU

Honolulu dominates Hawaii. Nearly 80 percent of the state's population and a matching percentage of its economy are concentrated in the city and its suburbs. The second largest city outside Oahu, Hilo on Hawaii, although increasing its population at a more rapid rate than Honolulu, has only one-twentieth Honolulu's metropolitan area population. No other state, no other region in this textbook, is so totally dominated by a single urban center.

Mountains, the ocean, and federally held land hug the city, crowding it into a series of narrow valleys and a fringe along the coast. Recent construction of two highways and tunnels have increased the city's connections with the northeast coast of the island (the Kilua-Kaneohe urbanized area on that coast has grown to over 100,000), and construction is also increasing along Oahu's central valley.

The problems of Honolulu are similar to those of other American cities, although somewhat magnified. Living costs are high, perhaps 20 percent above the national average. Among U.S. cities, only Anchorage, Alaska, matches the expense of living in Honolulu. Housing costs are also high. With so much land held by estates, residential land for sale is scarce. One result is that about 40 percent of the population lives in owner-occupied housing. Housing is noticeabley inadequate for low-income groups. The arrangement of the city, the many barriers provided by physical features and federal land, and the general high density of occupation all inhibit traffic flow and increase congestion. The city's leeward location, while providing a generally dry climate that encourages tourism, leads to occasional periods of little wind, and a genuine atmospheric pollution problem is evident.

All of this does not particularly damn Honolulu as a city. Instead, it emphasizes that it is a city with a set of problems and attractions much like that of any other American city. It may be in paradise, but it is a large city in paradise.

INTER-ISLAND DIVERSTIY

The major Hawaiian islands are part of the same state, they have similar geologic histories, and they are closely spaced in a vast ocean, yet each tends to have its own character. Oahu is densely populated and intensely used, and it offers a view of bustle and confusion common to urban America. The island of Hawaii, the Big Island, has by comparison an air of relative space and distance, with large ranches, high, barren volcanoes, and large stretches of almost treeless land. Its land area is dominated by five huge shield volcanoes. Mauna Loa alone covers half the island, and Mauna Kea almost another quarter. Sugar, cattle ranching, and tourism are its major industries. (see Table 19-2).

Table 19-2 Hawaiian Place Names

Common Name	Hawaiian Name	Meaning
Hawaii	Hawai'i	
Hiilawe	Hi'lawe	Lift (and) carry
Honolulu		Protected bay
Kahoolawe	Kaho'olawe	The carrying away
Kauai	Kaua'i	
Kilauea	Kilauea	Spewing, much spreading
Lanai	Lana'i	Day (of) conquest
Mauna Kea		White mountain
Mauna Loa		Long Mountain
Molokai	Moloka'i	
Niihau	Ni'ihau	
Oahu	O'ahu	
Pali		Cliff
Waialeale	Wai'ale'ale	Overflowing water
Waikiki	Waikiki	Sprouting water
Waimea		Reddish water
Waipio	Waipi'o	Curved water

Source: Joseph R. Morgan, *Hawaii* (Boulder, Colo.: Westview Press, 1983).

Kauai is heavily eroded into a spectacular scenery of mountains, canyons, cliffs, and waterfalls. As the northernmost of the islands, it is more exposed to frontal storms that travel eastward across the Pacific Ocean. Thus, it can be drenched by orographic precipitation from the trade winds as well as frontal precipitation. It is sometimes called the garden isle because of its lush tropical vegetation. Kauai is becomming increasingly popular with tourists because of its dramatic physical environment. Neigboring Niihau sits in the lea of the larger Kauai and is shielded from the prevailing trade winds. The result is a much drier climate. Niihau is privately owned and is operated as the Niihau Ranch Company. Most of its few hundred residents are native Hawaiians.

Maui, the second largest of the islands, offers a contrast between the plantations of its central lowlands and the rugged mountains to either side. Tourist development, concentrated along the western coastal strip, has been intense, with the result that Maui had the most rapid rate of population increase of any of the islands in the 1970s and 1980s. Still, much of the rest of the island remains little changed and sparsely populated.

Molokai is half ranchland and half rugged mountains. Its north coast is dominated by spectacular sea cliffs, while the south shore is a broad coastal plain. It is perhaps the least economically developed of the populated Hawaiian Islands. The island economy was dominated by pineapple production until the Del Monte pineapple plantation closed in 1983.

Lanai and Kahoolawe are both in the lea of much higher Maui. As a result, both are dry. Neither have any permanent streams. Pineapple production is the only important economic activity on Lanai. The U.S. navy administers Kahoolawe, and uses it for military exercises much to the dismay of many Hawaiians.

An Island Paradise?

It has become fashionable to talk of trouble in paradise when discussing Hawaii. Certainly the state has its share of problems. High living costs and high unemployment rates will continue. The combined impact of the flood of tourists, of economic development, of change of many kinds in its physical environment has Hawaii groping for a solution. The economy is too closely tied to tourism and the military, as both are given to rapid fluctuation resulting from recession or changing military spending programs.

Still, these should not overshadow Hawaii's positive attributes. No part of America, and few parts of the world, can match its level of racial and ethnic assimilation. All people are not treated equally in Hawaii, but they are far closer to that goal than is most of the rest of the world. Much of Hawaii remains a place of great beauty. Its climate is often superlative. Few other states in the country can match Hawaii's attempts at controlling growth to maintain the quality of their environment. Hawaii passed America's first statewide land-use program, and it has a statewide sign ordinance. Concern for tourist-oriented development on the outer islands has exerted a strong counterbalance to rapid growth. For many, Hawaii will remain an island paradise.

ADDITIONAL READINGS

Armstrong, R. Warwick (ed.). *Atlas of Hawaii,* 2nd ed. Honolulu: University Press of Hawaii, 1983.

Carlquist, Sherwin J. *Hawaii: A Natural History,* 2nd ed. Honolulu: Pacific Tropical Botanical Garden, 1980.

Farrell, Bryan. *Hawaii: The Legend That Sells.* Honolulu: University Press of Hawaii, 1980.

Glick, Clarence E. *Sojourners and Settlers: Chinese Migrants in Hawaii.* Honolulu: University of Hawaii Press, 1980.

Hazama, Dorothy O., and Jane O. Komeiji. *'Okage Sama De': The Japanese in Hawaii.* Honolulu: Bess Press, 1986.

Kent, Noel J. *Hawaii: Inlands Under the Influence.* New York: Monthly Review Press, 1983.

Lind, Andrew W. *Hawaii's People,* 4th ed. Honolulu: University Press of Hawaii, 1980.

Morgan, Joseph. *Hawaii: A Geography.* Boulder, Colo.: Westview Press, 1983.

Glossary

Absolute Humidity the mass of water vapor in the atmosphere per unit of volume of space.

Accessibility a locational characteristic that permits a place to be reached by the efforts of those at other places.

Accessibility Resource a naturally occurring landscape feature that facilitates interaction between places.

Acid Rain rain that has become more acidic than normal (a pH below 5.0) as certain oxides present as airborne pollutants are absorbed by the water droplets. The term is often applied generically to all acidic precipitation.

Air Mass a very large body of atmosphere defined by essentially similar horizontal air temperatures. Moisture conditions are also usually similar throughout the mass.

Alluvial Soils soils deposited through the action of moving water. These soils lack horizons and are usually highly fertile.

Antebellum before the war; in the United States, belonging to the period immediately prior to the Civil War.

Arete a sharp, narrow mountain ridge. It often results from the erosive activity of alpine glaciers flowing in adjacent valleys.

Arroyo a deep gully cut by a stream that flows only part of the year; a dry gulch. A term normally used only in desert areas

Badlands very irregular topography resulting from wind and water erosion of sedimentary rock.

Base Level the lowest level to which a stream can erode its bed. The ultimate base level of all streams is, of course, the sea.

Batholith a very large body of igneous rock, usually granite, that has been exposed by erosion of the overlying rock.

Bedrock the solid rock that underlies all soil or other loose material; the rock material that breaks down to eventually form soil.

Bilingual the ability to use either one of two languages, especially when speaking.

Biological Diversity a concept recognizing the variety of life forms in an area of the earth and the ecological interdependence of these life forms.

Biota the animal and plant life of a region considered as a total ecological entity.

Boll Weevil a small grayish beetle of Mexico and the southeastern United States with destructive larvae that hatch in and damage cotton bolls.

Break-in-Bulk Point commonly, a transfer point on a transport route where the mode of transport (or type of carrier) changes and where large-volume shipments are reduced in size. For example, goods may be unloaded from a ship and transferred to trucks at an ocean port.

Caprock a strata of erosion-resistant sedimentary rock (usually limestone) found in arid areas. Caprock forms the top layer of most mesas and buttes.

Carrying Capacity the number of people that can be supported in an area given the quality of the natural environment and the level of technology of the population.

CBD the central business district of an urban area, typically containing an intense concentration of office and retail activities.

Chinook a warm, dry wind experienced along the eastern side of the Rocky Mountains in Canada and the United States. Most common in winter and spring, it can result in a rise in temperature of 20° C (35° to 40° F) in a quarter of an hour.

Climax Vegetation the vegetation that would exist in an area if growth had proceeded undisturbed for an extended period. This would be the "final" collection of plant types that presumably would remain forever, or until the stable conditions were somehow disturbed.

Confluence the place at which two streams flow together to form one larger stream.

Coniferous bearing cones; from the conifer family.

Continental Climate the type of climate found in the interior of the major continents in the middle, or temperate, latitudes. Characterized by a great seasonal variation in temperatures, four distinct seasons, and a relatively small annual precipitation.

Continental Divide the line of high ground that separates the oceanic drainage basins of a continent; the river systems of a continent on opposite sides of a continental divide flow toward different oceans.

Coulee a dry canyon eroded by Pleistocene floods that cut into the lava beds of the Columbia Plateau.

Conurbation an extensive urban area formed when two or more cities, originally separate, coalesce to form a continuous metropolitan region.

Core Area the portion of a country that contains its economic, political, intellectual, and cultural focus. It is often the center of creativity and change (see **Hearth**).

Crop-lien System a farm financing scheme whereby money is loaned at the beginning of a growing season to pay for farming operations with the subsequent harvest used as collateral for the loan.

Culture the accumulated habits, attitudes, and beliefs of a group of people that define for them their general behavior and way of life; the total set of learned activities of a people.

Culture Hearth the area from which the culture of a group diffused (see **Hearth**).

Cut-and-Sew Industry the manufacture of basic ready-to-wear clothing. Such facilities usually have a small fixed investment in the manufacturing facility.

Deciduous Forest forests where the trees lose their leaves each year.

De Facto Segregation the spatial and social separation of populations that occurs without legal sanction.

Degree Day deviation of one degree temperature for one day from an arbitrary standard, usually the long-term average temperature for a place.

De Jure Segregation the spatial and social separation of populations that occurs as a consequence of legal measures.

Demography the systematic analysis of population.

Discriminatory Shipping Rates a transportation charge levied in a manner that is inequitable to some shippers, primarily because of those shippers' location.

Dome an uplifted area of sedimentary rocks with a downward dip in all directions. Often caused by molten rock material pushing upward from below. The sediments have often eroded away, exposing the rocks that resulted when the molten material cooled.

Dry Farming a type of farming practiced in semiarid or dry grassland areas without irrigation using such approaches as fallowing, maintaining a finely broken surface, and growing drought-tolerant crops.

Economies of Agglomeration the economic advantages that accrue to an activity by locating close to other activities; benefits that follow from complementarity or shared public services.

Economies of Scale savings achieved in the cost of production by larger enterprises because the cost of initial investment can be defrayed across a greater number of producing units.

Emergent Coastline a shoreline resulting from a rise in land surface elevation relative to sea level.

Enclave a tract or territory enclosed within another state or country.

Erratic a boulder that has been carried from its source by a glacier and deposited as the glacier melted. Thus, the boulder is often of a different rock type from surrounding types.

Estuary the broad lower course of a river that is encroached on by the sea and affected by the tides.

Evapotranspiration the water lost from an area through the combined effects of evaporation from the ground surface and transpiration from the vegetation.

Exotic Stream a stream found in an area that is too dry to have spawned such a flow. The flow originates in some moister section.

Extended Family a family that includes three or more generations. Normally, that would include grandparents, their sons or daughters, and their children, as opposed to a "nuclear family", which is only the married couple and their offspring.

Fall Line the physiographic border between the piedmont and coastal plain regions. The name derives from the river rapids and falls that occur as the water flows from hard rocks of the higher piedmont onto the softer rocks of the coastal plain.

Fallow agricultural land that is plowed or tilled but left unseeded during a growing season. Fallowing is usually done to conserve moisture.

Fault Block Mountain a mountain mass created either by the uplift of land between faults or the subsidence of land outside the faults.

Fault Zone a fault is a fracture in the earth's crust along which movement has occurred. The movement may be in any direction and involve material on either or both sides of the fracture. A "fault zone" is an area of numerous fractures.

Federation a form of government in which powers and functions are divided between a central government and a number of political subdivisions that have a significant degree of political autonomy.

Feral Animal a wild or untamed animal, especially one having reverted to such a state from domestication.

Fish Ladder a series of shallow steps down which water is allowed to flow; designed to permit salmon to circumvent artificial barriers such as power dams as the salmon swim upstream to spawn.

Focality the characteristic of a place that follows from its interconnections with more than one other place. When interaction within a region comes together at a place (i.e., when the movement focuses on that location), the place is said to possess "focality".

Functional Diversity the characteristic of a place where a variety of different activities (economic, political, social) occur; most often associated with urban places.

Geomorphology the study of the arrangement and form of the earth's crust and of the relationship between these physical features and the geologic structures beneath.

Ghetto originally, the section of a European city to which Jews were restricted. Today, commonly defined as a section of a city occupied by members of a minority group who live there because of social restrictions on their residential choice.

Glacial Till the mass of rocks and finely ground material carried by a glacier, then deposited when the ice melted. Creates an unstratified material of varying composition.

Great Circle Route the shortest distance between two places on the earth's surface. The route follows a line described by the intersection of the surface with an imaginary plane passing through the earth's center.

Hazardous Waste unwanted by-products remaining in the environment and posing immediate or potential hazard to human life.

Hearth the source area of any innovation. The source area from which an idea, crop, artifact, or good is diffused to other areas.

Heavy Industry manufacturing activities engaged in the conversion of large volumes of raw materials and partially processed materials into products of higher value; hallmarks of this form of industry are considerable capital investment in large machinery, heavy energy consumption, and final products of relatively low value per unit weight (see **Light Industry**).

Hinterland the area tributary to a place and linked to that place through lines of exchange, or interaction.

Horizon a distinct layer of soil encountered in vertical section.

Humus partially decomposed organic soil material.

Hydrography the study of the surface waters of the earth.

Hydroponics the growing of plants, especially vegetables, in water containing essential mineral nutrients rather than in soil.

Igneous Rock rock formed when molten (melted) materials harden.

Indentured Labor work performed according to a binding contract between two parties. During the early colonial period in America, this often involved long periods of time and a total work commitment.

Inertia Costs of Location costs born by an activity because it remains located at its original site, even though the distributions of supply and demand have changed.

Insular either of an island, or suggestive of the isolated condition of an island.

Intervening Opportunity the existence of a closer, less expensive opportunity for obtaining a good or service, or for a migration destination. Such opportunities lessen the attractiveness of more distant places.

Intracoastal Waterway System a waterway channel, maintained through dredging and sheltered for the most part by a series of linear offshore islands, that extends from New York City to Florida's southern tip and from Brownsville, Texas, to the eastern end of Florida's panhandle.

Isohyet a line on a map connecting points that receive equal precipitation.

Jurisdiction the right and power to apply the law; the territorial range of legal authority or control.

Karst an area possessing surface topography resulting from the underground solution of subsurface limestone or dolomite.

Kudzu a vine, native to China and Japan but imported into the United States; originally planted for decoration, for forage, or as a ground cover to control erosion. It now grows wild in many parts of the southeastern United States.

La Belle Province the former nickname of the province of Quebec, still commonly used.

Lacustrine Plain a nearly level land area that was formed as a lakebed.

Latitude a measure of distance north or south of the equator. One degree of latitude equals approximately 110 kilometers (69 miles).

Leaching a process of soil nutrient removal through the erosive movement and chemical action of water.

Legume a plant, such as the soybean, which bears nitrogen-fixing bacteria on its roots, and thereby increases soil nitrogen content.

Life-Cycle Stage a period of uneven length in which the relative dependence of an individual on others helps define a complex of basic social relations that remains relatively consistent throughout the period.

Light Industry manufacturing activities that use moderate amounts of partially processed materials to produce items of relatively high value per unit weight (see **Heavy Industry**).

Lignite a low-grade, brownish coal of relatively poor heat-generating capacity.

Loess a soil made up of small particles that were transported by the wind to their present location.

Longitude a measure of distance east and west of a line drawn between the North and South poles and passing through the Royal Observatory at Greenwich, England.

Maritime Climate a climate strongly influenced by an oceanic environment, found on islands and the windward shores of continents. It is characterized by small daily and yearly temperature ranges and high relative humidity.

Mediterranean Climate a climate characterized by moist, mild winters and hot, dry summers.

Metamorphic Rock rock that has been physically altered by heat and/or pressure.

Metes and Bounds a system of land survey that defines land parcels according to visible natural landscape features and distance. The resultant field pattern is usually very irregular in shape.

Metropolitan Coalescence the merging of the urbanized areas of separate metropolitan regions; Megalopolis is an example of this process.

Monadnock an isolated hill or mountain of resistant rock rising above an eroded lowland.

Moraine the rocks and soil carried and deposited by a glacier. An "end moraine," either a ridge or low hill running perpendicular to the direction of ice movement, forms at the end of a glacier when the ice is melting.

MSA Metropolitan Statistical Area. A statistical unit of one or more counties that focus on one or more central cities larger than a specified size, or with a total population larger than a specified size. A reflection of urbanization.

Multilingual the ability to use more than one language when speaking or writing (see **Bilingual**). This term often refers to the presence of more than two populations of significant size within a single political unit, each group speaking a different language as their primary language.

Municipal Waste unwanted by-products of modern life generated by people living in an urban area.

Nuclear Family see **Extended Family**.

Open Range a cattle or sheep ranching area characterized by a general absence of fences.

Orographic Rainfall precipitation that results when moist air is lifted over a topographic barrier such as a mountain range.

Outwash rocky and sandy surface material deposited by meltwater that flowed from a glacier.

Overburden material covering a mineral seam or bed

that must be removed before the mineral can be removed in strip mining.

Permafrost a permanently frozen layer of soil.

Physiographic Region a portion of the earth's surface with a basically common topography and common morphology.

Plural Society a situation in which two or more culture groups occupy the same territory but maintain their separate cultural identities.

Plate Tectonics geologic theory that the bending (folding) and breaking (faulting) of the solid surface of the earth results from the slow movement of large sections (plates) of that surface.

Platted Land land that has been divided into surveyed lots.

Pleistocene period in geologic history (basically the last 1 million years) when ice sheets covered large sections of the earth's land surface not now covered by glaciers.

Polynodal many-centered.

Postindustrial an economy that gains its basic character from economic activities developed primarily after manufacturing grew to predominance. Most notable would be quaternary economic patterns.

Precambrian Rock the oldest rocks, generally more than 600 million years old.

Presidio a military post. Spanish.

Primary Product a product that is important as a raw material in developed economies; a product consumed in its primary (i.e., unprocessed) state (see **Staple Product**).

Primary Sector that portion of a region's economy devoted to the extraction of basic materials (e.g., mining, lumbering, agriculture).

Pueblo a type of Indian village constructed by some tribes in the southwestern United States. A large community dwelling, divided into many rooms, up to five stories high, and usually made of adobe. Also, a Spanish word for town or village.

Quaternary Sector that portion of a region's economy devoted to informational and idea-generating activities (e.g., basic research, universities and colleges, and news media).

Rail Gauge the distance between the two rails of a railroad.

Rainshadow an area of diminished precipitation on the lee (downwind) side of a mountain or mountain range.

Rangs lines of long, narrow fields in French Canada, created to maximize access to rivers or roads for transportation.

Region an area having some characteristic or characteristics that distinguish it from other areas. A territory of interest to people and for which one or more distinctive traits are used as the basis for its identity.

Resource anything that is both naturally occurring and of use to humans.

Riparian Rights the rights of water use possessed by a person owning land containing or bordering a water course or lake.

Riverine located on or inhabiting the banks or the area near a river or lake.

Scarp (also **Escarpment**) A steep cliff or steep slope, formed either as a result of faulting or by erosion or inclined rock strata.

Scots-Irish The North American descendants of Protestants from Scotland who migrated to northern Ireland in the 1600s.

Secondary Sector that portion of a region's economy devoted to the processing of basic materials extracted by the primary sector.

Second Home a seasonally occupied dwelling that is not the primary resident of the owner. Such residences are usually found in areas with substantial opportunities for recreation or tourist activity.

Sedimentary Rock rock formed by the hardening of material deposited in some process; most commonly sandstone, shale, and limestone.

Seigneuries large land grants in French Canada possessing feudal privileges that were awarded by the kings of France to noblemen and to the Church. These lands were then parceled out to individual settlers.

Sharecropping a form of agricultural tenancy where the tenant pays for use of the land with a predetermined share of his or her crop rather than with a cash rent.

Shield a broad area of very old rocks above sea level. Usually characterized by thin, poor soils and low population densities.

Silage fodder (livestock feed) prepared by storing and fermenting green forage plants in a silo.

Silo usually a tall, cylindrical structure in which fodder (livestock feed) is stored; may be a pit dug for the same purpose.

Sinkhole crater formed when the roof of a cavern collapses; usually found in areas of limestone rock.

Smog mixture of particulate matter and chemical pollutants in the lower atmosphere, usually over urban areas.

Site features of a place related to the immediate environment on which the place is located (e.g., terrain, soil, subsurface geology, ground water).

Situation features of a place related to its location relative to other places (e.g., accessibility, hinterland quality).

Soluble capable of being dissolved; in this case, the characteristic of soil minerals that leads them to be carried away in solution by water (see **Leaching**).

Space Economy the locational pattern of economic activities and their interconnecting linkages.

Spatial Complementarity the occurrence of location pairing such that items demanded by one place can be supplied by another.

Spatial Interaction movement between locationally separate places.

Staple Product a product that becomes a major component in trade because it is in steady demand; thus, a product that is basic to the economies of one or more major consuming populations (see **Primary Product**).

Sustainable Yield the amount of a naturally self-reproducing community, such as trees or fish, that can be harvested without diminishing the ability of the community to sustain itself.

Temperature Inversion an increase in temperature with height above the earth's surface, a reversal of the normal pattern.

Territory a specific area or portion of the earth's surface, not to be confused with "region."

Tertiary Sector that portion of a region's economy devoted to service activities (e.g., transportation, retail and wholesale operations, insurance).

Threshold the minimum-sized market for an economic activity. The activity will not be successful until it can reach a population larger than this threshold size.

Time-distance a time measure of how far apart places are (how long does it take to travel from place A to place B?); this may be contrasted with other distance metrics such as geographic distance (how far is it?) and cost distance (how much will it cost to get there?).

Township and Range the rectangular system of land subdivision of much of the agriculturally settled United States west of the Appalachians; established by the Land Ordinance of 1785.

Transferability the extent to which a good or service can be moved from one location to another; the relative capacity for spatial interaction.

Transhumance the seasonal movement of people and animals in search of pasture. Commonly, winters are spent in snow-free lowlands and summers in the cooler uplands.

Tree Line either the latitudinal or elevational limit of normal tree growth. Beyond this limit, closer to the poles or at at higher or lower elevations, climatic conditions are too severe for such growth.

Tropics technically, the area between the Tropic of Cancer (21 1/2° north latitude) and the Tropic of Capricorn (21 1/2 ° south latitude), characterized by the absence of a cold season. Often used to describe any area possessing what is considered to be a hot, humid climate.

Underemployment a condition among a labor force such that a portion of the labor force could be eliminated without reducing the total output. Some individuals are working less than they are able or want to, or they are engaged in tasks that are not entirely productive.

Underpopulation economically, a situation where an increase in the size of the labor force will result in an increase in per worker productivity.

Water Table the level below the land surface at which the subsurface material is fully saturated with water. The depth of the water table reflects the minimum level to which wells must be drilled for water extraction.

Zoning the public regulation of land and building use to control the character of a place. Surface material is fully saturated with water.

Map Appendix

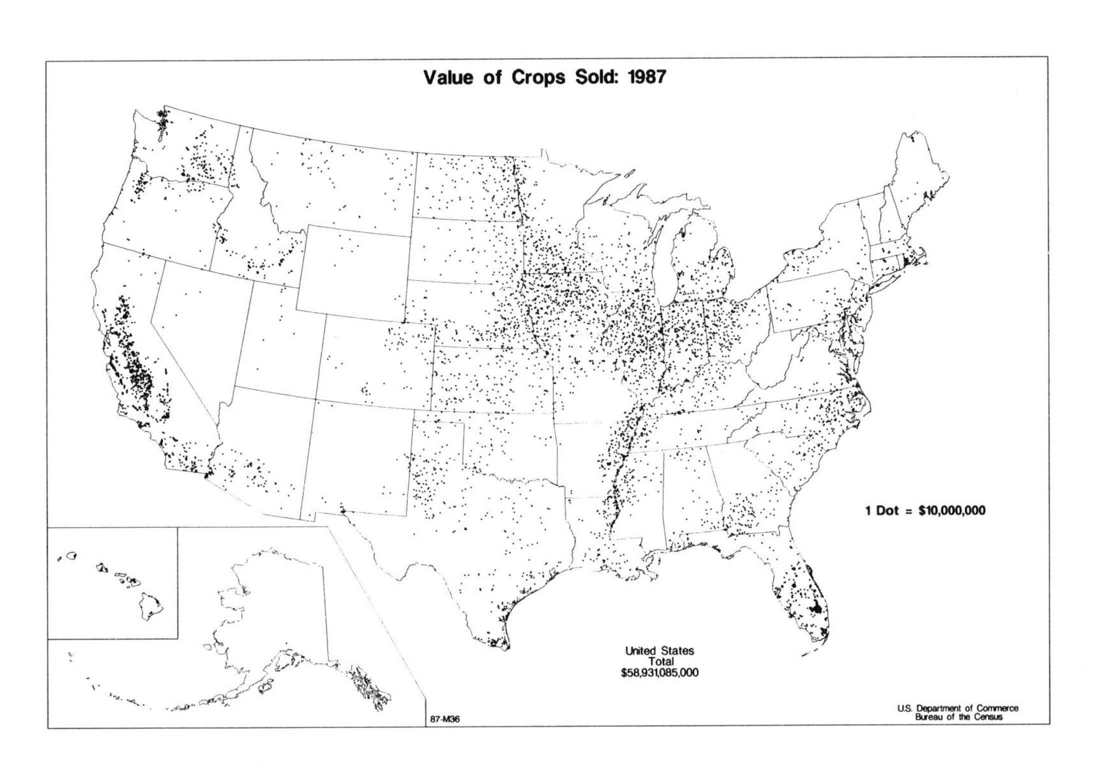

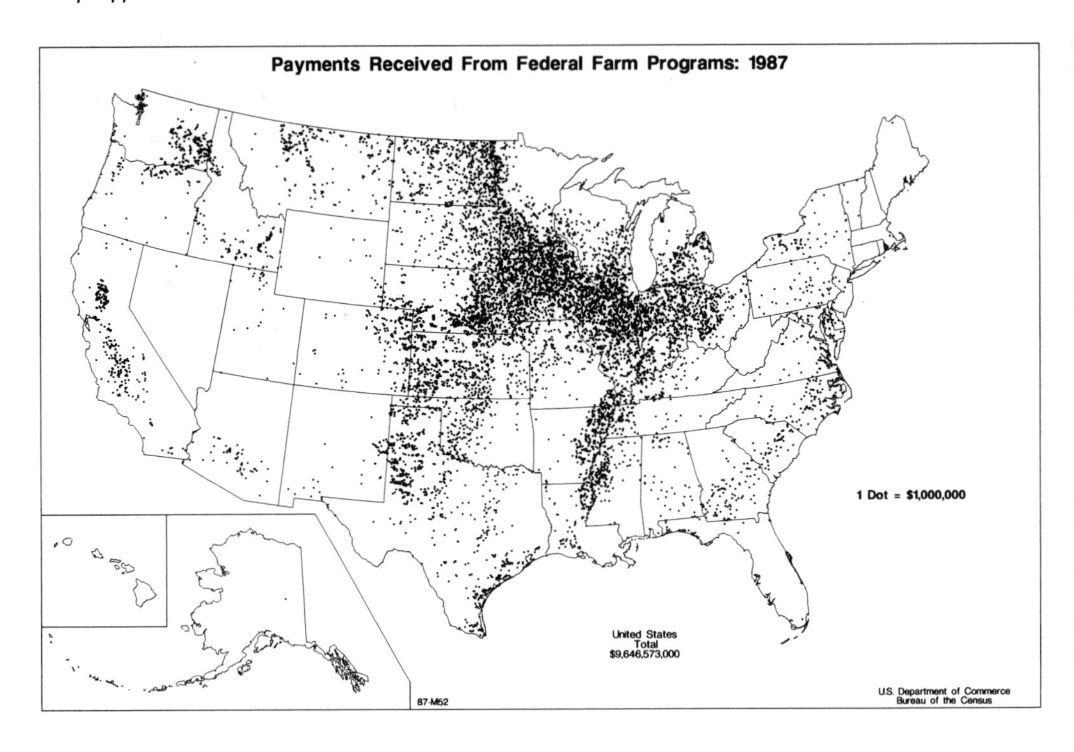
Payments Received From Federal Farm Programs: 1987
1 Dot = $1,000,000
United States
Total
$9,646,573,000
87-M52
U.S. Department of Commerce
Bureau of the Census

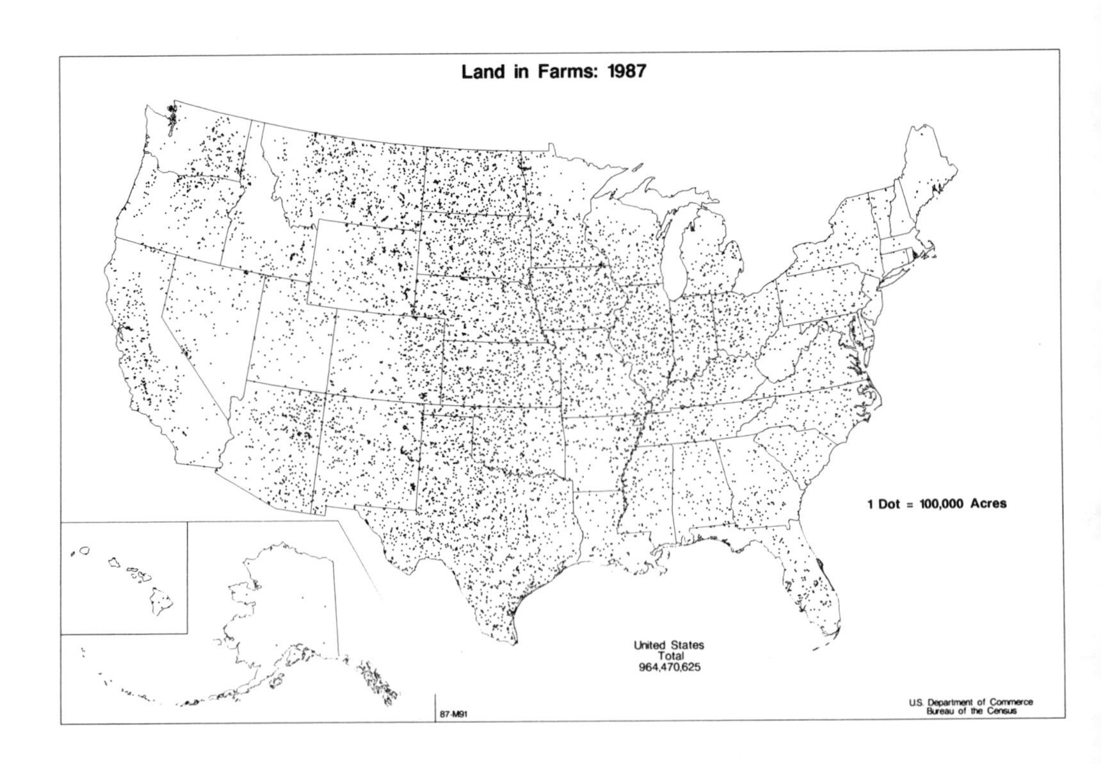
Land in Farms: 1987
1 Dot = 100,000 Acres
United States
Total
964,470,625
87-M91
U.S. Department of Commerce
Bureau of the Census

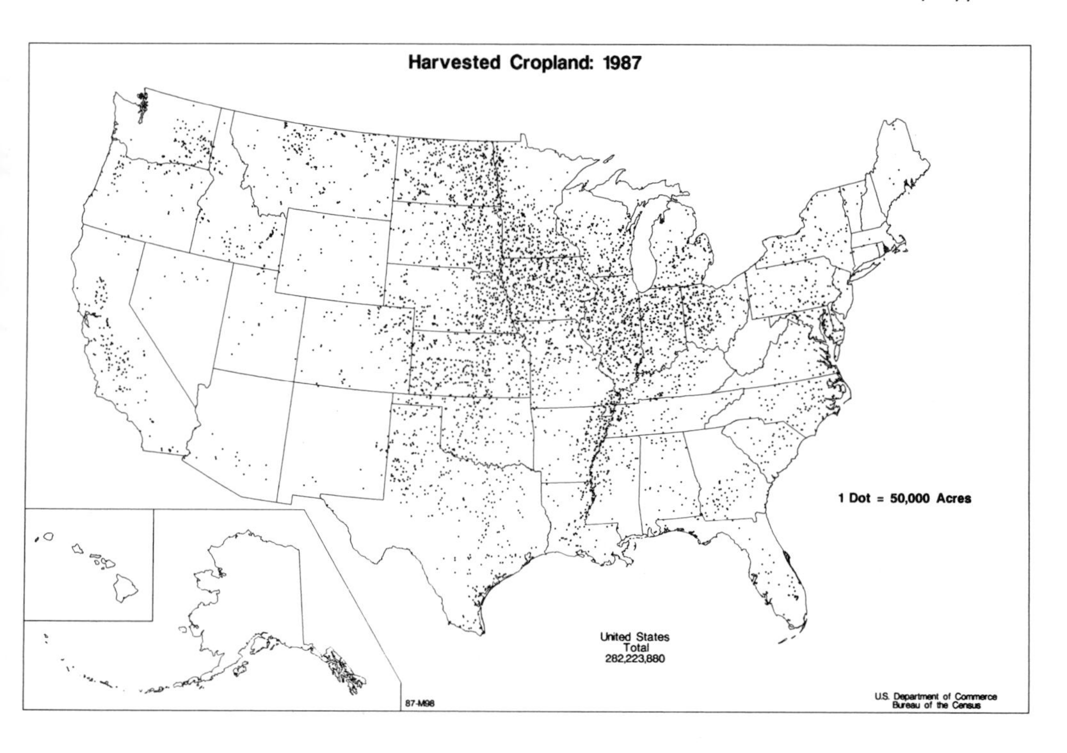
Harvested Cropland: 1987
1 Dot = 50,000 Acres
United States
Total
282,223,880
87-M98
U.S. Department of Commerce
Bureau of the Census

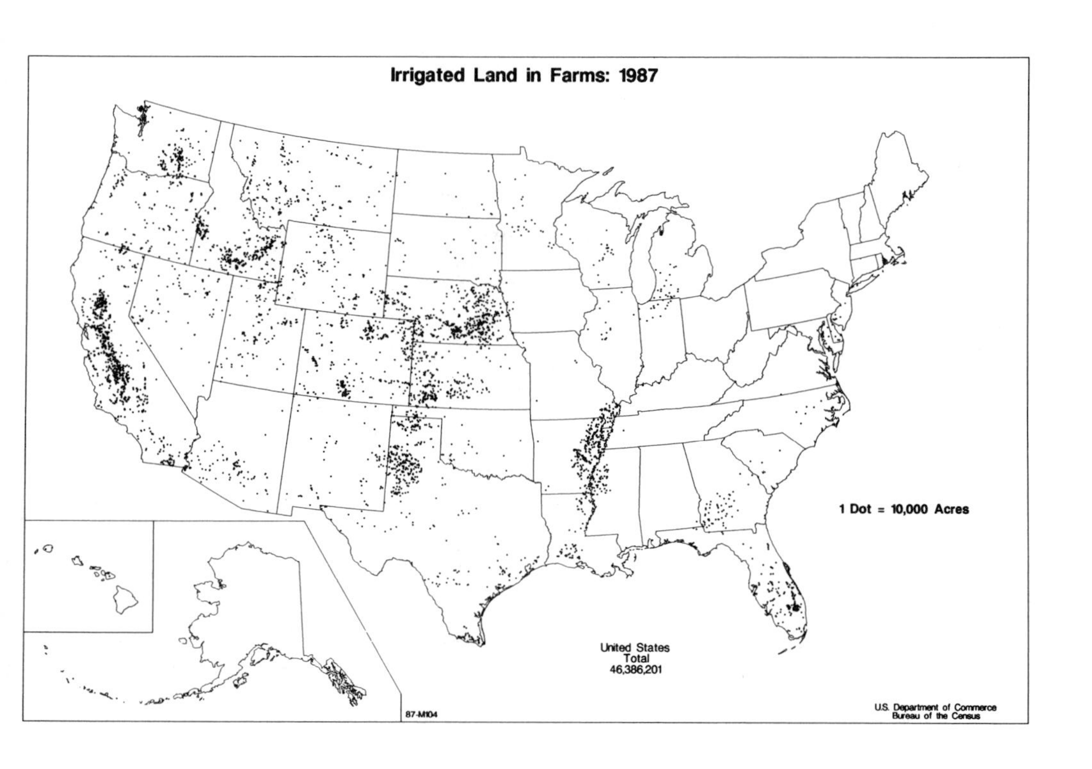
Irrigated Land in Farms: 1987
1 Dot = 10,000 Acres
United States
Total
46,386,201
87-M104
U.S. Department of Commerce
Bureau of the Census

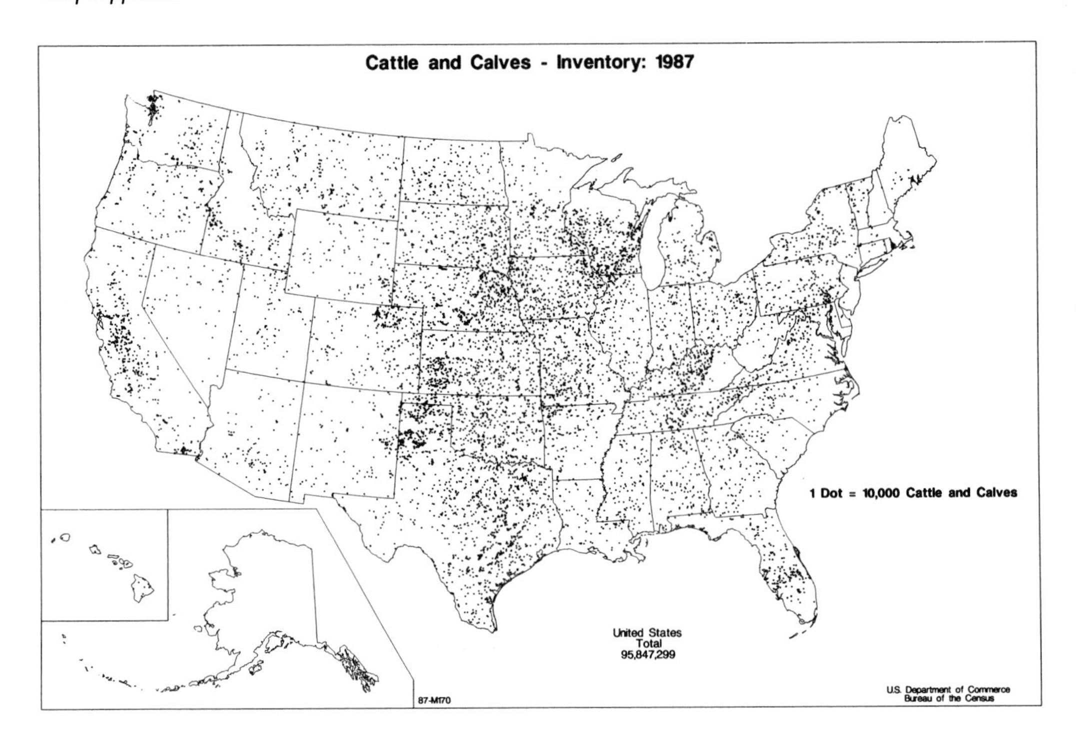
Cattle and Calves - Inventory: 1987
1 Dot = 10,000 Cattle and Calves
United States
Total
95,847,299
87-M170
U.S. Department of Commerce
Bureau of the Census

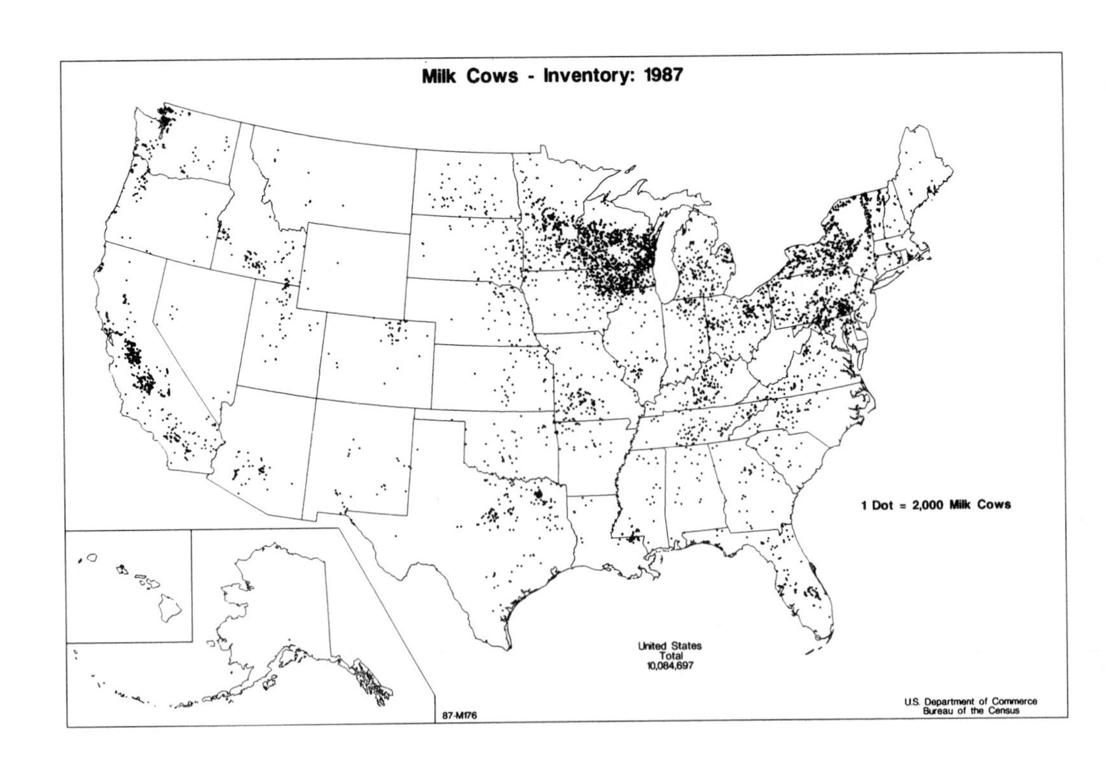
Milk Cows - Inventory: 1987
1 Dot = 2,000 Milk Cows
United States
Total
10,084,697
87-M176
U.S. Department of Commerce
Bureau of the Census

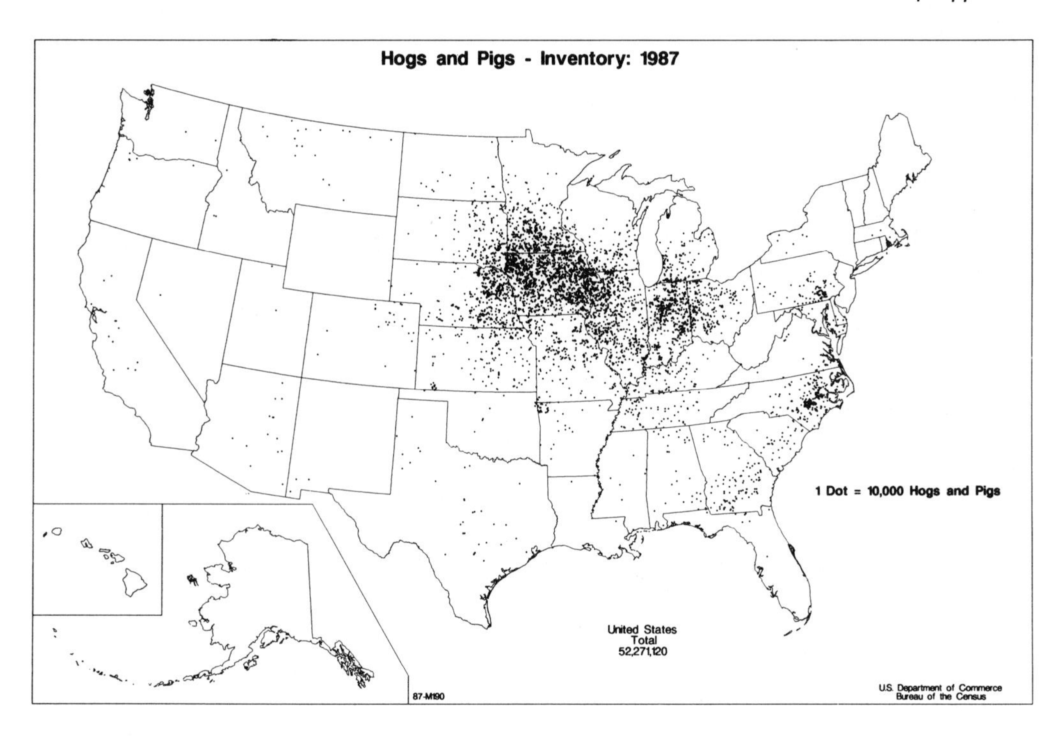
Hogs and Pigs - Inventory: 1987
1 Dot = 10,000 Hogs and Pigs
United States
Total
52,271,120
87-M190
U.S. Department of Commerce
Bureau of the Census

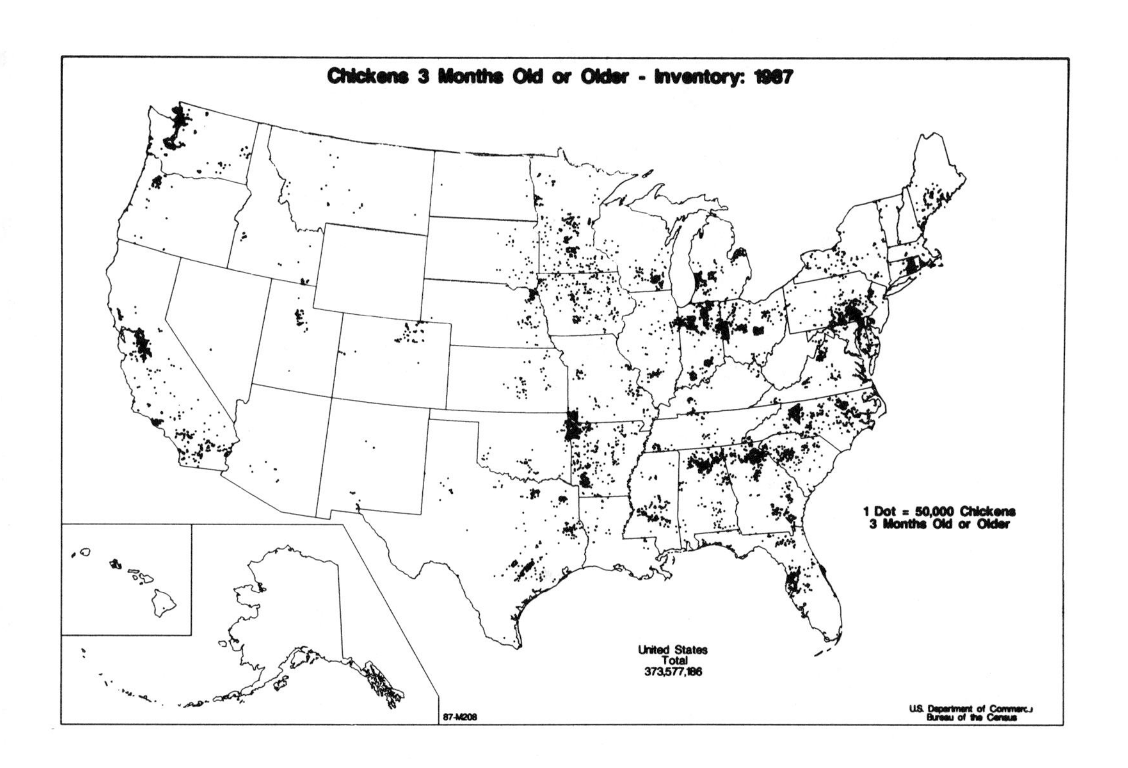
Chickens 3 Months Old or Older - Inventory: 1987
1 Dot = 50,000 Chickens
3 Months Old or Older
United States
Total
373,577,186
87-M208
U.S. Department of Commerce
Bureau of the Census

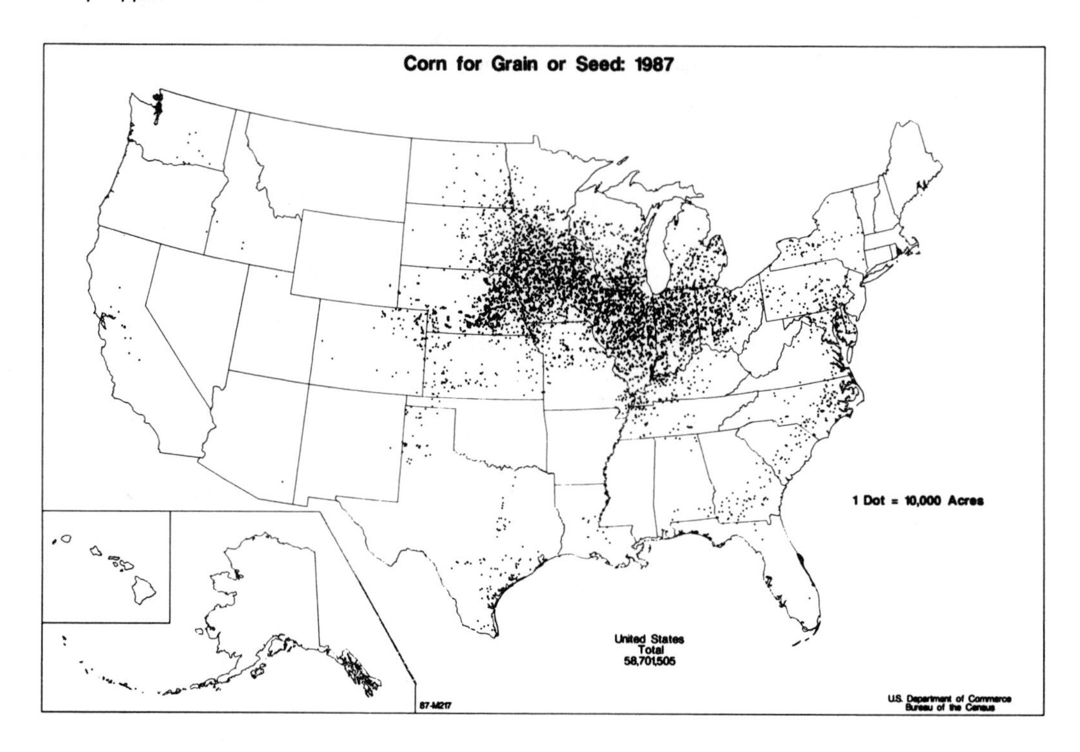
Corn for Grain or Seed: 1987
1 Dot = 10,000 Acres
United States
Total
58,701,505
87-M217
U.S. Department of Commerce
Bureau of the Census

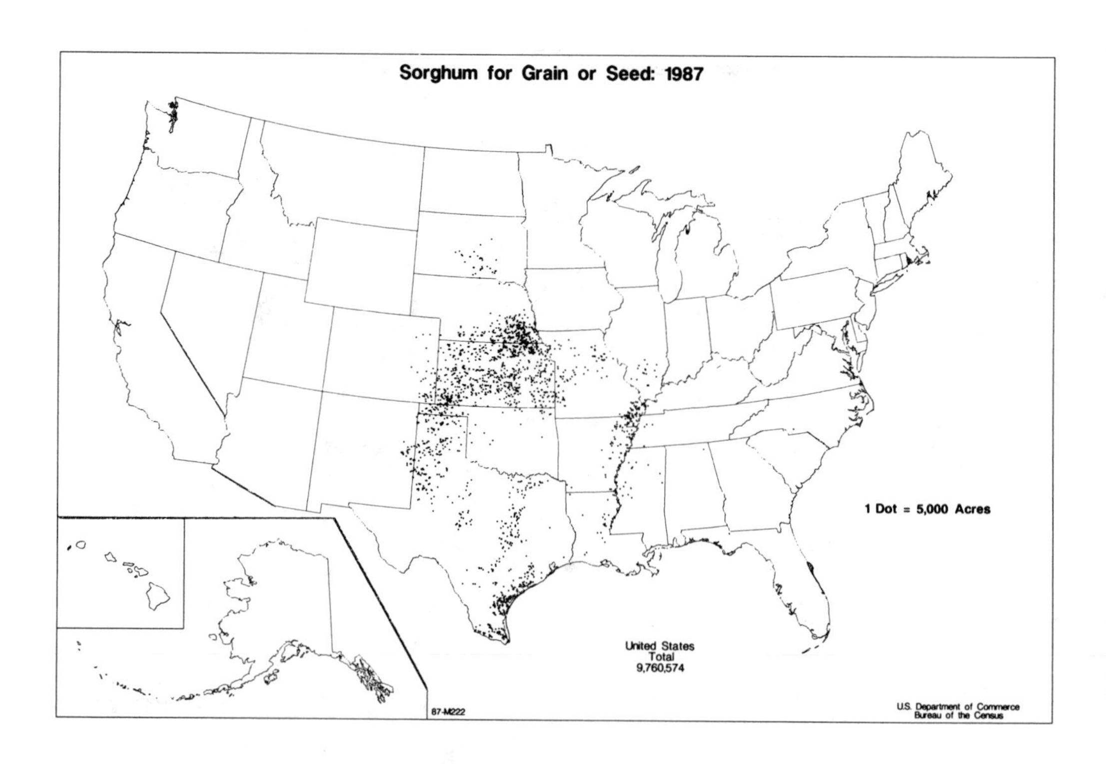
Sorghum for Grain or Seed: 1987
1 Dot = 5,000 Acres
United States
Total
9,760,574
87-M222
U.S. Department of Commerce
Bureau of the Census

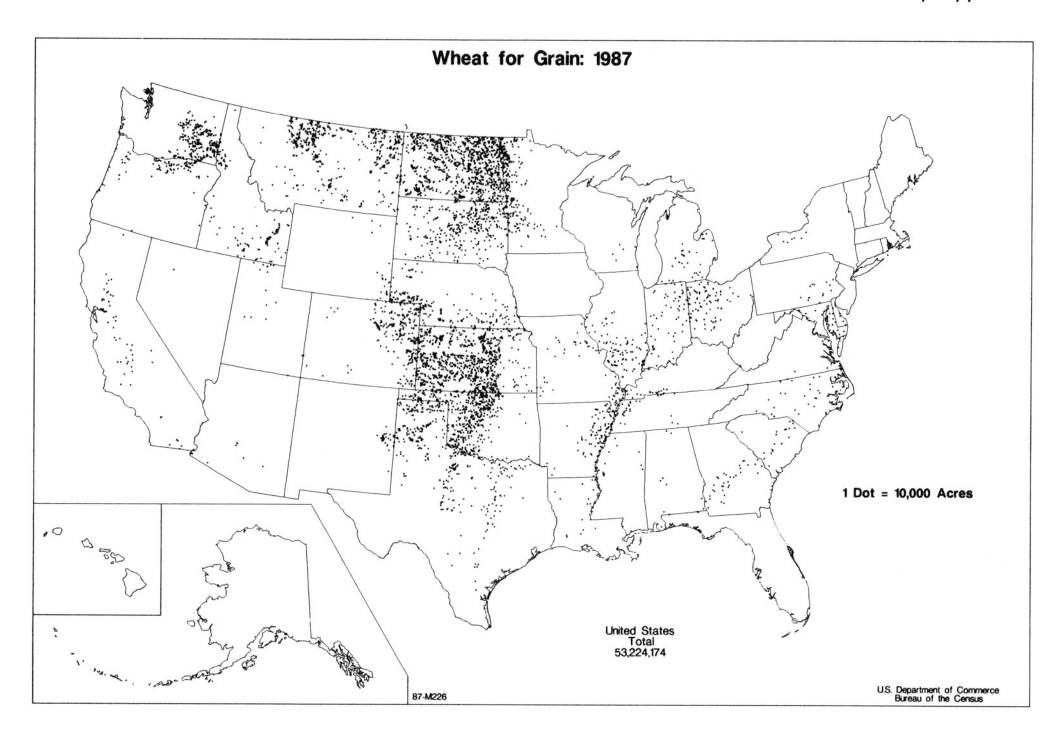
Wheat for Grain: 1987
1 Dot = 10,000 Acres
United States
Total
53,224,174
87-M226
U.S. Department of Commerce
Bureau of the Census

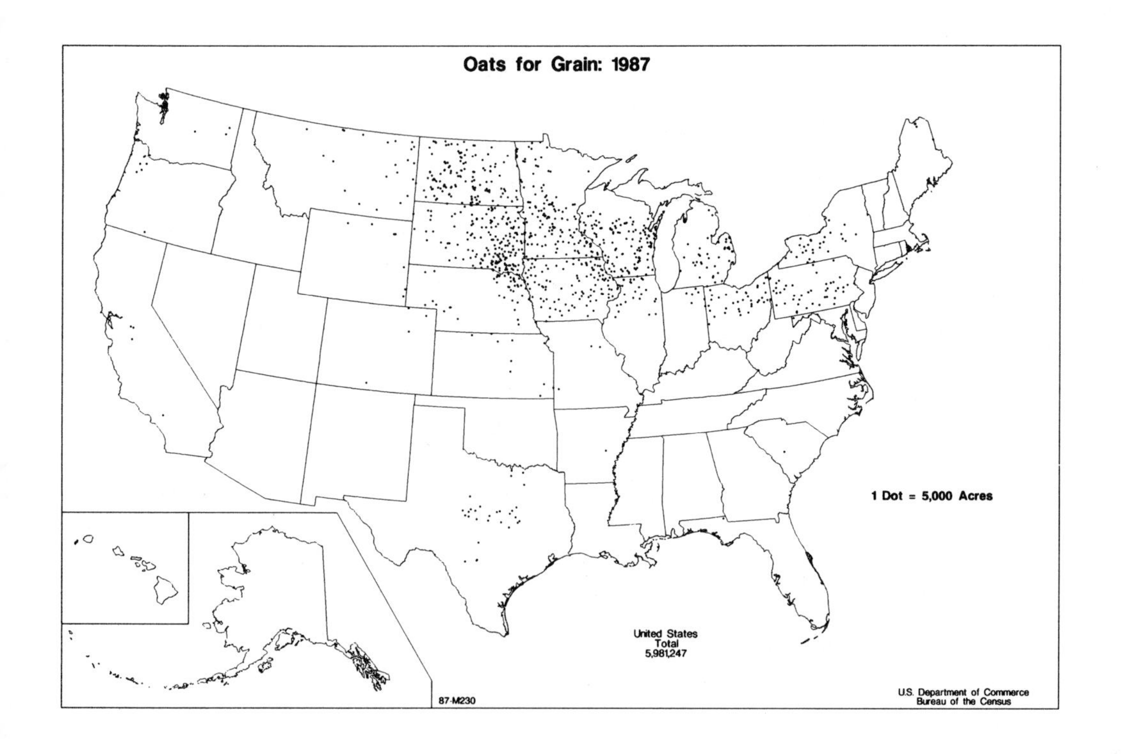
Oats for Grain: 1987
1 Dot = 5,000 Acres
United States
Total
5,981,247
87-M230
U.S. Department of Commerce
Bureau of the Census

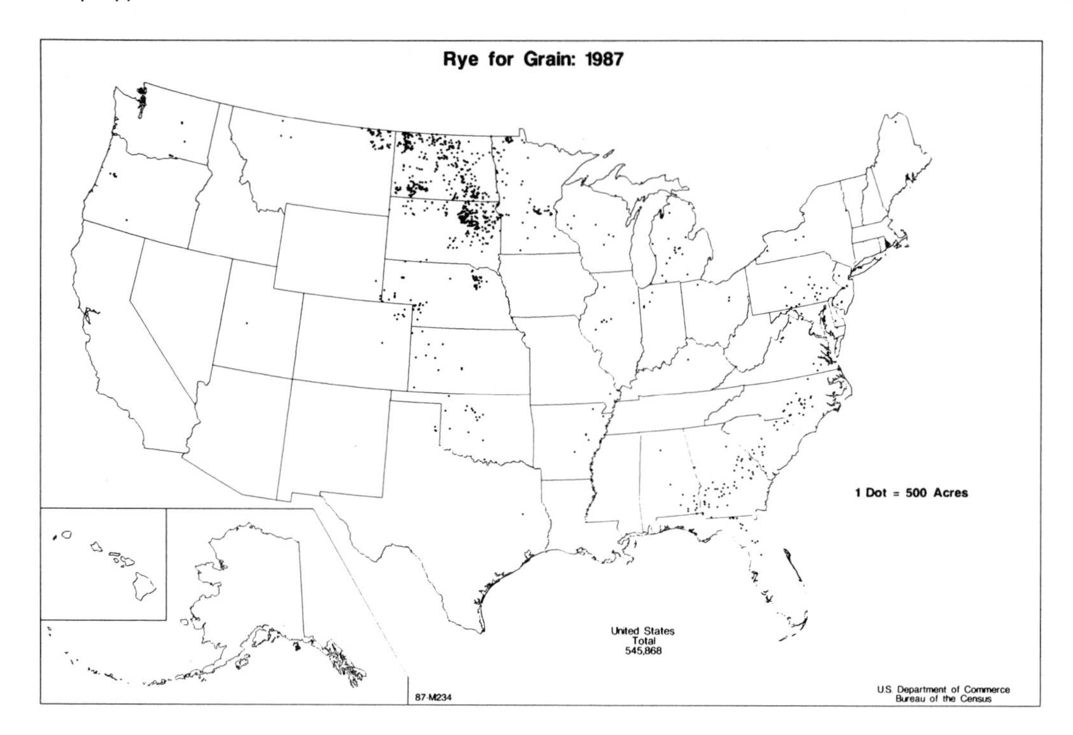
Rye for Grain: 1987
1 Dot = 500 Acres
United States
Total
545,868
87-M234
U.S. Department of Commerce
Bureau of the Census

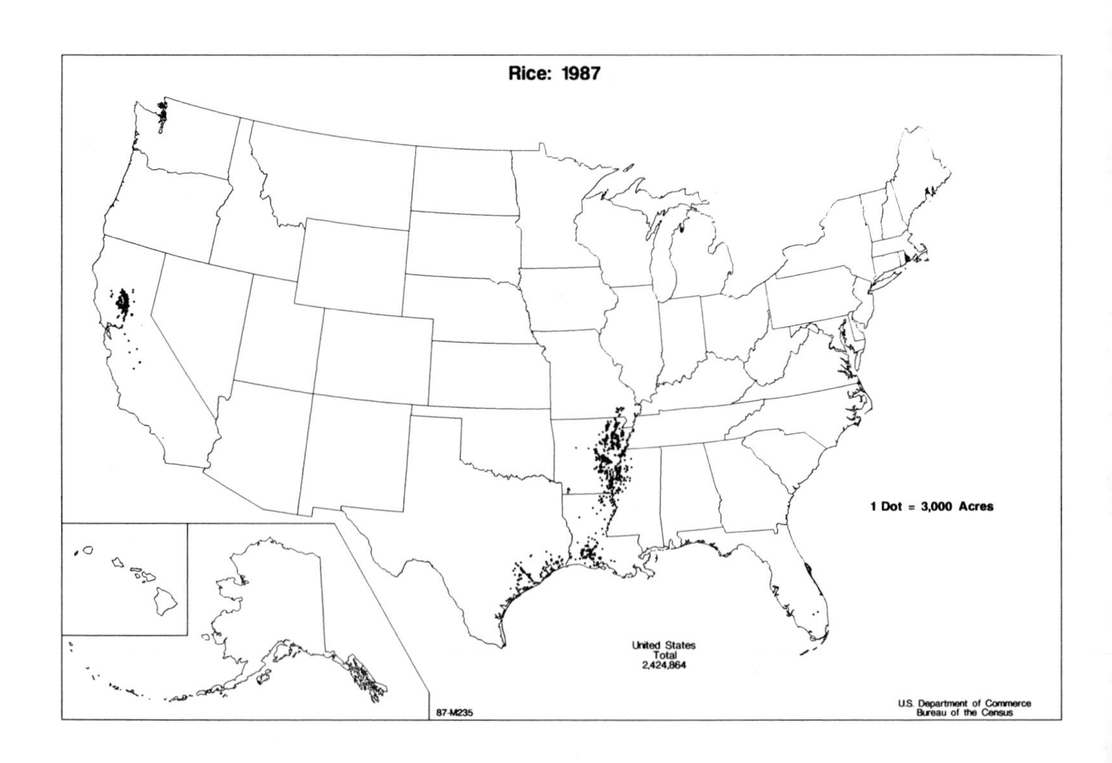
Rice: 1987
1 Dot = 3,000 Acres
United States
Total
2,424,864
87-M235
U.S. Department of Commerce
Bureau of the Census

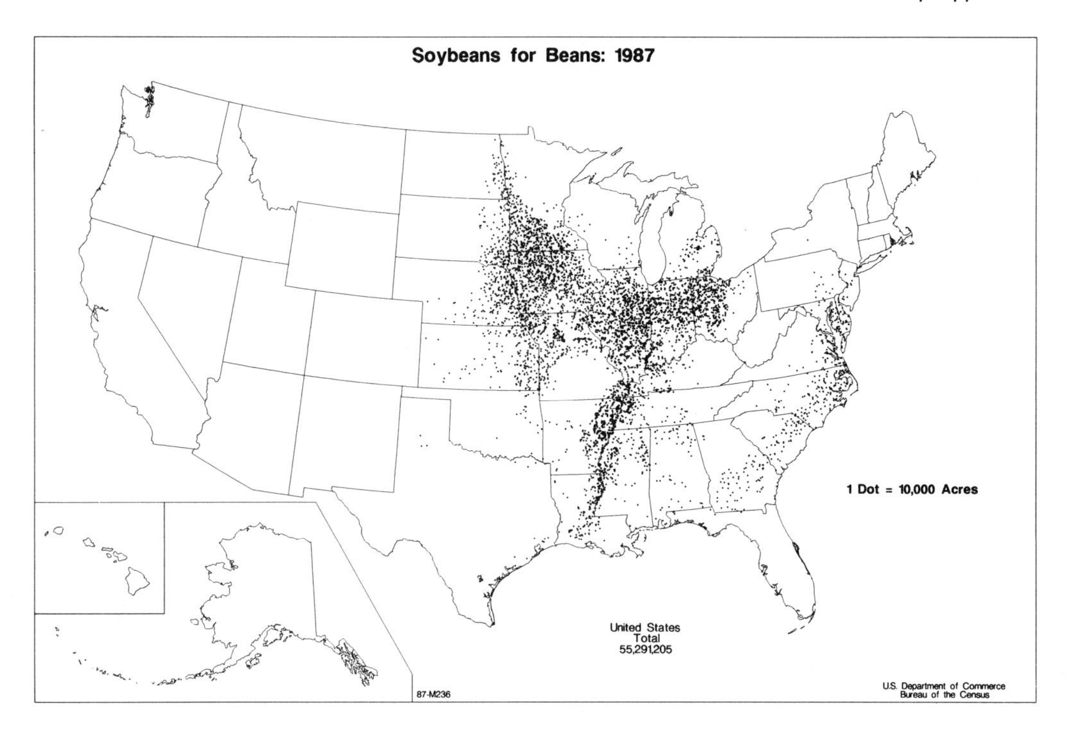
Soybeans for Beans: 1987
1 Dot = 10,000 Acres
United States
Total
55,291,205
87-M236
U.S. Department of Commerce
Bureau of the Census

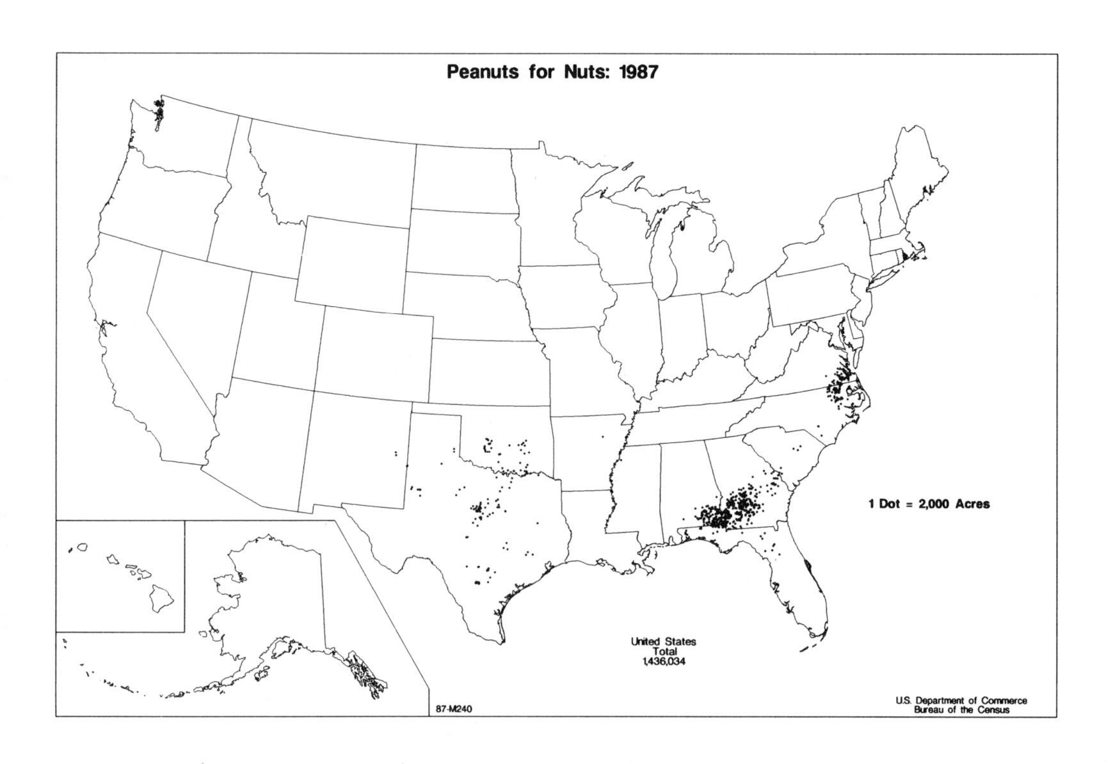
Peanuts for Nuts: 1987
1 Dot = 2,000 Acres
United States
Total
1,436,034
87-M240
U.S. Department of Commerce
Bureau of the Census

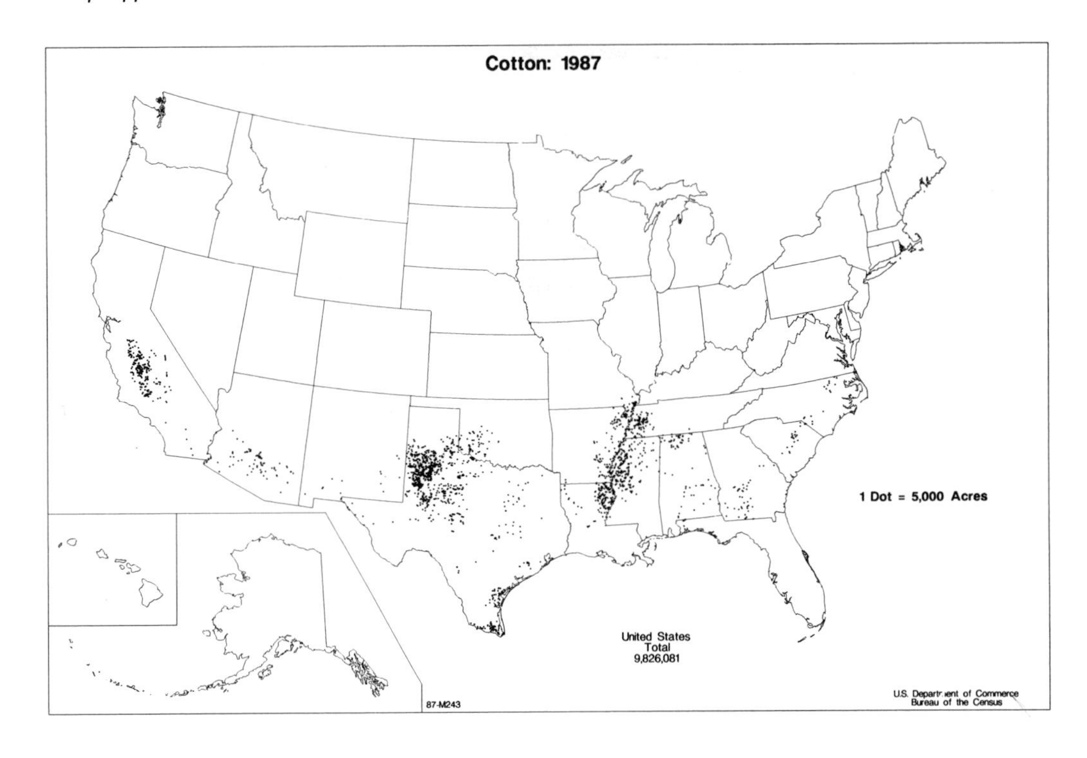
Cotton: 1987
1 Dot = 5,000 Acres
United States
Total
9,826,081
87-M243
U.S. Department of Commerce
Bureau of the Census

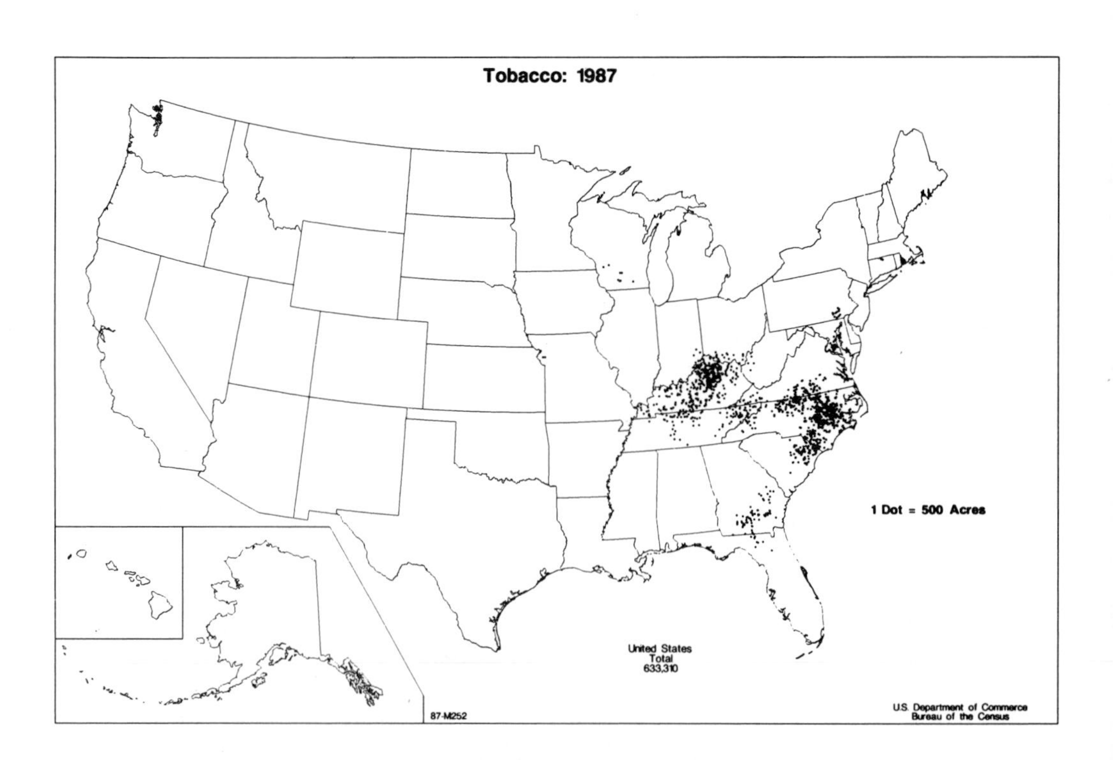
Tobacco: 1987
1 Dot = 500 Acres
United States
Total
633,310
87-M252
U.S. Department of Commerce
Bureau of the Census

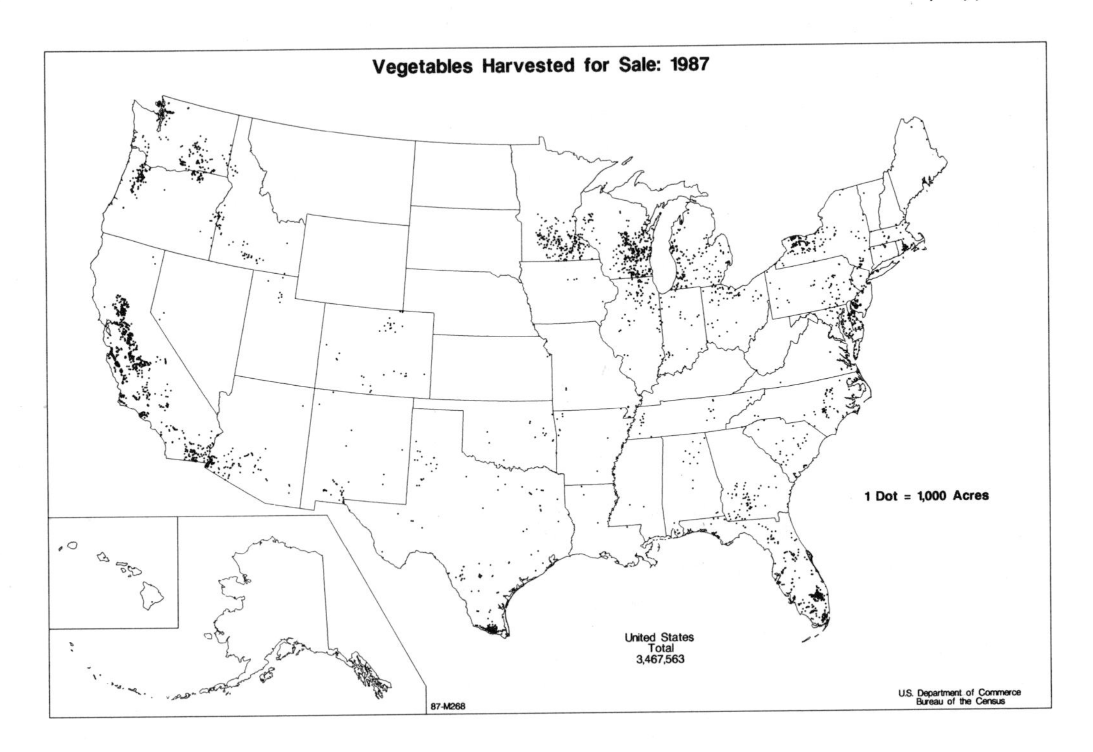
Vegetables Harvested for Sale: 1987
1 Dot = 1,000 Acres
United States
Total
3,467,563
87-M268
U.S. Department of Commerce
Bureau of the Census

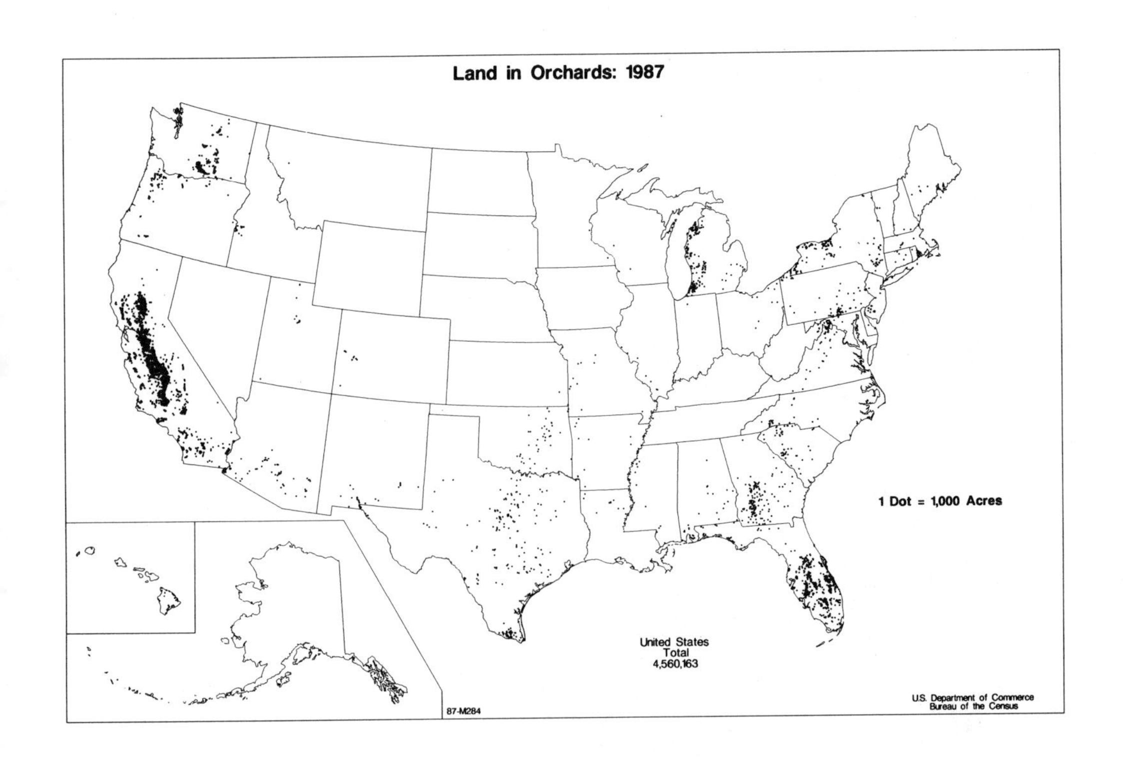
Land in Orchards: 1987
1 Dot = 1,000 Acres
United States
Total
4,560,163
87-M284
U.S. Department of Commerce
Bureau of the Census

Index